Distributed by
North-Holland Publishing Company
Amsterdam-London

ISBN 0 7204 2060 1

Responsible for publication Á. Császár,
Secretary general
János Bolyai Mathematical Society

Printed in Hungary

MTA KESZ Sokszorosító

COLLOQUIA MATHEMATICA
SOCIETATIS JÁNOS BOLYAI, 5.

HILBERT SPACE OPERATO

AND

OPERATOR ALGEBR

Edited by BÉLA SZ.-NAGY

DISTRIBUTED BY
NORTH-HOLLAND PUBLISHING COMPANY
AMSTERDAM - LONDON

PREFACE

The Bolyai János Mathematical Society sponsors in every year one or more international conferences in various fields of mathematics of general interest. The conference on "Linear Spaces and Operators" held in 1964 at Balatonföldvár was a success and this encouraged the Society to organize in 1970 again a conference in the domain of functional analysis, this time at Tihany, another picturesque part of the shore of lake Balaton. The subject now was specified as "Hilbert Space Operators and Operator algebras".

It is perhaps not only a personal preference to say that this is one of the most exciting fields in contemporary mathematics. Beyond its intrinsic interest, it connects areas of research that apparently lie widely apart, as shown by its applications in various parts of pure mathematics, probability theory, mathematical statistics, as well as in quantum mechanics and quantum field theory. So it is understandable that the rapid progress of the theory is followed by vivid interest not only of the specialists but of a wide group of nonspecialists also: mathematicians as well as physicists and scientists of various other areas.

The conference was attended by a considerable number of very distinguished scientists from various countries. Many interesting papers were presented and they were followed by high level and fruitful formal and informal discussions.

The work of the conference began in the afternoon on September 14 and ended at noon on September 18. The length of the lectures was generally 25 minutes including the time of discussion. Four or five lectures were given in the morning and a similar number in the afternoon session; there were no parallel sections.

The present volume of the proceedings of the conference contains the papers submitted for publication to the Organizing Committee. Shortened versions of most of these papers were read at the conference by their authors; a few other authors who were not able to attend for various reasons had their papers presented by other participants or by the Organizing Committee. In addition, the full programme of the conference, as well as a list of participants and their mailing addresses are included. This list also contains authors who had submitted papers but did not attend; their names are marked with an asterisk.

May this volume enhance a fruitful and widespread continuation of the work done in the Conference

The Editor

SCIENTIFIC PROGRAM

September 14. Monday

Afternoon session

Chairman: B. SZ.-NAGY

$3^{45} - 4^{10}$	P.R. Halmos: Quasi algebraic operators
$4^{15} - 4^{40}$	P. Rosenthal: Problems on invariant subspaces and operator algebras
$4^{45} - 5^{10}$	Ju.A. Rozanov: On the structure of measurable linear operators

Chairman: P.R. HALMOS

$5^{20} - 5^{45}$	I.M. Gel'fand: The cohomology of infinite-dimensional Lie algebras
$5^{50} - 6^{15}$	L. Cooper: Integral transforms and group representations
$6^{20} - 6^{50}$	H. Helson: Weighted shifts in Hilbert space

September 15. Tuesday

Morning session

Chairman: I.M. GEL'FAND

$9 - 9^{25}$	I.C. Gohberg: Banach algebras generated by singular integral operators
$9^{30} - 9^{55}$	Ch. Davis: Dilation and characteristic functions

Chairman: L. COOPER

$10^{10} - 10^{35}$	H. Langer: Erweiterung J-nichtnegativer Operatoren
$10^{40} - 11^{5}$	E. Durszt: Matrix representation of ϱ-dilations

Afternoon session

Chariman: G.W. MACKEY

$3^{45} - 4^{10}$	J. Dixmier: On the point spectrum of an operator
$4^{15} - 4^{40}$	H. Halpern: A generalized dual for a C^*-algebra
$4^{45} - 5^{10}$	L.A. Coburn: C^*-algebras generated by semigroups of isometries

Chairman: J. DIXMIER

$5^{20} - 5^{45}$	S. Sakai: On global type II_1 w^*-algebras
$5^{50} - 6^{15}$	G.W. Mackey: Products of subgroups and projective multipliers
$6^{20} - 6^{45}$	F.F. Bonsall: Hermitian operators on Banach spaces

September 16. Wednesday

Morning session

Chairman: P. MASANI

$9 - 9^{25}$	B. Sz.-Nagy: Quasi-similarity of operators of class C_0
$9^{30} - 9^{55}$	C. Foiaş: On models of operators with cyclic vectors
$10 - 10^{25}$	J. Bognár: A characterization of the conjugation of operators in spaces with indefinite metric

Chairman: F.F. BONSALL

$10^{30} - 10^{55}$	J. Szűcs: On the non-existence of unicellular operators on certain non-separable Banach spaces
$11 - 11^{25}$	J.D. Pincus: Almost commuting operators

Afternoon session

Chairman: CH. DAVIS

$3 - 3^{25}$	P. Masani: Some remarks on eigenpackets of self-adjoint operators
$3^{30} - 3^{55}$	P.L. Butzer: On the Cayley transform and semigroup operators
$4 - 4^{25}$	K. Scherer: Approximationssätze über lineare Operatoren im Dualen

September 17. Thursday

Morning session

Chairman: C. FOIAŞ

$9 - 9^{25}$	M.A. Kaashoek: On the peripheral spectrum
$9^{30} - 10$	Ju.M. Berezanskiĭ: Generalized power moment problems

Chairman: I.C. GOHBERG

$10 - 10^{25}$	V.I. Gorbačuk: On self-adjoint extensions of some hermitian operators in the space of indefinite metric
$10^{30} - 10^{55}$	J. Woods: On classification of ITPFI factors
$11 - 11^{25}$	L. Máté: On tensor products and the multiplier problem

Afternoon session

Chairman: JU.M. BEREZANSKIĬ

$3^{30} - 4^{10}$	M.S. Brodskiĭ – E. Kisilevskiĭ: On some invariant subspaces of dissipative operators of exponential type
$4^{15} - 4^{40}$	E. Denčev: Commutative self-adjoint extensions of differential operators

Chairman: S. SAKAI

$4^{50} - 5^{15}$	J.W. Helton: Operators with a representation as a multiplier on a Sobolev space
$5^{20} - 5^{40}$	L. Zsidó: Commutant and tensor product of von Neumann algebras

September 18. Friday

Morning session

Chairman: P.L. BUTZER

$8^{45} - 8^{55}$	I. Kovács: Power bounded operators and finite type von Neumann algebras
$9 - 9^{25}$	I. Suciu: On operator representations of function algebras
$9^{30} - 9^{55}$	H. Johnen: Lipschitzklassen auf kompakten Mannigfaltigkeiten

Chairman: H. LANGER

$10 - 10^{25}$	S. Stratilă presents the paper by S. Teleman: Representation of von Neumann algebras by sheaves
$10^{30} - 10^{55}$	R.A. Alexandrian: On the kernel of spectrum and construction of complete system of generalized eigenfunctions of self-adjoint operators
$11 - 11^{15}$	L. Coburn presents the paper by R.G. Douglas and J.L. Taylor: On Wiener-Hopf operators with measure kernel

LIST OF PARTICIPANTS

ALEXANDRIAN, R.A., Dept. of Math., Univ. of Yerevan, Yerevan, USSR

ARSENE, G., Inst. of Math. Acad. Sci., Calea Griviţei 21, Bucharest 12, Roumania

BADER, P., Battelle Inst., Advanced Studies Center, Carouge (Geneva) 7 Rte de Drize 1227, Switzerland

BEREZANSKIĬ, JU.M., Inst. of Math. Ukrainian Acad. Sci., ul. Repina 3, Kiev, USSR

BEREZIN, F.A.,* Dept. of Math., State Univ. of Moscow, Moscow, USSR

BOGNÁR, J., Inst. of Math. Acad. Sci., Reáltanoda u. 13-15, Budapest 5, Hungary

BONSALL, F.F., Dept. of Math. of the Univ., 20 Chambers Str., Edinbourgh, Great Britain

BRODSKIĬ, M.S.,* Dept. of Higher Math., Ped. Inst. of Odessa, Komsomolskaja 26, Odessa, USSR

BROISE, M., Dept. of Math., Univ. of Paris, 11 Rue Pierre Curie, Paris 5, France

BUTZER, P., Dept. of Math., Technical Univ., 51 Aachen, Templergraben 55, Federal Republic of Germany

COBURN, L., Dept. of Math., Yeshiva Univ., Amsterdam Ave. & 185th str., New York, N.Y. 10033, USA

COOPER, J.L.B., Dept. of Math., Chelsea College of Sci. and Techn., Manresa Road, London SW3, Great Britain

DAVIS, Ch., Dept. of Math., Univ. of Toronto, Toronto 181, Ont., Canada

DENČEV, R., Dept. of Math., Univ. of Sofia, Sofia, Bulgaria

DIXMIER, J., Dept. of Math., Univ. of Paris, 9 Quai St. Bernard, Paris 5, France

DOUGLAS, R.G.,* Dept. of Math., State Univ. of New York at Stony Brook, Stony Brook, Long Island, N.Y. 11790, USA

DURSZT, E., Dept. of Math., Univ. of Szeged, Szeged, Hungary

DŽAFAROV, A.S., Dept. of Math., Univ. of Baku, Baku, USSR

FOIAŞ, C., Inst. of Math. Acad. Sci., Calea Griviţei 21, Bucharest 12, Roumania

GEHÉR, L., Dept. of Math., Univ. of Szeged, Szeged, Hungary

GEL'FAND, I.M., Dept. of Math., State Univ. of Moscow, Moscow, USSR

GOHBERG, I.C., Inst. of Math. and Phys. Moldavian Acad. Sci., Akademičeskaja 3, Kišinev, USSR

GORBAČUK, V.I., Inst. of Math. Ukrainian Acad. Sci., ul. Repina 3, Kiev, USSR

HALMOS, P.R., Dept. of Math., Indiana Univ., Bloomington, Ind. 47401, USA

HALPERN, H., Dept. of Math., Illinois Inst. of Techn., Chicago, Ill., USA

HATVANI, L., Dept. of Math., Univ. of Szeged, Szeged, Hungary

HELSON, H., Dept. of Math., Univ. of California, Berkeley, Cal. 94720, USA

HELTON, J.W., Dept. of Math., State Univ. of New York at Stony Brook, Stony Brook, Long Island, N.Y. 11790, USA

IOHVIDOV, I.S.,* Dept. of Math., Univ. of Voronež, Voronež, USSR

JOHNEN, H., Dept. of Math., Technical Univ., 51 Aachen, Templergraben 55, Federal Republic of Germany

JUBERG, R.K.,* Dept. of Math., Univ. of California, Irvine, Cal. 92664, USA

KAASHOEK, M., Dept. of Math., Univ. of Amsterdam, De Boelelaan 1081, Amsterdam, Holland

KALISCH, R.G.,* Dept. of Math., Univ. of California, Irvine, Cal. 92664, USA

KISILEVSKIĬ, G.E., Dept. of Math., Ped. Inst., 44 ul. Karla Marksa, Žitomir, USSR

KOMLÓSI, S., Dept. of Math., Univ. of Szeged, Szeged, Hungary

KOVÁCS, I., Dept. of Math., Univ. of Szeged, Szeged, Hungary

KOVÁTS, SZ., Dept. of Math., Univ. of Szeged, Szeged, Hungary

KREĬN, M.G.,* Dept. of Math., Technical Univ. for Architecture of Odessa, Odessa, USSR

KRUPNIK, N.JA.,* Inst. of Math. and Phys. Moldavian Acad. Sci., Akademičeskaja 3, Kišinev, USSR

LANGER, H., Dept. of Math., Technical Univ. of Dresden, Zellescher Weg, Dresden, German Democratic Republic

MACKEY, G.W., Dept. of Math., Harvard Univ., Cambridge, Mass. 02139, USA

MASANI, P., Dept. of Math., Indiana Univ., Bloomington, Ind. 47401, USA

MÁTÉ, A., Dept. of Math., Univ. of Szeged, Szeged, Hungary

MÁTÉ, E., Dept. of Math., Univ. of Szeged, Szeged, Hungary

MÁTÉ, L., Dept. of Math., Faculty of Electic Engineering, Technical Univ., Sztoczek u. 2., Budapest 11, Hungary

MOORE III,B., Dept. of Math., Univ. of New Hampshire, Durham, N. H., USA

NAGY, P., Dept. of Math., Univ. of Szeged, Szeged. Hungary

NORDGREN, E., Dept. of Math., Univ. of New Hampshire, Durham, N. H., USA

PINCUS, J.D., Dept. of Math., State Univ. of New York at Stony Brook, Stony Brook, Long Island, N.Y. 11790, USA

PINTÉR, L., Dept. of Math., Univ. of Szeged, Szeged, Hungary

PONOMAREV, V.A.,* Dept. of Math., State Univ. of Moscow, Moscow, USSR

RÁCZ, A., Dept. of Math., Univ. of Timişoara, Timişoara, Roumania

ROSENTHAL, P., Dept of Math., Univ. of Toronto, Toronto 181, Ont., Canada

ROZANOV, JU.A., Steklov Inst. of Math. Acad. Sci., 47 ul. Vavilova, Moscow, V – 333, USSR

SAKAI, S., Dept. of Math., Univ. of Pennsylvania, Philadelphia, Penn. 19104, USA

SCHERER, K., Dept. of Math., Technical Univ., 51 Aachen, Templergraben 55, Federal Republic of Germany

STRĂTILĂ, S., Inst. of Math. Acad. Sci., Calea Griviţei 21, Bucharest 12, Roumania

ŠUBIN, M.A.,* Dept. of Math, State Univ. of Moscow, Moscow, USSR

SUCIU, I., Inst. of Math. Acad. Sci., Calea Griviţei 21, Bucharest 12, Roumania

SZÉP, A., Inst. of Math., Acad. Sci., Reáltanoda u. 13-15, Budapest 5, Hungary

SZÉP, L., Inst. of Math., Acad. Sci., Reáltanoda u. 13-15, Budapest 5, Hungary

SZ.-NAGY, B., Dept. of Math., Univ. of Szeged, Szeged, Hungary

TAYLOR, J.L.,* Dept. of Math., Univ. of Utah, Salt Lake City, Utah, USA

TELEMAN, S.,* Inst. of Math. Acad. Sci, Calea Griviţei 21, Bucharest 12, Roumania

WESTPHAL, U.,* Dept. of Math., Technical Univ., 51 Aachen, Templergraben 55, Federal Republic of Germany

WOODS, J., Dept. of Math., Queen's Univ., Kingston, Ont., Canada

ZSIDÓ, L., Inst. of Math. Acad. Sci., Calea Griviţei 21, Bucharest 12, Roumania

COLLOQUIA MATHEMATICA SOCIETATIS JÁNOS BOLYAI
5. HILBERT SPACE OPERATORS, TIHANY (HUNGARY), 1970

Generalized power moment problems

JU. M. BEREZANSKIĬ

In this lecture we shall talk about a generalization of the classical power moment problem, which arises while describing objects satisfying some of the axioms of quantum field theory (the axioms of positivity and local commutativity) [1,2,3].

In all that follows we shall consider only separable spaces; $\mathrm{l.h.}(A)$ and $\mathrm{r.l.h.}(A)$ denote the linear hull and the real linear hull, respectively, of the set A; $H^n = \underbrace{H \otimes \dots \otimes H}_{n \text{ times}}$ $(n = 1, 2, \dots;\ H^0 = C)$ is the nth tensorial power of the Hilbert space H.

1. We begin with the simplest generalization. As is well known, the central feature of the theory of the power moment problem is to find conditions on the numerical sequence $s_{\mathbf{n}}$, where $\mathbf{n} = (n_1, \dots, n_M)$ runs over all points of $\underbrace{[0,\infty) \times \dots \times [0,\infty)}_{n \text{ times}}$ with integral coordinates, ensuring the existence of a representation

$$(1) \qquad s_{\mathbf{n}} = \int\limits_{\mathbf{R}^M} \lambda^{\mathbf{n}} d\varrho(\lambda) \qquad (\lambda^{\mathbf{n}} = \lambda_1^{n_1} \dots \lambda_M^{n_M},\ \lambda = \lambda_1, \dots, \lambda_M))$$

with a non-negative finite measure $d\varrho(\lambda)$. In the one-dimensional case $(M = 1,\ \mathbf{n} = n)$

the necessary and sufficient condition for the existence of a representation (1) is that $(s_n)_{n=0}^{\infty}$ be a moment-type sequence, i.e. that the matrix $\{\mathbf{K}_{jk}\}_{j,k=0}^{\infty}$, $\mathbf{K}_{jk} = s_{j+k}$ be positive definite (p.d.). In the multi-dimensional case it is necessary to impose additional restrictions on the growth of $s_{\mathbf{n}}$ ensuring the self-adjointness of the relevant operators (case of the determined moment problem), or providing for the extendability of the occurring Hermitian operators to commuting self-adjoint ones (see, e.g., [4], Ch. VIII, §5, Sections 4-7). The situation is analogous in the infinite dimensional case, i.e. when $M = \infty$ [5]. In this section the representation (1) is generalized to the case in which the role of s_n is played by functionals over the nth tensorial power of a nuclear space.

Assume that $\mathfrak{H} = \bigcap_{\tau \in T} \mathfrak{H}_\tau$ is a nuclear space (where the spaces $\mathfrak{H}_\tau$ are the Hilbert spaces determining it), $\mathfrak{H}^n = \bigcap_{\tau \in T} \mathfrak{H}_\tau^n$ the nth tensorial power of $\mathfrak{H}$ $(n = 0, 1, \ldots)$, $(\mathfrak{H}^n)' = \bigcup_{\tau \in T} (\mathfrak{H}_\tau^n)'$ the corresponding adjoint space of antilinear functionals. Assume that in $\mathfrak{H}$ we are given an involution $\varrho \longrightarrow \bar{\varrho}$ that may be continuously extended to an involution in each $\mathfrak{H}_\tau$. By forming the tensorial powers and then by going over to adjoint operators, it can naturally be extended to involutions on the tensorial powers and on their adjoint spaces; for the notation of these involutions we retain the bar $^{-}$. The index Re will indicate that only those elements real with respect to the relevant involution of the given space are considered. For example, $\mathfrak{H}'_{Re}$ is the space formed by all $\alpha \in \mathfrak{H}'$ such that $\bar{\alpha} = \alpha$.

Consider the sequence $\mathbf{S} = (\mathbf{S}_n)_{n=0}^{\infty}$, where $\mathbf{S}_n \in (\mathfrak{H}^n)'$ is completely symmetric (i.e. $s_n(\varrho_1 \otimes \ldots \otimes \varrho_n)$ is a symmetric function of $\varrho_1, \ldots, \varrho_n \in \mathfrak{H}$). We call $\mathbf{S}$ a moment-type sequence if the condition of p.d. is satisfied: for any finite sequence $u = (u_j)_{j=0}^{\infty}$ $(u_j \in \mathfrak{H}^j)$ we have

$$\sum_{j,k=0}^{\infty} \mathbf{S}_{j+k}(u_j \otimes \bar{u}_k) \geq 0 \ . \tag{2}$$

The set of all finite sequence will be denoted by $\mathbf{C}_0$.

Now we formulate a condition of the type of determinedness of the classical moment problem. Since $\mathbf{S}_n \in (\mathfrak{H}^n)' = \bigcup_{\tau \in T} (\mathfrak{H}_\tau^n)'$, there exists a $\tau = \tau(n)$ such that $\mathbf{S}_n \in (\mathfrak{H}_{\tau(n)}^n)'$ $(n = 0, 1, \ldots)$. Fix a set $\mathfrak{L} \subseteq (\mathfrak{H}_{\tau(n)}^n)$ with $\ell.h.(\mathfrak{L})$ dense in $\mathfrak{H}$, and set $d(\tau, \mathfrak{L}) = \sup \|\varrho\|_{\mathfrak{H}_\tau}$. We say that $\mathbf{S}$ is determined if the class

(3) $$\mathbf{C}\{M_n\}, \quad M_n = d^n(\tau(4n), \mathfrak{k}) \|\mathbf{S}_{4n}\|^{1/4}_{(\mathfrak{H}^{4n}_{\tau(4n)})'}$$

is quasi-analytic. We shall also consider the more comprehensive case of a quasi-determined $\mathbf{S}$ in which the above condition is fulfilled with the modification that the closure of $l.h.(\mathfrak{k})$ has codimension one in $\mathfrak{H}$, and in (3) instead of $\mathbf{S}_{4n}$ its orthogonal projection onto $(\mathcal{L}^{4n}_{\tau(4n)})'$ occurs, where $\mathcal{L}_\tau$ denotes the closure of $l.h.(\mathfrak{k})$ in $\mathfrak{H}_\tau$.*

Theorem 1. *Suppose $\mathbf{S} = (\mathbf{S}_n)_{n=0}^{\infty}$ is a moment-type sequence satisfying the conditions of determinedness or quasi-determinedness. Then there exists a finite non-negative measure $d\rho(\lambda)$ defined on a σ-algebra of subsets of $\mathfrak{H}'_{Re}$ such that*

(4) $$\mathbf{S}_n = \int_{\mathfrak{H}'_{Re}} \underbrace{\lambda \otimes \ldots \otimes \lambda}_{n \text{ times}} \, d\rho(\lambda) \qquad (n = 0, 1, \ldots).$$

Here the integrand is the vector function $\mathfrak{H}_{Re} \ni \lambda \to \underbrace{\lambda \otimes \ldots \otimes \lambda}_{n \text{ times}} \in (\mathfrak{H}^n)'$ the integral being weakly convergent. If $\mathbf{S}$ is determined then the measure $d\rho(\lambda)$ is unique (in case of quasi-determinedness we do not, in general, have uniqueness). Conversely, every sequence of the form (4) is a moment-type sequence.

We make now some observations. a) From (4) it is easily seen that $\mathbf{S}_n$ is real – this immediately follows from (2) as well. b) While speaking of the uniqueness of the measure $d\rho(\lambda)$, we have in mind uniqueness in the class of all measures determined by their values on cylindrical sets of the form $C_\Delta = \{\lambda \in \mathfrak{H}'_{Re} : (\lambda(\varphi_1), \ldots \ldots \lambda(\varphi_n)) \in \Delta\}$, where $\varphi_1, \ldots, \varphi_n \in r.l.h.(\mathfrak{k})$ and Δ is a Borel set in $\mathbf{R}^n$. The measure whose existence is claimed by the above theorem belongs to this class. c) The measure $d\rho(\lambda)$ is actually concentrated on a Hilbert space $\mathfrak{H}'_{\tau_0, Re} \subset \mathfrak{H}'_{Re}$ depending on $\mathbf{S}$. d) It is well known that p.d. functions $f(\varphi) = \int_{\mathbf{R}^1} e^{i\lambda\varphi} d\rho(\lambda)$ $(\varphi \in R^1)$ and usual moment-type sequences $s = (s_n)_{n=0}^{\infty}$ are closely connected: if $f(\varphi)$ is p.d. then $s_n = (-i)^n \left(\frac{d^n f}{d\varphi^n}\right)(0)$ is a moment-type sequence. There is analogous connection between the moment -type sequences just considered and p.d. functions $f(\varphi)$ of $\varphi \in \mathfrak{H}$ and the theorem of R.A. Minlos and V.V. Sazonov (see,

*The conditions of determinedness and quasi-determinedness may be weakened. Thus, e.g. in the case of determinedness, M_n in (3) can be replaced by $M_n = d^n(\tau(2n), \mathfrak{k}) \|\mathbf{S}_{2n}\|^{1/2}_{(\mathfrak{H}^{2n}_{\tau(2n)})'}$.

e.g., [6], Ch. IV. §4) on their representation. We also observe that Theorem 1 describes the set of all moments of a generalized random process in an intrinsic way.

We prove Theorem 1 as follows. Introduce on $\mathbf{C}_0$ a quasi-scalar product by taking $\langle u,v\rangle_{\mathbf{S}} = \sum_{j,k=0}^{\infty} \mathbf{S}_{j+k}(v_j \otimes \bar{u}_k)$ $(u,v \in \mathbf{C}_0)$. After factorization with respect to $\{w \in \mathbf{C}_0 : \langle w,w\rangle_{\mathbf{S}} = 0\}$ and completion we obtain a Hilbert space $H_{\mathbf{S}}$. For an $e = \bar{e} \in \mathfrak{H}$ define the Hermitian operator A'_e in $H_{\mathbf{S}}$ with the aid of the correspondence

$$\mathbf{C}_0 \ni u = (u_0, u_1, u_2, \dots) \to (0, e \otimes u_0, e \otimes u_1, e \otimes u_2, \dots) \in \mathbf{C}_0 ;$$

A'_e and A'_ℓ commute with each other. Denote by A_e the closure of A'_e. In the determined case the operators A_e $(e \in \mathfrak{E})$ are self-adjoint and commute with each other. In the quasi-determined case one of them is non-self-adjoint, nevertheless it can be extended to a self-adjoint operator such that the system of operators obtained is commutative; this can be done with the aid of a procedure going back to M.S. Livšic [7]. Choosing for e vectors $e_1, e_2, \dots \in \mathfrak{E}$ such that $\ell.h.((e_n)_{n=0}^{\infty})$ is dense in $\mathfrak{H}$, we obtain a system of commuting self-adjoint operators. Then we expand this commuting operator system into a continual integral of common eigenvectors [8], and evaluate the respective generalized projection operators $P(\lambda)$. In this way we obtain (4).

Observe that in the determined case Theorem 1 in a somewhat less precise formulation can also be derived from the M-dimensional case by letting $M \to \infty$. This is done with the aid of R.A. Minlos's theorem on the extension of measures [9] along the same lines as the theorem of R.A. Minlos and V.V. Sazonov follows from the M-dimensional case of Bochner's theorem (cf. [6], Ch. IV, §4).

2. Theorem 1 is related to the so-called Euclidean field theory. Field theory in A.S. Wightman's formulation involves a theorem on the representation of more general objects than the moment-type sequence $\mathbf{S} = (\mathbf{S}_n)_{n=0}^{\infty}$. Now we proced towards an adequate formulation.

As in Section 1, let $\mathfrak{H}$ be a nuclear space, denote by $\mathfrak{H}^n \ni u_n \to \overleftarrow{u}_n \in \mathfrak{H}^n$ an involution depending on $n = 0, 1, \dots$ such that $\overleftarrow{u_j \otimes v_k} = \overleftarrow{v}_k \otimes \overleftarrow{u}_j$ $(u_j \in \mathfrak{H}^j;$ $v_k \in \mathfrak{H}^k;$ $j, k = 0, 1, \dots)$. A matrix $\mathbf{K} = (\mathbf{K}_{jk})_{j,k=0}^{\infty}$ $(\mathbf{K}_{jk} \in (\mathfrak{H}^{j+k})')$ is called p.d. if, similarly to (2), we have

$$(5)\qquad \sum_{j,k=0}^{\infty} \mathbf{K}_{jk}(u_j \otimes \bar{u}_k) \geq 0 \qquad (u \in \mathbf{C}_0)$$

Fix a real subspace $\mathcal{Y} \subseteq \mathcal{H}_{Re}$. A p.d. matrix $\mathbf{K}$ is called moment-type if

$$(6)\qquad \mathbf{K}_{j+1\,k}(u_j \otimes \varphi \otimes v_k) = \mathbf{K}_{j\,k+1}(u_j \otimes \varphi \otimes v_k),\ \mathbf{K}_{j+1\,k+1}(u_j \otimes \varphi \otimes \psi \otimes v_k) =$$
$$= \mathbf{K}_{j+1\,k+1}(u_j \otimes \psi \otimes \varphi \otimes v_k)\quad (\varphi, \psi \in \mathcal{Y};\ u_j \in \mathcal{H}^j, v_k \in \mathcal{H}^k;\ j,k = 0,1,\dots)$$

(in the case $\mathcal{Y} = \mathcal{H}_{Re}$ it follows from (6) that $\mathbf{K}_{jk}$ depends only on the sum of its indices: $\mathbf{K}_{jk} = \mathbf{S}_{j+k}$, and the sequence $\mathbf{S} = (\mathbf{S}_n)_{n=0}^{\infty}$ is moment-type).

The analogue of the condition of determinedness can be obtained in the following manner. Since $\mathbf{K}_{jk} \in (\mathcal{H}^{j+k})' = \bigcup_{\tau \in T} (\mathcal{H}_{\tau}^{j+k})'$, there exists a $\tau = \tau(j,k)$ such that $\mathbf{K}_{jk} \in (\mathcal{H}_{\tau(j,k)}^{j+k})'$ $(j,k = 0,1,\dots)$. Fix $\mathcal{L} \subseteq \mathcal{Y}$ such that $\mathrm{r.l.h.}(\mathcal{L})$ is dense in $\mathcal{Y}$. We shall say that $\mathbf{K}$ is determined if the class

$$\mathbf{C}\{M_n\},\quad M_n = d^n(\tau(n,n), \mathcal{L}) \|\mathbf{K}_{nn}\|^{1/2}_{(\mathcal{H}^{2n}_{\tau(n,n)})'}$$

is quasi-analytic (cf. footnote *). Analogously to Section 1 we may introduce the notion of quasi-determinedness of $\mathbf{K}$ as well, but we will not go into details.

In order to state the representation of type (4) it is convenient to pass from the matrix $\mathbf{K} = (\mathbf{K}_{jk})_{j,k=0}^{\infty}$ to the one $\mathbf{K}^{\circ} = (\mathbf{K}^{\circ}_{jk})_{j,k=0}^{\infty}$, where the elements $\mathbf{K}^{\circ}_{jk} \in \in (\mathcal{H}^{j+k})'$ are defined by the following relations: $\mathbf{K}^{\circ}_{jk}(u_j \otimes v_k) = \overline{\mathbf{K}_{jk}(\bar{u}_j \otimes \bar{v}_k)}$ $(u_j \in \mathcal{H}^j, v_k \in \mathcal{H}^k)$. It is clear that the matrices $\mathbf{K}$ and $\mathbf{K}^{\circ}$ uniquely determine each other. The following representation theorem containing Theorem 1 is valid:

Theorem 2. *Let $\mathbf{K} = (\mathbf{K}_{jk})_{j,k=0}^{\infty}$ be a moment-type matrix satisfying the condition of determinedness. Then the following representation holds:*

$$(7)\qquad \mathbf{K}^{\circ}_{jk} = \sum_{\alpha=0}^{j} \sum_{\beta=0}^{k} \int_{\mathcal{H}'_{Re}/\hat{\mathcal{Y}}} \underbrace{\lambda' \otimes \dots \otimes \lambda'}_{j-\alpha} \otimes d\rho_{\alpha\beta}(\lambda) \otimes \underbrace{\lambda' \otimes \dots \otimes \lambda'}_{k-\beta}\quad (j,k = 0,1,\dots).$$

Here $\hat{\mathcal{Y}}$ is the subspace of those functionals in $\mathcal{H}'_{Re}$ orthogonal to $\mathcal{Y}$, $\mathcal{H}'_{Re}/\hat{\mathcal{Y}} \ni \lambda \to \to \lambda' \in \mathcal{H}'_{Re}$ is a continuous left inverse mapping, fixed in advance, of the natural homomorphism $\mathcal{H}'_{Re} \to \mathcal{H}'_{Re}/\hat{\mathcal{Y}}$, and $\rho(\Delta) = (\rho_{\alpha\beta}(\Delta))_{\alpha,\beta=0}^{\infty}$ $(\rho_{\alpha\beta}(\Delta) \in (\mathcal{H}^{\alpha+\beta})')$

is a matrix measure defined on a σ *-algebra of subsets of* $\mathfrak{H}'_{Re}/\hat{\mathfrak{G}}$ *. Furthermore, the values of* $\varrho(\Delta)$ *are p.d. matrices in the sence of* (5), *and*

(8) $$(\varrho_{\alpha\beta}(\Delta))(\varphi \otimes u_{\alpha+\beta-1}) = (\varrho_{\alpha\beta}(\Delta))(u_{\alpha+\beta-1} \otimes \varphi) = 0$$

$$(\varphi \in \mathfrak{G}, \ u_{\alpha+\beta-1} \in \mathfrak{H}^{\alpha+\beta-1}).$$

holds. Such a measure is uniquely determined by (7). Conversely, all matrices of form (7) are moment-type.

We add that the integral under (7) converges in a weak sense defined in a natural way. As for uniqueness, a remark analogous to b) in Sect. 1 should be made. An observation of type c) in Sect. 1 also applies. Finally we mention that in the case $\mathfrak{G} = \mathfrak{H}_{Re}$ we have $\mathfrak{H}'_{Re}/\hat{\mathfrak{G}} = \mathfrak{H}'_{Re}$ and $\lambda' = \lambda$; in view of (8) $\varrho_{\alpha\beta}(\Delta) = 0$ holds for $\alpha+\beta > 0$, and the representation under (7) becomes identical to (4). The proof of Theorem 2 can be carried out analogously to the way presented in Sect. 1; in doing so, the vectors $e_1, e_2, \ldots \in \mathfrak{E}$ should be chosen such that $\text{r.l.h.}\,((e_\mu)_{\mu=0}^{\infty})$ is dense in $\mathfrak{G}$.

3. We give some examples. a) Let $T = \{0\}$, $\mathfrak{H}_0 = \mathbf{C}^M$ $(M = 1, 2, \ldots)$, and choose the usual transition to the complex conjugate as involution. Here $\mathbf{S}_n \in (\mathfrak{H}^n)' = \mathbf{C}^{nM}$, and owing to complete symmetry, to this vector there corresponds a collection of coordinates $s_{n_1,\ldots,n_M} = s_{\mathbf{n}}$, where $n_1, \ldots, n_M = 0, 1, \ldots$, $\mathbf{n} = (n_1, \ldots, n_M)$, $n_1 + \ldots + n_M = n$ (if $e_1, \ldots, e_M$ is a basis in $\mathbf{C}^M$, then $s_{n_1,\ldots,n_M}$ is the coordinate of $\mathbf{S}_n$ with respect to the basis vector $e_{\alpha_1} \otimes \ldots \otimes e_{\alpha_M}$ in $\mathbf{C}^{nM}$, provided in the collection $(\alpha_1, \ldots, \alpha_M)$ the index 1 occurs n_1 times, the index 2 n_2 times, etc; the order of occurence of the indices is unimportant because of the symmetry of $\mathbf{S}_n$). Condition 2 becomes the usual condition of being moment-type: $\sum_{j,k} s_{j+k} \bar{\xi}_j \xi_k \geq 0$, while representation (4) written coordinate-wise goes over into (1). Thus our discussion embraces also the usual M -dimensional moment problem with a certain assumption on the growth of $s_{\mathbf{n}}$ ensuring its solvability (as mentioned earlier, this assumption can be precisely formulated).

b) Let $T = \{0, 1, \ldots\}$, let $\mathfrak{H}_\tau$ be the Sobolev space $\mathbf{W}_2^\tau(\mathbf{R}^N, (1+|x|^2)^\tau dx)$ i.e. the completion of the class $\mathbf{C}_0^\infty(\mathbf{R}^N)$ of all infinitely differentiable functions with compact support, the completion being taken with respect to the scalar product

$$(9)\quad (\varphi,\psi)_{W_2^\tau(\mathbf{R}^N,(1+|x|^2)^\tau dx)} = \sum_{|\alpha|\le\tau} \int_{\mathbf{R}^N} (D^\alpha\varphi)(x)\overline{(D^\alpha\psi)(x)}(1+|x|^2)^\tau dx .$$

It is not difficult to show that the corresponding nuclear space $\mathfrak{H}$ coincides with the space $\mathcal{S}(\mathbf{R}^N)$ of. L. Schwartz consisting of those functions of the class $C^\infty(\mathbf{R}^N)$ vanishing at infinity more rapidly than any power of $|x|^{-1}$, and $\mathfrak{H}^n = \mathcal{S}(\mathbf{R}^{nN})$. The sequence S_n consists of tempered distributions. It can be shown that for a suitable $\mathcal{L}$ with linear hull dense in $\mathcal{S}(\mathbf{R}^N)$ we have $d(\tau,\mathcal{L}) \le \tau^{(N+\varepsilon)\tau}$ $(\varepsilon>0)$. Hence the condition for the determinedness of our S follows (the conditions for quasi-determinedness will not be considered here).

c) Analogously to b), we may consider the case in which $\mathfrak{H}$ is the S.L. Sobolev – L. Schwartz space $\mathcal{D}(\mathbf{R}^N)$ of functions with compact support. Now T consists of pairs $\tau=(\tau_1,\tau_2(x))$, where $\tau_1 = 0,1,\ldots$ and $C^\infty(\mathbf{R}^N)\ni\tau_2(x)\ge 1$ $(x\in\mathbf{R}^N)$ is a weight function. The Sobolev space $W_2^{\tau_1}(\mathbf{R}^N,\tau_2^2(x)dx)$ takes the role of the space $\mathfrak{H}_\tau$ (in the expression under (9) τ and $(1+|x|^2)^\tau$ should be replaced by τ_1 and $\tau_2^2(x)$, respectively.

d) Analogously to examples b) and c), we may choose for $\mathfrak{H}_\tau$ a family (countable or uncountable) of spaces with various weight functions. Then the role of $\mathfrak{H}'$ will be played by a certain sequence space, and considerations of the type occurring in Example a) lead to the infinite-dimensional moment problem. This enables us to obtain results that are close to Theorem 5.4 in [5].

e) Now we study the case of moment-type sequences consisting of elements of negative Hilbert spaces. Let $\mathfrak{H}_{-1}\supseteq\mathfrak{H}_0\supseteq\mathfrak{H}_1$ be a chain with involution; this may always be considered as part of a chain with involution $\ldots\supseteq\mathfrak{H}_{-2}\supseteq\mathfrak{H}_{-1}\supseteq\mathfrak{H}_0\supseteq$ $\supseteq\mathfrak{H}_1\supseteq\mathfrak{H}_2\supseteq\ldots$ determining a nuclear space $\mathfrak{H}$ with respect to these notions see e.g. [4], Ch. 1). So, if $S_n\in\mathfrak{H}_{-1}^n$, then $S_n\in(\mathfrak{H}^n)'$, and the representation (4) with a measure $d\rho(\lambda)$ concentrated on $\mathfrak{H}_{-\tau_0,\mathrm{Re}}$ is valid. Since there is sufficient freedom in the choice of $\mathfrak{H}$, as $\mathfrak{H}_{-\tau_0,\mathrm{Re}}$ we can take any space $\mathfrak{H}_{-2,\mathrm{Re}}$ such that the corresponding positive space $\mathfrak{H}_2$ embedded in $\mathfrak{H}_1$ is quasi-nuclear. So, if now we set $\tau(n)=1$ $(n=0,1,\ldots)$ then the condition of determinedness assumes the following form: the class $C\{\|S_{2n}\|^{1/2}_{\mathfrak{H}_{-1}^{2n}}\}$ is quasi-analytic.

f) In [10], on p. 18 it is shown that if the moment-type sequence

$S = (S_n)_{n=0}^{\infty}$ consists of sufficiently smooth functions $S_n = S_n(x_1, \dots, x_n)$ of the points $x_1, \dots, x_n \in \mathbf{R}^N$, then under certain conditions of the type of determinedness we have the representation

$$S_n(x_1, \dots, x_n) = \int_{C_{Re}(\mathbf{R}^N)} \lambda(x_1) \dots \lambda(x_n) \, d\varrho(\lambda) \quad (x_1, \dots, x_n \in \mathbf{R}^N;\ n = 0, 1, \dots),$$

where the domain of integration is the space of real continuous functions defined on $\mathbf{R}^N$, and the integral converges absolutely and uniformly whenever $(x_1, \dots, x_n)$ varies within a compact subset of $\mathbf{R}^{nN}$.

Similarly to a) – f), we may also consider examples of moment-type matrices. Here we only observe that in the case of function spaces the mapping

$$u_n(x_1, x_2, \dots, x_{n-1}, x_n) \to \overleftarrow{u}_n(x_1, x_2, \dots, x_{n-1}, x_n) = \overline{u_n(x_n, x_{n-1}, \dots, x_2, x_1)}$$

may serve as an involution $\leftarrow$, where the bar indicates transition to the complex conjugate. Then the conditions (6) are certain stipulations on the symmetry of $K_{jk} = K_{jk}(x_1, \dots, x_j, y_1, \dots, y_k)$. Examples of moment-type matrices are contained in the papers [10, 13-17].

4. Representation (7) in more special circumstances was obtained by the author in [11-15] (for detailed proofs see [10, 16, 17]). In these papers cases are considered both of a finite dimensional $\mathfrak{Y}$, when the integration in (7) is of finite multiplicity, and of infinite dimensional $\mathfrak{Y}$, leading to a continual integral in (7). Later V.P. Gačok [18-20], found that the particular case of representations such that $K_{jk}(x_1, \dots, x_j, y_1, \dots, y_k)$ is completely symmetric is of interest in the Euclidean theory of fields. In case S_n belongs to the nth tensorial power of a Hilbert space, and essentially in the determined case, with the aid of the scheme and techniques elaborated in [11-14], he makes an attempt [18-20] to obtain representations of type (4), though he is led to an erroneous answer (compare Theorem 5 in [19] with our formula (4)). The results of Sect. 1 have been obtained by the author jointly with S.N. Šifrin.

REFERENCES

[1] R.F. Streater and A.S. Wightman, *PCT, spin and statistics, and all that* (Russian translation of the English original. New York – Amsterdam, 1964) (Moscow, 1966).

[2] R. Jost, *The general theory of quantized fields* (reviewed Russian translation of the English original, Providence, R.I., 1965) (Moscow, 1967).

[3] N.N. Bogoljubov, A.A. Logunov, and I.T. Todorov, *Foundations of an axiomatic approach to quantum field theory* (Russian) (Moscow, 1969).

[4] Ju.M. Berezanskiĭ, *Eigenfunction expansions of selfadjoint operators* (Kiev, 1965). (Russian)

[5] A.G. Kostjučenko and B.S. Mitjagin, Positive-definite functionals on nuclear spaces (Russian), *Trudy Moskov. Mat. Obšč.*, 9 (1960), 283-316.

[6] I.M. Gel'fand and N.Ja. Vilenkin, *Generalized functions.* Vol. 4: *Some applications of harmonic analysis. Rigged Hilbert spaces* (Moscow, 1958). (in Russian)

[7] M.S. Livšic, Doctoral dissertation, Steklov Math. Inst., Moscow, 1945.

[8] Ju.M. Berezanskiĭ, Decomposition by generalized eigenvectors and integral representation of positive definite kernels in form of continual integrals, *Sibirsk. Mat. Ž.*, 9 (1968), 998-1013 (in Russian).

[9] R.A. Minlos, Generalized random processes and their extension to a measure *Trudy Moskov. Mat. Obšč.*, 8 (1959), 497-518 (in Russian).

[10] Ju.M. Berezanskiĭ, Representation of functionals of Wightman's type by continual integrals (Russian), *Funkcional. Anal. i Priložen.*, 3 (1969), 3-12.

[11] Ju.M. Berezanskiĭ, Self-adjointness of field operators and integral representations of Wightman-type functionals (Russian), *Ukrain. Mat. Ž.*, 18 (1966), 3-12.

[12] Ju.M. Berezanskiĭ and M.L. Gorbačuk, Integral representation of positive-definite Wightman-type functionals (Mathematical Congress, Moscow, 1966). (in Russian)

[13] Ju.M. Berezanskiĭ, A generalization of the power moment problem, *Dokl. Akad. Nauk SSSR,* 172 (1967), 514-517 (in Russian).

[14] Ju.M. Berezanskiĭ, Integral representation of positive definite functionals of Wightman's type, *Ukrain. Mat. Ž.,* 19 (1967), 89-95 (in Russian).

[15] Ju.M. Berezanskiĭ, Integral representation of Wightman-type functionals which are generalized functions of the space variables, *Uspehi Mat. Nauk,* 25 (1970), 261-262 (in Russian).

[16] Ju.M. Berezanskiĭ, Generalized power moment problem, *Trudy Moskov. Mat. Obšč.,* 21 (1970), 47-101 (in Russian).

[17] Ju.M. Berezanskiĭ, On the generalized power moment problem, *Ukrain. Mat. Ž.,* 22 (1970), 435-460 (in Russian).

[18] V.P. Gačok, Integral representations of vacuum expectation values involving local commutativity. Preprint (Inst. Teoret. Fiz. Akad. Nauk USSR, Kiev, 19 1967).

[19] V.P. Gačok, Integral representations in the euclidean quantum field theory and their consequences in relativistic case. Preprint (Inst. Teoret. Fiz. Akad. Nauk USSR, Kiev, 1967).

[20] V.P. Gačok, Moment problem and quantum field theory. (Self-review of doctoral dissertation, Inst. Mat. Akad. Nauk USSR, Kiev, 1968). (in Russian).

COLLOQUIA MATHEMATICA SOCIETATIS JÁNOS BOLYAI
5. HILBERT SPACE OPERATORS, TIHANY (HUNGARY), 1970

Symbols of operators and quantization

F. A. BEREZIN and M. A. ŠUBIN

Let $L_2(M)$ be the Hilbert space of functions square-integrable on a measure space M. It is convenient to define linear operators on $L_2(M)$ by means of functions of two variables. The best known way is by means of kernel:

$$(Af)(x) = \int K(x,y)f(y)dy .$$

In case of such a correspondence between operators A and functions $\varphi(x,y)$, we shall call the function $\varphi(x,y)$ the *symbol* of the corresponding operator A. The kernel $K(x,y)$ is the particular case of symbol. In this paper we give a survey on different sorts of symbols, considering mainly the case when M is the real n-dimensional euclidean space with the usual Lebesque measure. One of the sources of appearance of symbols is the expression of operators in algebras of operators through generators $\hat{p}_k$ and $\hat{q}_k$, where $\hat{p}_k$ and $\hat{q}_k$ are the "operators of impulse and coordinate" well known from quantum mechanics:

$$\hat{p}_k f = \frac{h}{i}\,\frac{\partial f}{\partial x_k}, \quad \hat{q}_k f = x_k f .$$

The expression of other operators through $\hat{p}_k, \hat{q}_k$ is not unique; for example,

$\hat{p}\hat{q} = \hat{q}\hat{p} + \frac{h}{i}E$. However, if we agree upon a standard entry, or, as one often calls it, a *normal form,* then expression becomes unique. For example, every polynomial of p_k, q_k may be written in form of a linear combination of operators of form

$$(*) \qquad \hat{p}_1^{k_1} \ldots \hat{p}_n^{k_n} \hat{q}_1^{k'_1} \ldots \hat{q}_n^{k'_n}$$

($\hat{p}_k$ stands left from $\hat{q}_k$).

To the operator under (*) we make correspond the function $p_1^{k_1} \ldots p_n^{k_n} q_1^{k_1} \ldots q_n^{k_n}$. This correspondence may be extended linearly to a one-to-one correspondence between operators having polynomial expression through $\hat{p}_k, \hat{q}_k$, and usual polynomials of $2n$ variables p_k, q_k which are the symbols of these operators. Furthermore the correspondence in question may be extended to operators of a more general form. In this paper we consider also the symbols that come from the analogous $\hat{q}\hat{p}$ -normal form, symmetrical normal form (definition in § 1), and the Wick normal form, which latter is similar to the $\hat{q}\hat{p}$ -form, but with other generators $\hat{a}_k = \frac{1}{\sqrt{2}}(\hat{p}_k + i\hat{q}_k)$, $\hat{a}_k^* = \frac{1}{\sqrt{2}}(\hat{p}_k - i\hat{q}_k)$. We also consider such Wick's normal form that $\hat{q}_k$ and $\hat{p}_k$ are not operators of coordinate and impulse, but generators of a Clifford algebra. (This case is important in Fermi's variant of second quantization.) In this case the symbol is an element of a Grassman algebra. The inverse correspondence: symbol $\longrightarrow$ operator is called quantization. Everything in this paper that considers methods of quantization (by means of different normal forms) keeps within the general scheme of linear quantization.

In this paper the symbols and the methods of quantization are marked in the same way as the corresponding normal forms (for example $\hat{q}\hat{p}$ -symbol, $\hat{q}\hat{p}$ -quantization etc.). The paper consists of two parts. The first part has in the main a heuristic character. We establish here the basic formulae (the symbol of composition of operators is expressed by the symbols of factors, the trace of operators is expressed by its symbol etc.). We deduce an expression for the symbol of the operator $e^{it\hat{H}}$ by the symbol of the operator $\hat{H}$ by means of Feynmann's integral on trajectories in the phase space. We establish also the connection between quantum and classical mechanics and the various quasiclassical asymptotics. Although the formulae obtained in this paper have no strong ground in most cases, nevertheless, in the authors opinion, the symbols are the most suitable mathematical methods for their derivation. The second part contains the strong results about one class of $\hat{q}\hat{p}$ -symbols. In particular, the indices of defect of an operator with symbol of this class are found. The symbols of hypoelliptic operators in this class are studied in detail. In conclusion we note that the algebra of operators with generators $\hat{p}_k, \hat{q}_k$ and the Clifford algebra are not the unique algebras the elements of which are defined by means of symbols; they are themselves parts of other more simple algebras. For example, if G is an arbitrary Lie algebra and $\mathfrak{U}$ is its associative cover, then it is often useful to define the elements of $\mathfrak{U}$ by means of symbols which are generalizations of symmetrical symbols described in reference [1]. It is essential for the needs of the theory of group representations to find the generalization of Wick's symbols in the same case.

The present paper is written on basis of papers [2], [3], [4], [5], except the considerations connected with the asymptotic of the Green function, which are published for the first time.

1. THE DEFINITION OF LINEAR QUANTIZATION AND EXAMPLES

Let L be the phase space of the classical mechanical system with n degrees of freedom, $q=(q_1,\dots,q_n)$, $p=(p_1,\dots,p_n)$ being the canonical coordinates in L. The expressions qp, qx, p^2 etc. are abbreviated notations for $qp=\sum q_i p_i$, $qx=\sum q_i x_i$, $p^2=\sum p_i^2$ etc. and dp, dq etc. are abbreviated notation for products of differentials $dp = dp_1 \dots dp_n$, $dq = dq_1 \dots dq_n$, respectively. The Hilbert space of all functions $f(x)$ with inner product $(f,g)=\int f\bar{g}\,dx$ will be denoted by L_2. The operators $\hat{p}_k$ and $\hat{q}_k$ in L_2 have the form

$$(\hat{p}_k f)(x) = \frac{h}{i}\frac{\partial f}{\partial x_k}, \quad (\hat{q}_k f)(x) = x_k f(x).$$

The following relations are fulfilled between them:

$$\hat{p}_k\hat{p}_{k'} - \hat{p}_{k'}\hat{p}_k = \hat{q}_k\hat{q}_{k'} - \hat{q}_{k'}\hat{q}_k = 0, \quad \hat{p}_k\hat{q}_{k'} - \hat{q}_{k'}\hat{p}_k = ihE\delta_{kk'}.$$

The problem of quantization is to associate with every classical observable, i.e. every real function $f(q,p)$, a quantum observable, i.e. a selfadjoint operator in a Hilbert space. It is assumed that the following conditions are fulfilled:

a) the correspondence $f \longrightarrow \hat{f}$ depends on the parameter h (Planck's constant), moreover the commutator $[\hat{f},\hat{g}]=\hat{f}\hat{g}-\hat{g}\hat{f}$ may be expressed as $[\hat{f},\hat{g}] = ih\hat{c}+o(h)$, where $\hat{c}$ is the operator corresponding to the function $c=[f,g]$, $[f,g]$ being the Poisson bracket;

b) The function $f(q,p)$ is the limit in some sense of the operator $\hat{f}$ when $h \to 0$.

Evidently, quantization is not, in considerable degree, a unique operation. We consider some variants of quantization below. We shall call the function $f(q,p)$ corresponding to the opertor $\hat{f}$ the symbol of this operator.

The correspondence between operators and symbols is completely defined by formulae expressing the symbols of the operators $\hat{p}\hat{A}$, $\hat{A}\hat{p}$, $\hat{q}\hat{A}$, $\hat{A}\hat{q}$ by the symbols of operator $\hat{A}$ and of the identity operator. We shall say that a linear quantization is defined if these formulae have the forms.

$$\hat{p}_i\hat{A} \longleftrightarrow L^1_{p_i}A, \; \hat{A}\hat{p}_i \longleftrightarrow L^2_{p_i}A, \; \hat{q}_i\hat{A} \longleftrightarrow L^1_{q_i}A, \; \hat{A}\hat{q}_i \longleftrightarrow L^2_{q_i}A,$$

where $L^1_{p_i}$ etc. are differential operators of first order whose coefficients by derivatives are constants and those by free members are linear. For example

$$L^1_{p_i} = \sum_j (\alpha^{(1)}_{ji}p_j + \beta^{(1)}_{ij}q_j + \gamma^{(1)}_{ij}\frac{\partial}{\partial p_j} + \delta^{(1)}_{ij}\frac{\partial}{\partial q_j})$$

(the matrices $\alpha^{(1)}_{ij}$ cannot be arbitrary: they must satisfy the conditions induced by the commutation formulae $[\hat{p}_i,\hat{q}_j] = -ih\delta_{ij}$).

1. Kernels of operators. The simplest linear quantization is to associate with every operator A its kernel $K(x,y)$, which is a distribution of two variables $x = (x_1,\dots,x_n)$ and $y = (y_1,\dots,y_n)$, $K(x,y) = \langle x|A|y\rangle$, i.e.

$$(\hat{A}f)(x) = \int K(x,y)f(y)\,dy .$$

For this correspondence

$$\text{(1.1)} \quad \hat{p}_k\hat{A} \longleftrightarrow -ih\frac{\partial K}{\partial x_k}, \; \hat{A}\hat{p}_k \longleftrightarrow ih\frac{\partial K}{\partial y_k}, \; \hat{q}_k\hat{A} \longleftrightarrow x_kA, \; \hat{A}\hat{q}_k \longleftrightarrow Ky_k$$

hold, so we have in fact a linear quantization.

2. $\hat{q}\hat{p}$-quantization. With every polynomial $f(q,p) = \sum f_{m_1\dots m_n, m'_1\dots m'_n} p_1^{m_1}\cdots p_n^{m_n} q_1^{m'_1}\cdots q_n^{m'_n}$ we associate the following operator in L_2 :

$$\text{(1.2)} \qquad \hat{f} = \sum f_{m_1\dots m_n, m'_1\dots m'_n}\hat{q}_1^{m'_1}\cdots q_n^{m'_n}p_1^{m_1}\cdots p_n^{m_n},$$

which is a differential operator with polynomial coefficients. We shall extend the correspondence between polynomials and polynomial differential operators to a correspondence between functions and operators of a more general form. To this purpose we note that if the operator f corresponds to the polynomial f then *

$$\hat{p}_k\hat{f} \leftrightarrow (p_k - ih\frac{\partial}{\partial q_k})f, \quad \hat{f}\hat{p}_k \leftrightarrow fp_k,$$

(1.3)

$$\hat{q}_k\hat{f} \leftrightarrow q_k f, \quad \hat{f}\hat{q}_k \leftrightarrow (q_k - ih\frac{\partial}{\partial p_k})f.$$

According to the definition of linear quantization we require these formulae to be valid also in case $f(q,p)$ is not a polynomial. Let f be an operator in L_2, and $K(x,y) = \langle x|\hat{f}|y\rangle$ its kernel. We shall look for the correspondence between functions and operators in the form

$$f(q,p) = \int L(q,p|x,y)K(x,y)\,dx\,dy$$

(1.4)

$$K(x,y) = \int L^*(x,y|q,p)f(q,p)\,dq\,dp\,.$$

From relations (1.1) and (1.3) for L and L^* we deduce the equations:

$$(p_k - ih\frac{\partial}{\partial q_k})L = ih\frac{\partial L}{\partial x_k}, \qquad p_k L = -ih\frac{\partial L}{\partial y_k};$$

(1.5)

$$q_k L = x_k L, \qquad (q_k - ih\frac{\partial}{\partial p_k})L = y_k L;$$

$$-ih\frac{\partial L^*}{\partial x_k} = (p_k + ih\frac{\partial}{\partial q_k})L^*; \qquad ih\frac{\partial L^*}{\partial y_k} = p_k L^*;$$

(1.5')

$$x_k L^* = q_k L^*; \qquad y_k L^* = (q_k + ih\frac{\partial}{\partial p_k})L^*.$$

All eight equations are obtained quite similarly, therefore we explain their

* Proof of formulae (1.3). Let $f = q_1^{m_1}\cdots q_n^{m'_n}p_1^{m'_1}\cdots p_n^{m_n}$. We note that if $m'_k > 0$, then

$$\hat{p}_k\hat{q}_k^{m'_k} = (\hat{p}_k\hat{q}_k - \hat{q}_k\hat{p}_k)\hat{q}_k^{m'_k-1} + \hat{q}_k\hat{p}_k\hat{q}_k^{m'_k-1} = -ih\hat{q}_k^{m'_k-1} + \hat{q}_k\hat{p}_k\hat{q}_k^{m'_k-1} = -mih\hat{q}_k^{m'_k-1} + \hat{q}_k^{m'_k}p_k.$$

Therefore $\hat{p}_k f \leftrightarrow (p_k - ih\frac{\partial}{\partial q_k})f$. The proof of the fourth formula under (1.3) is similar. The second and third formulae are evident.

derivation only in case of the first equation under (1.5).

From (1.1) and (1.3) we conclude that the operator $\hat{p}_k \hat{f}$ corresponds to the symbol $(p_k - ih\frac{\partial}{\partial q_k})f$ and to the kernel $-ih\frac{\partial K}{\partial x_k}$. By means of (1.4) we now deduce the following identities:

$$(p_k - ih\frac{\partial}{\partial q_k})f = \int (p_k - ih\frac{\partial}{\partial q_k}) LK\,dx\,dy =$$

$$= \int L(-ih\frac{\partial}{\partial x_k})\,dx\,dy = ih\int \frac{\partial L}{\partial x_k} K\,dx\,dy\,.$$

(The last equality is obtained by integration by parts.) Thus we have

$$\int (p_k - ih\frac{\partial}{\partial q_k}) LK\,dx\,dy = ih\int \frac{\partial L}{\partial x_k} K\,dx\,dy\,.$$

This relation must be satisfied for arbitrary K. Therefore the first equality in (1.5) must be fulfilled.

Solutions of equations (1.5), (1.5') are unique up to a scalar factor and are equal to the functions

(1.6) $$L = \delta(q-x)e^{\frac{i}{h}p(y-q)}, \quad L^* = (2\pi h)^{-n}\delta(q-x)e^{-\frac{i}{h}p(y-q)}.$$

The factors are defined by the condition that the identity operator corresponds to the kernel $\delta(x-y)$ on the one hand, and to the symbol $f \equiv 1$ on the other.

From (1.6) we finilly deduce the following relations between symbols and kernels

(1.7) $$f(q,p) = \int K(q,p)\,e^{-\frac{i}{h}p(y-q)}\,dy$$

(1.7') $$K(x,y) = (2\pi h)^{-n}\int f(x,p)\,e^{-\frac{i}{h}p(y-x)}\,dp\,.$$

Let $\varphi(x) \in L^2$. From (1.7') it follows that the function $(\hat{f}\varphi)(x)$ can be expressed as

(1.8) $$(\hat{f}\varphi)(x) = \int f(x,p)\,\tilde{\varphi}(p)\,e^{\frac{i}{h}px}\,dp$$

with

$$\tilde{\varphi}(p) = (2\pi h)^{-n}\int \varphi(y)\,e^{-\frac{i}{h}py}\,dy\,.$$

(1.8) is the basis of theory of pseudodifferential operators * [6].

(1.7) and (1.7') entails that if $\hat{f} = \hat{f}_1\hat{f}_2$ then the symbols of the operators $\hat{f}, \hat{f}_1, \hat{f}_2$ are connected by the following relation:

(1.9) $$f(q,p) = \frac{1}{(2\pi h)^n}\int f_1(q,p_1)\,f_2(q_1,p)\,e^{-\frac{i}{h}(p_1-p)(q_1-q)}\,dq_1\,dp_1\,.$$

From (1.7) and (1.7') the formula

(1.9') $$f(q,p) = f_1\left(q, p + \frac{h}{i}\frac{\partial}{\partial\tilde{q}}\right) f_2(\tilde{q},p)\Big|_{\tilde{q}=q}$$

is also easily derived. (1.9) is the basis for evaluating of Feynmann's functional integral on trajectories in the phase space.

Besides this formula we also mention two other remarkable ones:

(1.10) $$\mathrm{Sp}\,\hat{f} = \int K(x,x)\,dx = \frac{1}{(2\pi h)^n}\int f(q,p)\,dq\,dp\,;$$

(1.11) $$\mathrm{Sp}\,\hat{f}_1\hat{f}_2^* = \int K_1(x,y)\,K_2(x,y)\,dx\,dy = \frac{1}{(2\pi h)^n}\int f_1(q,p)\,f_2(q,p)\,dq\,dp.$$

Formula (1.11) plays the role of Plancherel's theorem for the transformation

* For the reader whom the reasoning given above seems inconvincing we suggest to verify, directly by using (1.8), that if $f = \sum f_{m_1\ldots m'_n}\,p_1^{m_1}\ldots q_n^{m'_n}$ then

$$(\hat{f}\varphi)(x) = \sum f_{m_1\ldots m'_n}\,x_1^{m'_1}\ldots x_n^{m'_n}\left(\frac{h}{i}\right)^{m_1+\cdots+m_n}\frac{\partial^{m_1+\cdots+m_n}\varphi}{\partial x_1^{m_1}\ldots\partial x_n^{m_n}}\,.$$

under (1.7).*

3. $\hat{p}\hat{q}$ - quantization. With the polynomial

$$f(q,p) = \sum f_{m_1 \dots m_n, m'_1 \dots m'_n} p_1^{m_1} \dots p_n^{m_n} q_1^{m'_1} \dots q_n^{m'_n}$$

we now associate the operator $\hat{f}$ in L_2 of form

$$f = \sum f_{m_1 \dots m_n, m'_1 \dots m'_n} \hat{p}_1^{m_1} \dots \hat{p}_n^{m_n} \hat{q}_1^{m'_1} \dots \hat{q}_n^{m'_n}, \tag{1.12}$$

which is a polynomial differential operator in L_2 as in the preceding case.

The continuation of the correspondence between polynomial differential operators and polynomials is based on considerations similar to those used in the case of $\hat{q}\hat{p}$ -quantization. The following formulae, analogous to those under (1.3), are fulfilled:

$$\begin{aligned} &\hat{p}_k\hat{f} \leftrightarrow p_k f, \quad \hat{f}\hat{p}_k \leftrightarrow (p_k + ih\frac{\partial}{\partial q_k})f, \\ &\hat{q}_k\hat{f} \leftrightarrow (q_k + ih\frac{\partial}{\partial p_k})f, \quad \hat{f}\hat{q}_k \leftrightarrow f q_k . \end{aligned} \tag{1.13}$$

We look for the correspondence between symbols and kernels in the form described under (1.4). From (1.1) and (1.13), for the functions L and L^* we derive some equations similar to those at (1.5) and (1.5')** :

$$\begin{aligned} &p_k L = ih\frac{\partial L}{\partial x_k}; \quad (p_k + ih\frac{\partial}{\partial q_k})L = -ih\frac{\partial L}{\partial y_k}; \\ &(q_k + ih\frac{\partial}{\partial p_k})L = x_k L; \quad q_k L = y_k L; \end{aligned} \tag{1.14}$$

* If the function $K(x,y)$ is smooth and rapidly decreasing then formulae (1.7') and (1.11) may be obtained immediately from (1.7) in a way similar to the derivation of the formula for the inverse Fourier transform and of Plancherel's theorem for the Fourier transform.

Thus, in case $K(x,y)$ is a distribution (for example corresponding to a polynomial $f(q,p)$), it becomes possible to approach formulae (1.7) and (1.7') from the positions from which one usually considers the Fourier transforms of distributions [7] .

** The proof is identical (see the footnote on p. 25)

(1.14')
$$-ih\frac{\partial L^*}{\partial x_k} = p_k L^*; \qquad ih\frac{\partial L^*}{\partial y_k} = \left(p_k - ih\frac{\partial}{\partial q_k}\right)L^*;$$
$$x_k L^* = \left(q_k - ih\frac{\partial}{\partial p_k}\right)L^*; \qquad y_k L^* = q_k L^*.$$

By (1.14) and (1.14') we can determine the functions L and L^* up to a scalar factor, which is defined by the condition that the identity operator corresponds to the kernel $K(x,y) = \delta(x-y)$ and the symbol $f \equiv 1$. We find:

(1.15) $$L = \delta(q-y)e^{-\frac{i}{h}p(x-q)}, \quad L^* = (2\pi h)^{-n}\delta(q-y)e^{\frac{i}{h}p(x-q)}.$$

Thus we have:

(1.16) $$f(q,p) = \int K(x,q)e^{-\frac{i}{h}p(x-q)}dx,$$

(1.16') $$K(x,y) = (2\pi h)^{-n}\int f(y,p)e^{\frac{i}{h}p(x-y)}dp.$$

From (1.16) and (1.16') we derive the composition formula: if $f = f_1 f_2$ then we have

(1.17) $$f(q,p) = \frac{1}{(2\pi h)^n}\int f_1(q_1,p)f_2(q,p_1)e^{\frac{i}{h}(q-q_1)(p-p_1)}dq_1dp_1.$$

The formula for trace and the Plancherel formula have the forms given under (1.10) and (1.11).*

We temporarily denote by $f_{\hat{q}\hat{p}}$ the symbol corresponding to the operator $\hat{f}$ by $\hat{q}\hat{p}$-quantization ($\hat{q}\hat{p}$-symbol) and by $f_{\hat{p}\hat{q}}$ the symbol corresponding to the operator f by $\hat{p}\hat{q}$-quantization ($\hat{p}\hat{q}$-symbol).

From (1.7) and (1.16) we obtain the relation between $f_{\hat{p}\hat{q}}$ and $f_{\hat{q}\hat{p}}$:

(1.18) $$f_{\hat{p}\hat{q}}(q,p) = \frac{1}{(2\pi h)^n}\int f_{\hat{q}\hat{p}}(q',p')e^{\frac{i}{h}(p'-p)(q'-q)}dq'dp',$$

*About the proof of the formulae under (1.16), (1.16'), and (1.17), the same arguments are true as in case of $\hat{q}\hat{p}$-quantization.

(1.18') $$f_{\hat{q}\hat{p}}(q,p) = \frac{1}{(2\pi h)^n}\int f_{\hat{p}\hat{q}}(q',p')e^{-\frac{i}{h}(p'-p)(q'-q)}dq'dp'.$$

From (1.11) we get the Plancherel formula for the transformation (1.18), (1.18'):

(1.19) $$\int f_{\hat{q}\hat{p}}(q,p)\overline{g_{\hat{q}\hat{p}}(q,p)}dq\,dp = \int f_{\hat{p}\hat{q}}(q,p)\overline{g_{\hat{p}\hat{q}}(q,p)}dq\,dp.$$

Finally we shall deduce an expression of the symbol of the adjoint of an operator by the symbol of this latter operator.

Put

$$\hat{f} = \sum f_{m_1\dots m_n, m_1'\dots m_n'}\hat{p}_1^{m_1}\dots\hat{p}_n^{m_n}\hat{q}_1^{m_1'}\dots\hat{q}_n^{m_n'}.$$

Then

$$\hat{f}^* = \sum \overline{f}_{m_1\dots m_n'}\hat{q}_n^{m_n'}\dots\hat{p}_1^{m_1}.$$

We denote the $\hat{p}\hat{q}$- and $\hat{q}\hat{p}$-symbols of the operator $\hat{f}^*$ by $f^*_{\hat{p}\hat{q}}$ and $f^*_{\hat{q}\hat{p}}$, respectively. It is obvious that

(1.20) $$f^*_{\hat{q}\hat{p}} = \overline{f}_{\hat{p}\hat{q}} = \frac{1}{(2\pi h)^n}\int \overline{f}_{\hat{q}\hat{p}}(q',p')e^{-\frac{i}{h}(p-p')(q-q')}dq'dp'.$$

Similarly

(1.20') $$f^*_{\hat{p}\hat{q}}(q,p) = \overline{f}_{\hat{q}\hat{p}}(q,p) = \frac{1}{(2\pi h)^n}\int \overline{f}_{\hat{p}\hat{q}}(q',p')e^{\frac{i}{h}(p-p')(q-q')}dq'dp'.$$

4. T h e s y m m e t r i c a l o r t h e W e y l q u a n t i z a t i o n. Let A and B be two noncommuting operators. Consider the operator $(\alpha A+\beta B)^n$ and decompose it by powers of α and β:

$$(\alpha A+\beta B)^n = \sum \frac{n!}{k!\,l!}\alpha^k\beta^l(A^kB^l).$$

The operator (A^kB^l) defined by this formula will be called the symmetrical product of the operators A^k and B^l.

Examples: $(AB) = \frac{1}{2}(AB+BA)$, $(A^2B) = \frac{1}{3}(A^2B+ABA+BA^2)$. Now consider the polynomial

$$f(q,p)=\sum f_{m_1\dots m_n, m_1'\dots m_n'} p_1^{m_1}\dots p_n^{m_n} q_1^{m_1'}\dots q_n^{m_n'}.$$

With this polynomial we associate the following operator in L_2 :

$$\hat{f}=\sum f_{m_1\dots m_n, m_1'\dots m_n'}(\hat{p}_1^{m_1}\hat{q}_1^{m_1'})\dots(\hat{q}_n^{m_n}\hat{q}_n^{m_n'}).$$

It is not difficult to verify by induction on the degree of the polynomial f that the correspondence thus constructed between polynomials and polynomial differential operators is one-to-one [3].

As in the previous cases, the basic part in further extending the described correspondence between polynomials and polynomial differential operators is played by the formulae expressing the symbols of the operators $\hat{p}_k\hat{f}, \hat{f}\hat{p}_k, \hat{q}_k\hat{f}, \hat{f}\hat{q}_k$ through that of the operator $\hat{f}$:

$$\hat{p}_k\hat{f} \leftrightarrow \left(p_k-\frac{ih}{2}\frac{\partial}{\partial q_k}\right)f, \quad \hat{f}\hat{p}_k \leftrightarrow \left(p_k+\frac{ih}{2}\frac{\partial}{\partial q_k}\right)f,$$

(1.23)

$$\hat{q}_k\hat{f} \leftrightarrow \left(q_k+\frac{ih}{2}\frac{\partial}{\partial p_k}\right)f, \quad fq_k \leftrightarrow \left(q_k-\frac{ih}{2}\frac{\partial}{\partial p_k}\right)f.$$

The proof of these formulae is contained in [3]. It is more difficult than the proofs of the formulae at (1.3) and (1.13). From (1.1) and (1.23) we can derive equations for the functions L and L^* relating symbols and kernels of operators.

$$\left(p_k-\frac{ih}{2}\frac{\partial}{\partial q_k}\right)L = ih\frac{\partial L}{\partial x_k}; \quad \left(p_k+\frac{ih}{2}\frac{\partial}{\partial q_k}\right)L = -ih\frac{\partial L}{\partial y_k};$$

$$\left(q_k+\frac{ih}{2}\frac{\partial}{\partial p_k}\right)L = x_k L; \quad \left(q_k-\frac{ih}{2}\frac{\partial}{\partial p_k}\right)L = y_k L;$$

(1.24)

$$-ih\frac{\partial L^*}{\partial x_k} = \left(p_k+\frac{ih}{2}\frac{\partial}{\partial q_k}\right)L^*; \quad ih\frac{\partial L^*}{\partial y_k} = \left(p_k-\frac{ih}{2}\frac{\partial}{\partial q_k}\right)L^*;$$

$$x_k L^* = \left(q_k-\frac{ih}{2}\frac{\partial}{\partial p_k}\right)L^*; \quad y_k L^* = \left(q_k+\frac{ih}{2}\frac{\partial}{\partial p_k}\right)L^*.$$

From (1.24) and the norming condition $f\equiv 1 \leftrightarrow K(x,y) = \delta(x,y)$ we obtain:

(1.25) $$L = \delta\left(q - \frac{x+y}{2}\right) e^{-\frac{1}{ih} p(y-x)}, \quad L^* = \frac{1}{(2\pi h)^n} \delta\left(q - \frac{x+y}{2}\right) e^{\frac{1}{ih} p(y-x)}$$

Hence

(1.26)
$$f(q,p) = \int K\left(q - \frac{\xi}{2}, q + \frac{\xi}{2}\right) e^{-\frac{p\xi}{ih}} d\xi ,$$
$$K(x,y) = \frac{1}{(2\pi h)^n} \int f\left(p, \frac{x+y}{2}\right) e^{\frac{1}{ih} p(y-x)} dp .$$

(1.26) entails the composition formula

(1.27) $$f(q,p) = \frac{1}{(\pi h)^{2n}} \int f_1(q_1,p_1) f_2(q_2,p_2) e^{-\frac{2}{ih}\begin{vmatrix} 1 & 1 & 1 \\ q_1 & q_2 & q \\ p_1 & p_2 & p \end{vmatrix}} dq_1\, dq_2\, dp_1\, dp_2 ;$$

we may also write this in the form:

(1.27') $$f(q,p) = f_1\left(q - \frac{ih}{2} \frac{\partial}{\partial \tilde{p}}, p - \frac{ih}{2} \frac{\partial}{\partial \tilde{q}}\right) f_2(\tilde{q}, \tilde{p}) \Bigg|_{\substack{\tilde{q} = q \\ \tilde{p} = p}} .$$

The formula (1.10) for trace and the Plancherel formula are fulfilled, as before.

For the Weyl quantization, unlike for the $\hat{q}\hat{p}$- and $\hat{p}\hat{q}$ -quantizations, the connection between the symbols f and f^* of the operator $\hat{f}$ and its adjoint $\hat{f}^*$ is very simple. Using the relation $K^*(x,y) = \overline{K(y,x)}$ between the corresponding kernels, from the first formula in (1.26) we get

(1.28) $$f^*(q,p) = \overline{f(q,p)} .$$

In particular, with a selfadjoint operator a real symbol is associated.

5. Second quantization, Bose case. The operations of creation and annihilation are the starting point:

$$\hat{a}^*_k = \frac{1}{\sqrt{2}}(\hat{q}_k - i\hat{p}_k), \qquad \hat{a}_k = \frac{1}{\sqrt{2}}(\hat{q}_k + i\hat{p}_k).$$

These operators satisfy the following relations:

$$\hat{a}_k\hat{a}_{k'} - \hat{a}_{k'}\hat{a}_k = \hat{a}^*_k\hat{a}^*_{k'} - \hat{a}^*_{k'}\hat{a}^*_k = 0, \quad \hat{a}_k\hat{a}^*_{k'} - \hat{a}^*_{k'}\hat{a}_k = hE\cdot\delta_{kk'}.$$

Several variants of correspondence between functions and operators are possible. We shall consider only the Wick normal form. To this end we rewrite the polynomials $f(q,p)$ by means of the variables $a_k = \frac{1}{\sqrt{2}}(q_k + ip_k)$, $a^*_k = \frac{1}{\sqrt{2}}(q_k - ip_k)$:

(1.29) $$f = \sum_{k,k'} \sum_{m_i, m'_i} \varphi_{m_1\dots m_k, m'_1\dots m'_{k'}} a^*_{m_1}\dots a^*_{m_k} a_{m'_1}\dots a_{m'_{k'}}.$$

With the polynomial f we associate the operator

(1.30) $$\hat{f} = \sum_{k,k'} \sum_{m_i m'_i} \varphi_{m_1\dots m_k, m'_1\dots m'_k} \hat{a}^*_{m_1}\dots\hat{a}^*_{m_k}\hat{a}_{m'_1}\dots\hat{a}_{m'_k}.$$

We suppose that the coefficients $\varphi_{m_1\dots m_k, m'_1\dots m'_k}$ are symmetrical separately in the first and second groups of indexes. The extension of this correspondence further, and the proofs of the basic formulae of the operator calculus may be carried out in the usual way.*

Basic formulae (see [2]):

if $\hat{f}_i \leftrightarrow f_i(a^*, a)$ then

(1.31) $$\hat{f} = \hat{f}_1\hat{f}_2 \leftrightarrow \int f_1(a^*, \alpha) f_2(\alpha^*, a)\, e^{-(\alpha^* - a^*)(\alpha - a)} \prod d\alpha^* d\alpha;$$

if $\hat{g} = f^*$ then

(1.32) $$g(a^*, a) = f^*(a^*, a),$$

where f^* is the function complex conjugate to f (the variables a^* and a are complex conjugate to each other, i.e. $(a^*)^* = a$); furthermore,

* The second quantization is studied much more thoroughly than described above. In particular, it is known that every bounded operator corresponds to a symbol that is an entire function of $2n$ complex variables a_k, a^*_k (which are considered as independent). This result and also the formulae below can be established in a more convenient way if we consider the realization of the operators $\hat{a}_k, \hat{a}^*_k$ in the Fock space, and do not appeal to their expressions by means of kernel in L_2 (see [2]).

$$\mathrm{Sp}\,\hat{f} = \int f(a^*, a) \prod da^* da \,. \tag{1.33}$$

In this formula we wrote $\prod da^* da = (2\pi)^{-n} dq_1 dp_1 \cdots dq_n dp_n$. The product $\prod d\alpha^* d\alpha$ in (1.31) has an analogous sense. (The normalizing factor is defined by the condition $\int e^{-a^* a} \prod da^* da = 1$. We recall that $a_k = \frac{1}{\sqrt{2}}(q_k + ip_k)$ and $a_k^* = \frac{1}{\sqrt{2}}(q_k - ip_k)$, where the q_k and p_k are real variables.)

6. Second quantization. Fermi case. As before, in this case any operator in Fock's space can be written in the Wick normal form (1.30) with coefficients $\varphi_{m_1 \cdots | \cdots m_k}$ antisymmetrical in the first and second groups of indices.

The operators $\hat{a}_k$ and $\hat{a}_k^*$ satisfy the following relations:

$$\hat{a}_k \hat{a}_{k'} + \hat{a}_{k'} \hat{a}_k = 0, \quad \hat{a}_k^* \hat{a}_{k'}^* + \hat{a}_{k'}^* \hat{a}_k^* = 0, \quad \hat{a}_k^* \hat{a}_{k'} + \hat{a}_{k'} \hat{a}_k^* = E\,\delta_{kk'},$$

i.e. they are generators of a Clifford algebra.

Let G be a Grassman algebra with $2n$ anticommuting generators $a_1, \ldots, a_n, a_1^*, \ldots, a_n^*$:

$$a_i a_j + a_j a_i = a_i^* a_j + a_j a_i^* = a_i^* a_j^* + a_j^* a_i^* = 0 \,.$$

The elements of G can be uniquely written in form (1.29) with coefficients $\varphi_{m_1 \cdots m_k | m_1' \cdots m_k'}$ antisymmetrical separately in the first and the second groups of indices. With any element of G of form (1.29) we associate the corresponding operator of form (1.30) in the Fock space of the Fermi system. Thus the symbol of the operator under (1.30) in Fermi's case is not a function of $2n$ complex variables as above, but an element of a Grassman algebra. The formula under (1.31) remains valid, and so does the one under (1.32) if we stipulate that now f^* is not the complex conjugate function, which is meaningless in the present case, but the element of G of form

$$f^* = \sum \varphi_{m_1 \cdots m_k | m_1' \cdots m_k'} a_{m_{k'}}^* \cdots a_{m_1}^* a_{m_k} \cdots a_{m_1} \,.$$

The formula under (1.33) is not now fulfilled; instead, we have the following:

$$\mathrm{Sp}\,\hat{f} = \int f(a^*, a)\, e^{2aa^*} \prod da^* da \,. \tag{1.34}$$

The integrals in (1.31) and (1.34) are often called integrals with anticommuting variables.*

In view of the great analogy between formal properties of functions and elements of the Grassman algebra G it is natural that elements of this algebra are called functions with anticommuting variables. The quantization by means of the Wick normal form (in both variants) is important because it can easily be extended to the case of an infinite number of degrees of freedom. In this latter case (1.29) becomes an infinite series (convergent for any a_k, a_k^* that are square summable** in Bose's case and formal in Fermi's case).

The formulae under (1.31), (1.33), and (1.34) are fulfilled as before, while integrals in these formulae become functional integrals, which can be interpreted as limits of finite-dimensional ones.

2. THE FUNCTIONAL INTEGRAL FOR $\exp\left(\frac{it}{h}\hat{H}\right)$

Let $\hat{H}$ be a Hamiltonian and $H(q,p)$ be its symbol. We shall determine the symbol of the operator $e^{\frac{it\hat{H}}{h}}$. First of all we note that $e^{\frac{it\hat{H}}{h}} = 1 + \frac{it\hat{H}}{h} + \frac{t^2}{h^2}\hat{R}$, where the operator $\hat{R}$ is an entire function of $\hat{H}$. Passing to symbols we find that

$$G(q,p|t) = 1 + \frac{itH}{h} + \frac{t^2}{h^2} R = e^{\frac{itH}{h}} + \frac{t^2}{h^2} r .$$

We denote by $\hat{u}(t)$ and $\hat{r}(t)$ the operators with symbols $e^{\frac{itH}{h}}$ and $r(t)$ respectively. The following identity is fulfilled:

$$e^{\frac{it\hat{H}}{h}} = \left(e^{\frac{it\hat{H}}{h}}\right)^N = \left[\hat{u}\left(\frac{t}{N}\right) + \frac{t^2}{N^2h^2}\hat{r}\left(\frac{t}{N}\right)\right]^N .$$

*Definition: $\int a_k da_k = \int a_k^* da_k^* = 1$, $\int da_k = \int da_k^* = 0$, the multiple integration is explained by successive integration, and the differentials da_k, da_k^* anticommute with each other and with a_k, a_k^* (see [2]).

**I.e. such that $\sum |a_k|^2 < \infty, \sum |a_k^*|^2 < \infty$; a_k, a_k^* are independent and not related variables.

Making $N \longrightarrow \infty$, we see that the second addend ceases to play role, and we obtain:

$$e^{\frac{it\hat{H}}{h}} = \lim_{N \to \infty} \left[\hat{u}\left(\frac{t}{N}\right)\right]^N . \tag{2.1}$$

The symbol of the operator $\left[\hat{u}\left(\frac{t}{N}\right)\right]^N$ will be denoted by $G_N(q, p|t)$.

$\hat{q}\hat{p}$- quantization. From formula (1.9) we find:

$$G_N(q,p|t) =$$

$$= \left(\frac{1}{2\pi h}\right)^{n(N-1)} \int e^{\frac{it}{Nh}\sum_1^N H(q_{k-1}, p_k) - \frac{i}{h}\sum_1^N p_k(q_k - q_{k-1})} dq_1 \dots dq_{N-1} dp_1 \dots dp_{N-1},$$

$$p_0 = p_N = p, \quad p_0 = q_N = q . \tag{2.2}$$

Here p_0 does not occur in the integrand, and is added only for reasons of symmetry between p and q . We set $p_k = p(t_k)$, $q_k = q(t_k)$, $t_k = \frac{kt}{N}$, and $\Delta = \frac{t}{N}$. Obviously the exponent in (2.2) is a sum approximating the integral S of action:

$$S = \int_0^t \left[H(q(\tau), p(\tau)) - p\frac{dq}{d\tau}\right] d\tau , \tag{2.3}$$

where $p(\tau)$, $q(\tau)$ is a closed trajectory passing through the point p, q: $q(0) = q(t) = q$, $p(0) = p(t) = p$. The factor $\left(\frac{1}{2\pi h}\right)^{n(N-1)}$ can be incorporated in the differentials, and we obtain:

$$G(q,p|t) = \int e^{\frac{i}{h}S} \prod dq\, dp . \tag{2.4}$$

The domain of integration contains all closed trajectories passing through the point p, q.

By applying the formula under (1.7) to the symbol $G_n(q,p|t)$ we may determine the kernel $K_N(x,y)$ of the operator $\left[\hat{u}\left(\frac{t}{N}\right)\right]^N$. It can be expressed as above by means of the integral in (2.2) with a difference in the conditions on the trajectories, which are now the following: $q_0 = x$, $q_N = y$ (p_N is a variable of integration). Passing to limit by making $N \to \infty$, we obtain Feynmann's formula [15]:

(2.5) $$\langle x | e^{\frac{it\hat{H}}{h}} | y \rangle = \int e^{\frac{i}{h}S} \prod dq\,dp .$$

(The integral must be taken on trajectories satisfying the conditions $q(0) = x$, $q(t) = y$.)

$\hat{p}\hat{q}$ – quantization. Using the formula under (1.17) we find:

$$G_N(q,p|t) =$$

(2.6) $$= \left(\frac{1}{2\pi h}\right)^{n(N-1)} \int e^{\frac{it}{hN}\sum_1^N H(q_k,p_{k-1}) - \frac{i}{h}\sum_1^N p_{k-1}(q_k - q_{k-1})} dq_1 \dots dq_{N-1} dp_1 \dots dp_{N-1}.$$

From (1.16') we obtain that the kernel of the operator $[\hat{u}(\frac{t}{N})]^N$ can be expressed by a formula that differs from (2.6) only in the boundary conditions $q_0 = x$, $q_N = y$ and in the constant factor $\left(\frac{1}{2\pi h}\right)^{nN}$.

The formal expressions for the symbol and the kernel of the operator $e^{\frac{it\hat{H}}{h}}$ are of the same form as those given under (2.4) and (2.6), respectively.

Weyl's quantization. By using (1.27) we obtain: *

$$G_N(q,p|t) =$$

(2.7) $$= \left(\frac{1}{\pi h}\right)^{2nN} \int e^{\frac{it}{hN}\sum H(q_k,p_k) - \frac{2i}{h}\sum[(p_k-\xi_k)(\eta_{k+1}-\eta_k) - (q_k-\eta_k)(\xi_{k+1}-\xi_k)]} \cdot$$

$$\cdot dq_1 \dots dq_N dp_1 \dots dp_N d\xi_1 \dots d\xi_N d\eta_1 \dots d\eta_N ;$$

$$\xi_1 = p_0, \quad \eta_1 = q_0, \quad \xi_N = p, \quad \eta_N = q .$$

*In proving (2.7) the following identity is used:

$$\begin{vmatrix} 1 & 1 & 1 \\ \eta_k & \xi_k & \eta_{k+1} \\ \xi_k & p_k & \xi_{k+1} \end{vmatrix} = -(p_k - \xi_k)(\eta_{k+1} - \eta_k) + (q_k - \eta_k)(\xi_{k+1} - \xi_k) .$$

If N is odd then by means of integration with respect to the variables ξ_i, η_j it is possible to express the symbol G_N in an equivalent form:

$$G_N(q,p|t) =$$

(2.8) $$= \left(\frac{1}{\pi h}\right)^N \int e^{\frac{it}{hN}\sum H(q_k,p_k) - \frac{i}{h}\sum \tilde{p}_k(q_k - q_{k-1})} dq_1 \dots dq_N dp_1 \dots dp_N$$

$$\tilde{p}_k = p_k + \sum (-1)^s p_{k+s}, \quad p_N = p, \quad p_{N+s} = p_s \text{ if } s > 1 .$$

Suppose that $p_k = p(t_k)$, $p(t)$ has continuous second derivative, and $p(0) = p(t) = p$. Then we have $\sum (-1)^s p_{k+s} = \Delta r$, where $\Delta = \frac{t}{N}$ and $|r| <$ $<$ const. For all k this is verified in the same way, therefore we consider only the case $k = 1$. In this case we have:

$$p_2 - p_3 + \dots + p_N - p = \frac{1}{2}[(p_1 - p_2) + (p_2 - p_3) + \dots + (p_N - p)] +$$
$$+ \frac{1}{2}[\{(p_2 - p_3) - (p_1 - p_2)\} + \{(p_4 - p_5) - (p_3 - p_4)\} + \dots + \{(p_N - p) - (p_{N-1} - p_N)\}] .$$

We note that $(p_k - p_{k+1}) - (p_{k-1} - p_k) = \Delta^2 p''(\bar{t}_k)$, where $t_{k-1} \le \bar{t}_k \le$ $\le t_{k+1}$. Therefore the expression between the second pair of square brackets is equal to $\Delta^2 \sum p''(t_k) = \Delta r_1$, where $r_1 = \Delta \sum p''(\bar{t}_k)$ is a sum approximating $\int_0^t p''(s)\,ds$.

The expression between the first pair of square brackets is equal to $\frac{1}{2}(p_1 - p) = \frac{\Delta}{2} p''(\bar{t})$, $0 \le \bar{t} \le t_1$.

Returning to the integral under (2.8), we obtain that the expression in the exponent is, as before, a sum approximating the integral of action under (2.3). Therefore we get the previous expression under (2.4) for $G(q,p|t)$.

We note that it is possible to write one more expression for the Weyl symbol of $G(q,p|t)$, obtained from (2.7) by the formal limit passage $N \to \infty$:

$$G(q,p|t) =$$

(2.9) $$= \int e^{\frac{i}{h}\int_0^t \left[H(q(\tau),p(\tau)) + 2(p(\tau)-\xi(\tau))\frac{d\eta}{d\tau} - 2(q(\tau)-\eta(\tau))\frac{d\xi}{d\tau}\right]d\tau} dq\,dp\,d\xi\,d\eta$$

$$\xi(0) = p(0),\quad \eta(0) = q(0),\quad \xi(t) = p,\quad \eta(t) = q.$$

Passing from symbols to kernels by (2.6), we can derive an expression for the kernels $K_N(x,y)$ of the operators $[\hat{u}(\frac{t}{N})]^N$. It has not, however, a form so elegant as in the previous cases and we omit it. Formally, ba making $N \to \infty$ we arrive at the expression given under (2.5).

S e c o n d   q u a n t i z a t i o n . Using (1.31) we obtain the expression

(2.10) $$G_N(a^*,a|t) = \int e^{\frac{it}{hN}\sum_0^{N-1} H(a_k^*, a_{k+1}) + \frac{1}{h}\sum (a_k^* - a_{k+1}^*)a_{k+1}} \prod_1^{N-1} da_s^* da_s$$

$$a_0^* = a_N^* = a^*,\qquad a_0 = a_N = a$$

for the symbol $G_N(a^*,a|t)$ of the operator $[\hat{u}(\frac{t}{N})]^N$.

The formula under (2.10) is equally fulfilled in the Bose and the Fermi cases. Formal limit passage by making $N \to \infty$ gives the following formula for G :

(2.11) $$G(a^*,a|t) = \int e^{\frac{i}{h}\int_0^t H(a^*(\tau),a(\tau))d\tau + \frac{1}{h}\int_0^t \frac{da^*}{d\tau} a(\tau)d\tau} \prod da^* da .$$

To conclude study of functional integrals for $\exp(\frac{it}{h}\hat{H})$, we note that the expression under (2.4) can be obtained also by means of the Wick formula (see [4]). Expressions analogous to (2.4) can be obtained also for S -matrices by using arguments simular to those given above [4].

In conclusion of this section we note that formal limit passage for the cases $\hat{q}\hat{p}$ -, $\hat{p}\hat{q}$ - and Weyl's quantizations give us the same functional integral (2.4). This circumstance shows that the functional integral itself has no exact sense. It must be understood as a limit of integrals of finite multiplicity and the result may depend on the way of approximation (for example, the limits of (2.2) and (2.6) are equal to the $\hat{q}\hat{p}$ - and $\hat{p}\hat{q}$-symbols of the operator $e^{\frac{it\hat{H}}{h}}$, which are different already in case $\hat{H} = \hat{p}^2 + \hat{q}^2$).

We note here that Daletsky's paper [8] turned attention to the dependence of the functional integral on the method of its approximation. For stochastic integrals, which are very similar to functional integrals, this dependence also exists and is well known.

3. THE CONNECTION BETWEEN QUANTUM AND CLASSICAL MECHANICS. QUASICLASSICAL ASYMPTOTICS

We use here the symmetrical or the Weyl quantization to explain the connections between quantum and classical mechanics. The purpose of this section is to turn the attention of the reader to one possible method of obtaining of various quasiclassical asymptotics, which we consider very natural, although it is comparatively little known.

Cauchy problem for operators. We recall that the evolution of a physical quantity with time is given by the formula:

(3.1)
$$\hat{f}(t) = e^{-\frac{it}{h}\hat{H}} f e^{\frac{it}{h}\hat{H}},$$

where $\hat{f}(t)$ is the operator corresponding to the physical quantity at the moment t, $\hat{f} = \hat{f}(0)$, and $\hat{H}$ is the operator of energy (Hamiltonian). The differential equation of motion corresponding to (3.1) has the form:

(3.2)
$$ih \frac{\partial \hat{f}}{\partial t} = [\hat{H}, \hat{f}],$$

where $[\hat{H},\hat{f}] = \hat{H}\hat{f} - \hat{f}\hat{H}$ is the commutator of the operators $\hat{H}$ and $\hat{f}$.

We now recall that in classical mechanics the equation of motion for an observable quantity f that is a function of a point in the phase space and of time (i.e. of the variables q, p, t) has the form:

(3.3)
$$\frac{\partial f}{\partial t} = [H, f] = \sum_n \left(\frac{\partial H}{\partial q_n} \frac{\partial f}{\partial p_n} - \frac{\partial H}{\partial p_n} \frac{\partial f}{\partial q_n} \right),$$

where $H(q,p)$ is Hamilton's function, which can physically be interpreted as the energy at the point q,p. On the right-hand side of (3.3), the so-called Poisson bracket stands of the functions $H(q,p)$ and $f(t|q,p)$. The equation (3.3) may be solved by using the characteristics:

(3.4) $$f(t|q,p) = f(0|\tilde{q}(q,p,-t), \tilde{p}(q,p,-t)),$$

where

(3.5) $$\tilde{q}(q,p,t), \tilde{p}(q,p,t), \tilde{q}(q,p,0) = q, \quad \tilde{p}(q,p,0) = p$$

are the phase coordinates of a moving point; thus

(3.6) $$\frac{\partial \tilde{q}_n}{\partial t} = \frac{\partial H}{\partial \tilde{p}_n}, \quad \frac{\partial \tilde{p}_n}{\partial t} = \frac{\partial H}{\partial \tilde{q}_n}, \quad H = H(\tilde{q}_n, \tilde{p}_n).$$

The typical form of the operator $\hat{H}$ in quantum mechanics is as follows:

(3.7) $$H = \frac{1}{2m}(\hat{p}_1^2 + \dots + \hat{p}_N^2) + U(\hat{q}_1, \dots, \hat{q}_N),$$

The corresponding Weyl symbol obviously has the form:

(3.7') $$H(q,p) = \frac{1}{2m}(p_1^2 + \dots + p_N^2) + U(q_1, \dots, q_N);$$

and so it does not depend on h. $H(q,p)$ is the Hamiltonian of the classical system into which the quantum system is transformed as $h \to 0$.

Let us consider such a quantum quantity $\hat{f}$ that is an operator with a symmetrical symbol

$$f(h|q,p) = \sum_0^\infty h^n f_n(q,p).$$

We denote by $\hat{f}(t)$ the quantity given by (3.1). We denote the corresponding function by $f(t,h|q,p)$. From (3.2), (3.7') and (1.27') it follows that $f(t,h|q,p)$ satisfies the equation:

(3.8) $$\frac{\partial f}{\partial t} = [H,f] + \sum_1^\infty \left(\frac{ih}{2}\right)^{2k} \frac{1}{(2k+1)!} \sum \frac{\partial^{2k+1} U}{\partial q_{i_1} \cdots \partial q_{i_{2k+1}}} \cdot \frac{\partial^{2k+1} f}{\partial p_{i_1} \cdots \partial p_{i_{2k+1}}}.$$

We now expand $f(t,h|q,p)$ into a power series of h:

(3.9) $$f(t,h|q,p) = f(t|q,p) + hf_1(t|q,p) + h^2 f_2(t|q,p) + \dots$$

Substituting this expansion into (3.8) and comparing the coefficients by equal powers of h, we obtain the following system of equations:

$$\frac{\partial f}{\partial t} = [H, f]$$

(3.10) $$\frac{\partial f_1}{\partial t} = [H, f_1]$$

$$\frac{\partial f_2}{\partial t} = [H, f_2] - \frac{1}{4.3!}\sum \frac{\partial^3 U}{\partial q_{i_1}\partial q_{i_2}\partial q_{i_3}} \cdot \frac{\partial^3 f}{\partial p_{i_1}\partial p_{i_2}\partial p_{i_3}}$$

. . . .

The initial conditions for this system are the following:

(3.11) $$f(0|q,p) = f_0(q,p), \quad f_i(0|q,p) = f_i(q,p), \qquad i \geq 1 .$$

In particular, if $f(h|q,p) = f_0(q,p)$ does not depend on h then $f_i(0|q,p) = 0$. From (3.3), (3.4) and (3.9) it follows:

$$f(t,h|q,p) = f(0,0|\tilde{q}(q,p,-t), \tilde{p}(q,p,-t)) + O(h),$$

where $\tilde{q}(q,p,t), \tilde{p}(q,p,t)$ are the phase coordinates of a classical point. These functions satisfy the classical equations of motion and the initial conditions under (3.5).

It is not difficult to express all the terms of the formal expansion of $f(t,h|q,p)$ by powers of h. We consider in particular the cases when $\hat{f} = \hat{p}_k$ and $\hat{f} = \hat{q}_k$. We denote the symbol of the operator $\hat{f}(t)$ in this case by $P_k(q,p,t|k)$ and $Q_k(q,p,t|h)$, respectively. According to (3.11), the expansions of P_k and Q_k by powers of h have the forms

$$P_k = \tilde{p}_k(q,p,t) + O(h) \quad \text{and} \quad Q_k = \tilde{q}_k(q,p,t) + O(h) .$$

T h e a s y m p t o t i c o f t h e G r e e n f u n c t i o n . We are going to determine a group of equations defining the asymptotic, as $h \to 0$, of the symbol $G(q,p|t)$ of the operator $\hat{G} = e^{\frac{it\hat{H}}{h}}$. We look for G in the form

$$G = e^{\frac{i}{h}F_{-1} + F_0 + hF_1 + \cdots}$$

Passing in the equation $\frac{h}{i}\frac{\partial \hat{G}}{\partial t} = \hat{H}\hat{G}$ to symbols, we immediately obtain:

$$\frac{\partial F_{-1}}{\partial t} = H\left(p - \frac{1}{2}\frac{\partial F_{-1}}{\partial q},\ q + \frac{1}{2}\frac{\partial F_{-1}}{\partial p}\right),\quad F(0,q,p) = 0\ . \tag{3.12}$$

This equation can be solved by a known method using characteristics [9]. It is interesting that the characteristics of this equation have an immediate operator sense. We denote by $\hat{P}$ and $\hat{Q}$ the operators $\hat{P} = \hat{G}\hat{p}\hat{G}^{-1}$, $\hat{Q} = \hat{G}\hat{q}\hat{G}^{-1}$. Obviously, we have $\hat{P}\hat{G} = \hat{G}\hat{q}$ and $\hat{Q}\hat{G} = \hat{G}\hat{q}$. Passing in this equalities from operators to symbols, and using the formulae under (1.23), and then keeping in this formulae only the terms of zero degree in h, we obtain:

$$\left(\frac{\partial P_0}{\partial p}\cdot\frac{\partial}{\partial q} - \frac{\partial P_0}{\partial q}\cdot\frac{\partial}{\partial p} + \frac{\partial}{\partial q}\right)F_{-1} = P - p\,;$$

(3.13)

$$\left(\frac{\partial Q_0}{\partial q}\cdot\frac{\partial}{\partial p} - \frac{\partial Q_0}{\partial p}\cdot\frac{\partial}{\partial q} + \frac{\partial}{\partial p}\right)F_{-1} = q - Q\,;$$

here P_0, Q_0 have the same meaning as in (3.6), i.e. they are solutions of the classical Hamilton equations.

We note that the system of equations under (3.13) can also be obtained from (3.12) by the method of characteristics.

T h e a s y m p t o t i c o f p r o p e r v a l u e s. We suppose that the potential $U(q_1,\ldots,q_N)$ in (3.7') satisfies the condition

$$U(q_1,\ldots,q_N) \longrightarrow +\infty \quad \text{as} \quad |q| = \sqrt{q_1^2 + \cdots + q_N^2} \longrightarrow \infty\ .$$

As is well known, this condition ensures that the operator $\hat{H}$ given by (3.7) has a purely discrete spectrum. We denote by $N(E,h)$ the number of the proper values less than or equal to E of the operator $\hat{H}$. We are going to determine the asymptotic of $N(E,h)$ as $h \longrightarrow 0$.

We denote by $G(\beta,h|q,p)$ the Weyl symbols of the operator $e^{-\beta\hat{H}}$, and by E_n the proper values of operators $\hat{H}$. We note that:

$$\text{(3.14)} \qquad Sp\, e^{-\beta\hat{H}} = \sum e^{-E_n\beta} = \int_0^\infty e^{-\beta E} d_E N(E,h)$$

(here Sp indicates trace). On the other hand, according to the formula under (1.10), which can also be verified for symmetrical symbols, we have:

$$Sp\, e^{-\beta\hat{H}} = \frac{1}{(2\pi h)^N} \int G(\beta, h | q, p)\, d^N q\, d^N p \, .$$

From (1.27') it formally follows that *

$$\lim_{h\to 0} G(\beta, h | q, p) = e^{-\beta H(q,p)} \, .$$

Therefore $Sp\, e^{-\beta\hat{H}}$ has the following asymptotic as $h \to 0$**:

$$\text{(3.15)} \qquad \begin{aligned} Sp\, e^{-\beta\hat{H}} &\approx \frac{1}{(2\pi h)^N} \int e^{-\beta H(q,p)} d^N q\, d^N p = \\ &= \frac{1}{(2\pi h)^N} \int e^{-\beta E} \delta(H(q,p) - E)\, d^N q\, d^N p\, dE = \frac{1}{(2\pi h)^N} \int_0^\infty e^{-\beta E} \varrho'(E)\, dE \, , \end{aligned}$$

where $\varrho(E)$ is the volume of the domain in the phase space with energy less than:

$$\varrho(E) = \int_{H(q,p)\le E} d^N q\, d^N p \, .$$

*In case $H = \frac{1}{2}(\hat{p}^2 + \hat{q}^2)$ (harmonical oscillator) the function $G(\beta, h | q, p)$ can be easily calculated:

$$G(\beta, h | q, p) = \left(\mathrm{ch}\, \frac{\beta h}{2}\right)^{-1} \exp\left\{ \frac{-\mathrm{th}\, \frac{\beta h}{2}}{h} (q^2 + p^2) \right\} .$$

Thus the function $G(\beta, h | q, p)$ is analytic in the neighbourhood of $h = 0$, and, as one may expect:

$$G(\beta, 0 | q, p) = \exp\left\{ -\frac{\beta}{2}(q^2 + p^2) \right\} .$$

Apparently, the analyticity in h of the functions $G(\beta, h | q, p)$ is a general property of many Hamiltonians.

**We note that, apart from the asymptotical formula under (3.15), the inequality

$$(*) \qquad Sp\, e^{-\beta\hat{H}} \le \frac{1}{(2\pi h)^N} \int e^{-\beta H(q,p)} d^N q\, d^N p$$

is also fulfilled. (This is the so-called Feynmann inequality). The trace on the left-hand side exists if the integral on the right-hand side exists. So this inequality may be used for the proof of the discreteness of the spectrum of the operator $\hat{H}$ with increasing potential $U(q)$. (The proof of (*) is given only for Hamiltonians of type $H = p^2/2m + U(q)$ [10].)

A comparaison of (3.14) and (3.15) gives:

(3.16) $$d_E N(E,h) \approx \frac{\varrho'(E)}{(2\pi h)^N}\, dE .$$

Observe that $N(0,h) = \varrho(0) = 0$. Therefore

(3.17) $$N(E,h) \approx \frac{\varrho(E)}{(2\pi h)^N} .$$

From (3.16) it follows that the number of proper values lying in any interval (a,b), $b > a > 0$, tends to ∞ as $h \to 0$. Hence the density of proper values must increase with decreasing h.*

B o h r f o r m u l a . When the number N of degrees of freedom is equal to 1 we can find an asymptotic, as $h \to 0$, with individual proper values. We shall suppose that the proper values $E_n(h)$ are enumerated in an increasing order.

Simultaneously with the operator under (3.7) we also shall consider the operator $\hat{H}_0$ (harmonical oscillator):

(3.18) $$\hat{H}_0 = \frac{1}{2}(\hat{q}^2 + \hat{p}^2) .$$

The proper values of H_0 are well known (see for example [11]): $E^0_n(h) = (n+\frac{1}{2})h$. We consider a function $f(x,h) = f_0(x) + f_1(x,h)$ satisfying the following property: $f\left[(n+\frac{1}{2})h, h\right] = E_n(h)$. As we saw it above, the levels of energy draw together as $h \to 0$ and we have $E_1(h) \to 0$. Therefore $f(0,0) = 0$, and so $f_0(0) = 0$. We suppose that $f_1(cx, x)$ does not tend to 0 slower than $f_0(x)$ as $x \to 0$. In this case we have $E_n(h) \sim f_0((n+\frac{1}{2})h)$ as $h \to 0$.

Let us consider the inverse function $F(x,h)$ of $f(x,h)$: $F(f(x,h),h) = x$. Similarly to $f(x,h)$, the function $F(x,h)$ has the form $F(x,h) = F_0(x) + F_1(x,h)$, where $F_0(x)$ is the inverse function of $f_0(x)$ and $\lim_{h\to 0} F_1(x,h) = 0$.

We note that the function $f(x,h)$ possesses the property $\hat{H} = \hat{U}(h) \cdot f(\hat{H}_0,h)\hat{U}^{-1}(h)$, where $\hat{U}(h)$ is a unitary operator. Therefore, for all $\beta > 0$ we have:

(3.19) $$\mathrm{Sp}\, e^{-\beta F(\hat{H},h)} = \mathrm{Sp}\, e^{-\beta \hat{H}_0} .$$

* (see ** footnote on preceding page)

We shall denote by $G_0(\beta, h|q,p)$ the symmetrical symbol of the operator $e^{-\beta\hat{H}_0}$, and by $G(\beta, h|q,p)$ that of $e^{-\beta\hat{H}}$. Obviously, we have:

$$G_0(\beta,0|q,p) = e^{-\frac{\beta}{2}(q^2+p^2)}, \quad G(\beta,0|q,p) = e^{-\beta F_0(\frac{p^2}{2m}+U(q))} .$$

We use the formula under (1.10). Multiplying both sides of (3.19) by $2\pi h$ and passing to the limit $h \to 0$, we obtain:

$$(3.20) \qquad \int e^{-\beta F_0(\frac{p^2}{2m}+U(q))} dq\,dp = \int e^{-\frac{\beta}{2}(q^2+p^2)} dq\,dp .$$

This is the equation for the function F_0 we were aiming at.

We transform the first of these integrals according to (3.15):

$$(3.21) \qquad \int e^{-\beta F_0(\frac{p^2}{2m}+U(q))} dq\,dp = \int e^{-\beta F_0(E)} \varrho'(E)\,dE ,$$

where

$$\varrho(E) = \int_{\frac{p^2}{2m}+U(q)\leq E} dq\,dp .$$

We recall that the proper values $E_n(h)$ of operator $\hat{H}$ are enumerated in an increasing order. Therefore the function $F(x,h)$ is not decreasing by x. Obviously, the same property is satisfyed by $F_0(x) = \lim_{h\to 0} F(x,h)$. Hence in the last integral of (3.21) we may make the following substitution of variable: $F_0(E) = \lambda$. Finally, we obtain:

$$\int e^{-\beta F_0(\frac{p^2}{2m}+U(q))} dq\,dp = \int_0^\infty e^{-\beta\lambda} \frac{\varrho'(E)}{F_0'(E)} d\lambda, \qquad F_0(E) = \lambda .$$

Transforming the second integral in (3.20) in a similar way, we find:

$$\int e^{-\frac{\beta}{2}(q^2+p^2)} dq\,dp = 2\pi \int_0^\infty e^{-\beta\lambda} d\lambda .$$

Therefore

$$\frac{\varrho'(E)}{F_0'(E)} = 2\pi, \quad F_0(E) = \frac{1}{2\pi}\varrho(E) + C.$$

The function $F_0(x)$ is the inverse of $f_0(x)$. Therefore we have $F_0(0) = f_0(0) = 0$. As $\varrho(0) = 0$, here $C = 0$ holds. Taking the direct form of $\varrho(E)$ into consideration, we obtain at last:

$$F_0(E) = \frac{1}{2\pi} \int_{\frac{p^2}{2m} + U(q) \le E} dq\, dp .$$

From here we can now derive the equation for the proper values $E_n(h)$ of the operator $\hat{H}$:

(3.22) $$\frac{1}{2\pi} \int_{\frac{p^2}{2m} + U(q) \le E_n(h)} dq\, dp = (n + \frac{1}{2}) h .$$

This well-known formula is due to Bohr (see any textbook of quantum mechanics, for example [11]).

4. PSEUDODIFFERENTIAL OPERATOR IN $\mathbf{R}^n$

We introduce a class of symbols for which the greater part of the heuristic arguments of the first section relating to $\hat{q}\hat{p}$ -, $\hat{p}\hat{q}$ -, and the Weyl symbols can be transformed into exact proofs. The corresponding operators form a subclass of the class of pseudodifferential operators in $\mathbf{R}^n$, and therefore the proofs can be modelled on ideas of L. Hörmander. Similar arguments are contained in Grušin [12].

By y we shall denote a point of phase space; i.e. $y = (q, p)$. Let $f(y)$ be a complex-valued function on the phase space $\mathbf{R}^{2n}$; let further m and ϱ be real numbers, $0 < \varrho \le 1$. We shall write $f(y) \in G_\varrho^m$ if $f(y) \in C^\infty(\mathbf{R}^{2n})$ and if for any multiindex γ there exists a constant C_γ such that the following inequality is fulfilled:

(4.1) $$|\partial^{|\gamma|} f(y)/\partial y^\gamma| \le C_\gamma (1+|y|)^{m-\varrho|\gamma|}, \qquad y \in \mathbf{R}^{2n} .$$

With every function $f(y) \in G_\varrho^m$ we associate an operator $\hat{f}$ the

$\hat{q}\hat{p}$ -symbol of which is $f(y)$ (for the sake of simplicity we consider only this case). We shall define this operator by means of the Fourier transformation, i.e. by the formula under (1.8).

We shall denote by $S(\mathbf{R}^n)$ the space of L. Schwartz of all complex-valued functions $\varphi(x)$ such that they and all their derivatives decrease more rapidly than every power of $|x|$ as $|x| \to \infty$. $S'(\mathbf{R}^n)$ will denote the space of distributions adjoint to $S(\mathbf{R}^n)$.

Then, analogously to some results of Hörmander, the following proposition can be proved:

Proposition. *If* $f(y) \in G_\varrho^m$ *then the operator* $\hat{f}$ *is a continuous operator from* $S(\mathbf{R}^n)$ *to* $S(\mathbf{R}^n)$ *and it can be extended to a continuous map from* $S'(\mathbf{R}^n)$ *to* $S'(\mathbf{R}^n)$. *Its distribution kernel, understood in the sense of L. Schwartz, belongs to* $S'(\mathbf{R}^n \times \mathbf{R}^n)$ *and is infinitely differentiable outside the diagonal in* $\mathbf{R}^n \times \mathbf{R}^n$.

We note here that, for any $u \in S'$, $\hat{f}u$ is a defined by the equality:

$$(\hat{f}u, \varphi) = (u, \hat{f}^*\varphi), \quad \varphi \in S,$$

where $\hat{f}^*$ is the formal adjoint of $\hat{f}$.

Theorem 1 (composition formula). *If* $f_1(y) \in G_\varrho^{m_1}$, $f_2(y) \in G_\varrho^{m_2}$ *then* $\hat{f}_1\hat{f}_2 = \hat{f}_h$, *where* $f_h(y) \in G_\varrho^{m_1+m_2}$ *and* $f_h(y)$ *is defined by the formula under* (1.9'), *and for every natural number* N *it has the following expansion:*

$$(4.2) \quad f_h(q,p) = \sum_{|\alpha| \leq N-1} \frac{(ih)^{|\alpha|}}{\alpha!} \frac{\partial^{|\alpha|} f_1(q,p)}{\partial p^\alpha} \frac{\partial^{|\alpha|} f_2(q,p)}{\partial q^\alpha} + h^N r_N(q,p,h),$$

where $r_N(q,p,h) \in G_\varrho^{m-\varrho N}$ *with the constants in the inequality under* (4.1) *not depending on* h .

Remark. The formula under (1.9') must be understood here in following way:

$$(4.3) \qquad f_h(q_0,p_0) = \hat{f}_1(q,p_0+p) f_2(q,p_0)\Big|_{q=q_0},$$

where, on the right-hand side, we have the result, which is a distribution in $S'(\mathbf{R}^n_q)$, of applying the operator with the $\hat{q}\hat{p}$ -symbol $f_1(q,p_0+p)$ to a function $f_2(q,p_0)$ of the variable q .

Proof of Theorem 1. We shall derive the desired estimate for the remainder r_N . From the properties of the formula under (4.2) it follows that it is sufficient to find for any N_0 an N such that $r_N \in G_\varrho^{m-\varrho N_0}$ with constants not depending on in the inequality under (4.1). Expanding the function f_1 under (4.3) by the Taylor formula about the point (q_0, p_0), and writing the remainder term in integral form, we see that it is sufficient to estimate the integrals

$$I = \iint e^{i(q-z)\zeta} f_1(q, p+t\zeta) f_2(z,p)\, dz\, d\zeta \, , \tag{4.4}$$

where $f_1 \in G_\varrho^{m'}$, $f_2 \in G_\varrho^{m''}$, and m' and m'' are sufficiently small.

Lemma. *Let* $\tilde{f}_2(\zeta, p) = \int e^{-iz\zeta} f_2(z,p)\, dz$. *Then with* $m''+n-\varrho N < 0$ *the following estimate holds:*

$$|\tilde{f}_2(\zeta,p) \le C_N |\zeta|^{-N} (1+p)^{m''+n-\varrho N} \, .$$

Proof of the lemma. If γ is a multiindex with $|\gamma| = N$ then we have

$$|\zeta^\gamma \tilde{f}_2(\zeta,p)| = |\int e^{-iz\zeta} \partial_z^\gamma f_2(z,p)\, dz| \le$$

$$\le \int (1+|z|+|p|)^{m''-\varrho N} dz = C_N (1+|p|)^{m''+n-\varrho N} \, ,$$

where $C_N = \int (1+|\eta|)^{m''-\varrho N} d\eta$. The assertion of the lemma follows.

Let us estimate now the integral under (4.4). We decompose it into two parts:

$$I = I_1 + I_2 = \int_{|\zeta| \le |y|/2} \cdots + \int_{|\zeta| \ge |y|/2} \cdots \, ,$$

where $y = (q,p)$. In the integral I_1 we have:

$$|f_1(q, p+t\zeta)| \le C(1+|y|)^{m'} .$$

Therefore, for $m'' < -n$ we obtain:

$$|I_1| \le C(1+|y|)^{m'+n} (1+|p|)^{m''+n} \le C(1+|y|)^{m'+n} \, .$$

To estimate I_2 we use the lemma. For $m' < 0$ we obtain:

$$|I_2| \le C \int_{|\zeta| \ge |y|/2} |\tilde{f}_2(\zeta, p)\, d\zeta \le C \int_{|\zeta| \ge |y|2} |\zeta|^{-N} d\zeta \cdot (1+|p|)^{m''+n-\varrho N} \le$$

$$\le C|y|^{-N+n} (1+|p|)^{m''+n-\varrho p} \le C_p |y|^{-N+n},$$

provided N is sufficiently large.

The obtained estimates give the required estimate for I in case $|y| \ge 1$. In case $|y| \le 1$ the estimate may be obtained from the obvious equality:

$$I = \iint \frac{f_1(q, p+t\zeta)}{(1+|\zeta|^2)^M} [(1-\Delta_z)^M b(z,p)]\, e^{i(q-z)\zeta} dz\, d\zeta ,$$

where Δ_z is the result of the Laplace operator performed on z. If $m'' < -n$ and M is sufficiently large then the integral is convergent, and we have the following estimate uniformly in t:

$$|I| \le C \quad \text{for} \quad |y| \le 1 .$$

Theorem 1 is proved.

We may handle the question of the symbol of the adjoint of an operator and its expansion into a power series of h in a similar way.

We now consider the question of the inverse of an operator and its regularizator.

Definition. Γ_ϱ^m will denote the subset of G_ϱ^m containing such symbols $f(y)$ as satisfy the following estimates:

(4.5) $$|\partial^\gamma{}_y f(y)/f(y)| \le C_\gamma |y|^{-\varrho|\gamma|}, \quad |y| \ge M .$$

This condition is an obvious analogue of hypoellipticity of Hörmander and it may be written in terms of matrix symbols. We obtain

Theorem 2. *If* $f(y) \in \Gamma_\varrho^m$ *then there exists a symbol* $g(y) \in G_\varrho^M$ *such that the following relations are fulfilled:*

$$\hat{g}\hat{f} = I + T_1 , \qquad \hat{f}\hat{g} = I + T_2 ,$$

where the operators T_1 *and* T_2 *have kernels belonging to* $S(\mathbf{R}^n \times \mathbf{R}^n)$*, and therefore they map the whole of* $S'(\mathbf{R}^n)$ *into* $S(\mathbf{R}^n)$.

Corollary. *If* $f(y) \in \Gamma_\varrho^m$ *then the operator* $\hat{f}: S \longrightarrow S$ *has finite dimensional kernel and cokernel, and the relation*

$$\hat{f}u = \varphi ,$$

where $\varphi \in S$, $u \in S'$ *implies* $u \in S$ *. In particular, if* $u \in S'$ *and* $\hat{f}u = 0$ *then* $u \in S$.

Now we shall study the influence of a small parameter h .

Theorem 3. *Let* $f(y) \in \Gamma_\varrho^m$ *and* $f(y) \neq 0$ *for all* $y \in \mathbf{R}^{2n}$ *. Then, for sufficiently small* h *, the operator* $\hat{f}$ *is invertible and the inverse operator has form* $\hat{g}_h$ *with* $g_h(y) \in G_\varrho^M$ *. where* $g_h(y)$ *has the following asymptotic expansion into a power series of* h *:*

$$g_h(y) = g_0(y) + hg_1(y) + \dots + h^{N-1} g_{N-1}(y) + h^N r_N(y;h) ,$$

where $g_0(y) = (f(y))^{-1}$ *and* $r_N(y;h) \in G_\varrho^{S(N)}$ *with constants in the inequality under* (4.1) *not depending on* h *and with* $S(N)$ *tending to* $-\infty$ *as* $N \longrightarrow +\infty$.

This theorem may be obtained from Theorems 1 and 2. We only note the symbol $g_h(q,p)$ may be constructed by means of successive approximations starting with $g_0(y) = (f(y))^{-1}$.

Let now $f(y) \in G_\varrho^m$ be the symbol of a formally selfadjoint operator $\hat{f}$ (i.e. of an operator symmetric on $C_0^\infty(\mathbf{R}^n)$). We are going to establish a theorem giving sufficient conditions for the closure of the operator $\hat{f}$ to be selfadjoint:

Theorem 4. *Let* $f(y) \pm i \in \Gamma_\varrho^m$ *. Then the indexes of defects of the operator* $\hat{f}$ *are equal to* 0 .

Proof. We suppose that $u \in L^2(\mathbf{R}^n)$ and $\hat{f}u = \pm iu$ (the operator on the left-hand side is to be understood as being adjoint, to the corresponding operator on C_0^∞ , in the sense of distributions specified above for all $u \in S'$). From the corollary of Theorem 2 we derive that $u \in S$; and from here it follows that $u = 0$

because the operator $\hat{f}$ is simmetrical on S, as shown by the continuity of $\hat{f}$ on S and by the density of C_0^∞ in S. Theorem 4 is proved.

In conclusion we note that the results given above may be extended to a larger class of symbols defined by the inequalities

$$|\partial_q^\beta \partial_p^\alpha f(q,p)| \le C_{\alpha\beta}(1+|p|)^{m_1-\varrho_1|\alpha|+\delta_1|\beta|}(1+|q|)^{m_2-\varrho_2|\beta|+\delta_2|\alpha|},$$

where $\delta_1 < \varrho_1$ and $\delta_2 < \varrho_2$.

REFERENCES

[1] F.A. Berezin, *Functional Anal. i Priložen.*, 1 (1967), 1-14.

[2] F.A. Berezin, *Method of second quantization* (Russian) (Moscow, 1965). English translation: Academic Press, London, 1966.

[3] F.A. Berezin, *Trudy Moskov. Mat. Obšč.*, 17 (1967), 118-184.

[4] F.A. Berezin, *Žurnal Teoret. i. Mat. Fiz.*, (1970) (to appear).

[5] M.A. Šubin, *Dokl. Akad. Nauk SSSR*, (1970) (to appear).

[6] Pseudodifferential operators (collection of papers; (Russian) Moscow, 1968).

[7] A.N. Kolmogorov and S.V. Fomin, *Elements of the theory of functions and of functional analysis* (Moscow. 1968).

[8] Ju.L. Daleckiĭ, *Uspehi Mat. Nauk*, 17 (1962), 3-115.

[9] R. Courant and D. Hilbert, *Methoden der mathematischen Physik.* I und II (Berlin, 1931 and 1937).

[10] K. Simanzik, *Journ. of Math. Phys.*, 6 (1965).

[11] L.D. Landau and E.M. Lifšic, *Quantum Mechanics* (Russian) (Moscow, 1963).

[12] V.D. Grušin, *Dokl. Akad. Nauk SSSR*, (1970) (to appear).

[13] R. Feynmann, *Phys. Rev.*, 84 (1965), 108.

COLLOQUIA MATHEMATICA SOCIETATIS JÁNOS BOLYAI
5. HILBERT SPACE OPERATORS. TIHANY (HUNGARY). 1970

Involution as operator conjugation

J. BOGNÁR

1. INTRODUCTION

Let $\mathbf{K}$ stand for the real number field $\mathbf{R}$ or the complex number field $\mathbf{C}$. Let $\mathfrak{E}$ be a Banach space over $\mathbf{K}$, and $\mathcal{L}(\mathfrak{E})$ the Banach algebra of all bounded linear operators on $\mathfrak{E}$. Both on $\mathfrak{E}$ and on $\mathcal{L}(\mathfrak{E})$ the norm will be denoted by $|\cdot|$.

We shall use also the following notations.

$\mathfrak{N}(T)$: the null-space of the operator $T \in \mathcal{L}(\mathfrak{E})$.

$\mathfrak{R}(T)$: the range of the operator $T \in \mathcal{L}(\mathfrak{E})$.

$\langle x \rangle$: the 1-dimensional subspace spanned by the vector $x \in \mathfrak{E}$.

A *symmetric* (resp. *skew-symmetric) bilinear form* $(\cdot,\cdot)$ on $\mathfrak{E}$ is, by definition, a mapping of $\mathfrak{E} \times \mathfrak{E}$ to $\mathbf{K}$ such that

(1) $$(\alpha_1 x_1 + \alpha_2 x_2, y) = \alpha_1 (x_1, y) + \alpha_2 (x_2, y)$$

$$(\alpha_1, \alpha_2 \in \mathbf{K}; \quad x_1, x_2 \in \mathfrak{E})$$

and

(2) $$(y,x) = \overline{(x,y)} \qquad (x,y \in \mathfrak{X})$$

(resp.

(3) $$(y,x) = -\overline{(x,y)} \qquad (x,y \in \mathfrak{X})).$$

A symmetric or skew-symmetric bilinear form $(\cdot\,,\cdot)$ is said to be *non-degenerate* if

$$(x,y) = 0 \quad (y \in \mathfrak{X}) \quad \text{implies} \quad x = 0,$$

and *bounded* if

$$|(x,y)| \leqq \beta |x| |y| \qquad (x,y \in \mathfrak{X})$$

for a suitable $\beta > 0$.

For particular spaces $\mathfrak{X}$ and non-degenerate symmetric or skew-symmetric bilinear forms $(\cdot\,,\cdot)$ it may happen that to each $T \in \mathcal{L}(\mathfrak{X})$ there is a $T^* \in \mathcal{L}(\mathfrak{X})$ satisfying the relation

(4) $$(Tx,y) = (x,T^*y) \qquad (x,y \in \mathfrak{X}).$$

(Owing to non-degeneracy T^* is unique.) We say that T^* is the *adjoint* of T relative to the form $(\cdot,\cdot)$. The mapping $*: \mathcal{L}(\mathfrak{X}) \to \mathcal{L}(\mathfrak{X})$ will be called the *conjugation* relative to $(\cdot,\cdot)$.

It is easy to see that the following three identities hold for any conjugation:

(5) $$(\alpha_1 T_1 + \alpha_2 T_2)^* = \bar{\alpha}_1 T_1^* + \bar{\alpha}_2 T_2^* \qquad (\alpha_1,\alpha_2 \in \mathbf{K};\ T_1,T_2 \in \mathcal{L}(\mathfrak{X})),$$

(6) $$(T_1 T_2)^* = T_2^* T_1^* \qquad (T_1,T_2 \in \mathcal{L}(\mathfrak{X})),$$

(7) $$T^{**} = T \qquad (T \in \mathcal{L}(\mathfrak{X})).$$

A mapping $*: \mathcal{L}(\mathfrak{X}) \to \mathcal{L}(\mathfrak{X})$ with properties (5), (6), (7) is said to be an *involution* on $\mathcal{L}(\mathfrak{X})$. Thus every conjugation is an involution.

In the present note we prove that every involution on $\mathcal{L}(\mathfrak{X})$ is a conjugation relative to a suitable form $(\cdot,\cdot)$.

The classical result in this direction is a theorem of Kakutani and

Mackey [1], [2] (see also Kawada [3], Rickart [4; Corollary 4.10.8]) characterizing conjugations relative to definite forms $(.,.)$. We propose a proof of this theorem which at some points seems to be more elementary and straightforward than the known ones. Then we indicate the modifications needed in order to treat the case of general symmetric or skew-symmetric forms and, in particular, symmetric forms with a finite rank of indefiniteness.

2. CONJUGATION RELATIVE TO DEFINITE FORMS

Let the symmetric bilinear form $(.,.)$ be *definite:*

(8) $$(x,x) \neq 0 \qquad (x \in \mathfrak{E},\ x \neq 0).$$

Suppose that the conjugation $*$ relative to $(.,.)$ exists. (According to the theory of Hilbert spaces and Theorem 1 below this is the case if and only if the norm $\|\cdot\|$ defined by the relation $\|x\| = (x,x)^{1/2}$ $(x \in \mathfrak{E})$ is equivalent to $|\cdot|$.)

If for some T in $\mathcal{L}(\mathfrak{E})$ $T^*T = 0$, then $(T^*Tx,x) = (Tx,Tx) = 0$ for every x in $\mathfrak{E}$, so that in view of (8) $T = 0$.

Theorem 1 (Kakutani and Mackey). *Let* $*$ *be an involution on* $\mathcal{L}(\mathfrak{E})$ *with the additional property*

(9) $$T^*T \neq 0 \qquad (T \in \mathcal{L}(\mathfrak{E}),\ T \neq 0).$$

Then there is a definite symmetric bilinear form $(.,.)$ *on* $\mathfrak{E}$ *such that* $*$ *concides with conjugation relative to* $(.,.)$. *Moreover,* $(.,.)$ *is apart from a multiplicative real non-zero constant uniquely determined, and the norm* $\|\cdot\|$ *given by the relation*

(10) $$\|x\| = |(x,x)|^{\frac{1}{2}} \qquad (x \in \mathfrak{E})$$

is equivalent to $|\cdot|$.

In order to establish Theorem 1 we need a lemma which (in a more general form) is due to Eidelheit [5]. The proof of this lemma, given below for the sake of completeness, belongs to Rickart (see [4; Theorem 2.4.14]).

Lemma 1. *Let* $*$ *be any involution on* $\mathcal{L}(\mathfrak{E})$. *Then the formula*

$$|T|^* = |T^*| \qquad (T \in \mathcal{L}(\mathfrak{E}))$$

defines a Banach algebra norm $|\cdot|^*$ *equivalent to* $|\cdot|$ *on* $\mathcal{L}(\mathfrak{E})$.

Proof of Lemma 1. It is evident that $|\cdot|^*$ is a norm on $\mathcal{L}(\mathfrak{X})$ satisfying $|T_1T_2|^* \leqq |T_1|^*|T_2|^*$ for every T_1, T_2 in $\mathcal{L}(\mathfrak{X})$. To prove completeness, let $|T_m^* - T_n^*| \to 0$ $(m,n \to \infty)$. Then, for some T, $|T_n^* - T| \to 0$ i.e. $|T_n - T^*|^* \to 0$ $(n \to \infty)$.

To prove equivalency it is sufficient, according to the closed graph theorem, to show that the identity mapping of the Banach space $(\mathcal{L}(\mathfrak{X}), |\cdot|^*)$ onto the Banach space $(\mathcal{L}(\mathfrak{X}), |\cdot|)$ is closed.

So let

$$|T_n|^* \to 0, \quad |T_n - T| \to 0 \qquad (n \to \infty).$$

Consider the operator $F \in \mathcal{L}(\mathfrak{X})$ defined by the relation

$$F(x) = \varphi(x)y \qquad (x \in \mathfrak{X}),$$

where y is a fixed element of $\mathfrak{X}$ and φ is a fixed bounded linear form on $\mathfrak{X}$. For every $S \in \mathcal{L}(\mathfrak{X})$ we have

$$(SF)^2 = \varphi(Sy)SF.$$

Hence

$$|\varphi(Sy)| \leqq |S|^* |F|^*.$$

Substituting $S = T_n$ $(n = 1,2,\dots)$ we obtain that $\varphi(T_n y) \to 0$ $(n \to \infty)$. On the other hand, $\varphi(T_n y) \to \varphi(Ty)$. Thus $\varphi(Ty) = 0$ or, φ and y being arbitrary, $T = 0$.

Proof of Theorem 1. We choose an operator $T_0 \in \mathcal{L}(\mathfrak{X})$ of rank 1, and a vector $e \in \mathfrak{X}$ satisfying the relations

(11) $$T_0 e \neq 0, \qquad |e| = 1.$$

Then

(12) $$\mathfrak{X} = \langle e \rangle \dotplus \mathfrak{N}(T_0).$$

For every $x \in \mathfrak{X}$ the relations

(13) $$P_x e = x, \qquad P_x \mathfrak{N}(T_0) = 0$$

define a linear operator P_x on $\mathfrak{X}$. As the components in (12) are closed, each P_x belongs to $\mathcal{L}(\mathfrak{X})$. Moreover, for $\alpha \in \mathbf{K}$, $y \in \mathfrak{N}(T_0)$ we have $|P_x(\alpha e + y)| =$

$= |\alpha| |x| = |P_e(\alpha e + y)| |x| \leqq |P_e| |\alpha e + y| |x|$, so that

(14) $$|P_x| \leqq |P_e| |x| \qquad (x \in \mathfrak{E}) .$$

Put

(15) $$T_0^* T_0 e = f .$$

In view of (12) and (9) $f \neq 0$.

We claim

(16) $$\mathcal{R}(P_x^*) \subset \langle f \rangle \qquad (x \in \mathfrak{E}).$$

In fact, $P_x = Q T_0^* T_0$, where Q is any bounded linear operator such that $Qf = x$. Hence $P_x^* = T_0^* T_0 Q^*$, $\mathcal{R}(P_x^*) \subset \mathcal{R}(T_0^* T_0)$.

Relation (16) implies the existence of a bounded linear form φ_x satisfying

(17) $$P_x^* y = \varphi_x(y) f \qquad (x, y \in \mathfrak{E}).$$

By (12), (13) and (15) this can be rewritten as

(18) $$P_x^* P_y = \varphi_x(y) T_0^* T_0 \qquad (x, y \in \mathfrak{E}).$$

Now suppose that $(\cdot, \cdot)$ is a definite symmetric bilinear form on $\mathfrak{E}$ such that (4) is satisfied for every $T \in \mathcal{L}(\mathfrak{E})$. Then $(P_x e, y) = (e, P_x^* y)$ yields

(19) $$(x, y) = \overline{\varphi_x(y)} (e, f) \qquad (x, y \in \mathfrak{E}),$$

where $(e,f) = (e, T_0^* T_0 e) = (T_0 e, T_0 e)$ is a real non-zero constant from $\mathbf{K}$. This proves uniqueness.

To prove existence, choose any real non-zero number λ and put

(20) $$(x, y) = \lambda \overline{\varphi_x(y)} \qquad (x, y \in \mathfrak{E}) .$$

From (13) it follows that

(21) $$P_{\alpha_1 x_1 + \alpha_2 x_2} = \alpha_1 P_{x_1} + \alpha_2 P_{x_2} \qquad (\alpha_1, \alpha_2 \in \mathbf{K};\ x_1, x_2 \in \mathfrak{E}).$$

Hence, by relations (17) and (5),

(22) $$\varphi_{\alpha_1 x_1 + \alpha_2 x_2} = \bar{\alpha}_1 \varphi_{x_1} + \bar{\alpha}_2 \varphi_{x_2} \qquad (\alpha_1, \alpha_2 \in \mathbf{K};\ x_1, x_2 \in \mathfrak{E}).$$

Thus the definition (20) conforms with (1).

From (18), applying the involution $*$ to both sides and interchanging x and y we obtain

$$P_x^* P_y = \overline{\varphi_y(x)}\, T_0^* T_0 \qquad (x, y \in \mathcal{E}). \tag{23}$$

Comparing (23) with (18) we find

$$\varphi_y(x) = \overline{\varphi_x(y)} \qquad (x, y \in \mathcal{E}). \tag{24}$$

Therefore (20) satisfies relation (2).

Let $\varphi_x(x) = 0$ for some x. Then (18), (9) and (13) yield $x = 0$. Therefore the symmetric bilinear form (20) is definite.

According to (13),

$$P_{Tx} = TP_x \qquad (x \in \mathcal{E};\ T \in \mathcal{L}(\mathcal{E})). \tag{25}$$

Hence $P_{Tx}^*(y) = (TP_x)^* y = P_x^* T^* y$ so that in view of (17)

$$\varphi_{Tx}(y) = \varphi_x(T^* y) \qquad (x, y \in \mathcal{E};\ T \in \mathcal{L}(\mathcal{E})). \tag{26}$$

Thus the involution $*$ coincides with conjugation relative to the form (20).

By Lemma 1 and relation (14) we have

$$|P_x^* y| \leq \gamma |x| |y| \qquad (x, y \in \mathcal{E}) \tag{27}$$

with some $\gamma > 0$. Taking the definition (17) of φ_x into account one finds

$$|\varphi_x(y)| \leq \gamma_1 |x| |y| \qquad (x, y \in \mathcal{E}), \tag{28}$$

where $\gamma_1 = \gamma |f|^{-1}$. Therefore the norm $\|\cdot\|$ given by (10) and (20) is continuous relative to $|\cdot|$.

In order to complete the proof suppose that $|\cdot|$ is discontinuous relative to $\|\cdot\|$. Then for some sequence $\{x_n\}_1^\infty \subset \mathcal{E}$ the scalar sequence $\{\|x_n\|\}_1^\infty$ is bounded while $\{|x_n|\}_1^\infty$ is unbounded. From (13) and (11) we obtain $|P_{x_n}| \geq$ $\geq |x_n|$. Hence $\{|P_{x_n}|\}_1^\infty$ and by Lemma 1 also $\{|P_{x_n}^*|\}_1^\infty$ is unbounded. Making use of the principle of uniform boundedness one finds a y in $\mathcal{E}$ such that $\{|P_{x_n}^* y|\}_1^\infty$ is unbounded. On the other hand, the latter sequence must be bounded since $|P^*_{x_n} y| = |\varphi_{x_n}(y)|\,|f| = |\lambda|^{-1}|(x_n, y)|\,|f| \leq |\lambda|^{-1}|f|\,\|y\|\,\|x_n\|$ for every n .

3. CONJUGATION RELATIVE TO GENERAL SYMMETRIC OR SKEW–SYMMETRIC FORMS

Let $(\cdot,\cdot)$ be any non-degenerate symmetric (resp. skew-symmetric) bilinear form on $\mathfrak{X}$. Suppose that the conjugation $*$ relative to this form exist. (In this case $(\cdot,\cdot)$ must be bounded as will be seen from the next theorem.)

If $\mathbf{K} = \mathbf{C}$, then $*$ is conjugation also relative to the skew-symmetric (resp. symmetric) form $(\cdot,\cdot)'$ defined by $(x,y)' = i(x,y)$ $(x,y \in \mathfrak{X})$ and it is an arbitrary involution (see Theorem 2 below).

If $\mathbf{K} = \mathbf{R}$, then for symmetric resp. skew-symmetric forms the involution $*$ behaves in two different ways. Namely, in the skew-symmetric case (3) implies

$$(29) \qquad (x,x) = 0 \qquad (x \in \mathfrak{X}),$$

hence for every operator T of rank 1 we have $(T^*Tx,y) = (Tx,Ty) = 0$ $(x,y \in \mathfrak{X})$, so that by non-degeneracy $T^*T = 0$. In the symmetric case, however, $T_0^*T_0 \neq 0$ holds for at least one $T_0 \in \mathfrak{L}(\mathfrak{X})$ of rank 1 ; otherwise, in view of (29), the form $(\cdot,\cdot)$ would be semi-definite and we would have $(x,y)^2 \leqq (x,x)(y,y) = 0$ $(x,y \in \mathfrak{X})$ contradicting non-degeneracy.

Theorem 2. *Let $*$ be any involution on $\mathfrak{L}(\mathfrak{X})$. Then there is one, and apart from a multiplicative constant only one, non-degenerate symmetric or skew--symmetric bilinear form $(\cdot,\cdot)$ on $\mathfrak{X}$ such that $*$ coincides with conjugation relative to $(\cdot,\cdot)$. Moreover, $(\cdot,\cdot)$ is bounded. In the case $\mathbf{K} = \mathbf{R}$ the form $(\cdot,\cdot)$ can be chosen to be symmetric if and only if there is a $T_0 \in \mathfrak{L}(\mathfrak{X})$ satisfying the relations*

$$(30) \qquad \operatorname{rank} T_0 = 1, \qquad T_0^* T_0 \neq 0,$$

and skew-symmetric if and only if such a T_0 does not exist. In the case $\mathbf{K} = \mathbf{C}$ both the symmetric and skew-symmetric possibilities can always be realized.

Proof. I) First we let $\mathbf{K}$ to be any of the fields $\mathbf{R}, \mathbf{C}$, and assume that there exists a $T_0 \in \mathfrak{L}(\mathfrak{X})$ with properties (30). Starting from this T_0 the construction which yields the uniqueness part of Theorem 1 can be repeated almost invariantly. We obtain that any non-degenerate symmetric or skew-symmetric bilinear form $(\cdot,\cdot)$

satisfying (4) for every $T \in \mathcal{L}(\mathcal{E})$ must be given by the equality (19), where the non-zero constant $(e,f) = (T_0 e, T_0 e)$ is real for a symmetric form and pure imaginary for a skew-symmetric one. Hence the relations $\mathbf{K} = \mathbf{R}$ and (30) are incompatible with skew-symmetry.

To prove existence, we define $(.,.)$ by (20), where λ is real and non-zero. In the same way as in the proof of Theorem 1 it can be verified that $(.,.)$ is a bounded symmetric bilinear form on $\mathcal{E}$ satisfying (4) for every $T \in \mathcal{L}(\mathcal{E})$. Under the present assumptions instead of definiteness we have non-degeneracy. Really, $(x,y) = 0 \ (y \in \mathcal{E})$ together with relations (20) and (17) implies $P_x^* = 0$. Therefore, in view of (7), $P_x = 0$. Making use of (13) it follows that $x = 0$.

If $\mathbf{K} = \mathbf{C}$, in (20) we can take $\bar{\lambda} = -\lambda$, $\lambda \neq 0$. Then we obtain a skew-symmetric form.

II) Next let $\mathbf{K}$ be any of $\mathbf{R}, \mathbf{C}$, and suppose that there does not exist any $T_0 \in \mathcal{L}(\mathcal{E})$ satisfying (30).

Let $T_1 \in \mathcal{L}(\mathcal{E})$, rank $T_1 = 1$. Owing to (7), $T_1^* \neq 0$. Choose a $g \notin \mathcal{N}(T_1^*)$, and a $T_2 \in \mathcal{L}(\mathcal{E})$ such that $\mathcal{N}(T_2) = \mathcal{N}(T_1)$, $\mathcal{R}(T_2) = \langle g \rangle$. Then

(31) $$T_1^* T_2 \neq 0 .$$

Since the operators $T_1, T_2, T_1 + T_2$ are of rank 1 and the condition (30) is not fulfilled for any element of $\mathcal{L}(\mathcal{E})$, we have

$$T_1^* T_1 = T_2^* T_2 = (T_1 + T_2)^* (T_1 + T_2) = 0 .$$

Therefore

(32) $$T_1^* T_2 = - T_2^* T_1 .$$

Let

(33) $$e \notin \mathcal{N}(T_2), \qquad |e| = 1 .$$

Then

(34) $$\mathcal{E} = \langle e \rangle \dotplus \mathcal{N}(T_2) .$$

Put

(35) $$T_1^* T_2 e = f .$$

According to (31) and (34) $f \neq 0$.

For every $x \in \mathfrak{L}$ we define a linear operator P_x by the relations

(36) $$P_x e = x, \qquad P_x \mathfrak{N}(T_2) = 0 .$$

From this point on the reasoning imitates part I) of the present proof. The role of $T_0^* T_0$ will be played by $T_1^* T_2$, and (32) must be taken into account.

4. CONJUGATION RELATIVE TO SYMMETRIC FORMS WITH A FINITE RANK OF INDEFINITENESS

Let $(\cdot,\cdot)$ be a non-degenerate symmetric bilinear form on $\mathfrak{L}$ having the property that $\mathfrak{L}$ is the direct sum of two subspaces,

(37) $$\mathfrak{L} = \mathfrak{L}^+ \dot{+} \mathfrak{L}^- ,$$

where

(38) $$(x,x) > 0 \qquad (x \in \mathfrak{L}^+, \ x \neq 0),$$

(39) $$(x,x) < 0 \qquad (x \in \mathfrak{L}^-, \ x \neq 0),$$

(40) $$(x,y) = 0 \qquad (x \in \mathfrak{L}^+, \ y \in \mathfrak{L}^-)$$

and

(41) $$0 < \min\{\dim \mathfrak{L}^+, \dim \mathfrak{L}^-\} = k < \infty .$$

We say, $(\cdot,\cdot)$ has a *finite rank of indefiniteness* k.

The relation

(42) $$[x,y] = (x^+, y^+) - (x^-, y^-) \qquad (x = x^+ + x^-, \ y = y^+ + y^-; \ x^+, y^+ \in \mathfrak{L}^+; \ x^-, y^- \in \mathfrak{L}^-)$$

defines a definite form $[\cdot,\cdot]$. The respective norm $\|\cdot\|$ is given by

(43) $$\|x\| = [x,x]^{\frac{1}{2}} \qquad (x \in \mathfrak{L}).$$

If, in addition to (37)-(41), the condition

(44) $$\mathfrak{E}^+, \mathfrak{E}^- \text{ are complete relative to } \|\cdot\|$$

is fulfilled, $\mathfrak{E}$ is said to be a *Pontrjagin space* with rank of indefiniteness k. (For a comprehensive treatment of Pontrjagin spaces see [6]).

In a Pontrjagin space the norms $\|\cdot\|$ corresponding to different decompositions (37) are equivalent [6]. They will be called the *natural norms* belonging to $(\cdot,\cdot)$.

Lemma 2. *A non-degenerate symmetric bilinear form* $(\cdot,\cdot)$ *on* $\mathfrak{E}$ *has a finite rank of indefiniteness* k $(0<k<\infty)$ *if and only if the maximal dimension of subspaces* $\mathcal{L}\subset\mathfrak{E}$ *with* $(x,x)=0$ *for every* $x\in\mathcal{L}$ *(or, equivalently,* $(x,y)=0$ *for every* $x,y\in\mathcal{L}$ *) equals* k.

Proof. If $(x,x)>0$ $(<0,=0)$ for every non-zero element of a subspace $\mathcal{L}$, we say that $\mathcal{L}$ is positive definite (negative definite, neutral).

Let $(\cdot,\cdot)$ have a finite rank of indefiniteness k. One can find two systems $\{e_n^+\}_1^k\subset\mathfrak{E}^+, \{e_n^-\}_1^k\subset\mathfrak{E}^-$ satisfying the relations $(e_m^+,e_n^+)=\delta_{mn}$, $(e_m^-,e_n^-)=-\delta_{mn}$ $(m,n=1,\dots,k)$. The span of the vectors $e_n^++e_n^-$ $(n=1,\dots k)$ is a k-dimensional neutral subspace. On the other hand, a $(k+1)$-dimensional subspace cannot be neutral since, owing to the definition of k, it has a non-zero intersection with at least one of the subspaces $\mathfrak{E}^+, \mathfrak{E}^-$.

Suppose, conversely, that the maximal dimension of neutral subspaces in $\mathfrak{E}$ equals k, where $0<k<\infty$. First we observe that $\mathfrak{E}$ cannot contain both infinite-dimensional positive definite and infinite-dimensional negative definite subspaces.

Otherwise let $\dim\mathfrak{M}=k+1$, $\mathfrak{M}$ positive definite. Set

$$\mathfrak{M}^\perp=\{x\in\mathfrak{E}:\ (x,y)=0\quad (y\in\mathfrak{M})\}.$$

Any infinite-dimensional subspace of $\mathfrak{E}$ meets $\mathfrak{M}^\perp$ in an infinite-dimensional subspace. Therefore $\mathfrak{M}^\perp$ contains $(k+1)$-dimensional negative definite subspaces. Consequently, using the construction presented at the beginning of this proof, we could find a $(k+1)$-dimensional neutral subspace in $\mathfrak{E}$, contrary to the hypothesis.

It follows [6; Lemma 1.1] that $(\cdot,\cdot)$ has a finite rank of indefiniteness k_1. By what has already been proved, $k_1=k$.

Now let $(\cdot,\cdot)$ denote any non-degenerate symmetric bilinear form on $\mathfrak{E}$ with a finite rank of indefiniteness k. Suppose that the conjugation $*$ relative to $(\cdot,\cdot)$ exists. (By the theory of Pontrjagin spaces and Theorem 3 below this is the case if and only if $\mathfrak{E}$ is a Pontrjagin space with natural norms equivalent to $|\cdot|$.) By Lemma 2 the maximal rank of operators $T \in \mathcal{L}(\mathfrak{E})$ such that $(Tx, Ty) = 0$ for every x, y in $\mathfrak{E}$, i.e. $T^*T = 0$, is equal to k. Moreover, according to the discussion at the beginning of Section 3, in the case $\mathbf{K} = \mathbf{R}$ there exists a $T_0 \in \mathcal{L}(\mathfrak{E})$ of rank 1 satisfying $T_0^* T_0 \neq 0$.

Theorem 3. *Let* $*$ *be an involution on* $\mathcal{L}(\mathfrak{E})$ *such that*

$$0 < \max\{\operatorname{rank} T : T \in \mathcal{L}(\mathfrak{E}),\ T^*T = 0\} = k < \infty . \tag{45}$$

For $\mathbf{K} = \mathbf{R}$ *we require also the existence of a* $T_0 \in \mathcal{L}(\mathfrak{E})$ *satisfying (30). Then there is one, and apart from an arbitrary non-zero real multiplicative constant only one, non--degenerate symmetric bilinear form* $(\cdot,\cdot)$ *on* $\mathfrak{E}$ *such that* $*$ *coincides with conjugation relative to* $(\cdot,\cdot)$. *The form* $(\cdot,\cdot)$ *turns* $\mathfrak{E}$ *into a Pontrjagin space with rank of indefiniteness* k *and with natural norms equivalent to* $|\cdot|$.

Proof. From Theorem 2 we obtain the uniqueness conclusion and the existence of a bounded non-degenerate symmetric bilinear form $(\cdot,\cdot)$ satisfying (4) for every $T \in \mathcal{L}(\mathfrak{E})$. From condition (45) it follows that the maximal dimension of subspaces $\mathcal{L} \subset \mathfrak{E}$ with $(x,y) = 0$ for every $x, y \in \mathcal{L}$ equals k. Therefore, according to Lemma 2, $(\cdot,\cdot)$ has rank of indefiniteness k.

Consider a direct decomposition (37) having the properties (38)-(41). The subspaces $\mathfrak{E}^+, \mathfrak{E}^-$ are closed. Really, if e.g. $\mathfrak{E}^+$ is not closed, then its closure contains a non-zero vector x in common with $\mathfrak{E}^-$. By (39) we have $(x, x) < 0$. On the other hand, (38) and the boundedness of $(\cdot,\cdot)$ imply $(x, x) \geqq 0$.

Hence the linear operators P^+, P^-, J defined by the relations $P^+x^+ = x^+$, $P^+x^- = 0$ $(x^+ \in \mathfrak{E}^+, x^- \in \mathfrak{E}^-)$, $P^- = I - P^+$, $J = P^+ - P^-$ belong to $\mathcal{L}(\mathfrak{E})$. We set

$$T^{[*]} = JT^*J \qquad (T \in \mathcal{L}(\mathfrak{E})).$$

From (4) and (42) it follows that

$$[Tx, y] = [x, T^{[*]}y] \qquad (x, y \in \mathfrak{E};\ T \in \mathcal{L}(\mathfrak{E})).$$

Applied to the involution $[*]$ Theorem 1 says, in particular, that the norm $\|\cdot\|$ given by (43) is equivalent to $|\cdot|$. As $\mathfrak{L}^+, \mathfrak{L}^-$ are closed, also condition (44) is satisfied.

5. Remarks. In the special case $\dim \mathfrak{L} < \infty$ Theorems 1 and 2 have been proved recently [7] by a similar, but a little more complicated method.

The characterization (45) of Pontrjagin spaces was proposed to the author by M.G. Kreĭn.

The question of characterizing conjugations in J-spaces, i.e. vector spaces $\mathfrak{L}$ with a symmetric bilinear form $(\cdot,\cdot)$ satisfying (37)-(40) and (44) but not necessarily (41), seems to be unsolved. H. Langer pointed out that in the special case where $\mathfrak{L}$ is a Hilbert space by assumption, the symmetric form given by Theorem 2 always turns $\mathfrak{L}$ into a J-space. This follows from the spectral theorem and the proof of Theorem 3.

I express my deep gratitude to I.S. Louhivaara for making the necessary literature available, and to B. Sz.–Nagy for helpful criticism which led to the final version of the paper.

LITERATURE

[1] S. Kakutani and G.W. Mackey, Two characterizations of real Hilbert space, *Ann. Math.*, 45(1944), 50-58.

[2] S. Kakutani and G.W. Mackey, Ring and lattice characterizations of complex Hilbert space, *Bull. Amer. Math. Soc.*, 52(1946), 727-733.

[3] Y. Kawada, Über den Operatorenring Banachscher Räume, *Proc. Imp. Acad. Tokyo*, 19(1943), 616-621.

[4] G.E. Rickart, General theory of Banach algebras (Princeton-Toronto-London--New York, 1960).

[5] M. Eidelheit, On isomorphisms of rings of linear operators., *Studia Math.*, 9(1940), 97-105.

[6] I.S. Iohvidov and M.G. Kreĭn, Spectral theory of operators in spaces with indefinite metrics. I, *Trudy Mosk. Mat. Obšč.*, 5(1956), 367-432 (in Russian).

[7] J. Bognár, Operator conjugation with respect to symmetric and skew-symmetric forms, *Acta Sci. Math.*, 31(1970), 69-73.

COLLOQUIA MATHEMATICA SOCIETATIS JÁNOS BOLYAI
5. HILBERT SPACE OPERATORS, TIHANY (HUNGARY), 1970

Hermitian operators on Banach spaces

F. F. BONSALL

1. INTRODUCTION

Some recent results on Hermitian operators on Banach spaces are briefly surveyed, the principal theorems being stated without proof, and these operators are compared with self-adjoint operators on a Hilbert space. Some examples and applications are given, including a short discussion of the spectral decomposition of compact Hermitian operators. In particular, it is proved that a compact Hermitian operator can be expressed in terms of its spectral projections if the eigenvalues are of order $o(\frac{1}{n})$. The final section is concerned with the perturbation of Hermitian operators by compact operators.

I am grateful to A.M. Sinclair for prepublication access to his results mentioned below, in particular Theorem 2 of which I make repeated use.

Throughout this article X will denote a complex Banach space, X' its dual space, $B(X)$ the Banach algebra of all bounded linear operators on X ; A will denote a complex unital Banach algebra, and $Sp(a)$, $\varrho(a)$ will denote the spectrum and spectral radius respectively of an element $a \in A$. The closed convex hull of a set E will be denoted by $\overline{co}\, E$.

2. NUMERICAL RANGES

Of the several alternative definitions of Hermitian operators, the most natural is in terms of the reality of the numerical range.

Let S denote the unit sphere $\{x \in X : \|x\| = 1\}$ in X, S' the unit sphere in X', π_1 the natural projection of $X \times X'$ onto X, and let

$$\Pi = \Pi(X) = \{(x,f) \in S \times S' : f(x) = 1\} .$$

Given $T \in B(X)$, the *numerical range* $V(T)$ of T is defined by

$$V(T) = \{f(Tx) : (x,f) \in \Pi\} .$$

More generally, given $\Gamma \subset \Pi$, let

$$V_\Gamma(T) = \{f(Tx) : (x,f) \in \Gamma\} .$$

The following elementary lemma is essentially due to Lumer [8].

Lemma 1. *Let* Γ *be a subset of* Π *such that* $\pi_1\Gamma$ *is dense in* S. *Then*

$$\sup \operatorname{Re} V_\Gamma(T) = \inf\{\alpha^{-1}(\|I + \alpha T\| - 1) : \alpha > 0\} =$$
$$= \lim_{\alpha \to 0+} \alpha^{-1}(\|I + \alpha T\| - 1) .$$

It follows at once from this lemma that

(1) $$\overline{\mathrm{co}}\, V_\Gamma(T) = \overline{\mathrm{co}}\, V(T)$$

for all choices of Γ such that $\pi_1\Gamma$ is dense in S. In particular the choice of a semi-inner-product on X that reproduces the norm of X is equivalent to the choice of Γ containing exactly one (x,f) for each $x \in S$, and then $V_\Gamma(T)$ becomes the numerical range $W(T)$ in the sense of Lumer [8]. Also, for operators $T \in B(X')$, Lemma 1 together with the theorem of Bishop and Phelps [2] shows that $\overline{\mathrm{co}}\, V(T)$ can be obtained from $\Pi(X)$ instead of $\Pi(X')$,

(2) $$\overline{\mathrm{co}}\, V(T) = \overline{\mathrm{co}}\{(Tf)(x) : (x,f) \in \Pi(X)\} \qquad (T \in B(X')) .$$

Given an element $a \in A$, it is natural to define its numerical range $V(A,a)$ by

$$V(A,a) = \{f(ax) : (x,f) \in \Pi(A)\} .$$

However, since A has a unit element 1 with $\|1\| = 1$, a simpler expression is available for $V(A,a)$. Let $D(1)$ denote the set of all normalized states on A, i.e. functionals $f \in A'$ such that

$$\|f\| = f(1) = 1 .$$

Then it is very easily verified that

(3) $$V(A,a) = \{f(a) : f \in D(1)\},$$

from which it is apparent that $V(A,a)$ is compact and convex.

Given an operator $T \in B(X)$, we may regard T as an element of the unital Banach algebra $B(X)$, and have available the compact convex numerical range $V(B(X),T)$. Lemma 1 shows that

(4) $$\overline{co}\, V(T) = V(B(X),T) .$$

For a fuller account of numerical ranges with proofs, see Bonsall and Duncan [5].

3. HERMITIAN OPERATORS

An operator $T \in B(X)$ or an element $h \in A$ is said to be *Hermitian* if its numerical range is real. We denote by $H(A)$ the set of all Hermitian elements of A.

The above remarks, and particularly the equalities (1), (4) show that all choices of numerical range determine the same Hermitian operators. In particular $T \in B(X)$ is a Hermitian operator if and only if it is a Hermitian element of the Banach algebra $B(X)$. It can occur, however, that it is convenient to verify the Hermitian property in terms of a particular choice of $\Gamma \subset \ (X)|$. For example, by (2), $T \in B(X')$ is Hermitian if and only if

(5) $$(Tf)(x) \in \mathbf{R} \qquad ((x,f) \in \ (X)) .$$

It is also a corollary of Lemma 1 that $h \in A$ is Hermitian if and only if

$$\lim_{\alpha \to 0\ (\alpha \in \mathbf{R})} \alpha^{-1}(\|1 + i\alpha h\| - 1) = 0,$$

which is the definition introduced by Vidav [13]. It was shown by Vidav that this is equivalent to

(6) $$\|\exp(i\alpha h)\| = 1 \qquad (\alpha \in \mathbf{R}),$$

which is often the most convenient characterization.

When X is a Hilbert space, $V(T)$ concides with the classical numerical range $W(T)$ and $T \in B(X)$ is Hermitian if and only if it is self-adjoint. Likewise an element h of a B^*-algebra is Hermitian if and only if $h^* = h$. The metric characterization of B^*-algebras due to Vidav [13] has recently been improved by Berkson [1], Glickfeld [6], and Palmer [10], so that it now takes the following highly satisfactory form.

Theorem 1. *There exists an involution* A *making* A *a* B^**-algebra if and only if*

(7) $$A = H(A) + iH(A).$$

This theorem can now be given an elementary and reasonably short proof, and so it has become a convenient tool for the study of B^*-algebras. For example, Theorem 1 together with (5) gives an immediate proof that the second dual of a B^*-algebra with the Arens multplication and the natural involution is a B^*-algebra.

A crucial step in the proof of Theorem 1 is the proof that condition (7) implies that $\rho(h) = \|h\|$ for Hermitian elements h. Since the square of a Hermitian operator need not be Hermitian, it was very surprising when Sinclair [11] proved the following theorem showing that this equality is true for all unital Banach algebras A (i.e. without condition (7)).

Theorem 2. $\rho(h) = \|h\| \quad (h \in H(A))$.

An elementary and intuitive proof of this theorem has been given recently by Bonsall and Crabb [4], based on the following lemma.

Lemma 2. *Let* $h \in H(A)$ *and* $\rho(h) \leq \pi/2$. *Then*

$$h = \arcsin(\sin h).$$

Here $\arcsin a$ is defined, for $a \in A$ such that $\{\|a^n\|\}$ is bounded, by

$$\arcsin a = a + \frac{1}{2} \cdot \frac{1}{3} a^3 + \frac{1}{2} \cdot \frac{3}{4} \cdot \frac{1}{5} a^5 + \dots .$$

If we take Lemma 2 for granted, Theorem 2 is immediate, for (6) gives $\|\sin h\| \leq 1$. Therefore given a Hermitian element h with $\rho(h) \leq \pi/2$, we have

$$\|h\| \leq \|\sin h\| + \frac{1}{2}\cdot\frac{1}{3}\|\sin h\|^3 + \ldots \leq 1 + \frac{1}{2}\cdot\frac{1}{3} + \ldots = \frac{\pi}{2} .$$

Example 1. Theorem 2 yields a simple proof that differentiation is a bounded linear operator on the space of all uniformly almost periodic functions with their Fourier exponents lying in a compact set. Let τ be a positive real number, let

$$u_\lambda(s) = \exp(i\lambda s) \qquad (\lambda, s \in \mathbf{R}).$$

Let E_0 be the linear hull of $\{u_\lambda : -\tau \leq \lambda < \tau\}$, and let E be the closure of E_0 in the space of bounded complex functions on $\mathbf{R}$ with respect to the uniform norm; i.e. E is the space of all uniformly almost periodic functions with their Fourier exponents in $[-\tau, \tau]$.

Let T_0 be the operator defined on E_0 by $T_0 f = -if'$, where f' is the derivative of f. Obviously $T_0 u_\lambda = \lambda u_\lambda$, and so

$$\exp(i\alpha T_0)\, u_\lambda(s) = \exp(i\alpha\lambda)\, u_\lambda(s) = u_\lambda(s+\alpha).$$

Let $\lambda_1, \ldots, \lambda_n \in [-\tau, \tau]$, let E_1 be the linear hull of $\{u_{\lambda_1}, \ldots, u_{\lambda_n}\}$, and let T_1 be the restriction of T_0 to E_1.

We have

$$\exp(i\alpha T_1) f(s) = f(s+\alpha) \qquad (f \in E_1,\ \alpha, s \in \mathbf{R}),$$

and so $\|\exp(i\alpha T_1)\| = 1$, T_1 is Hermitian. Evidently, $Sp(T_1) = \{\lambda_1, \ldots, \lambda_n\}$, and so, by Theorem 2,

$$\|T_1\| = \rho(T_1) = \max\{|\lambda_1|, \ldots, |\lambda_n|\} \leq \tau .$$

Since $\lambda_1, \ldots, \lambda_n$ are arbitrary, we have now proved that $T_0 \in B(E_0)$ and $\|T_0\| = \tau$. Therefore T_0 has a unique extension $T \in B(E)$, and $\|T\| = \tau$.

It is now easy to see that every $f \in E$ is differentiable and that $Tf = -if'$. For, given $f \in E$, there exist $f_n \in E_0$ uniformly convergent to f. Then $\{T_0 f_n\}$ convergens uniformly to Tf, and so

$$\int_0^t (Tf)(s)\,ds = \lim_{n\to\infty} \int_0^t (T_0 f_n)(s)\,ds = \lim_{n\to\infty} -i \int_0^t f_n'(s)\,ds =$$

$$= \lim_{n \to \infty} -i(f_n(t) - f_n(0)) = -i(f(t) - f(0)),$$

from which the result follows.

It is clear that this operator T is Hermitian. It provides a natural example of a Hermitian operator the square of which is not Hermitian. To check that T^2 is not Hermitian, it is enough to consider $\exp(i\frac{\pi}{2}T^2)(u_0 + iu_1 + u_2)$.

Given a self-adjoint operator T on a Hilbert space with $\|T\| \leq 1$, it is easy to express T as a convex combination of operators of the form $\exp(iU)$ with U self-adjoint. It is enough to take $R = \arccos T$, and then $T = \frac{1}{2}(\exp(iR) + \exp(-iR))$ is an expression of the required kind. This argument fails for a Hermitian operator T on a Banach space since we have no reason to suppose that $\arccos T$ is Hermitian. However, if we allow infinite 'convex' combinations we can obtain a similar result.

Theorem 3. *Let* $h \in H(A)$ *and* $\|h\| \leq 1$. *Then*

$$h = \sum_{-\infty}^{\infty} 4\pi^{-2}(2n+1)^{-2} \exp\{(n + \frac{1}{2})\pi i(h-1)\}.$$

Proof. By Lemma 2, $\frac{\pi}{2}h = \arcsin(\sin\frac{\pi}{2}h)$; and we have

$$\sin\frac{\pi}{2}h = \frac{1}{2}\{\exp(\frac{\pi}{2}i(h-1)) + \exp(-\frac{\pi}{2}i(h-1))\}.$$

Since the sum of the coefficients in the $\arcsin$ series is $\frac{\pi}{2}$, it follows that

$$h = \sum_{-\infty}^{\infty} \alpha_n \exp(n\frac{\pi}{2}i(h-1)),$$

where $\alpha_n \geq 0$ and $\sum_{-\infty}^{\infty} \alpha_n = 1$. Moreover, the coefficients α_n are absolute constants independent of h and of the Banach algebra A. We may therefore evaluate them by taking $A = \mathbf{C}$ and $h = t \in \mathbf{R}$ with $-1 \leq t \leq 1$. This presents no difficulty.

Theorem 3 gives an explicit expression for Hermitian elements in the unit ball of A as infinite 'convex' combinations of vertices of the ball.

4. COMPACT HERMITIAN OPERATORS

The following theorem gives a partial analogue of the classical spectral decomposition of a compact self-adjoint operator on a Hilbert space.

Theorem 4. *Let* $T \in B(X)$ *be compact and Hermitian, let* $\{\lambda_n\}$ *be the non-zero (distinct) eigenvalues of* T *arranged so that* $|\lambda_{n+1}| \leq |\lambda_n|$ $(n = 1, 2, \dots)$, *and let* P_n *be the spectral projection corresponding to* λ_n. *Then the following statements hold.*

(i) *Each* λ_n *has ascent* 1.

(ii) $\|P_n\| = 1$.

(iii) *If each* P_n *is Hermitian, then* $T = \sum_n \lambda_n P_n$.

(iv) *If* $\lim_{n\to\infty} n\lambda_n = 0$, *then* $T = \sum_n \lambda_n P_n$.

Proof. (i) Clearly $\lambda_n \in V(T) \subset \mathbf{R}$, and so λ_n is a boundary point of $\overline{\text{co}}\, V(T)$. A theorem of Nirschl and Schneider [9] now shows that λ_n has ascent 1.

(ii) By (i), $P_n X$ is the null space and $(I - P_n)X$ is the range of $\lambda_n I - T$. Sinclair [12] has proved that if an eigenvalue λ of an operator $U \in B(X)$ is a boundary point of $V(B(X), U)$, then the null space of $\lambda I - U$ is orthogonal (in the sense of R.C. James) to the range of $\lambda I - U$. This theorem is applicable to λ_n and T, and gives

$$\|P_n x + (I - P_n)y\| \geq \|P_n x\| \quad (x, y \in X),$$

from which $\|x\| \geq \|P_n x\|$.

(iii) Consider the case in which there are infinitely many λ_n, the finite case being easier. Assume that each P_n is Hermitian, and let $S_n = \sum_{k=1}^{n} \lambda_k P_k$. Then $T - S_n$ is Hermitian and so Theorem 2 gives

$$\|T - S_n\| = \rho(T - S_n) = |\lambda_{n+1}|.$$

But $\lim_{n\to\infty} \lambda_n = 0$ by the compactness of T.

(iv) With S_n as above and $R_n = \sum_{k=1}^{n} P_k$, we have

$$T - S_n = T_n(I - R_n),$$

where T_n denotes the restriction of T to the subspace $(I-R_n)X$. The restriction of a Hermitian operator to an invariant subspace is evidently Hermitian, and therefore, by Theorem 2,

$$\|T_n\| = \rho(T_n) = |\lambda_{n+1}|.$$

But, by (ii), $\|R_n\| \le n$ and so $\|T-S_n\| \le (n+1)|\lambda_{n+1}|$.

Since $I - R_n = (I-P_n)\cdots(I-P_1)$, this proof also shows that the required conclusion also holds whenever $\|I - P_n\| \le 1$ for all n.

Example 2. It is easy to verify that $T \in B(\ell_p)$ with $1 \le p \le \infty$, $p \ne 2$, is Hermitian if and only if there is a bounded real sequence $\{\lambda_n\}$ such that $T\{\xi_n\} = \{\lambda_n \xi_n\}$ for every $\{\xi_n\} \in \ell_p$. T is also compact if and only if $\lim_{n\to\infty} \lambda_n = 0$. Let e_n denote the sequence with 1 for the nth term and 0 for all other terms, and let the λ_n be distinct. Then the spectral projection P_n is given by $P_n\{\xi_n\} = \xi_n e_n (\{\xi_n\} \in \ell_p)$. Plainly P_n is multiplication by the bounded real sequence e_n, and so P_n is Hermitian.

Example 3. With the notation of Example 1, take E_1 to be the linear hull of $\{u_0, u_1, u_2\}$, and consider the operator T_0 of Example 1 restricted to this space. Let P be the spectral projection corresponding to the eigenvalue 2. Then P is not Hermitian and $\|I - P\| > 1$.

To see this, note that $P(\xi_0 u_0 + \xi_1 u_1 + \xi_2 u_2) = \xi_2 u_2$. Consider the point $x = 6u_0 + 6u_1 - u_2$. We have

$$\|(I-P)x\| = \sup_{\Theta} |6 + 6e^{i\Theta}| = 12,$$

but

$$\|x\| = \sup_{\Theta} |6 + 6e^{i\Theta} - e^{2i\Theta}| = \sup_{\Theta} |6e^{-i\Theta} + 6 - e^{i\Theta}| =$$

$$= \sup_{\Theta} |6 + 5\cos\Theta - 7i\sin\Theta| = 11.$$

This shows that $\|I - P\| > 1$, and therefore, by Theorem 2, $I - P$ is not Hermitian, and so P is not Hermitian.

5. PERTURBATION OF HERMITIAN OPERATORS BY COMPACT OPERATORS

$K(X)$ will denote the set of all compact linear operators on X, and π will denote the canonical mapping of $B(X)$ onto $B(X)/K(X)$. Given $T \in B(X)$, $WSp(T)$ will denote the *Weyl spactrum* given by

$$WSp(T) = \bigcap \{Sp(T+C) : C \in K(X)\} .$$

Definition. $T \in B(X)$ is said to be *almost Hermitian* if $T = U + C$ with $U \in H(B(X))$ and $C \in K(X)$, it is said to be *quasi-Hermitian* if $\pi T \in H(B(X)/K(X))$.

Lemma 3. *Let* K *be a closed two-sided ideal of a unital Banach algebra* B, *and let* π *denote the canonical mapping of* B *onto* B/K. *Then*

$$V(B/K, \pi a) \subset V(B, a) \qquad (a \in A) .$$

Proof. Immediate.

Remarks. (i) It is obvious that Hermitian operators are almost Hermitian, and Lemma 3 shows that almost Hermitian operators are quasi-Hermitian.

(2) If X is a Hilbert space, $T \in B(X)$ is quasi-Hermitian if and only if its imaginary part is compact. That operators with compact imaginary part are almost Hermitian and therefore quasi-Hermitian is obvious. On the other hand if T is quasi-Hermitian, then πT is a Hermitian and therefore self-adjoint element of the B^*-algebra $B(X)/K(X)$, so $\pi(T-T^*) = 0$, $T - T^* \in K(X)$.

(3) All quasi-Hermitian Riesz operators are compact. For if T is a quasi-Hermitian Riesz operator, then πT is a quasi-nilpotent Hermitian element of $B(X)/K(X)$. Therefore, by Theorem 2, $\|\pi T\| = \rho(\pi T) = 0$, from which $T \in K(X)$.

In particular, all Hermitian Riesz operators are compact.

(4) If T_1, T_2 are quasi-Hermitian (almost Hermitian), then $i(T_1T_2 - T_2T_1)$ is quasi-Hermitian (almost Hermitian). This follows at once from the corresponding statement for Hermitian elements (Vidav [13]) and the fact that $K(X)$ is a two-sided ideal.

Lemma 4. *In the notation of Lemma 3,*

$$V(B/K, \pi a) = \bigcap \{V(B, a+k) : k \in K\} .$$

Proof. Let $E(a) = \cap\{V(B, a+k): k \in K\}$. By Lemma 3,

$$V(B/K, \pi a) = V(B/K, \pi(a+k)) \subset V(B, a+k),$$

and therefore $V(B/K, \pi a) \subset E(a)$. The sets $V(B/K, \pi a)$ and $E(a)$ are compact and convex, and $V(B/K, \pi(\lambda a)) = \lambda V(B/K, \pi a)$, $E(\lambda a) = \lambda(E)a$. Therefore it is enough to prove that

$$\sup \operatorname{Re} E(a) \leq \sup \operatorname{Re} V(B/K, \pi a). \tag{8}$$

Let $\mu = \sup \operatorname{Re} V(B/K, \pi a)$ and let $\varepsilon > 0$. By Lemma 1, there exists $\beta > 0$ such that $\|\pi(1+\beta a)\| < 1 + \beta(\mu + \varepsilon)$. Then, by definition of the canonical norm on B/K, there exists $k_0 \in K$ such that $\|1 + \beta a + k_0\| < 1 + \beta(\mu + \varepsilon)$. Write $k = k_0/\beta$. Then

$$\frac{1}{\beta}\{\|1 + \beta(a+k)\| - 1\} < \mu + \varepsilon,$$

and so $\sup \operatorname{Re} V(B, a+k) < \mu + \varepsilon$. This proves (8) and completes the proof.

Theorem 5. *Let* $T \in B(X)$. *Then* T *is quasi-Hermitian if and only if*

$$\cap\{V(B(X), T+C): C \in K(X)\} \subset \mathbf{R}.$$

Proof. Immediate from Lemma 4.

Corollary 1. *Let* T *be quasi-Hermitian. Then* $WSp(T) \subset \mathbf{R}$ *and all points of* $Sp(T) \setminus WSp(T)$ *are eigenvalues of* T *with finite dimensional eigenspaces.*

Proof. We have $Sp(T+C) \subset V(B(X), T+C)$, and so $WSp(T) \subset \mathbf{R}$ by Theorem 5. The rest follows by Weyl's theorem (Halmos [7], Problem 143).

It is evident that the set of quasi-Hermitian operators is a closed subset of $B(X)$, and this raises the question whether the set of almost Hermitian operators is closed. The following example shows that this is the case when $X = \ell_p$ $(1 \leq p < \infty)$.

Example 4. Let $1 \leq p < \infty$, $p \neq 2$, and use the notation of Example 2. Given $T \in B(\ell_p)$, let τ_j denote the jth term of the sequence Te_j, let $\varrho_j = \operatorname{Re} \tau_j$, and for each $x = \{\xi_j\} \in \ell_p$, let $R(T)x = \{\varrho_j \xi_j\}$. We have $|\varrho_j| \leq |\tau_j| \leq \|Te_j\| \leq \|T\|$, and so $R(T)x \in \ell_p$, and $\|R(T)x\| \leq \|T\| \, \|x\|$. Thus $R(T) \in B(\ell_p)$. By Example 2, R projects $B(X)$ onto $H(B(X))$. It is easy to prove that R maps $K(X)$ into $K(X)$. It then follows that $T \in B(X)$ is almost Hermitian if and only if $T - R(T) \in K(X)$. The continuity of R now shows that set of almost Hermitian operators on ℓ_p is closed.

This result suggests that perhaps on the spaces ℓ_p, $1 \le p < \infty$ quasi-Hermitian and almost Hermitian operators may coincide.

REFERENCES

[1] E. Berkson, Some characterizations of C^*-algebras, *Ill. J. Math.*, 10 (1966), 1-8.

[2] E. Bishop and R.R. Phelps, A proof that every Banach space is subreflexive, *Bull. Amer. Math. Soc.*, 67 (1961), 97-98.

[3] H.F. Bohnenblust and S. Karlin, Geometrical properties of the unit sphere of Banach algebras, *Ann. of Math.*, 62 (1955), 217-229.

[4] F.F. Bonsall and M.J. Crabb, The spectral radius of a Hermitian element of a Banach algebra, *Bull. London Math. Soc.*, 2 (1970), 178-180.

[5] F.F. Bonsall and J. Duncan, *Numerical ranges of operators on normed spaces and of elements of normed algebras,* London Math. Soc. Lecture Notes 2 (1970).

[6] B.W. Glickfeld, A metric characterization of $C(X)$ and its generalization to C^*-algebras, *Ill. J. Math.*, 10 (1966), 547-566.

[7] P.R. Halmos, *Hilbert space problem book* (Princeton – Toronto – London, 1967).

[8] G. Lumer, Semi-inner-product spaces, *Trans. Amer. Math. Soc.*, 100 (1961), 29-43.

[9] N. Nirschl and H. Schneider, The Bauer fields of values of a matrix, *Numer. Math.*, 6 (1964), 355-365.

[10] T.W. Palmer, Characterizations of C^*-algebras, *Bull. Amer. Math. Soc.*, 74 (1968), 538-540.

[11] A.M. Sinclair, The norm of a Hermitian element in a Banach algebra (to appear).

[12] A.M. Sinclair, Eigenvalues in the boundary of the numerical range (to appear).

[13] I. Vidav, Eine metrische Kennzeichnung der selbstadjungierten Operatoren, *Math. Zeit.*, 66 (1956), 121-128.

COLLOQUIA MATHEMATICA SOCIETATIS JÁNOS BOLYAI
5. HILBERT SPACE OPERATORS, TIHANY (HUNGARY), 1970

On certain invariant subspaces of dissipative operators of exponential type

M. S. BRODSKIĬ

We shall say that the bounded linear operator A acting in the separable Hilbert space $\mathfrak{H}$ belongs to the class $\Lambda^{(exp)}$, provided: 1) A is dissipative;* 2) the spectrum of A does not contain points different from zero; 3) $(I-\lambda A)^{-1}$ is a function of exponential type. The type of growth of the function $(I-\lambda A)^{-1}$ will be denoted by $\sigma(A)$. If A satisfies the conditions 1), 2) and $\tau(A) = \operatorname{sp} A_I < \infty$ (or $\dim A_I \mathfrak{H} = 1$, respectively), then we shall say that A belongs to the class Λ_v (resp. Λ_1).

We have the relations $\Lambda_1 \subset \Lambda_v \subset \Lambda^{(exp)}$. For every operator $A \in \Lambda^{(exp)}$ the inequality $\sigma(A) \leqq 2\tau(A)$ holds. Completely non-selfadjoint** operators of

*The operator A is said to be dissipative if $A_I = \frac{1}{2i}(A - A^*) \geqq 0$.

**The operator A is said to be completely non-selfadjoint, if there is no non-zero subspace invariant for A and A^*, in which the induced operator is selfadjoint.

the class Λ_1 are unicellular and satisfy the condition $\sigma(A) = 2\tau(A)$ [1].

According to a theorem of G.E. Kisilevskiĭ [2], the space $\mathfrak{H}$ where the completely non-selfadjoint operator $A \in \Lambda_V$ acts can be represented as the approximative sum* of a finite or countable number of subspaces $\mathfrak{H}_j$ invariant for A so that in each of them the induced operator A_j is unicellular; moreover, the numbers $\sigma(A_j)$ are uniquely defined by the operator A. Obviously, in order to carry over completely the Jordan theory of finite-dimensional operators to the class Λ_V, one has to add to the theorem of G. E. Kisilevskiĭ the solution of the inverse problem that consists of describing the process of constructing arbitrary operators in the class Λ_V from unicellular operators of the same class. An idea of the difficulties arising here is given already by the simplest case where the space $\mathfrak{H}$ is represented as the approximative sum of two of its subspaces $\mathfrak{H}_1$ and $\mathfrak{H}_2$. The point is that if one selects in them unicellular operators $A_1 \in \Lambda_V$ and $A_2 \in \Lambda_V$ arbitrarily, then in $\mathfrak{H}$ this defines an operator A that may turn out to be unbounded or non-dissipative.

In the present paper we investigate the relative position of those invariant subspaces for an operator of class $\Lambda^{(exp)}$ in which the induced operator belongs to Λ_1. In particular, we solve the inverse problem mentioned above for the operators of class Λ_V admitting a decomposition into unicellular operators of class Λ_1.

We shall make use of the following results [1].

Theorem 0.1. *If* A *is a completely non-selfadjoint operator of class* $\Lambda^{(exp)}$ *acting in* $\mathfrak{H}$ *and* $\sigma(A) = 0$, *then* $\mathfrak{H} = 0$.

Theorem 0.2. *If the operator* A_0 *is induced by the operator* $A \in \Lambda^{(exp)}$ *in a subspace invariant for* A, *then* $A_0 \in \Lambda^{(exp)}$ *and* $\sigma(A_0) \leqq \sigma(A)$.

Theorem 0.3. *Let* A *be a completely non-selfadjoint operator of class* $\Lambda^{(exp)}$ *acting in the space* $\mathfrak{H}$, *and let* A_γ $(\gamma \in \Gamma)$ *denote the operators induced in some invariant subspaces* $\mathfrak{H}_\gamma$ *of* A. *If* $\mathfrak{H} = \tilde{\bigcup}_{\gamma\in\Gamma} \mathfrak{H}_\gamma$, *then* $\sigma(A) = \sup_{\gamma\in\Gamma} \sigma(A_\gamma)$.

Theorem 0.4. *Let, under the assumptions of the preceding theorem, the subspaces* $\mathfrak{H}_\gamma$ *be ordered by inclusion. If* $A \in \Lambda_V$, *then* $\inf \sigma(A_\gamma) = \sigma(A_0)$,

*The space $\mathfrak{H}$ is said to be the approximative sum of its subspaces $\mathfrak{H}_\gamma$ $(\gamma \in \Gamma)$ if 1) $\tilde{\bigcup}_{\gamma\in\Gamma} \mathfrak{H}_j = \mathfrak{H}$; 2) $(\tilde{\bigcup}_{\gamma\in\Gamma_1} \mathfrak{H}_j) \cap (\tilde{\bigcup}_{\gamma\in\Gamma_2} \mathfrak{H}_\gamma) = 0$ for every $\Gamma_1, \Gamma_2 \subset \Gamma$ satisfying the conditions $\Gamma_1 \cup \Gamma_2 = \Gamma$, $\Gamma_1 \cap \Gamma_2 = 0$. The symbol $\tilde{\bigcup}_{\gamma\in\Gamma} \mathfrak{H}_\gamma$ stands for the closure of the linear hull of the subspaces $\mathfrak{H}_\gamma$ $(\gamma \in \Gamma)$.

where A_0 *stands for the operator induced in the subspace* $\mathfrak{H}_0 = \bigcap_{\gamma \in \Gamma} \mathfrak{H}_\gamma$.

1. Let $\mathfrak{G}$ be a separable Hilbert space of infinite dimension. Let us construct the Hilbert space $L^{(2)}_{\mathfrak{G}}(0, \ell)$ $(0 < \ell < \infty)$, consisting of all weakly measurable vector functions $f(x)$ $(0 \leq x \leq \ell)$ with values in $\mathfrak{G}$ such that

$$\|f\|^2 = \int_0^\ell \|f(x)\|^2_{\mathfrak{G}}\, dx < \infty .$$

The scalar product in $L^{(2)}_{\mathfrak{G}}(0,\ell)$ is defined by the formula

$$(f,g) = \int_0^\ell (f(x), g(x))_{\mathfrak{G}}\, dx .$$

We define an operator J in $L^{(2)}_{\mathfrak{G}}(0,\ell)$ putting

$$(Jf)(x) = 2i \int_x^\ell f(y)\, dy ,$$

and associate with each vector $g \in \mathfrak{G}$ the vector function $\hat{g} = \hat{g}(x) \equiv g$ $(0 \leq x \leq \ell)$. It is easy to see that the set of all vector functions $\hat{g}(x)$ coincides with the range of the operator J_I. The operator J is completely non-selfadjoint and belongs to the class $\Lambda^{(exp)}$ [1].

Lemma 1.1. *If* P *is the orthoprojector onto the invariant subspace* $L \subset L^{(2)}_{\mathfrak{G}}(0,\ell)$ *of* J, *then*

(1) $$\int_0^x ((f-Pf)(x-y), (Pg)(\ell-y))_{\mathfrak{G}}\, dy = 0$$

$$(f(x), g(x) \in L^{(2)}_{\mathfrak{G}}(0,\ell); \quad 0 \leq x \leq \ell) .$$

Proof. As

$$(J^{*n} h)(t) = \frac{(-2i)^n}{(n-1)!} \int_0 s^{n-1} h(t-s)\, ds$$

$$(h(x) \in L^{(2)}_{\mathfrak{G}}(0,\ell); \quad n = 1, 2, \ldots)$$

and

$$P\mathcal{J}^{*n}(I-P)=0 \qquad (n=1,2,\dots),$$

we have

$$\int_0^\ell s^{n-1}\int_s^\ell ((f-Pf)(t-s),(Pg)(t))_{\mathfrak{g}}\,dt\,ds =$$

$$=\int_0^\ell\int_0^t (s^{n-1}(f-Pf)(t-s),(Pg)(t))_{\mathfrak{g}}\,ds\,dt =$$

$$=\int_0^\ell\Big(\int_0^t s^{n-1}(f-Pf)(t-s)\,ds,(Pg)(t)\Big)_{\mathfrak{g}}dt=\frac{(n-1)!}{(-2i)^n}(\mathcal{J}^{*n}(I-P)f,Pg)=0.$$

Therefore

$$\int_s^\ell ((f-Pf)(t-s),(Pg)(t))_{\mathfrak{g}}\,dt=0. \tag{2}$$

The equality (1) follows from (2) with the help of the substitutions $s=\ell-x, t=\ell-y$. The lemma is proved.

Let $h\in\mathfrak{g}$ $(h\neq 0)$ and $\tau\in(0,\ell]$. Let $L(h,\tau)$ denote the set of all vector functions of the form $\varphi(x)h$ $(0\leqq x\leqq\ell)$, where $\varphi(x)$ is any scalar function having square-summable absolute value and vanishing almost everywhere on the interval $[\tau,\ell]$. Evidently, $L(h,\tau)$ is a subspace in $L^{(2)}_{\mathfrak{g}}(0,\ell)$, invariant with respect to the operator $\mathcal{J}$. The operator induced in $L(h,\tau)$ will be denoted by $\mathcal{J}(h,\tau)$. It is not difficult to verify that $\mathcal{J}(h,\tau)\in\Lambda_1$. The range of the operator $(\mathcal{J}(h,\tau))_I$ is spanned by the vector function

$$h(\tau;x)=\begin{cases} h, & 0\leqq x<\tau, \\ 0, & \tau\leqq x\leqq\ell. \end{cases}$$

Moreover, $(\mathfrak{I}(h,\tau))_I h(\tau;x) = \tau h(\tau;x)$.

Lemma 2.1. *Every invariant subspace* $L \subset L^{(2)}_{\mathfrak{g}}(0,\ell)$ *of* $\mathfrak{I}$ *in which the induced operator is of class* Λ_1, *coincides with one of the subspaces* $L(h,\tau)$.

Proof. Let $\mathfrak{I}_L$ be the operator induced in L, and $e(x)$ a unit vector belonging to the one-dimensional subspace $(\mathfrak{I}_L)_I L$. The projection on L of any vector function $\hat{f}(x)$ is collinear with $e(x)$. Substituting in (1) $\hat{f}(x)$ and $e(x)$ for $f(x)$ and $g(x)$, respectively, we obtain:

$$\int_0^x (\hat{f}, e(\ell-y))_{\mathfrak{g}}\, dy = \int_0^x ((\hat{f},e)\, e(x-y), e(\ell-y))_{\mathfrak{g}}\, dy,$$

$$\int_0^x e(\ell-y)\, dy = \int_0^x (e(\ell-y), e(x-y))_{\mathfrak{g}}\, dy \int_0^\ell e(y)\, dy. \tag{3}$$

Differentiating both sides of the latter relation and setting

$$h = \int_0^\ell e(y)\, dy,$$

we conclude that $e(x) = \varphi(x)h$, where $\varphi(x)$ is a scalar function. Now from (3) it follows that

$$\int_0^x (1 - \|h\|^2_{\mathfrak{g}}\, \overline{\varphi(x-y)})\, \varphi(\ell-y)\, dy = 0$$

On account of Titchmarsh's theorem on convolution [3], there exists a number such that

$$\varphi(x) = \begin{cases} \dfrac{1}{\|h\|^2_{\mathfrak{g}}}, & 0 \leqq x < \tau, \\ 0, & \tau \leqq x \leqq \ell. \end{cases}$$

It remains to observe that L is the closure of the linear hull of vector functions of the form $\mathfrak{I}^n(\varphi(x)h)$ $(n = 0,1,\ldots)$.

Lemma 3.1. *If* $0<\tau\leqq\tau_0\leqq\ell$, *then the projection on the subspace* $L(h,\tau)$ *of the vector function* $h_0(\tau_0;x)$ *is collinear with the vector function* $h(\tau;x)$.

Proof. Let P be the orthoprojector onto $L(h,\tau)$. Then

$$Ph_0(\tau_0;x)=\psi(x)h \qquad (\psi(x)\in L^{(2)}(0,\ell);\quad \psi(x)=0 \quad (\tau\leqq x\leqq\ell))$$

and, consequently,

$$\int_0^\tau (h_0(\tau_0;x),\varphi(x)h)_{\mathfrak{G}}\,dx = \int_0^\tau (\psi(x)h,\varphi(x)h)_{\mathfrak{G}}\,dx$$

$$(\varphi(x)\in L^{(2)}(0,\tau)).$$

Thus

$$(h_0(\tau_0;x),h)_{\mathfrak{G}}=\psi(x)(h,h)_{\mathfrak{G}}, \qquad \psi(x)=\frac{(h_0,h)_{\mathfrak{G}}}{(h,h)_{\mathfrak{G}}}$$

$$(0\leqq x<\tau).$$

Theorem 1.1. *Let* $h_j\in\mathfrak{G}$ $(j=1,2,\dots,n)$ *and* $\tau_j\in(0,\ell]$ $(j=1,2,\dots,n)$. *Then either*

$$\widetilde{\bigcup_{j=1}^{n}} L(h_j,\tau_j)=L(h_1,\tau_1)\dotplus L(h_2,\tau_2)\dotplus\cdots\dotplus L(h_n,\tau_n) \tag{4}$$

and at the same time

$$\widetilde{\bigcup_{j=1}^{n}} L(h_j,\tau_j)=\bigoplus_{j=1}^{n} L(e_j,\tau_j), \tag{5}$$

where $\{e_j\}_1^n$ *is an orthonormal system, or there exists a number* j_0 $(1\leqq j_0\leqq n)$ *such that*

$$L(h_{j_0},\tau_{j_0})\subset \widetilde{\bigcup_{j\neq j_0}} L(h_j,\tau_j).$$

Proof. Without loss of generality one may assume that

(6) $$\tau_1 \geqq \tau_2 \geqq \cdots \geqq \tau_n \,.$$

Let the vectors h_j $(j = 1,2,\ldots,n)$ be linearly independent. Then forming an orthonormal system

$$e_j = \alpha_{j1} h_1 + \alpha_{j2} h_2 + \cdots + \alpha_{jj} h_j \qquad (j = 1,2,\ldots,n)$$

and making use of the relations (6) we find that

(7) $$L(e_j, \tau_j) \subset \tilde{\bigcup_{k=1}^{j}} L(h_k, \tau_k)\,.$$

We have also

$$h_j = \beta_{j1} e_1 + \beta_{j2} e_2 + \cdots + \beta_{jj} e_j \qquad (j = 1,2,\ldots,n)$$

and, consequently,

(8) $$L(h_j, \tau_j) \subset \bigoplus_{k=1}^{n} L(e_k, \tau_k)\,.$$

The relations (7) and (8) imply (5). Since every vector function contained in the right-hand side of the equality (5) can be uniquely represented as the sum of n terms belonging to the subspaces $L(h_j, \tau_j)$ $(j = 1,2,\ldots,n)$ respectively, the equality (4) is also valid.

In case the vector h_{k_0} is a linear combination of the vectors $h_1, h_2, \ldots, h_{k_0-1}$, we have

$$L(h_{k_0}, \tau_{k_0}) \subset \tilde{\bigcup_{k=1}^{k_0-1}} L(h_k, \tau_k)\,.$$

Theorem 2.1. *Let* $h^{(\gamma)} \in \mathcal{G}$ *and* $\tau^{(\gamma)} \in (0, \ell]$ $(\gamma \in \Gamma)$. *If the induced operator* A *in the subspace* $L = \tilde{\bigcup_{\gamma \in \Gamma}} L(h^{(\gamma)}, \tau^{(\gamma)})$ *is of class* Λ_v, *then*

(9) $$L = \bigoplus_{j=1}^{\omega} L(e_j, \tau_j) \qquad (\omega \leqq \infty)\,,$$

where $\{e_j\}_1^{\omega}$ *is an orthonormal system and the numbers* τ_j *satisfy the relations*

(10) $$\tau_1 = \tau_2 = \cdots = \tau_{p_1} > \tau_{p_1+1} = \tau_{p_1+2} = \cdots = \tau_{p_2} > \tau_{p_2+1} = \cdots\,.$$

The representation (9) *is unique in the sense that if*

(11) $$L = \bigoplus_{j=1}^{\omega'} L(e'_j, \tau'_j) \qquad (\omega' \leqq \infty),$$

where $\{e'_j\}_1^{\omega'}$ *is a new orthonormal system, and*

(12) $$\tau'_1 = \tau'_2 = \ldots = \tau'_{p'_1} > \tau'_{p'_1+1} = \tau'_{p'_1+2} = \ldots = \tau'_{p'_2} > \tau'_{p'_2+1} = \cdots,$$

then $\omega = \omega'$, $\tau_j = \tau'_j$ $(j = 1, 2, \ldots, \omega)$ and

(13) $$\bigoplus_{j=p_k+1}^{p_{k+1}} L(e_j, \tau_j) = \bigoplus_{j=p_k+1}^{p_{k+1}} L(e'_j, \tau_j) \qquad (k = 0, 1, \ldots;\ p_0 = 0).$$

Proof. In view of Theorem 0.3, $\tau_1 = \sup_{\gamma \in \Gamma} \tau^{(\gamma)} < \infty$. We are going to show that there exists a vector $e_1 \in \mathcal{G}$ $(\|e_1\|_{\mathcal{G}} = 1)$ such that $L(e_1, \tau_1) \subset L$. This assertion is evident if τ_1 coincides with one of the numbers $\tau^{(\gamma)}$. On the other side, if every $\tau^{(\gamma)}$ is different form τ_1, then by Theorem 1.1 and the condition $A \in \Lambda_v$ there exists a finite-dimensional subspace $\mathcal{G}_0 \subset \mathcal{G}$ and a sequence $L(h^{(\gamma_j)}, \tau^{(\gamma_j)})$ so that 1) $\tau^{(\gamma_1)} \leqq \tau^{(\gamma_2)} \leqq \cdots$, 2) $\tau^{(\gamma_j)} \longrightarrow \tau_1$, 3) the vectors $h^{(\gamma_{kn+1})}$, $h^{(\gamma_{kn+2})}, \ldots, h^{(\gamma_{kn+n})}$ form a basis in G_0 for every non-negative integer k. Hence it follows that for any $e_1 \in \mathcal{G}_0$ $(\|e_1\|_{\mathcal{G}} = 1)$ the subspaces $L(e_1, \tau^{(\gamma_{kn+1})})$ $(k = 0, 1, \ldots)$ are contained in L. Therefore $L(e_1, \tau_1) \subset L$.

According to Theorem 1.1 the linear hull of the subspaces $L(e_1, \tau_1)$ and $L(h^{(\gamma_0)}, \tau^{(\gamma_0)})$ $(\gamma_0 \in \Gamma)$ either coincides with $L(e_1, \tau_1)$ or it can be represented in the form $L(e_1, \tau_1) \oplus L(g^{(\gamma_0)}, \tau^{(\gamma_0)})$. Thus $L = L(e_1, \tau_1) \oplus L_1$, where L_1 is the closure of the linear hull of some subspaces $L(g^{(\delta)}, \tau^{(\delta)})$ $(\delta \in \Delta)$ such that $\tau_2 = \sup_{\delta \in \Delta} \tau^{(\delta)} \leqq \tau_1$. Representing L_1 analogously in the form $L(e_2, \tau_2) \oplus L_2$ and continuing the process we obtain one of the following relations:

$$L = L(e_1, \tau_1) \oplus L(e_2, \tau_2) \oplus \ldots \oplus L(e_\omega, \tau_\omega),$$

(14) $$L = \bigoplus_{j=1}^{\infty} L(e_j, \tau_j) \oplus L'.$$

Consider the equality (14) and denote by A_k the operator induced in the subspace

$$L_k = L \ominus \Big[\bigoplus_{j=1}^{k} L(e_j, \tau_j) \Big].$$

Since

(15) $$\sum_{j=1}^{\infty} \tau_j < \infty, \qquad \sigma(A_k) = 2\tau_k \longrightarrow 0$$

and $L' = \bigcap_{k=1}^{\infty} L_k$, Theorems 0.4 and 0.1 imply $L' = 0$.

We have shown that the subspace L can be represented in the form (9). Moreover, owing to the first of the relations (15), condition (10) is fulfilled. If at the same time the representation (11) holds and for the numbers τ'_j the conditions (12) are satisfied then, according to the theorem of G. E. Kisilevskiĭ [2], $\omega = \omega'$ and $\tau_j = \tau'_j$ $(j = 1, 2, \ldots, \omega)$. By lemma 3.1 we have

$$e'_j(\tau_j; x) = \sum_{k=1}^{\omega} c_{jk} e_k(\tau_k; x), \qquad e_j(\tau_j; x) = \sum_{k=1}^{\omega} d_{jk} e'_k(\tau_k; x)$$

$$(j = 1, 2, \ldots, p_1; \quad 0 \leqq x \leqq \ell),$$

whence it follows easily that

$$e'_j = \sum_{k=1}^{p_1} c_{jk} e_k, \qquad e_j = \sum_{k=1}^{p_1} d_{jk} e'_k \qquad (j = 1, 2, \ldots, p_1).$$

Applying Theorem 1.1 we find:

$$\bigoplus_{j=1}^{p_1} L(e_j, \tau_j) = \bigoplus_{j=1}^{p_1} L(e'_j, \tau_j).$$

The other equalities (13) can be proved in a similar way.

2. In the space $\mathfrak{h}$ let us be given a completely non-selfadjoint operator of class $\Lambda^{(\exp)}$. In $L^{(2)}_{\mathfrak{g}}(0, \ell)$ $(2\ell = \sigma(A))$ there exists a subspace L invariant for $\mathfrak{J}$ and such that the operator $\mathfrak{J}_L$ induced in L satisfies the condition $UA = \mathfrak{J}_L U$, where U is an isometric mapping of $\mathfrak{h}$ onto L [4]. This result enables us to reformulate Theorems 1.1 and 2.1 as follows.

Theorem 1.2. *If* A *is a completely non-selfadjoint operator of class* $\Lambda^{(\exp)}$ *and* H_j $(j = 1,2,\dots,n)$ *are invariant subspaces of* A *with the property that the operators induced in them belong to the class* Λ_1 *, then either*

$$\tilde{\bigcup_{j=1}^{n}} \mathfrak{H}_j = \mathfrak{H}_1 \dot{+} \mathfrak{H}_2 \dot{+} \dots \dot{+} \mathfrak{H}_n ,$$

or there is a number j_0 $(1 \leqq j_0 \leqq n)$ *such that* $H_{j_0} \subset \tilde{\bigcup}_{j \neq j_0} \mathfrak{H}_j$.

Theorem 2.2. *Let* A *be a completely non-selfadjoint operator of class* Λ_V *and let* $\mathfrak{H}^{(\gamma)}$ $(\gamma \in \Gamma)$ *be invariant subspaces of* A *, the induced operators belonging to the class* Λ_1 *.*
Then

$$\tilde{\bigcup_{\gamma \in \Gamma}} \mathfrak{H}^{(\gamma)} = \bigoplus_{j=1}^{\omega} \mathfrak{H}_j \qquad (\omega \leqq \infty),$$

where the subspaces $\mathfrak{H}_j$ *are invariant with respect to* A*, while the operators* A_j *induced in them belong to* Λ_1 *and satisfy the conditions**

$$\tau(A_1) = \tau(A_2) = \dots = \tau(A_{p_1}) > \tau(A_{p_1+1}) = \tau(A_{p_1+2}) = \dots =$$
$$= \tau(A_{p_2}) > \tau(A_{p_2+1}) = \dots \; .$$

If, at the same time,

$$\tilde{\bigcup_{\gamma \in \Gamma}} \mathfrak{H}^{(\gamma)} = \bigoplus_{j=1}^{\omega'} \mathfrak{H}'_j \qquad (\omega' \leqq \infty),$$

where the subspaces $\mathfrak{H}'_j$ *are also invariant relative to* A*, and the operators* A_j *induced in them belong to* Λ_1 *and satisfy the relations*

$$\tau(A'_1) = \tau(A'_2) = \dots = \tau(A'_{p'_1}) > \tau(A'_{p'_1+1}) = \tau(A'_{p'_1+2}) = \dots =$$
$$= \tau(A'_{p'_2}) > \tau(A'_{p'_2+1}) = \dots \; .$$

then $\omega = \omega'$, $\tau(A_j) = \tau(A'_j)$ $(j = 1,2,\dots,\omega)$ *and*

*The orthogonal decomposition into operators of class Λ_1 was obtained by G.E. Kisilevskiĭ [5] for any completely non-selfadjoint operator $A \in \Lambda_V$ having an n-dimensional imaginary component and satisfying the condition $\sigma(A) = \dfrac{2\tau(A)}{n}$.

$$\bigoplus_{j=p_k+1}^{p_{k+1}} \mathfrak{h}_j = \bigoplus_{j=p_k+1}^{p_{k+1}} \mathfrak{h}'_j \qquad (k = 0, 1, \dots ;\ p_0 = 0).$$

BIBLIOGRAPHY

[1] M.S. Brodskiǐ, *Triangular and Jordan type representations of linear operators* (Moscow, 1969).

[2] G.E. Kisilevskiǐ, Generalization of the Jordan theory to a class of linear operators in Hilbert space, *Dokl. Akad. Nauk SSSR*, 176 (1967), 768-770.

[3] E.C. Titchmarsh, *Introduction to the theory of Fourier integrals* (Oxford, 1948).

[4] L.E. Isaev, On a class of operators with spectrum concentrated in zero, *Dokl. Akad. Nauk SSSR*, 178 (1968), 783-785.

[5] G.E. Kisilevskiǐ, Conditions of unicellularity for dissipative Volterra operators with finite-dimensional imaginary component, *Dokl. Akad. Nauk SSSR*, 159 (1964), 505-508.

COLLOQUIA MATHEMATICA SOCIETATIS JÁNOS BOLYAI
5. HILBERT SPACE OPERATORS, TIHANY (HUNGARY), 1970

On the Cayley transform and semigroup operators

P. L. BUTZER and U. WESTPHAL

The Cayley transform V of the infinitesimal generator A associated with a contraction semigroup $\{T(t);\ t \geq 0\}$ of operators mapping a Hilbert space H into itself plays a fundamental role in the theory of Hilbert space operators. In this respect, the basic relations between the *discrete* semigroup $\{V^n;\ n \geq 0\}$ and the given *continuous* semigroup $\{T(t);\ t \geq 0\}$ are of interest. Thus, for example, Sz.-Nagy – Foiaş [14,15] have shown that a necessary and sufficient condition that $T(t)$ be for every t (i) a normal, (ii) selfadjoint, (iii) isometric or (iv) unitary operator is that V^n has the corresponding property. Furthermore,

(1) $$\lim_{t\to\infty} \|T(t)f\| = \lim_{n\to\infty} \|V^n f\| \qquad (f \in H).$$

If $\{T(t);\ t \geq 0\}$ is a semigroup of isometries, P. Masani [9,10] found an intimate connection between the "remote" subspaces of the two semigroups. Indeed,

(2) $$\bigcap_{t \geq 0} T(t)H = \bigcap_{n \geq 0} V^n H .$$

Another proof of the latter result is due to Sz.-Nagy [13].

The purpose of this lecture* is to present a further basic relationship between $T(t)$ and V^n, and to interpret the result in the light of saturation theory and in terms of the mean ergodic theorem.

Concerning the concepts, $\{T(t);\ t \geq 0\}$ is a contraction semigroup of class (C_0) on H into H provided (i) $T(t_1+t_2) = T(t_1)T(t_2)$ $(t_1, t_2 \geq 0)$ (ii) $\|T(t)\| \leq 1$, (iii) $\lim_{t\to 0+} \|T(t)f - f\| = 0$ for each $f \in H$. The infinitesimal generator A, defined by $\lim_{t\to 0+} \|\frac{T(t)-I}{t} f - Af\| = 0$, is a closed, in general unbounded operator with domain $D(A)$ dense in H. The resolvent $R(\lambda; A)$ of A, representable for $\lambda > 0$ by

$$(3) \qquad R(\lambda;A)f = \int_0^\infty e^{-\lambda t}\, T(t) f\, dt \qquad (f \in H),$$

has the properties that

$$\|\lambda R(\lambda;A)\| \leq 1 \quad (\lambda > 0), \qquad \lim_{\lambda\to\infty} \|\lambda R(\lambda;A)f - f = 0 \qquad (f \in H).$$

The Cayley transform of A, also called the cogenerator of the semigroup $\{T(t);\ t \geq 0\}$, is the operator V defined by $V = (A+I)(A-I)^{-1}$ (cf. [15]). In contrast to A, V is a bounded operator with $\|V\| \leq 1$ and $D(V) = H$.

The following equivalence theorem is crucial for the statement and proof of the main result of this lecture.

If $f \in H$, *the following assertions are equivalent:*

(a) $\|T(t)f - f\| = \mathcal{O}(t) \qquad (t \to 0+)$,

(b) $f \in D(A)$,

(c) $\|\lambda R(\lambda;A)f - f\| = \mathcal{O}(\lambda^{-1}) \quad (\lambda \to \infty)$,

(d) *there exists* $g \in H$ *such that*

$$\lim_{\lambda\to\infty} \|\lambda[\lambda R(\lambda;A)f - f] - g = 0 .$$

In this event, $Af = g$.

*Delivered by the first-named author at the Conference on Hilbert Space Operators and Operator Algebras, Tihany, on Sept. 16, 1970. This author is grateful to Pesi Masani for kindling an interest in the matter during a short visit at Indiana University in April of 1970.

For the implications (a) $\Leftrightarrow$ (b), see Butzer [3]; for the connections with the resolvent operators given by (c), (d) and the material as a whole, see the detailed treatment in Butzer–Berens [4].

Theorem 1. *If* $f \in H$, *the following four assertions are equivalent:*

(a) $\|T(t)f - f\| = \mathcal{O}(t) \qquad (t \to 0+)$,

(b) $f \in D(A)$,

(c) $\|(1-r)\sum_{n=0}^{\infty} r^n V^n f - 0\| = \mathcal{O}(1-r) \qquad (r \to 1-)$,

i.e., the sequence $\{V^n f\}$ *is Abel-summable to the zero element* 0 *for* $n \to \infty$; *and, in fact, the Abel-sum* $(1-r)\sum_{n=0}^{\infty} r^n V^n f$ *approximates* 0 *with order* $\mathcal{O}(1-r)$ *for* $r \to 1-$,

(d) *there exists* $h \in H$ *such that*

$$\lim_{r \to 1-} \|\sum_{n=0}^{\infty} r^n V^n f - h\| = 0,$$

i.e., the series $\sum_{n=0}^{\infty} V^n f$ *is Abel-summable to* h. *In this event,* $1/2(I-A)f = h$.

The proof rests upon the identity given in

Lemma 1. *For each* $f \in H$ *and* $0 < r < 1$.

(4) $$\sum_{n=0}^{\infty} r^n V^n f = f - \frac{r}{2}\,\frac{2}{1-r}\left[\frac{2}{1-r} R\left(\frac{2}{1-r}; A+I\right)f - f\right].$$

To establish this lemma, we need the formula

(5) $$V^n f = f - 2\int_0^{\infty} L_{n-1}^{(1)}(2t)\, e^{-t}\, T(t) f\, dt \qquad (f \in H;\ n = 1, 2, \ldots),$$

due to Masani [9], where

$$L_n^{(1)}(x) = \sum_{k=0}^{n} \frac{(-1)^k}{k!} \binom{n+1}{n-k} x^k$$

is the n th generalized Laguerre polynomial with index 1. (The proof, to be found in Masani–Robertson [11] for a semigroup of isometries, carries over immediately to contraction semigroups.) By (5) it follows that

$$\sum_{n=0}^{\infty} r^n V^n f = \frac{1}{1-r} f - 2r \int_0^{\infty} \sum_{n=0}^{\infty} r^n L_n^{(1)}(2t)\, e^{-t}\, T(t) f\, dt \qquad (f \in H).$$

Indeed, the interchange of sum and integral on the right follows by sharp estimates for Laguerre polynomials, the second one being derived in Askey–Wainger [1] as a consequence of an asymptotic representation of the Laguerre polynomials due to A. Erdélyi [7]:

(6) $$|L_n^{(1)}(x)| \leq \sqrt{n+1}\, e^{x/2} \qquad (x \geq 0;\ n = 0,1,2,\dots),$$

(7) $$|L_n^{(1)}(x)| = \mathcal{O}(e^{-\xi x + x/2}) \qquad (x \geq 8(n+1);\ n \text{ large enough};\ 0 < \xi < 1/2).$$

An application of the power series expansion (see [12])

$$\sum_{n=0}^{\infty} r^n L_n^{(1)}(x) = (1-r)^{-2} \exp\left(-\frac{xr}{1-r}\right) \qquad (|r| < 1)$$

of the generating function of the Laguerre polynomials then yields, putting $x = 2t$,

$$\sum_{n=0}^{\infty} r^n V^n f = \frac{1}{1-r} f - \frac{2r}{(1-r)^2} \int_0^{\infty} \exp\left(\frac{-2t}{1-r}\right) e^t\, T(t) f\, dt .$$

In view of the hypothesis upon $\{T(t); t \geq 0\}$, $\{e^t T(t);\ t \geq 0\}$ is a semigroup of operators of class (C_0) on H with generator $(A+I)$, its resolvent being representable for all $\lambda > 1$ in the form

$$R(\lambda; A+I) f = \int_0^{\infty} e^{-\lambda t} e^t\, T(t) f\, dt \qquad (f \in H).$$

Setting $\lambda = 2(1-r)^{-1}$ completes the proof of identity (4).

The proof of the theorem now follows readily by applying the equivalence theorem of saturation theory stated above, to the semigroup $\{e^t T(t);\ t \geq 0)$ (note that it is also true for semigroups of class (C_0) which need not necessarily consist of contractions). Here one also uses the fact that the statement $\|T(t)f - f\| = \mathcal{O}(t)$ is equivalent to $\|e^t T(t) f - f\| = \mathcal{O}(t)$ for $t \to 0+$.

The assertions of the theorem may be interpreted from several points of view.

I. From the point of view of the relationship between the $T(t)$ and V^n result (1) relates the behaviour of the semigroup for $n \to \infty$ to that of the powers for $n \to \infty$ and result (2) an intersection for all $t \geq 0$ to one for all $n \geq 0$. In contrast, the assertions (a) and (c) of our main theorem link the behaviour of $T(t)$ for $t \to 0+$, the state of this process at instant $t = 0$, with that (of Abel-summability) of the sequence $\{V^n\}$ for $n \to \infty$. i.e. to the behaviour of the process in the remote future.

II. From the angle of ergodic theory, let us first recall the

Mean Ergodic Theorem. *Let* S *be a power-bounded operator on* H *into* H, *i.e.,* $\|S^n\| \leq C$, $n = 0,1,2,\dots$. *Then*

$$\text{(8)} \qquad \operatorname*{s-lim}_{r \to 1-} (1-r) \sum_{n=0}^{\infty} r^n S^n f = Pf$$

for each $f \in H$, *where* P *is the projection (i.e.,* $P^2 = P$*) on the null space* $N(P) = \overline{R(I-S)}$ *along the range* $R(P) = N(I-S)$, *thus* $H = N(I-S) \oplus \overline{R(I-S)}$.

In the literature this theorem is usually formulated for the $(C,1)$-means $(1/n+1) \sum_{k=0}^{n} S^k f$ (compare [6, p. 661]); the above Abel version is then a corolarry. Taking for S the cogenerator V of the semigroup, then $R(P) = \{\theta\}$, thus is the zero operator, and $R(I-S) = D(A)$; in other words, the sequence $\{V^n\}$ is Abel-summable to the zero operator. Whereas the mean ergodic theorem is actually a *convergence* theorem, our main theorem asserting equivalences to (c) even gives a result for the *rate* of convergence (in the particular instance that $S = V$).

The question whether an order of approximation can be built into the mean ergodic theorem for arbitrary power-bounded operators S will not be considered here. In terms of the $(C,1)$-means it is conjectured that under general assumptions upon S one has

$$\left\| \frac{1}{n+1} \sum_{k=0}^{n} S^k f - Pf \right\| = O\left(\frac{1}{n}\right) \qquad (n \to \infty)$$

if and only if $f \in N(I-S) + R(I-S)$. The authors are considering problems of this type in a futher paper [5].

III. Concerning the interpretation in terms of saturation theory, assertions (c) or (d) of the theorem give new caharcterizations of the saturation class $D(A)$ of the process $T(t)$. Whether they are of practical interest will not be gone into here.

IV. The problem arises whether the Abel-sum of the sequence $\{V^n\}$ is saturated. For a positive solution to this question we need a "small-o" result.

Lemma 2. *If for* $f \in H$

$$\| (1-r) \sum_{n=0}^{\infty} r^n V^n f - 0 \| = o(1-r) \qquad (r \to 1-),$$

then $f = 0$.

The proof again follows by Lemma 1 provided one makes use of a result for the resolvent; see [4. p. 132]. The theorem as well as Lemma 2 yield

Corollary. *The Abel-sum of the operator sequence* $\{V^n\}$ *is saturated in with order* $O(1-r)$, $r \to 1-$, *and the saturation class is equal to that for the process* $\{T(t);\ t \geq 0\}$.

Remark. A proof of the implication (b) $\Rightarrow$ (d) of the Theorem by means of the theory of unitary dilations [15] was kindly communicated to the authors by Béla Sz.–Nagy. Ciprian Foiaş believes it would also be possible to prove the equivalence of (b), (c) and (d) with these methods. Whereas these proofs would be restricted to operators on Hilbert space, the methods of this lecture – which depend upon the formula (5) of Masani and saturation theory – can also be carried over immediately to reflexive Banach spaces. As a matter of fact, the assertions (a) and (c) are equivalent for any Banach space. Indeed, the cogenerator V of a contraction semigroup of class (C_0) is well defined on any Banach space and bounded with $\|V\| \leq 3$. The problem whether V is even a contraction operator, or at least power-bounded for any non-Hilbert space does not seem to have been considered. Nevertheless, the series $\sum_{n=0}^{\infty} r^n V^n f$, $0 < r < 1$, is convergent for any element f of a Banach space since by the Masani formula (5), that also remains valid, and the estimates (6), (7) one obtains that

$$\| V^n \| = O(n^{3/2}) \qquad (n \to \infty),$$

which implies the convergence as

$$\sum_{n=0}^{\infty} n^{3/2} r^n \leq (d^2/dr^2)(1/(1-r)).$$

Thus the identity (4) of Lemma 1 is also true in any Banach space.

As remarked, the assertions (c) and (b) of Theorem 1 may be interpreted as a saturation theorem or, in other words, as a theorem on *optimal* approximation for the Abel-sum of the operator sequence $\{V^n\}$. Counterparts in the situation of *non-optimal* approximation are also valid. For this purpose we need the following result on resolvent operators for which one may consult Hündgen [8] in case $0 < \alpha < 1$, $q = \infty$ and Berens [2, p. 50] in the setting of the theory of intermediate spaces. We state it immediately for an arbitrary Banach space X.

If $f \in X$ *and* $0 < \alpha < 1$, $1 \leq q \leq \infty$ *or* $\alpha = 1, q = \infty$, *the following two are equivalent:*

(i) $$\int_0^{\infty} (\lambda^{\alpha} \| \lambda R(\lambda; A) f - f \|)^q \frac{d\lambda}{\lambda} < \infty \; *$$

(ii) $$f \in (X, D(A))_{\alpha,q;K}, \quad \textit{i.e.} \quad \int_0^{\infty} (t^{-\alpha} K(t, f; X, D(A))^q \frac{dt}{t} < \infty .$$

Moreover, in the saturation case $\alpha = 1$, $q = \infty$ *the intermediate space* $(X, D(A))_{1,\infty;K}$ *of* X *and* $D(A)$ *is equal to the completion of* $D(A)$ *relative to* X *which, if* X *is reflexive, is the space* $D(A)$ *itself.*

For the concepts intermediate space, K-functional (which in the abstract setting plays the role of the modulus of smoothness), relative completion etc. see Butzer–Berens [4] and [2]. Two spaces are said to be equal if they are algebraically equal and their norms equivalent.

Using formula (4) we can apply this general result on resolvent operators to the discrete semigroup $\{V^n;\; n \geq 0\}$ giving the following theorem on non-optimal as well as optimal approximation.

*The modification for $q = \infty$ is obvious. The integral $\int_0^{\infty}$ may be replaced by $\int_{\varepsilon}^{\infty}$ for any $\varepsilon > 0$; the singularity of the integrand lies at infinity.

Theorem 2. *If* $f \in X$, $0 < \alpha < 1$, $1 \leq q \leq \infty$ *or* $\alpha = 1$, $q = \infty$, *the following are equivalent:*

(i) $$\int_{\varepsilon}^{1} \Big[(1-r)^{-\alpha} \Big\| (1-r) \sum_{n=0}^{\infty} r^n V^n f - 0 \Big\|\Big]^q \frac{dr}{1-r} < \infty \qquad (\varepsilon > 0)$$

(ii) $$\int_{0}^{\infty} [t^{-\alpha} K(t, f; X, D(A))]^q \frac{dt}{t} < \infty .$$

Note that in the optimal case $\alpha = 1$, $q = \infty$ this theorem generalizes the equivalence of the assertions (b) and (c) of Theorem 1 to arbitrary Banach spaces which need not necessarily be reflexive.

REFERENCES

[1] R. Askey and S. Wainger, Mean convergence of expansions in Laguerre and Hermite series, *Amer. J. Math.*, 87 (1965), 695-708.

[2] H. Berens, Interpolationsmethoden zur Behandlung von Approximationsprozessen auf Banachräumen (Lecture Notes, Berlin, 1968).

[3] P.L. Butzer, Über den Grad der Approximation des Identitätsoperators durch Halbgruppen von linearen Operatoren und Anwendungen auf die Theorie der singulären Integrale, *Math. Ann.*, 133 (1957), 410-425.

[4] P.L. Butzer and H. Berens, *Semi-groups of Operators and Approximation* (Berlin, 1967).

[5] P.L. Butzer and U. Westphal, On the mean ergodic theorem and saturation (in preparation).

[6] N. Dunford and J. Schwartz, *Linear Operators*. Part I. General Theory (New York, 1958).

[7] A. Erdélyi, Asymptotic forms for Laguerre polynomials, *J. Indian Math. Soc.*, 24 (1960), 235-250.

[8] W. Hündgen, Über gebrochene Potenzen der Resolventen von Operatoren mit Halbgruppeneigenschaft (Staatsexamensarbeit, TH Aachen 1968).

[9] P. Masani, On isometric flows on Hilbert space, *Bull. Amer. Math. Soc.*, 68 (1962), 624-632.

[10] P. Masani, Quasi-isometric measures and their applications, *Bull. Amer. Math. Soc.*, 76 (1970), 427-528.

[11] P. Masani and J. Robertson, The time-domain analysis of a continuous parameter weakly stationary stochastic process, *Pacific J. Math.*, 12 (1962), 1361-1378.

[12] G. Sansone, *Orthogonal Functions* (New York, 1959).

[13] B. Sz.-Nagy, Isometric flows in Hilbert space, *Proc. Cambridge Philos. Soc.*, 60 (1964), 45-49.

[14] B. Sz.-Nagy and C. Foiaş, Sur les contractions de l'espace de Hilbert. III, *Acta Sci. Math.*, 19 (1958), 26-45.

[15] B. Sz.-Nagy and C. Foiaş, *Analyse harmonique des opérateurs de l'espace de Hilbert* (Budapest, 1967).

COLLOQUIA MATHEMATICA SOCIETATIS JÁNOS BOLYAI
5. HILBERT SPACE OPERATORS. TIHANY (HUNGARY), 1970

C^*-algebras generated by semi-groups of isometries

L. A. COBURN

Introduction. In this expository paper, I discuss some recent results in the field of "concrete C^*-algebras". Some of these results were obtained jointly with C. Berger, R. Douglas, D. Schaeffer, and I.M. Singer.

The general setting is that one starts with a toplogical semi-group S and a strongly continuous representation $s \longrightarrow U_s$ of S as isometries on a Hilbert space H. One is then interested in determining the structure of $\mathfrak{A}(U_s: s \in S)$, the C^*-algebra generated by $\{U_s : s \in S\}$. In case $S = R^+$, the positive reals, Berger and I [2] were able to get a result analogous to my earlier result [5] for $S = Z^+$, the positive integers, that (up to $*$-isomorphism) there is only *one* such C^*-algebra which is non-commutative (i.e. for which not all the U_s are unitary). The algebra for for $S = R^+$ had previously been studied in a peper with R. Douglas [6,7]. For this algebra, and certain related C^*-algebras of the type described above, Douglas, Schaeffer, Singer and I have just completed a rather exotic "index theorem" with real-valued index [8].

*Research supported by a grant of the National Science Foundation

More recently, Singer and I [9] have applied the above results to the study of C^*-algebras of singular integral operators on R^n corresponding to partial differential operators with almost-periodic coefficients. Our results, which are based on estimates in [7], provide an example of "ellipticity" on a non-compact manifold.

In the next section, I describe the above-mentioned results in some detail. Needless to say, suspension of disbelief is required since no proofs are provided.

Main results. Let us recall that for $S = Z^+$, we have the result [5] that if $s \to U_s$ is not unitary then $\mathfrak{A}(U_s : s \in Z^+)$ is $*$-isomorphic to the C^*-algebra $\mathfrak{A}(S)$ generated by the simple unilateral shift, S. Moreover, if we take the standard representation of S as multiplication by z on the Hardy space $H^2(T)$ of the unit circle T and define for ϕ in $C(T)$, the continuous functions on T, the Toeplitz operators T_ϕ on $H^2(T)$ by

$$T_\phi f = P(\phi \cdot f),$$

where P is the orthogonal projection from the space of Lebesgue square-integrable functions, $L^2(T)$, onto $H^2(T)$, then we have [5]

Theorem 1. $\mathfrak{A}(T_z) = \{T_\phi + K : \phi \in C(T), K \in \mathcal{K}\}$ *where* $\mathcal{K}$ *is the collection of all compact operators on* H^2. *Moreover, the representation as sums is unique and the map* $\phi \to T_\phi + \mathcal{K}$ *is a* $*$*-isomorphism from* $C(T)$ *onto* $\mathfrak{A}(T_z)/\mathcal{K}$.

If A is a bounded operator on Hilbert space, A is called a *Fredholm operator* provided that A is invertible mod $\mathcal{K}$. It is well-known by now that [1] A is Fredholm if and only if $\operatorname{Ker} A$ and $\operatorname{Ker} A^*$ are finite-dimensional and $\operatorname{Im} A$ is closed. The collection of all Fredholm operators Fred is an open semigroup and the mapping

$$\operatorname{ind}(A) = \dim \operatorname{Ker} A^* - \dim \operatorname{Ker} A$$

is a continuous homomorphism from Fred onto Z with the discrete topology and

$$\operatorname{ind}(A+K) = \operatorname{ind}(A)$$

for $A \in \mathrm{Fred}$ and $K \in \mathcal{K}$. Using the fact [11] that if ϕ is an invertible element in $C(T)$ then

$$\phi(z) = z^n e^{\psi(z)}$$

for some n in Z and ψ in $C(T)$ so that $\phi_t(z) = z^n e^{t\psi(z)}$ is an arc of invertible elements of $C(T)$ joining ϕ to z^n, we see that by Theorem 1 we have $T_\phi + K$ in $\mathfrak{A}(T_z)$ is in Fred if and only if ϕ is invertible and moreover [10]

$$\mathrm{ind}(T_\phi+K) = \mathrm{ind}(T_\phi) = \mathrm{ind}(T_{z^n}) = \dim \mathrm{Ker}(T^*_{z^n}) = n = \frac{1}{2\pi}\int_0^{2\pi} d\arg\phi .$$

Now for $S = R^+$, we consider the Hardy space $H^2(R)$ consisting of all square-integrable functions, f, on R whose Fourier transforms, $\mathfrak{F}f$, are supported on R^+. Then for $s \in R^+$ we have a strongly continuous one-parameter semigroup of isometries given by

$$(\tilde{T}_s f)(x) = e^{isx} f(x).$$

Clearly, $\mathfrak{F}\tilde{T}_s\mathfrak{F}^{-1} = T_s$ is translation by s on $L^2(R^+)$. Now let P be the projection from $L^2(R)$ onto $H^2(R)$ and for ϕ almost-periodic on R define an operator W_ϕ on $H^2(R)$ by

$$W_\phi f = P(\phi f).$$

Then we have [6,7]

Theorem 2. *$\mathfrak{A}(\tilde{T}_s\colon s\in R^+) = \{W_\phi + C\colon \phi \in AP(R)$ and $C \in \tilde{\mathfrak{C}}\}$ where $AP(R)$ is the algebra of almost-periodic functions on R and $\tilde{\mathfrak{C}}$ is the commutator ideal in $\mathfrak{A}$. Moreover, the representation as sums is unique and the map $\phi \to W_\phi + \tilde{\mathfrak{C}}$ is a $*$-isomorphism from $AP(R)$ onto $\mathfrak{A}(\tilde{T}_s\colon s\in R^+)/\tilde{\mathfrak{C}}$. Further, if $s \to U_s$ is any non-unitary strongly continuous representation of R by isometries then [2] $\mathfrak{A}(U_s\colon s\in R^+)$ is $*$-isomorphic with $\mathfrak{A} = \mathfrak{A}(\tilde{T}_s\colon s\in R^+)$.*

Theorem 2 suggests that a "Fredholm-theory" of the type discussed for Z^+ is possible for R^+. Fortunately, the closed ideal $\mathfrak{C}$ does *not* consist of compact operators [6] so the problem is rather difficult. The key step is to notice that there is a a well-developed "Fredholm-theory" for operators in an arbitrary von Neumann factor V [4]. This theory is due to Breuer and in this theory, the role of the compact operators is taken over by the unique closed two-sided ideal $\mathfrak{J}$ in V. An operator A in V is said to be in $\mathrm{Fred}(V,\mathfrak{J})$ if and only if A is invertible mod $\mathfrak{J}$. If V is a II_∞ factor, operators in $\mathrm{Fred}(V,\mathfrak{J})$ are characterized in a manner analogous to ordinary Fredholm operators except that the appropriate dimension is the intrinsic real-valued von Neumann-Murray dimension in V [4]. In particular, if A is in $\mathrm{Fred}(V,\mathfrak{J})$ then $\mathrm{Ker}\,A$ and $\mathrm{Ker}\,A^*$ have finite von-Neumann-Murray dimension, dim. Further $\mathrm{Fred}(V,\mathfrak{J})$ is an open semi-group and

$$\operatorname{ind}(A) = \dim \operatorname{Ker}(A^*) - \dim \operatorname{Ker}(A)$$

is a continuous homomorphism from $\operatorname{Fred}(V, \mathfrak{J})$ onto R with the discrete toplogy and

$$\operatorname{ind}(A+C) = \operatorname{ind}(A)$$

for $A \in \operatorname{Fred}(V, \mathfrak{J})$ and $C \in \mathfrak{J}$.

To apply the theory of the above paragraph to $\mathfrak{A}(\tilde{T}_s: s \in R^+)$, in [8] we constructed a representation

$$\Phi: \mathfrak{A}(\tilde{T}_s: s \in R^+) \rightarrow V$$

with the property that $\Phi(C) \subset \mathfrak{J}$ for V a II_∞ factor. It is easy to see that if $A = W_\phi + C$ with ϕ invertible then by Theorem 2, $\Phi(A)$ is in $\operatorname{Fred}(V, \mathfrak{J})$. We use the fact that if ϕ is invertible in $AP(R)$, then

$$\phi(x) = e^{isx} e^{\psi(z)}$$

for some s in R and ψ in $AP(R)$. It then follows, as on T, from the construction of Φ that for ϕ invertible in $AP(R)$

$$\operatorname{ind}(\Phi(W_\phi + C)) = \operatorname{ind} \Phi(W_\phi) = \operatorname{ind} \Phi(W_{e^{isx}}) = \operatorname{ind} \Phi(\tilde{T}_s) =$$

$$= \dim \operatorname{Ker} \Phi(\tilde{T}_s^*) = s = \lim_{T \rightarrow \infty} \frac{1}{2T} \int_{-T}^{T} d \arg \phi .$$

This "index theorem" gives a useful necessary condition for invertibility. It should be observed that the real number s is just the Bohr mean motion [3] of ϕ.

For more general semigroups S, one does not expect uniqueness of the C^*-algebra generated by strongly continuous, isometric, non-unitary representations of S. On the other hand, it would be interesting to know exactly what the situation is for some examples. Theorems analogous to Theorems 1 and 2 have been obtained for special representations of more general semigroups than Z^+, R^+ [7].

To conclude, I will describe the application of the above results to algebras of singular integral operators on R^n. For smplicity, I will only consider in detail the case $n = 1$, although the general case is analogous. The results described below were obtained jointly with I.M. Singer [9]. We consider an arbitrary linear partial differential operator D on $L^2(R^n)$. It is well-known that for x in R^n and

$$\Lambda = \mathfrak{F}^{-1} M_{1+|x|} \mathfrak{F}$$

where $M_\psi f \equiv \psi \cdot f$ for f in $L^2(R^2)$, we can write

$$D = A_m \Lambda^m + A_{m-1} \Lambda^{m-1} + \ldots + A_0$$

where the A_j are bounded operators. If we assume that the coefficients of D are almost-periodic then for $k(x)$ homogeneous of degree 0 on R^n and continuous on S^{n-1} we see that the A_j have the form

$$M_\phi \mathfrak{F}^{-1} M_{k(x)} \mathfrak{F}$$

where ϕ is almost-peridoic. The C^*-algebra generated by all such operators is called the algebra of singular integral operators with almost-periodic coefficients and denoted by $\mathfrak{A}(R^n)$. We write

$$\tilde{\mathfrak{A}}(R^n) = \mathfrak{F} \mathfrak{A}(R^n) \mathfrak{F}^{-1}.$$

Now let χ be the characteristic function of the non-negative reals R^+. We have the following results for $n = 1$ [9]

Theorem. *$\tilde{\mathfrak{A}}(R)$ is just the C^*-algebra generated by all translations T_λ on $L^2(R)$ and M_χ. This algebra contains and is the uniform closure of the set of all sums*

$$(*) \qquad \sum_{j=1}^{m} M_{\psi_j} T_{\lambda_j}$$

where the ψ_j are step-functions, continuous from the right and having finitely-many discontinuities. Moreover, the commutator ideal $\mathcal{C}$ in $\tilde{\mathfrak{A}}(R)$ is the closed two-sided ideal generated by those sums (*) for which the ψ_j all vanish at $\pm\infty$. For ϕ any continuous almost-periodic function on R, $\tilde{\mathfrak{A}}(R)$ contains

$$A_\phi^+ \equiv M_\chi \mathfrak{F} M_\phi \mathfrak{F}^{-1} M_\chi, \quad A_\phi^- \equiv M_{1-\chi} \mathfrak{F} M_\phi \mathfrak{F}^{-1} M_{1-\chi}$$

and every element of $\tilde{\mathfrak{A}}(R)$ can be represented uniquely in the form

$$A_\phi^+ + A_\psi^- + C$$

*for C in $\mathcal{C}$. The map $(\phi,\psi) \to A_\phi^+ + A_\psi^- + \mathcal{C}$ is a *-isomorphism from*

$AP(R) \oplus AP(R)$ *onto* $\tilde{\mathfrak{A}}(R)/\mathcal{C}$.

Now in a manner similar to our treatment of $\mathfrak{A}(\tilde{T}_s : s \in R^+)$, we can construct a representation

$$\Phi : \tilde{\mathfrak{A}}(R) \longrightarrow V'$$

where V' is a II_∞ factor in such a way that $\Phi(\mathcal{C}) \subset \mathcal{J}'$. It then turns out that for (ϕ, ψ) invertible in $AP(R) \oplus AP(R)$, $\Phi(A_\phi^+ + A_\psi^- + C)$ is in $\mathrm{Fred}(V^1, \mathcal{J}^1)$ and

$$\mathrm{ind}(\Phi(A_\phi^+ + A_\psi^- + C)) = \mathrm{ind}\,\Phi(A_\phi^+ + A_\psi^-) = \lim_{T \to \infty} \frac{1}{2T} \int_{-T}^{T} d\arg\phi - \lim_{T \to \infty} \frac{1}{2T} \int_{-T}^{T} d\arg\psi \; .$$

This result gives a useful criterion for invertibility of singular integral operators with almost-periodic coefficients as well as for difference operators in $\tilde{\mathfrak{A}}(R)$.

In particular, for $D = A_m \Lambda^m + \ldots$ as described above, we say D is elliptic (II_∞) if A_m has the form $\mathcal{F}^{-1}(A_\phi^+ + A_\psi^- + C)\mathcal{F}$ where ϕ, ψ are invertible in $AP(R)$ and we define for such D

$$\mathrm{ind}\, D \equiv \mathrm{ind}\, \Phi(A_\phi^+ + A_\psi^- + C) \, .$$

For $n = 1$, and $H = \mathcal{F}^{-1} M_{2\chi - 1} \mathcal{F}$, it is easy to check that A_m has the form M_ϕ or $M_\phi H$ for ϕ almost-periodic and these operators have the forms

$$\mathcal{F}^{-1}(A_\phi^+ + A_\phi^- + C)\mathcal{F}, \quad \mathcal{F}^{-1}(A_\phi^+ + A_{-\phi}^- + C)\mathcal{F}$$

respectively so that D is elliptic (II_∞) if and only if ϕ is invertible and then

$$\mathrm{ind}\, D = 0 \; .$$

The case $n > 1$, together with proofs of the above results will appear in [9].

REFERENCES

[1] F.V. Atkinson, Normal solubility of linear equations in normal spaces, *Math. Sbornik*, 28 (70) (1951).

[2] C.A. Berger, and L.A. Coburn, One-parameter semigroups of isometries, to appear in *Bull. Amer. Math. Soc.*

[3] H. Bohr, Für Theorie der fastperiodischen funktionen. II, *Acta Math.*, 46 (1925), 101-214.

[4] M. Breuer, Fredholm theories in von Neumann algebras. I, II, *Math. Annalen,* 178 (1968), 243-254; 180 (1969), 313-325.

[5] L.A. Coburn, The C*-algebra generated by an isometry. I, II, *Bull. Amer. Math. Soc.,* 73 (1967), 722-726; *Trans. Amer. Math. Soc.,* 137 (1969), 211-217.

[6] L.A. Coburn, and R.G. Douglas, Translation operators on the half-line, *Proc. Nat. Acad. Sci.,* 62 (1969), 1010-1013.

[7] L.A. Coburn, and R.G. Douglas, C*-algebras of operators on a half-space. I (to appear).

[8] L.A. Coburn, R.G. Douglas, D.G. Schaeffer and I.M. Singer, C*-algebras of operators on a half-space. II: index theory (to appear).

[9] L.A. Coburn, and I.M. Singer, Algebras of singular integral operators with almost-periodic coefficients (to appear).

[10] R.G. Douglas, On the spectrum of a class of Toeplitz operators, *J. Math. Mech.,* 17 (1968) 433-436.

[11] H.L. Royden, Function algebras, *Bull. Amer. Math. Soc.,* 69 (1963), 281-298.

COLLOQUIA MATHEMATICA SOCIETATIS JÁNOS BOLYAI
5. HILBERT SPACE OPERATORS, TIHANY (HUNGARY), 1970

Group representations and integral transforms

J. L. B. COOPER

The integral transforms important in analysis are those which have particularly simple behaviour when the spaces on which the transformed functions are defined are subjected to certain transformation groups. This suggests that a profitable method of studying and classifying integral transforms is to investigate just what transforms have specific group transformation properties. It is trivially obvious that if T is a transformation from a function space A to a space B then if A undergoes transformation under a group so do the transforms TA. The study of such transformations becomes important when the groups involved take the form of combinations of multiplication by functions and substitution groups; these are the simplest groups one can define on a function space, and we call them appropriate groups. Specifically, an *appropriate group* on a function space $A(X)$ of functions defined on a space X consists of transformations $W(g)$ of the form $W(g)f(x) = Q(x,g)f(V(g)x)$, where $Q(x,g)$ is a continuous complex valued function in $X \times G$ and $V(g)$ is a representation of W by continuous transformations of X. A transformation T on a function space $A(X)$ to a space $B(U)$ on a set U obeys an *appropriate equation* if there are appropriate groups $W(g)$ on $A(X)$ and $W^*(g)$ on $B(U)$ such that $TW^*(g) = W^*(g)T$.

The simplest and most important case is that in which X and U are intervals of the real line and $\{W(g)\} = \{W(\alpha)\}$ is a representation of the additive group of the real line. In this case one can show [1] that under quite general assumptions the possible appropriate equations reduce to a few canonical forms.

If $W(g)$ acts on X continuously, then X splits into a countable set of intervals $\{I_n\}$ and a residual set E such that $V(\alpha)x = x$ for all $x \in E$ and all α, and for each I_n we can make a change of variables, mapping X homeomorphically on R, so that in the new variables $V(\alpha)x = x + \alpha$. In E, $W(\alpha)f(x) = e^{\alpha\psi(x)}f(x)$. In I_n, $W(\alpha)f(x) = \frac{O(x+\alpha)}{O(x)} f(x+\alpha)$. If we change the function space, replacing $f(x)$ by $O(x)f(x)$, then $W(\alpha)$ takes the simpler form $W(\alpha)f(x) = f(x+\alpha)$. A similar decomposition holds in U. If the function spaces are such that T can be decomposed into its actions on the restrictions of each function to the I_n and E, then by considering the maps from the restriction functions on the I_n or the E to the corresponding sets in $B(U)$, we find that there are four possible appropriate equations:

(I) $\quad I \to I \qquad [Tf(x+\alpha)](u) = [Tf(x)](u+\alpha)$;

(II) $\quad I \to E \qquad [Tf(x+\alpha)](u) = e^{\alpha\psi(u)}[Tf(x)](u)$;

(III) $\quad E \to I \qquad [T\{e^{\alpha\psi(u)}f(x)\}](u) = [Tf(x)](u+\alpha)$;

(IV) $\quad E \to E \qquad [T\{e^{\alpha\psi(u)}f(x)\}](u) = e^{\alpha\psi(u)}[Tf(x)](u)$.

The next question is that of the forms of solutions of these equations. This depends to some extent on the properties of the spaces A and B. The interesting cases are the equations (I) and (II), since (III) is in effect simply (II) for the inverse of T, and (IV) leads to the conclusion that T is substantially a substitution operator.

Under quite general conditions, the solutions of (I) are convolution transforms,

$$[Tf(x)](u) = \int k(x-u)f(x)\,dx,$$

where k is a generalized function; more generally, we can have solutions of the form

$$\int_{-\infty}^{u} e^{\lambda v} Tf(x)(v)\,dv = \int k(x-u)\, e^{\lambda x} f(x)\,dx .$$

The solutions of the equation (II) are of the nature of exponential transforms:

$$[Tf(x)](u) = \zeta(u) \int e^{x\varphi(u)} f(x)\,dx ,$$

where $\zeta(u)$ is an arbitrary function of u. This is the general form if the characteristic functions of intervals are in A and their span is dense in A. For spaces for which this is not the case this solution may not hold. For instance, an equation of the form (II) with $\varphi(u) = -u$ holds for the inverse Laplace transform acting on the space of integral functions of exponential type: but this is not representable by any real integral transform.

A further question is this: given two specific function spaces, does a non-trivial solution of these equations exist which acts from one of these spaces to the other? This depends critically on the topologies of the spaces concerned: some results for the particular case of the L^p spaces are given in [1].

When we go beyond the case of the real line and the real group, it does not seem possible to give an exhaustive classification of all possible functional equations. To take the next simplest case, the modes in which the translation group of the plane can act on subsets of the plane are very varied – the variety of ways in which the plane can be transformed by conformal maps is evidence of this. Some results of interest can be obtained by considering specific representations.

In particular, this method seems to offer an interesting way of studying and characterising exponential transforms in normed spaces.

Let M be the set of bounded measures on a Hilbert space H. We give M a topology in which a basis of neighbourhoods of $\mu \in M$ are the sets $N(\mu, \delta, \varepsilon)$ consisting of all $\mu' \in M$ such that there is a countable set $\{S_r, r = 0, 1, \ldots\}$ with $\cup S_r = H$, $\operatorname{diam} S_r < \delta$ if $r > 0$, $\mu(S_0) < \varepsilon$ and a translation t_h, $\|h\| < \delta$, so that

$$\sum \left(|t_h \mu'_+(S_r) - \mu_+(S_r)| + |t_h \mu'_-(S_r) - \mu_-(S_r)| \right) < C ,$$

where μ_+, μ_- are the positive and negative parts of μ. Here we write t_h for the translation of μ defined by

$$\int f(x+h)\mu(dx) = \int f(x) t_h \mu(dx).$$

Let T be a linear map from M to the complex valued functions on H such that the map $\mu \to T\mu(u)$ is continuous for each u, and $(Tt_h\mu)(u) = Q(h,u)(T\mu)(u)$. Then since $t_{h_1+h_2} = t_{h_1}t_{h_2}$, $Q(h_1+h_2, u) = Q(h_1,u)Q(h_2,u)$. Since $t_h\mu u \to \mu u$ on M as $h \to 0$, $Q(h,u)$ is continuous; and since $Q(h_1,u) = 0$ for one h_1 would imply that $Q(h,u) = 0$ for all h, either $Q(h,u)$ is nowhere zero or T is trivial; hence $\log Q(h,u)$ is a continuous linear function of h, so that $Q(h,u) = e^{\langle h,\varphi(u)\rangle}$.

Now consider a disjoint collection (S_r) satisfying the conditions of the definition of the topology on M, and choose $a_r \in S_r$. Write $\mu_{\{a\}}$ for the unit measure concentrated at a, and let $\mu' = \sum \mu(S_r)\mu_{\{a_r\}}$. Then as $\delta \to 0$, $T\mu'(u) \to$ $\to T\mu(u)$ for all u.

Now let $\zeta(u) = T\mu_{\{0\}}(u)$. Then $T\mu_{\{a_r\}} = \zeta(u)e^{(a_r,\varphi(u))}$. Thus $T\mu'(u) = \zeta(u)\sum e^{(a_r,\varphi(u))}\mu(S_r)$, so that on proceeding to the limit,

$$T\mu(u) = \zeta(u)\int e^{(x,\varphi(u))}\mu(dx).$$

The Fourier transform is a particular case of this form, with $\varphi(u) = iu$, but these equations are not sufficient to characterize it because of the arbitrariness of $\zeta(u)$. To characterize the Fourier transform we need to consider further functional equations.

For any real $\lambda > 0$ and any unitary U let m_λ, m_U be the transform of M defined by

$$\int f(x)\, m_\lambda \mu(dx) = \int f(\lambda x)\mu(dx), \quad \int f(x)\, m_U \mu(dx) = \int f(Ux)\mu(dx).$$

Suppose that T obeys the additional equations,

$$Tm_\lambda(u) = T\mu(\lambda u), \quad Tm_U\mu(u) = T\mu(U^* u).$$

Then we have:

$$\zeta(u)\, e^{(x,\varphi(u))} = \zeta(\lambda u)\, e^{(x,\varphi(\lambda u))}$$

$$\zeta(u)\, e^{(Ux,\varphi(u))} = \zeta(U^* u)\, e^{(x,\varphi(U^* u))}$$

for all x and u and for all U and $\lambda > 0$; from this it follows that $\zeta(u)$ is constant, and that $\varphi(\lambda u) = \lambda\varphi(u)$, $\varphi(U^*u) = U^*\varphi(u)$ and so that $\varphi(u) = cu$ for some constant c. T is therefore an exponential transform. If, in addition, it is a bounded transform on M, c must be purely imaginary and T the Fourier transform.

REFERENCES

[1] J.L.B. Cooper, Functional equations for linear transformations, *Proc. London Math. Soc.*, (1970), 1-26.

COLLOQUIA MATHEMATICA SOCIETATIS JÁNOS BOLYAI
5. HILBERT SPACE OPERATORS, TIHANY (HUNGARY), 1970

On commutative self-adjoint extensions of differential operators

R. DENČEV

Let a bounded domain Ω with piecewise smooth boundary be given in the n -dimensional Euclidean space E_n .

The following notations will be used: $C^\infty(\bar{\Omega})$ – the space of infinitely differentiable functions on $\bar{\Omega}$; $C_0^\infty(\Omega)$ – the space of infinitely differentiable functions with compact support on Ω ; $W^s(\Omega)$ – the Sobolev space of functions on Ω , having square-summable derivatives of order s with the corresponding metrics; $\mathring{W}^s(\Omega)$ – the closure of $C_0^\infty(\Omega)$ in the metrics of $W^s(\Omega)$; $\underset{0}{W}^s(\Omega) = \mathring{W}^1(\Omega) \cap W^s(\Omega)$ with the metrics of $W^s(\Omega)$.

We are given elliptic self-adjoint differential expressions of the second order with constant coefficients.

$$\mathfrak{A}u = \sum_{j,k=1}^{n} a_{jk} \frac{\partial^2 u}{\partial x_j \partial x_k} - au$$

$$\mathcal{B}u = \sum_{j,k=1}^{n} b_{jk} \frac{\partial^2 u}{\partial x_j \partial x_k} - bu \qquad (a>0,\ b>0,\ \mathcal{A} \neq \mathcal{B}).$$

We define the operators

$$(1) \qquad \begin{aligned} A_0 &: C_0^\infty(\Omega) \ni u \longrightarrow \mathcal{A}u \in C_0^\infty(\Omega) \\ B_0 &: C_0^\infty(\Omega) \ni u \longrightarrow \mathcal{B}u \in C_0^\infty(\Omega). \end{aligned}$$

The operators A_0 and B_0 are positive and symmetric in $L_2(\Omega)$. Denote by $\tilde{A}$ and $\tilde{B}$ their Friedrichs self-adjoint extensions, respectively [1].

Evidently, A_0 and B_0 commute, that is

$$A_0 B_0 u = B_0 A_0 u$$

for each $u \in C_0^\infty(\Omega)$. The present paper concerns the following problem: for what region Ω do the operators $\tilde{A}$ and $\tilde{B}$ commute?

According to a well-known theorem of the linear algebra [2], the matrices (a_{jk}), (b_{jk}) can be diagonalized simultaneously, i.e. there exist a non-degenerated matrix Z such that $Z'(a_{jk})Z = (\alpha_k \delta_{jk})$, $Z'(b_{jk})Z = (\delta_{jk}) = 1$. The transformation $x' = Zx$ maps the domain Ω into a domain Ω'.
Let

$$T:\ L_2(\Omega) \ni u(x) \longrightarrow \frac{1}{\sqrt{|\det Z|}} u(Z^{-1}x') = \hat{u}(x') \in L_2(\Omega').$$

The operator T is an isometric mapping of $L_2(\Omega)$ onto $L_2(\Omega')$. It is easy to see, that

$$A_0 = T^{-1} A_0' T - aI$$

$$B_0 = T^{-1} B_0' T - bI$$

where

$$A_0':\ C_0^\infty(\Omega) \ni u(x') \longrightarrow \sum_{k=1}^{n} \alpha_k \frac{\partial^2 u}{\partial x_k'^2} = \mathcal{A}' u$$

$$B_0': C_0^\infty(\Omega) \ni u(x') \to \sum_{k=1}^{n} \frac{\partial^2 u}{\partial x_k'^2} = \mathcal{B}'u .$$

It may also be proved that

$$(2) \qquad \begin{aligned} \tilde{A} &= T^{-1}\tilde{A}'T - aI \\ \tilde{B} &= T^{-1}\tilde{B}'T - bI , \end{aligned}$$

where $\tilde{A}$, $\tilde{B}$ are the Friedrichs extensions of the operators A_0' , B_0' respectively. It follows from [2] that the operators $\tilde{A}$ and $\tilde{B}$ commute if and only if the operators $\tilde{A}'$ and $\tilde{B}'$ do. Thus it is sufficient to study the conditions of commutation of $\tilde{A}'$ and $\tilde{B}'$.

We shall call the domain Ω' regular if it is bounded, has a piecewise smooth boundary and is convex in some neighbourhood of every point at which the boundary is not smooth.

Let the set of the coefficients α_k be decomposed into the subsets of equal coefficients:

$$(3) \qquad \begin{aligned} &\alpha_{i_1} = \alpha_{i_2} = \cdots = \alpha_{i_p} \\ &\alpha_{j_1} = \alpha_{j_2} = \cdots = \alpha_{j_q} \\ &\text{-----------------} \\ &\alpha_{l_1} = \alpha_{l_2} = \cdots = \alpha_{l_r} . \end{aligned}$$

The elements of two different subsets are different. Let Ω_i be a regular region in the p -dimensional subspace determined by the coordinate axis $x_{i_1}, x_{i_2}, \ldots, x_{i_p}$. Similarly $\Omega_j, \ldots, \Omega_l$ are the regions corresponding to the other groups of equal coefficients. Consider the cylinder

$$(4) \qquad \Omega' = \Omega_i \times \Omega_j \times \cdots \times \Omega_l .$$

The following theorem is true:

Theorem 1. *If the domain* Ω' *is regular and the operators* $\tilde{A}'$ *and* $\tilde{B}'$ *commute, then* Ω' *is a cylinder of the kind* (4).

To prove this we shall use the following.

Lemma. *Let the operators* A *and* B *in* $L_2(\Omega)$ *be self-adjoint, commuting and* $D(A)=D(B)$ *. *Then* $D(A^2) = D(B^2)$.

Proof of the lemma. Let $\mu(\xi)$ be a measure on E_1 . Denote by $L_2(E_1,\mu)$ the Hilbert space of all square-summable complex-valued functions $\hat{u}(\xi)$ on E_1 with the inner product

$$<\hat{u},\hat{v}> = \int \hat{u}(\xi)\overline{\hat{v}(\xi)}\,d\mu(\xi).$$

By the spectral theorem [5], since A and B commute, there exist a measure μ , a unitary transformation $\mathcal{U}: L_2(\Omega) \rightarrow L_2(E_1,\mu)$ and functions $f(\xi)$ and $g(\xi)$ such that

(5) $$A = \mathcal{U}^{-1}\hat{A}\,\mathcal{U}, \qquad B = \mathcal{U}^{-1}\hat{B}\,\mathcal{U},$$

where

$$\hat{A}: L_2(E_1,\mu) \ni \hat{u}(\xi) \rightarrow f(\xi)\hat{u}(\xi) \in L_2(E_1,\mu)$$

$$\hat{B}: L_2(E_1,\mu) \ni \hat{u}(\xi) \rightarrow g(\xi)\hat{u}(\xi) \in L_2(E_1,\mu).$$

From $D(A) = D(B)$ and (5) it follows

(6) $$D(\hat{A}) = D(\hat{B}).$$

We have

(7) $$D(\hat{A}) = \{\hat{u}: \int|\hat{u}|^2 d\mu < \infty,\ \int|f\hat{u}|^2 d\mu < \infty\}$$
$$D(\hat{B}) = \{\hat{u}: \int|\hat{u}|^2 d\mu < \infty,\ \int|g\hat{u}|^2 d\mu < \infty\}.$$

Prove that conditions $\int|\hat{u}|^2 d\mu < \infty$ and $\int|f^2\hat{u}|^2 d\mu < \infty$ imply $\int|g^2\hat{u}|^2 d\mu < \infty$. First we prove that

(8) $$\left(\int|\hat{u}|^2 d\mu < \infty,\ \int|f^2\hat{u}|^2 d\mu < \infty\right) \Rightarrow \int|f\hat{u}|^2 d\mu < \infty\ ^{**}$$

* By $D(A)$ we denote the domain of definition of the operator A .

** The symbol $\Longrightarrow$ means "implication".

This follows from the Cauchy inequality

(9) $$\int|f\hat{u}|^2 d\mu \leqq \sqrt{\int|f^2\hat{u}|^2 d\mu}\sqrt{\int|\hat{u}|^2 d\mu} < \infty .$$

From (6) and (7), we obtain

(10) $$\left(\int|\hat{u}|^2 d\mu < \infty, \int|f\hat{u}|^2 d\mu < \infty\right) \Leftrightarrow \left(\int|\hat{u}|^2 d\mu < \infty, \int|g\hat{u}|^2 d\mu < \infty\right).$$

Consequently, by substituting $f\hat{u}$ and $g\hat{u}$ in (10) instead of $\hat{u}$ we get

(11) $$\left(\int|f\hat{u}|^2 d\mu < \infty, \int|f^2\hat{u}|^2 d\mu < \infty\right) \Leftrightarrow \left(\int|f\hat{u}|^2 d\mu < \infty, \int|fg\hat{u}|^2 d\mu < \infty\right)$$

(12) $$\left(\int|g\hat{u}|^2 d\mu < \infty, \int|fg\hat{u}|^2 d\mu < \infty\right) \Leftrightarrow \left(\int|g\hat{u}|^2 d\mu < \infty, \int|g^2\hat{u}|^2 d\mu < \infty\right).$$

It follows from (8), (10) and (11) that

(13) $$\left(\int|\hat{u}|^2 d\mu < \infty, \int|f^2\hat{u}|^2 d\mu < \infty\right) \Rightarrow \left(\int|g\hat{u}|^2 d\mu < \infty, \int|fg\hat{u}|^2 d\mu < \infty\right).$$

From (12) and (13), we get

(14) $$\left(\int|\hat{u}|^2 d\mu < \infty, \int|f^2\hat{u}|^2 d\mu < \infty\right) \Rightarrow \int|g^2\hat{u}|^2 d\mu < \infty$$

and this is what we had to prove. The implication

(15) $$\left(\int|\hat{u}|^2 d\mu < \infty, \int|g^2\hat{u}|^2 d\mu < \infty\right) \Rightarrow \int|f^2\hat{u}|^2 d\mu < \infty$$

can be similarly proved.

It follows from (14) and (15) that

$$\left(\int|\hat{u}|^2 d\mu < \infty, \int|f^2\hat{u}|^2 d\mu < \infty\right) \Leftrightarrow \left(\int|\hat{u}|^2 d\mu < \infty, \int|g^2\hat{u}|^2 d\mu < \infty\right)$$

i.e.

$$D(A^2) = D(B^2).$$

Proof of Theorem 1. According to (3-4), since Ω' is regular we have

$$D(\tilde{A}') = D(\tilde{B}') = \overset{\circ}{W}{}^2(\Omega') = \{u: u \in W^2(\Omega'),\ u/\partial\Omega' = 0\}.$$

The conditions of the theorem and the lemma give

$$D(\tilde{A}'^2) = D(\tilde{B}'^2). \tag{16}$$

Obviously, we have

$$\begin{aligned} D(\tilde{A}'^2) &= \{u: u \in \overset{\circ}{W}{}^2(\Omega'),\ \tilde{A}'u \in W^2(\Omega'),\ \tilde{A}'u/\partial\Omega' = 0\} \\ D(\tilde{B}'^2) &= \{u: u \in \overset{\circ}{W}{}^2(\Omega'),\ \tilde{B}'u \in W^2(\Omega'),\ \tilde{B}'u/\partial\Omega' = 0\}. \end{aligned} \tag{17}$$

Let $u \in W^4(\Omega')$. The conditions $\tilde{A}'u/\partial\Omega' = 0$, $\tilde{B}'u/\partial\Omega' = 0$ will be written in local coordinates on the surface $\partial\Omega'$. We introduce a coordinate system in a neighbourhood of a smooth piece ω_1 of the surface $\partial\Omega'$

$$\begin{aligned} \xi_1 &= \xi_1(x_1, \dots, x_n) \\ &\cdots\cdots\cdots \\ \xi_n &= \xi_n(x_1, \dots, x_n) \end{aligned}$$

such that ω_1 is given by the equation $\xi_1 = 0$.

After changing coordinates, the expressions $\mathfrak{A}'u$ and $\mathfrak{B}'u$ get the form

$$\mathfrak{A}'u = \sum_{j,\ell=1}^{n} \frac{\partial^2 u}{\partial\xi_j \partial\xi_\ell} \sum_{k=1}^{n} \alpha_k \frac{\partial\xi_j}{\partial x_k} \frac{\partial\xi_\ell}{\partial x_k} + \sum_{j=1}^{n} \frac{\partial u}{\partial\xi_j} \sum_{k=1}^{n} \alpha_k \frac{\partial^2 \xi_j}{\partial x_k^2}.$$

$$\mathfrak{B}'u = \sum_{j,\ell=1}^{n} \frac{\partial^2 u}{\partial\xi_j \partial\xi_\ell} \sum_{k=1}^{n} \frac{\partial\xi_j}{\partial x_k} \frac{\partial\xi_\ell}{\partial x_k} + \sum_{j=1}^{n} \frac{\partial u}{\partial\xi_j} \sum_{k=1}^{n} \frac{\partial^2 \xi_j}{\partial x_k^2}.$$

Since $u/\omega_1 = 0$, then $\frac{\partial u}{\partial\xi_2}, \dots, \frac{\partial u}{\partial\xi_n}$ vanish on the boundary, and the conditions $\tilde{A}'u/\omega_1 = 0$, $\tilde{B}'u/\omega_1 = 0$ will mean

$$\left(\sum_{\ell=1}^{n} \frac{\partial^2 u}{\partial \xi_1 \partial \xi_\ell} \sum_{k=1}^{n} \alpha_k \frac{\partial \xi_1}{\partial x_k} \frac{\partial \xi_\ell}{\partial x_k} + \frac{\partial u}{\partial \xi_1} \sum_{k=1}^{n} \frac{\partial^2 \xi_1}{\partial x_k^2}\right)\Big/_{\omega_1} = 0 \tag{18}$$

$$\left(\sum_{\ell=1}^{n} \frac{\partial^2 u}{\partial \xi_1 \partial \xi_\ell} \sum_{k=1}^{n} \frac{\partial \xi_1}{\partial x_k} \frac{\partial \xi_\ell}{\partial x_k} + \frac{\partial u}{\partial \xi_1} \sum_{k=1}^{n} \frac{\partial^2 \xi_1}{\partial x_k^2}\right)\Big/_{\omega_1} = 0. \tag{19}$$

Let Ω_1 be a subdomain of Ω', the boundary $\partial\Omega_1$ of which satisfies the condition $\partial\Omega' \cap \partial\Omega_1 = \omega_1$. Denote by Ω_1^δ the set of points of Ω_1 the distance of which from $\partial\Omega_1 \setminus \omega_1$ is not less than δ. Let $\omega_1^\delta = \partial\Omega' \cap \partial\Omega_1^\delta \subset \omega_1$. We take n functions

$$\phi_k(\xi') = \phi_k(\xi_2, \dots, \xi_n) \in C^\infty(\omega_1), \qquad (k = 1, \dots, n)$$

which vanish outside ω_1^δ and such that

$$\begin{vmatrix} \dfrac{\partial \phi_1}{\partial \xi_2} & \cdots & \dfrac{\partial \phi_1}{\partial \xi_n} & \phi_1 \\ \dfrac{\partial \phi_2}{\partial \xi_2} & \cdots & \dfrac{\partial \phi_2}{\partial \xi_n} & \phi_2 \\ \multicolumn{4}{c}{\dotfill} \\ \dfrac{\partial \phi_n}{\partial \xi_2} & \cdots & \dfrac{\partial \phi_n}{\partial \xi_n} & \phi_n \end{vmatrix} \neq 0 \tag{20}$$

on ω_1^δ.

Let

$$\Psi_k(\xi') = \frac{-1}{\sum_{k=1}^{n} \left(\frac{\partial \xi_1}{\partial x_k}\right)^2} \left(\phi_k \sum_{k=1}^{n} \frac{\partial^2 \xi_1}{\partial x_k^2} + \sum_{\ell=2}^{n} \frac{\partial \phi_k}{\partial \xi_\ell} \sum_{k=1}^{n} \frac{\partial \xi_1}{\partial x_k} \frac{\partial \xi_\ell}{\partial x_k}\right).$$

Now construct functions $v_k(x) \in W^2(\Omega_1)$ satisfying the relations

$$v_k/\omega_1 = 0, \quad \frac{\partial v_k}{\partial \xi_1}/\omega_1 = \phi_k(\xi'), \quad \frac{\partial^2 v_k}{\partial \xi_1^2}/\omega_1 = \Psi_k(\xi')$$

and vanishing outside Ω_1^δ. Put

$$u_k(x) = \begin{cases} v_k(x) & (x \in \Omega_1), \\ 0 & (x \in \Omega' \setminus \Omega_1). \end{cases}$$

Obviously $u_k \in W^4(\Omega')$, $u_k/\partial\Omega' = 0$. Besides, the functions u_k satisfy (19) on ω_1 and vanish in a neighbourhood of the rest of the boundary $\partial\Omega'$. Thus $B'u_k/\partial\Omega' = 0$. Therefore $u_k \in D(\tilde{B}'^2)$. Then $u_k \in D(\tilde{A}'^2)$ and consequently u_k also satisfies (18). It follows from (20) that the $n-1$-dimensional vectors

$$\chi_k(\xi') = \left(\frac{\partial^2 u_k}{\partial \xi_1^2}, \frac{\partial^2 u_k}{\partial \xi_1 \partial \xi_2}, \dots, \frac{\partial^2 u_k}{\partial \xi_1 \partial \xi_n}, \frac{\partial u_k}{\partial \xi_1}\right), \qquad (k = 1, \dots, n)$$

are linearly independent at every point of ω_1^δ. The relations (18) and (19) mean that the vectors χ_k are orthogonal to the vectors

$$\pi(\xi') = \left(\sum_{k=1}^n \alpha_k \frac{\partial \xi_1}{\partial x_k}\frac{\partial \xi_1}{\partial x_k}, \dots, \sum_{k=1}^n \alpha_k \frac{\partial \xi_1}{\partial x_k}\frac{\partial \xi_n}{\partial x_k}, \sum_{k=1}^n \alpha_k \frac{\partial \xi_1}{\partial x_k^2}\right)$$

$$\rho(\xi') = \left(\sum_{k=1}^n \frac{\partial \xi_1}{\partial x_k}\frac{\partial \xi_1}{\partial x_k}, \dots, \sum_{k=1}^n \frac{\partial \xi_1}{\partial x_k}\frac{\partial \xi_n}{\partial x_k}, \sum_{k=1}^n \frac{\partial^2 \xi_1}{\partial x_k^2}\right).$$

It follows that $\pi(\xi')$ and $\rho(\xi')$ are collinear at every point of ω_1^δ, i.e. there exist a function $\varkappa(x)$ and $x \in \omega_1^\delta$ for which

$$\left(\sum_{k=1}^n (\alpha_k - \varkappa(x)) \frac{\partial \xi_1}{\partial x_k}\frac{\partial \xi_\ell}{\partial x_k}\right)/\omega_1^\delta = 0 \qquad (\ell = 1, \dots, n) \tag{21}$$

holds. Let

$$P_k(x) = (\alpha_k - \varkappa(x)) \frac{\partial \xi_1}{\partial x_k} . \tag{22}$$

The relations (17) take the form

$$\left(\sum_{k=1}^{n} P_k(x) \frac{\partial \xi_\ell}{\partial x_k} \right) \Big/ \omega_1^\delta = 0 \qquad (\ell = 1, \dots, n) . \tag{23}$$

It can be supposed, without essential restriction, that the equation $\xi_1(x) = 0$ is solved with respect to some of $x_1, \dots, x_n$, for instance to x_{i_1} (the index i_1 is the same as in (3)). Then

$$\xi_1(x) = x_{i_1} - \zeta(x') , \tag{24}$$

where

$$x' = (x_1, \dots, x_{i_1 - 1}, x_{i_1 + 1}, \dots, x_n) .$$

Denote by O_1^δ the orthogonal projection of ω_1^δ on the coordinate hyperplane $(x_1, \dots, x_{i_1-1}, x_{i_1+1}, \dots, x_n)$. If $x' \in O_1^\delta$ then the point $(x_1, \dots, x_{i_1-1}, \zeta(x'), x_{i_1+1}, \dots, x_n) \in \omega_1^\delta$ will be denoted by $M(x')$. Substituting $\ell = 1$ in (23) and taking into account (24) we get

$$P_{i_1}(M(x')) - \sum_{\substack{k=1 \\ k \neq i_1}}^{n} P_k(M(x')) \frac{\partial \xi}{\partial x_k} = 0 . \tag{25}$$

If $\ell = i_1$ we put

$$\eta_\ell(x') = \xi_\ell(M(x')) , \qquad x' \in O_1^\delta .$$

By differentiating we obtain

$$\frac{\partial \eta_\ell}{\partial x_k} = \frac{\partial \xi_\ell}{\partial x_k} + \frac{\partial \xi_\ell}{\partial x_{i_1}} \frac{\partial \zeta}{\partial x_k} .$$

Let us determine herefrom $\frac{\partial \xi_\ell}{\partial x_k}$ and substitute it in (23). We receive for

(26) $$\sum_{\substack{k=1\\k\neq i_1}}^{n} P_k(M(x'))\frac{\partial \eta_\ell}{\partial x_k} - \frac{\partial \xi_\ell}{\partial x_{i_1}}\sum_{\substack{k=1\\k\neq i_1}}^{n} P_k(M(x'))\frac{\partial \zeta}{\partial x_k} + P_{i_1}(M(x'))\frac{\partial \xi_\ell}{\partial x_{i_1}} = 0.$$

It follows from (25) and (26) that

(27) $$\sum_{\substack{k=1\\k\neq i_1}}^{n} P_k(M(x'))\frac{\partial \eta_\ell}{\partial x_k} = 0, \quad x'\in O_1^\delta, \quad \ell\neq i_1.$$

The functions $\eta_\ell(x')$, $\ell=1,\dots,n$, $\ell\neq i_1$ are independent. Thus the linear partial differential equation of the first order (27) has $n-1$ independent solutions. But this is possible only if [6] all the coefficients of the equation are equal to zero i.e.

(28) $$P_k(M(x'))=0, \quad k=1,\dots,n, \quad k\neq i_1, \quad x'\in O_1^\delta$$

It follows from (25) and (28) that

(29) $$P_{i_1}(M(x')=0, \quad x'\in O_1^\delta$$

also holds true.

From (28) and (29), taking in to account $\dfrac{\partial \xi_1}{\partial x_{i_1}} = 1$ and (3) we get the equalities

$$\varkappa(x') = \alpha_{i_1} = \alpha_{i_2} = \dots = \alpha_{i_p}$$

Then, if $k = i_1,\dots,i_p$, it follows from (22) and (28) that

$$\frac{\partial \xi_1}{\partial x_k}(M(x')) = 0, \quad x'\in O_1^\delta$$

i.e. ξ_1 does not depend on x_k. Thus ω_1^δ is a piece of the cylinder

(30) $$\xi_1(x_{i_1}, x_{i_2}, \dots, x_{i_p}) = 0.$$

So we have obtained that the boundary $\partial\Omega'$ is composed from pieces of cylinders of kind (30). Since Ω' is regular the assertion of the theorem follows.

It is easy to prove the converse of it.

Theorem 2. *If* Ω' *is a cylinder of kind* (4) *then the operators* $\tilde{A}'$ *and* $\tilde{B}'$ *commute.*

Proof. Let $u^i_\mu(x_{i_1},\dots,x_{i_p})$, $\mu = 1,2,\dots$ be a complete system of eigenfunctions of the Laplace operator $\Delta_i = \frac{\partial^2}{\partial x_{i_1}^2} + \dots + \frac{\partial^2}{\partial x_{i_p}^2}$ on the domain vanishing on the boundary. We construct the eigenfunctions $u^j_\nu(x_{j_1},\dots,x_{j_q}),\dots,u^\ell_\sigma(x_{\ell_1},\dots,x_{\ell_r})$, $\nu,\sigma = 1,2,\dots$ similarly. Then the set of functions

$$u_{\mu\nu\dots\sigma}(x) = u^i_\mu(x_{i_1},\dots,x_{i_p})\,u^j_\nu(x_{j_1},\dots,x_{j_q})\dots u^\ell_\sigma(x_{\ell_1},\dots,x_{\ell_r})$$

$$(\mu,\nu,\dots,\sigma = 1,2,\dots)$$

is a complete system of eigenfunctions both of the operators $\tilde{A}'$ and $\tilde{B}'$. Indeed $u_{\mu\nu\dots\sigma}(x) \in W^2(\Omega')$ because of

$$u^i_\mu \in W^2(\Omega_i),\quad u^j_\nu \in W^2(\Omega_j),\ \dots,\ u^\ell_\sigma \in W^2(\Omega_\ell).$$

The functions $u_{\mu\dots\sigma}(x)$ vanish on the boundary of Ω' since $u^i_\mu, u^i_\nu, \dots, u^i_\sigma$ vanish on the boundary of $\Omega_j, \dots, \Omega_\ell$, respectively. Hence $u_{\mu\dots\sigma} \in \overset{\circ}{W}{}^2(\Omega') = D(\tilde{A}') = D(\tilde{B}')$. We have

$$\tilde{A}' u_{\mu\nu\dots\sigma} = \mathfrak{A}'_{\mu\dots\sigma} = (\alpha_{i_1}\lambda^i_\mu + \alpha_{j_1}\lambda^j_\nu + \dots + \alpha_{\ell_1}\lambda^\ell_\sigma)\,u_{\mu\dots\sigma}$$

$$\tilde{B}' u_{\mu\nu\dots\sigma} = \mathfrak{B}'_{\mu\dots\sigma} = (\lambda^i_\mu + \lambda^j_\nu + \dots + \lambda^\ell_\sigma)\,u_{\mu\dots\sigma}\,.$$

Thus $u_{\mu\dots\sigma}$ are really common eigenvectors of the operators $\tilde{A}'$ and $\tilde{B}'$. The completeness of the system of functions $u_{\mu\dots\sigma}$ in $L_2(\Omega')$ is a consequence of the completeness of the systems $u^i_\mu, u^j_\nu, \dots, u^\ell_\sigma$ in $L_2(\Omega_i), L_2(\Omega_j), \dots, L_2(\Omega_\ell)$, respectively.

It follows that the operators $\tilde{A}'$ and $\tilde{B}'$ commute since they have a complete system of common eigenfunctions.

Up to now we have considered only regular domains. Consider now some nonregular two-dimensional domains.

Theorem 3. *Let Ω' be a two-dimensional domain with piecewise smooth boundary having some angles bigger than π. Suppose that in some neighbourhood of the vertex of such an angle the sides of the angle are straight. For such a domain the operators $\tilde{A}'$ and $\tilde{B}'$ do not commute.*

Proof. Suppose the operators $\tilde{A}'$ and $\tilde{B}'$ commute. Then [1] the operator $\tilde{A}' + \tilde{B}'$ is essentially self-adjoint.

We have

$$D(\tilde{A}'+\tilde{B}') = D(\tilde{A}') \cap D(\tilde{B}') .$$

Prove that

$$D(\tilde{A}' + \tilde{B}') = \overset{\circ}{W}{}^2(\Omega'). \tag{31}$$

It is necessary only to prove that $D(\tilde{A}') \cap D(\tilde{B}') \subset W^2(\Omega')$. Let $u(x) \in D(\tilde{A}') \cap D(\tilde{B}')$ and P be a vertex of the domain Ω' . Let $\zeta_p(x) \in C^\infty(\bar{\Omega}')$ vanish outside some neighbourhood of P containing no other vertexes and let $\zeta_p(x) \equiv 1$ in an other neighbourhood contained in this one. Since $u(x) \in D(\tilde{A}') \cap D(\tilde{B}')$ then $\zeta_p(x)\, u(x) \in \in D(\tilde{A}') \cap D(\tilde{B}')$. According to the results of [3-4] it follows from $\zeta_p u \in D(\tilde{B}')$ that

$$\zeta_p u = C r^{\frac{\pi}{\omega}} \sin \frac{\pi}{\omega} \varphi + u_1 ,$$

where r and φ are polar coordinates in a neighbourhood of P , C is a constant and $u_1 \in W^2(\Omega')$. On the other hand, since $\zeta_p u \in D(\tilde{A}')$, we have $\zeta_p u \in \in D(A_0^*)$ and consequently the function

$$\mathcal{A}'(\zeta_p u) = \left(\alpha_1 \frac{\partial^2}{\partial x_1^2} + \alpha_2 \frac{\partial^2}{\partial x_2^2}\right)(\zeta_p u) =$$

$$= C(\alpha_1 - \alpha_2) \frac{\pi}{\omega}\left(\frac{\pi}{\omega} - 1\right) r^{\frac{\pi}{\omega}-2} \sin\left(\frac{\pi}{\omega} - 2\right)\varphi + v_1, \quad v_1 \in L_2(\Omega')$$

belongs to $L_2(\Omega')$. This is possible only if $C = 0$. Consequently $\zeta_p u = u_1 \in W^2(\Omega')$. Thus $u \in W^2$ in a neighbourhood of every vertex. It is well known that $u \in W^2$ in a subdomain of Ω' containing no vertices. It follows from this that $u \in W^2(\Omega')$.

The relation (31) is proved.

Let us prove now that the operator $\tilde{A}' + \tilde{B}'$ is closed. Indeed, suppose that* $u \in D(\overline{\tilde{A}'+\tilde{B}'})$ and $(\overline{\tilde{A}'+\tilde{B}'})u = f$. Then there exists such a sequence $u_k \in D(\tilde{A}'+\tilde{B}')$ that $u_k \to u$ and $(\tilde{A}'+\tilde{B}')\, u_k \to f$. Since $u_k \in \overset{\circ}{W}{}^2(\Omega')$ then

* We denote by $\bar{A}$ the closure of A .

[3-4] the following estimate is true

$$\| u_k - u_\ell \|_{W^2(\Omega')} \leq C (\| (\tilde{A}' + \tilde{B}')(u_k - u_\ell) \| + \| u_k - u_\ell \|).$$

It is seen from the last inequality that $u \in \overset{\circ}{W}{}^2(\Omega')$ and $\tilde{A}' + \tilde{B}'$ is closed. Thus the operator $\tilde{A}' + \tilde{B}'$ is not only essentially selfadjoint but even self-adjoint. Consider the symmetric operator $K_0 = A'_0 + B'_0$ defined on $C_0^\infty(\Omega')$. The operator $\tilde{A}' + \tilde{B}'$ is a self-adjoint extension of the operator K_0 and

$$D(\tilde{A}' + \tilde{B}') \subset \overset{\circ}{W}{}^1(\Omega') = H_{K_0},$$

where H_{K_0} is the closure of $C_0^\infty(\Omega')$ in the sense of K_0-convergence (see [7], p. 388). According to a theorem of M.G. Kreĭn (see [7], p. 389) the operator $\tilde{A}' + \tilde{B}'$ is the Friedrichs extension $\tilde{K}$. This however is not possible because, according to [4] $D(\tilde{K}) \neq \overset{\circ}{W}{}^2(\Omega')$. The contradiction is due to the assumption that $\tilde{A}'$ and $\tilde{B}'$ commute. The theorem is proved.

From the theorem we obtain.

Corollary. *The only two dimensional domains with piecewise smooth boundary for which the operators $\tilde{A}'$ and $\tilde{B}'$ commute are the rectangles with sides parallel to the coordinate axes.*

The author thanks Prof. H. Hristov and Dr. G. Lassner for suggesting discussion.

REFERENCES

[1] K. Moren, *Hilbert space methods* (Moscow, 1965). (in Russian)

[2] F.R. Gantmaher, *Theory of matrices* (Moscow, 1967). (in Russian)

[3] V.A. Kondrat'ev, Boundary value problems for elliptic equations in a domain with conical or angular points, *Trudy Moskov. Mat. Obsc.*, 16 (1967), 209-292. (in Russian)

[4] N.S. Birman and E.E. Skvorcov, On the square summability of higher derivatives of the solution of Dirichlet's problem in a domain with piece-wise smooth boundary, *Matematika*, 30 (1962). (in Russian)

[5] L. Bers, F. John and M. Schechter, *Partial differential equations* (Moscow, 1966). (in Russian)

[6] E. Kamke, *Differential equations.* II. (Moscow, 1966). (in Russian)

[7] I.N. Ahiezer and I.M. Glazman, *Theory of linear operators in Hilbert space* (Moscow, 1966). (in Russian)

COLLOQUIA MATHEMATICA SOCIETATIS JÁNOS BOLYAI
5. HILBERT SPACE OPERATORS, TIHANY (HUNGARY), 1970

Sur le spectre ponctuel d'un opérateur

J. DIXMIER et C. FOIAŞ

Introduction. Tous les espaces de Banach considérés dans cet article sont supposés complexes. Si H est un espace de Banach, on notera $L(H)$ l'ensemble des endomorphismes continus de H. Si $u \in L(H)$, on notera $\sigma(u)$ le spectre de u et $\sigma_p(u)$ le spectre ponctuel de u.

On sait que $\sigma(u)$ est une partie compacte non vide de $\mathbf{C}$ (si $H \neq 0$). Réciproquement, si B est une partie compacte non vide de $\mathbf{C}$ et si H est un espace hilbertien de dimension infinie, il existe $u \in L(H)$ tel que $\sigma(u) = B$.

Soit H un espace hilbertien dont la dimension hilbertienne est au moins égale à la puissance du continu. Si A est une partie bornée de $\mathbf{C}$, on voit aussitôt qu'il existe un élément normal u de $L(H)$ tel que $\sigma_p(u) = A$. Nous verrons que la la situation est différente si H est un espace hilbertien séparable: dans ce cas, si $u \in L(H)$, $\sigma_p(u)$ *est un* F_σ *borné* (la prop. 1 donne un résultat plus général). Cela pose le problème suivant: si A est un F_σ borné dans $\mathbf{C}$, existe-t-il $u \in L(H)$ tel que $\sigma_p(u) = A$? La réponse est positive et on a même le résultat plus précis que voici (prop.3): *soient* H *un espace hilbertien séparable de dimension infinie,* A *un* F_σ *dans* $\mathbf{C}$, B *une partie compacte non vide de* $\mathbf{C}$ *contenant* A *; alors il existe* $u \in L(H)$ *tel* *que* $\sigma_p(u) = A$, $\sigma(u) = B$.

Soient H un espace de Banach et $u \in L(H)$. Il est bien connu que si $\lambda_1, \dots, \lambda_n$ sont des éléments de $\sigma_p(u)$ deux à deux distincts, et $x_1, \dots, x_n$ des vecteurs propres non nuls correspondants, alors les x_i sont linéairement indépendants. Cherchons à généraliser cette remarque lorsqu'on considère une infinité d'éléments de $\sigma_p(u)$. Pour tout $\lambda \in \sigma_p(u)$, soit x_λ un vecteur propre non nul correspondant. Soit μ une mesure sur $\sigma_p(u)$. Alors, est-il exact que la relation $\int x_\lambda d\mu(\lambda) = 0$ entraîne $\mu = 0$? Nous obtiendrons à ce sujet des résultats positifs et dès résultats négatifs (prop. 4, remarques 1 et 2). Pour donner un sens à la question, il faut d'ailleurs prouver qu'on peut choisir les x_λ dépendant assez régulièrement de λ . La prop. 3 permet de lever cette difficulté dans le cas hilbertien.

Faisons quelques rappels sur les endomorphismes. Soient H un espace de Banach et $u \in L(H)$. On dit que u possède la propriété de prolongement analytique unique si, pour toute partie ouverte G de $\mathbf{C}$ et toute application analytique f de G dans H telle que $(\lambda - u)f(\lambda)$ pour tout $\lambda \in G$, on a $f = 0$. Si u possède cette propriété, il existe, pour tout $x \in H$, une plus grande partie ouverte G_x de $\mathbf{C}$ dans laquelle est définie une application analytique $g : G_x \longrightarrow H$ telle que $(\lambda - u)g(\lambda) = x$ pour tout $\lambda \in G_x$. On pose $\mathbf{C} - G_x = \sigma(u, x)$. Si u possède une distribution spectrale (cf. [2]), alors u possède la propriété de prolongement analytique unique ([2], th. 1); pour toute partie fermée F de $\mathbf{C}$ notons $H(u,F)$ l'ensemble des $x \in H$ tels que $\sigma(u,x) \subset F$; alors $H(u,F)$ est un sous-espace vectoriel fermé de H stable par u , et $\sigma(u | H(u,F)) \subset F$; si H' est un sous-espace vectoriel fermé de H stable par u et tel que $\sigma(u | H') \subset F$, on a $H' \subset H(u,F)$ (cf. [2], p.151).

Proposition 1. *Soient* H *un espace de Banach réflexif séparable, et* $u \in L(H)$. *Alors* $\sigma_p(u)$ *est un* F_σ *borné.*

(Nous ignorons si la prop. 2 reste valable quand H est un espace de Banach séparable quelconque).

Soient B la boule unité de H , et P l'ensemble des vecteurs propres de u appartenant à B . On sait que B est compacte métrisable pour la topologie faible. Soit (x_n) une suite de points de P tendant faiblement vers un point x de B . Il existe des $\lambda_n \in \mathbf{C}$ tels que $|\lambda_n| \leq \|u\|$ et $ux_n = \lambda_n x_n$. Puis il existe une suite partielle (λ_{n_i}) qui tend vers une limite λ . On a $ux = \lambda x$, d'où $x \in P$.

Ainsi, P est faiblement fermé dans B et par suite compact métrisable pour la topologie faible.

Pour tout $x \in P - \{0\}$, soit $v(x)$ la valeur propre correspondante. Soit (y_n) une suite de points de $P - \{0\}$ tendant faiblement vers un point y de $P - \{0\}$. Pour toute suite partielle $(n(1), n(2), \dots)$ d'entiers, il existe, d'après le raisonnement ci-dessus, une suite partielle $(n(p_1), n(p_2), \dots)$ telle que $v(y_{n(p_i)}) \to v(y)$. Donc $v(y_n) \to v(y)$. La fonction v est donc continue sur $P - \{0\}$.

L'ensemble $P - \{0\}$ est réunion d'une suite $(K_1, K_2, \dots)$ de parties faiblement compactes. Alors $v(K_i)$ est une partie compacte de $\mathbf{C}$, donc la réunion des $v(K_i)$ est un F_σ. Or cet ensemble n'est autre que $\sigma_p(u)$.

Proposition 2. *Soient* H *un espace hilbertien séparable de dimension infinie,* A *un* F_σ *dans* $\mathbf{C}$, B *une partie compacte de* $\mathbf{C}$ *contenant* A. *Alors il existe* $u \in L(H)$ *tel que* $\sigma_p(u) = A$, $\sigma(u) = B$.

a) Supposons d'abord que B soit l'adhérence de A. Soit Ω un disque ouvert non vide dans $\mathbf{C}$ contenant B. Soit K l'espace de Sobolev formé par les $f \in L^2(\Omega)$ telles que $D^\alpha f \in L^2(\Omega)$ pour $|\alpha| \leq 2$ (les dérivées sont prises au sens des distributions); muni du produit scalaire

$$(f|g) = \sum_{|\alpha| \leq 2} \iint_\Omega D^\alpha f \cdot \overline{D^\alpha g}\, dx\, dy \quad .$$

l'espace K est un espace hilbertien séparable de dimension infinie. Soit $\lambda \in \Omega$. D'après le théorème de Sobolev (cf. par exemple [3] th. 5. I.) l'application $f \mapsto f(\lambda)$ de K dans $\mathbf{C}$ a un sens, et c'est une forme linéaire continue sur K ; il existe donc $g_\lambda \in K$ tel que

(1) $$f(\lambda) = (f | g_\lambda)$$

pour toute $f \in K$. Soit v l'endomorphisme continu de K défini par $(vf)(\lambda) = \lambda f(\lambda)$ pour tout $\lambda \in \Omega$. Pour $f \in K$ et $\lambda \in \Omega$, on a, compte tenu de (1),

$$(f | v^* g_\lambda) = (vf | g_\lambda) = (vf)(\lambda) = \lambda f(\lambda) = \lambda (f | g_\lambda) = (f | \bar{\lambda} g_\lambda)$$

donc

(2) $$v^* g_\lambda = \lambda g_\lambda \qquad \text{pour tout } \lambda \in \Omega .$$

Nous noterons v' l'endomorphisme v^* considéré comme opérant dans l'espace hilbertien K' conjugué de K . Alors, d'après (2),

(3) $$v' g_\lambda = \lambda g_\lambda$$

L'endomorphisme v admet la distribution spectrals $\varphi \mapsto t_\varphi$ définie par

$$(t_\varphi f)(\lambda) = \varphi(\lambda) f(\lambda) \qquad (\varphi \in C^\infty(\mathbf{R}^2)) .$$

Par suite, v' admet la distribution spectrale $\varphi \mapsto t'_\varphi$ où t'_φ est l'adjoint de t_φ considéré comme opérant dans K' (il est immédiat en effet que l'aplication $\varphi \mapsto t'_\varphi$ de $C^\infty(\mathbf{R}^2)$ dans $L(K')$ est linéaire continue).

Ecrivons A sous la forme $A_1 \cup A_2 \cup \ldots$, où $A_1, A_2, \ldots$ sont compacts. Soient $K_n = K'(v', A_n)$ et $K_0 = K'(v', B)$. Posons $v_n = v' | K_n$ pour $n \geq 0$. On a (cf. Introduction)

(4) $$\sigma(v_n) \subset A_n \quad \text{pour} \quad n > 0 .$$

Soient $\lambda \in A_n$ et L le sous-espace $\mathbf{C} g_\lambda$ de K' . D'après (3) et les rappels de l'introduction, on a $L \subset K_n$. Donc $g_\lambda \in K_n$ et $\lambda \in \sigma_p(v_n)$. Par suits, $A_n \subset \sigma_p(v_n)$. Compte tenu de (4), on a

(5) $$\sigma_p(v_n) = \sigma(v_n) = A_n .$$

Soit $u = v_1 \oplus v_2 \oplus \ldots$. On a $\sigma_p(u) = \sigma_p(v_1) \cup \sigma_p(v_2) \cup \ldots = A$, et par suite $\sigma(u) \supset B$. D'autre part, si $\lambda \in \mathbf{C} - B$, on a $(\lambda - v_n)^{-1} = (\lambda - v_0)^{-1} | K_n$ pour $n \geq 1$, donc $\sup_{n \geq 1} \| (\lambda - v_n)^{-1} \| < +\infty$ et par suite $\lambda \notin \sigma(u)$. Ainsi $\sigma(u) = B$.

b) Passons au cas général. Compte tenu de a), et grâce à la construction d'une somme hilbertienne, il suffit de construire, dans un espace hilbertien séparable, un endomorphisme continu u tel que $\sigma(u) = B$, $\sigma_p(u) = \phi$. On a $B = B' \cup B''$ avec B' parfait et B'' dénombrable. D'après la théorie spectrale, il existe un endomorphisme continu normal u' dans un espace hilbertien séparable tel que $\sigma(u') = B'$, $\sigma_p(u') = \phi$. Posons $B'' = \{\lambda_1, \lambda_2, \ldots\}$. Soit w un endomorphisme quasi-nilpotent injectif dans un espace hilbertien séparable (par exemple l'opérateur de Volterra). On a $\sigma(w) = \{0\}$, $\sigma_p(w) = \phi$. Soit u la somme hilbertienne $u' \oplus (w + \lambda_1) \oplus (w + \lambda_2) \oplus \ldots$ L'endomorphisme u est continu parce que la suite (λ_i) est bornée. On a $\sigma_p(u) = \phi$ et $\sigma(u) \supset B' \cup \{\lambda_1, \lambda_2, \ldots\} = B$. Si $\lambda \in \mathbf{C} - B$, il existe $\delta > 0$ tel que $|\lambda - \lambda_i| \geq \delta$ pour tout i ; alors

$$\| (w + \lambda_i - \lambda)^{-1} \| \leq \sup_{|\mu| \geq \delta} \| (w - \mu)^{-1} \| < +\infty$$

donc $\lambda \notin \sigma(u)$. Finalement, $\sigma(u) = B$.

Proposition 3. *Soient* H *un espace hilbertien séparable, et* $u \in L(H)$. *Pour tout* $\lambda \in \mathbf{C}$ *, soient* E_λ *le sous-espace propre correspondant à* λ , P_λ *le projecteur sur* E_λ , $d(\lambda)$ *la dimension hilbertienne de* E_λ.

(i) *Pour tout* $x \in H$ *, la fonction* $\lambda \mapsto \| P_\lambda x \|$ *est semi-continue supérieurement.*

(ii) *Il existe une suite* $(x_1, x_2, \dots)$ *d'application boréliennes de* $\mathbf{C}$ *dans* H *possédant les propriétés suivantes:*

a) *Si* $d(\lambda) = \aleph_0$, $(x_1(\lambda), x_2(\lambda), \dots)$ *est une base orthonormale de* E_λ.

b) *Si* $d(\lambda) = \aleph_0$, $(x_1(\lambda), x_2(\lambda), \dots, x_{d(\lambda)}(\lambda))$ *est une base orthonormale de* E_λ *, et* $x_i(\lambda) = 0$ *pour* $i > d(\lambda)$.

Si (i) est inexact, il existe $\lambda_0, \lambda_1, \lambda_2, \dots \in \mathbf{C}$ et $x \in H$ avec $\lambda_n \to \lambda_0$ et $\liminf_{n \to +\infty} \| P_{\lambda_n} x \| > \| P_{\lambda_0} x \|$. Nous allons en déduire une contradiction. Par extraction de suite partielle, on peut supposer que $P_{\lambda_n} x$ tend faiblement vers un élément z de H . Alors $x - P_{\lambda_n} x$ tend faiblement vers $x - z$, d'où

$$\| x - z \|^2 \le \liminf \| x - P_{\lambda_n} x \|^2 = \| x \|^2 - \limsup \| P_{\lambda_n} x \|^2 <$$

$$< \| x \|^2 - \| P_{\lambda_0} x \|^2 = \| x - P_{\lambda_0} x \|^2$$

ce qui est absurde puisque $z \in E_{\lambda_0}$.

Considérons le champ constant $E : z \mapsto H$ d'espaces hilbertiens sur $\mathbf{C}$. L'assertion (i) entraîne que le champ $z \mapsto E_z$ est un sous-champ borélien de ([1], p. 150, prop. 9). Alors (ii) résulte de [1], p. 144, prop. 1 (cf. p. 146, remarque).

Proposition 4. *Soient* H *un espace de Banach séparable,* u *un élément de* $L(H)$ *possédant la propriété de prolongement analytique unique. Soit* $\lambda \mapsto x_\lambda$ *une application borélienne bornée de* $\sigma_p(u)$ *dans* $H - \{0\}$ *telle que* $u x_\lambda = \lambda x_\lambda$ *pour tout* $\lambda \in \sigma_p(u)$. *Soit* μ *une mesure sur* $\mathbf{C}$ *concentrée sur une partie compacte* K *de* $\sigma_p(u)$ *et telle que* $\int x_\lambda \, d\mu(\lambda) = 0$. *Soit* α *la mesure de Lebesgue dans* $\mathbf{C}$ *Si* $\mathbf{C} - K$ *a un nombre fini de composantes connexes, ou si* $\alpha K = 0$ *, on a* $\mu = 0$.

Soit $\zeta \in \mathbf{C} - K$. La fonction $\lambda \mapsto (\lambda - \zeta)^{-1} x_\lambda$ est μ -intégrable. Posons

$$f(\zeta) = \int (\lambda-\zeta)^{-1} x_\lambda d\mu(\lambda).$$

Alors f est une application holomorphe de $\mathbf{C}-K$ dans H . En outre

$$uf(\zeta) = \int (\lambda-\zeta)^{-1} ux_\lambda d\mu(\lambda) = \int (\lambda-\zeta)^{-1} \lambda x_\lambda d\mu(\lambda) =$$

$$= \int (\lambda-\zeta)(\lambda-\zeta)^{-1} x_\lambda d\mu(\lambda) + \int \zeta(\lambda-\zeta)^{-1} x_\lambda d\mu(\lambda) =$$

$$= \int x_\lambda d\mu(\lambda) + \zeta \int (\lambda-\zeta)^{-1} x_\lambda d\mu(\lambda) = \zeta f(\zeta).$$

Donc $f = 0$ puisque u a la propriété de prolongement analytique unique. Soit $(e_1, e_2, \dots)$ une suite faiblement dense dans le dual de H . On a, pour $\zeta \in \mathbf{C}-K$,

$$0 = \langle f(\zeta), e_n \rangle = \int (\lambda-\zeta)^{-1} \langle x_\lambda, e_n \rangle d\mu(\lambda).$$

D'après les hypothèses faites sur K , les combinaisons linéaires des fonctions $\lambda \mapsto (\lambda-\zeta)^{-1}$ sur K sont denses dans l'espace des fonctions complexes continues sur K ([4], p. 78, th. 7.7 et 7.8). Donc la mesure $\langle x_\lambda, e_n \rangle d\mu(\lambda)$ est nulle. Autrement dit, il existe une partie μ-négligeable A_n de K telle que $\langle x_\lambda, e_n \rangle = 0$ pour $\lambda \in K - A_n$. Soit A la réunion des A_n , qui est μ-négligeable. On a $x_\lambda = 0$ pour $\lambda \notin A$. Donc $A = K$, d'où $\mu = 0$.

Remarque 1. Soient H un espace de Banach, et $u \in L(H)$. Si $\sigma_p(u)$ est d'intérieur vide, il est clair que u possède la propriété de prolongement analytique unique.

Remarque 2. Soient H un espace de Banach, u un élément de $L(H)$ qui ne possède *pas* la propriété de prolongement analytique unique (par exemple le "backwards shift operator" dans ℓ^2). Alors il existe une partie ouverte non vide G de $\mathbf{C}$ et une application analytique non nulle $\lambda \mapsto x_\lambda$ de G dans H telle que $ux_\lambda = \lambda x_\lambda$ pour tout $\lambda \in G$. Soit Γ un cercle contenu dans G et soit μ une mesure non nulle invariante par rotation sur Γ . Alors $\int x_\lambda d\mu(\lambda) = 0$, ce qui montre que la conclusion de la prop. 4 est en défaut.

BIBLIOGRAPHIE

[1] J. Dixmier, *Les algèbres d'operateurs dans l'espace hilbertien*, 2e éd. (Paris, 1969).

[2] C. Foiaş, Une application des distributions vectorielles à la théorie spectrale, *Bull. Sci. Math.*, 84(1960), 147-158.

[3] E. Gagliardo, Proprietà di alcune classi di funzioni in piu variabili, *Ric. di Mat.*, 7(1958), 102-137.

[4] J. Wermer, Banach algebras and analytic functions, *Advances in Mathematics*, 1(1961), 51-102.

COLLOQUIA MATHEMATICA SOCIETATIS JÁNOS BOLYAI
5. HILBERT SPACE OPERATORS, TIHANY (HUNGARY), 1970

Wiener–Hopf operators with measure kernels

R. G. DOUGLAS and J. L. TAYLOR

Let $L^p(\mathbf{R})$ $[L^p(\mathbf{R}^+)]$ denote the complex Lebesgue space for Lebesgue measure defined on the [non negative] real numbers. For k a function in $L^1(\mathbf{R})$ and λ a non-zero complex number the classical integral equation of Wiener and Hopf is

$$\lambda f(x) + \int_0^\infty k(x-t) f(t)\,dt = g(x),$$

where g is the given function and f the unknown function to be determined. In [11] Wiener and Hopf introduced a technique for solving this equation for a certain class of kernels k which utilized the Fourier transform and a certain factorization technique. Since then many authors (cf. [7], [4]) have considered analogues of this equation on various spaces and with more general kernels. Quite recently a rather different class of kernels has been investigated.

*Sloan Foundation Fellow.

**Research partially supported by the United States Air Force, Air Force Office of Scientific Research under AFOSR grant 1313-67.

In order to discuss these results we rewrite the equation in the form

$$\int_0^\infty f(t)\, d\nu(x-t) = g(x),$$

where ν is the Borel measure on $\mathbf{R}$ obtained by adding the absolutely continuous measure with derivative k to λ times the point mass δ_0 concentrated at the origin. This equation is completely equivalent to the preceding and at the same time suggests further generalizations. The equation was considered for purely atomic measures ν by Coburn and Douglas [2] on $L^2(\mathbf{R}^+)$ and Gohberg and Feldman [6] on $L^p(\mathbf{R}^+)$. Measures having no continuous singular part were also considered by Gohberg and Feldman [6] on $L^p(\mathbf{R}^+)$ as well as by the first author (unpublished) on $L^2(\mathbf{R}^+)$. In this note we consider an arbitrary Borel measure on $\mathbf{R}$ basing our results on a rather deep theorem on the convolution algebra of measures on $\mathbf{R}$ recently obtained by the second author [10]. It may be of interest to note that this latter work was motivated by this application to the study of Wiener-Hopf equations.

Let $M(\mathbf{R})$ denote the Banach algebra of complex Borel measures on $\mathbf{R}$ with convolution multiplication. For μ in $M(\mathbf{R})$ we define the Wiener-Hopf operator W_μ on $L^p(\mathbf{R}^+)$ such that $(W_\mu f)(x) = \int_0^\infty f(t)\, d\mu(x-t)$ foı f in $L^p(\mathbf{R}^+)$. This operator is bounded on each $L^p(\mathbf{R}^+)$ and has norm less than or equal to the total variation of μ. We can now write the Wiener-Hopf equation very succinctly $W_\mu f = g$ and the study of this equation is clearly equivalent to the study of the corresponding operator.

In this study of the operators W_μ our aim is quite modest. Recall that an operator T on a Banach space $\mathcal{H}$ is said to be a Fredholm operator if the range of T is closed and both the kernel and cokernel (that is the quotient space $\mathcal{H}/\mathrm{ran}\, T$) are finite dimensional. Moreover, the analytical index of T is defined $i_a(T) = \dim \ker T - \dim \operatorname{coker} T$. In this note we determine necessary and sufficient conditions on a measure μ for the operator W_μ on $L^1(\mathbf{R}^+)$ to be a Fredholm operator and provide a method for computing the index. As in previous studies (cf. [4]) this enables one to determine when such operators are invertible. Similar results are obtained for the operators W_μ on $L^p(\mathbf{R}^+)$ for $1 < p < \infty$ with the imposition of an additional hypothesis. In this case the so-called Wiener-Pitt phenomena intrudes at a key step. We discuss this and other questions raised by our results at the end of the note.

We begin by introducing enough notation to state the theorem for $M(\mathbf{R})$ which forms the basis for our results. For c in $\mathbf{R}$ let δ_c denote the unit mass concentrated at c and let ϱ denote the measure in $M(\mathbf{R})$ such that $d\varrho(x) = d\delta_0(x) + 2\chi_{[0,\infty)}(x)e^{-x}dx$. (Note that the Fourier transform of ϱ is the function $\frac{1+it}{1-it}$ and that ϱ is invertible in $M(\mathbf{R})$.)

Theorem A. ([10]) *If* μ *is an invertible measure in* $M(\mathbf{R})$, *then there exists an integer* n, *a real number* c, *and a measure* ν *in* $M(\mathbf{R})$ *such that* $\mu = \varrho^n * e^\nu * \delta_c$.

In other words up to a factor of the form $\varrho^{-n} * \delta_{-c}$, the measure μ has a logarithm, namely ν, in the algebra $M(\mathbf{R})$. We denote the pair (n,c) by $i_t(\mu)$. (For the connection of this result with the first Čech cohomology group of the maximal ideal space of $M(\mathbf{R})$ and other relations see [10].)

In order to make use of this result we need the following easy lemma concerning the behavior of Wiener-Hopf operators under composition. Although the mapping $\mu \longrightarrow W_\mu$ is not multiplicative we do have the following.

Lemma B. *If* μ, ν *and* ω *are measures in* $M(\mathbf{R})$ *such that* ν *is supported on* $[0,\infty)$ *and* ω *on* $(-\infty,0]$, *then* $W_\mu W_\nu = W_{\mu*\nu}$ *and* $W_\omega W_\mu = W_{\omega*\mu}$ *for the operators defined on* $L^p(\mathbf{R}^+)$ *for* $1 \leq p < \infty$.

Proof. Compute or compare [7].

This lemma along with Theorem A enables us to reduce the study of W_μ to several special cases which we consider in the following lemmas.

Lemma C. *The operator* W_{ϱ^n} *is a Fredholm operator on* $L^p(\mathbf{R}^+)$ *for* $1 \leq p < \infty$ *with index* $-n$.

Proof. Compare [7].

Lemma D. *The operator* W_{δ_c} *is a Fredholm operator on* $L^p(\mathbf{R}^+)$ *for* $1 \leq p < \infty$ *if and only if* $c = 0$.

Proof. If $c = 0$, then W_{δ_0} is the identity. Otherwise, W_{δ_c} is translation by c and hence for $c > 0$ the range has infinite dimensional defect or W_μ has an infinite dimensional kernel for $c < 0$.

Lemma E. *If* ν *is a measure on* $\mathbf{M}(\mathbf{R})$, *then the operator* W_{e^ν} *is invertible on* $L^p(\mathbf{R}^+)$ *for* $1 \leq p < \infty$.

Proof. If we write $\nu = \nu_1 + \nu_2$, where ν_1 has support on $[0,\infty)$ and ν_2 on $(-\infty, 0]$, then $W_{e^\nu} = W_{e^{\nu_2}} \cdot W_{e^{\nu_1}}$ by Lemma B, which further yields that $W_{e^{-\nu_1}} \cdot W_{e^{-\nu_2}}$ is the inverse $W^{-1}_{e^\nu}$ of W_{e^ν}.

We now state and prove our main result concerning the operators W_μ on $L^1(\mathbf{R}^+)$.

Theorem 1. *If* μ *is a measure in* $\mathbf{M}(\mathbf{R})$, *then* W_μ *is a Fredholm operator on* $L^1(\mathbf{R}^+)$ *if and only if* μ *is invertible in* $\mathbf{M}(\mathbf{R})$ *and* $i_t(\mu)$ *has the form* $(n,0)$ *for some integer* n. *Moreover, in this case* $i_a(W_\mu) = -n$.

Proof. Suppose μ is an invertible measure in $\mathbf{M}(\mathbf{R})$ and that we have factored $\mu = \varrho^n * e^\nu * \delta_c$ for some measure ν in $\mathbf{M}(\mathbf{R})$. If we assume that $n \geq 0$ and $c \leq 0$, then we can write $W_\mu = W_{\delta_c} W_{e^\nu} W_{\varrho^n}$ by Lemma B. Moreover since W_{e^ν} is invertible by Lemma E and W_{ϱ^n} is a Fredholm operator by Lemma C we see that W_μ is a Fredholm operator if and only if W_{δ_c} is and hence if and only if $c = 0$ by Lemma D. Moreover, since $i_a(W_\mu) = i_a(W_{\delta_c}) + i_a(W_{e^\nu}) + i_a(W_{\varrho^n})$ and the first two terms of the right hand side are 0 we see that $i_a(W_\mu) = -n$. in this case by Lemma C. Similar arguments are valid for other values of n and c.

Thus to complete the proof of the theorem we must show that the assumption that W_μ is a Fredholm operator implies that μ is an invertible measure in $\mathbf{M}(\mathbf{R})$. By multiplying on the right hand side of W_μ by a factor W_{ϱ^n} for sufficiently large n, we can reduce to the case where W_μ is left invertible. Thus assume that W_μ is a left invertible operator on $L^1(\mathbf{R}^+)$. Let F_μ denote the convolution operator defined on $L^1(\mathbf{R})$ by μ and T_c denote the operator translation by c on $L^1(\mathbf{R})$. If f is a function in $L^1(\mathbf{R})$ which vanishes on $(-\infty, 0)$, then $\| F_\mu f \| \geq \| W_\mu f \| \geq \varepsilon \| f \|$, where $\frac{1}{\varepsilon}$ is the norm of a left inverse for W_μ. Moreover, since F_μ commutes with T_c and T_c is an isometry, if f is supported on $[-c, \infty)$, then $\| F_\mu f \| = \| T_c F_\mu f \| = \| F_\mu (T_c f) \| \geq \varepsilon \| T_c f \| = \varepsilon \| f \|$. Finally, since the functions in $L^1(\mathbf{R})$ which are supported on some interval $[-c, \infty)$ are norm dense in $L^1(\mathbf{R})$, it follows that $\| F_\mu f \| \geq \varepsilon \| f \|$ for f in $L^1(\mathbf{R})$ and hence the range of F_μ is closed in $L^1(\mathbf{R})$.

Since the range of F_μ is translation invariant it will be all of $L^1(\mathbf{R})$ by the Wiener Tauberian Theorem if we knew that the Fourier-Stieltjes transform of μ didn't vanish; this follows in our situation from a result analogous to that in [8, Theorem 2.6.3]. Summarizing, we have shown that the operator F_μ is bounded below and is onto, and hence is invertible. Since the inverse F_μ^{-1} commutes with the translation operators, it now follows from a result due to Wendel and Helson [8, p. 75], that there exists a measure ω in $\mathbf{M}(\mathbf{R})$ such that $F_\mu^{-1} = F_\omega$. But ω must be the inverse of μ and the result is proved.

The assertion " W_μ invertible on $L^p(\mathbf{R}^+)$ implies μ is invertible in $\mathbf{M}(\mathbf{R})$" is false for $p = 2$ as we show later with an example. The proof just given breaks down in the last paragraph since the statement for operators on $L^p(\mathbf{R})$ that commute with the translation operators is false for $1 < p < \infty$. However, since the first paragraph does not use any results special to the case $p = 1$, we obtain the following.

Theorem 2. *If* μ *is an invertible measure in* $\mathbf{M}(\mathbf{R})$, *then* W_μ *is a Fredholm operator on* $L^p(\mathbf{R}^+)$ *for* $1 \leq p < \infty$ *if and only if* $i_t(\mu)$ *has the form* $(n, 0)$ *for some integer* n. *Moreover, in this case* $i_a(W_\mu) = -n$.

Lastly, we consider the problem of replacing the hypotheses of these theorems by conditions which can be verified. We do not consider that aspect of the question which concerns determining when a measure is invertible but assume this problem has somehow been solved. We shall not give detailed proofs since they are easy involving only the Fourier-Stieltjes transform.

Theorem 3. *If* μ *is an invertible measure on* $\mathbf{M}(\mathbf{R})$, *then* W_μ *is a Fredholm operator on* $L^p(\mathbf{R}^+)$ *for* $1 \leq p < \infty$ *if and only if a logarithm of the Fourier-Stieltjes transform of* μ *is bounded.*

Proof. Express μ in the form $\delta_c * \varrho^n * e^\nu$ and compute the Fourier-Stieltjes transform of this product.

If one wants to explicitly compute the factors we can give a procedure for measures belonging to a certain dense subset of $\mathbf{M}(\mathbf{R})$.

Let $\mathbf{M}_1(\mathbf{R})$ denote the measures μ in $\mathbf{M}(\mathbf{R})$ whose Fourier-Stieltjes transform $\hat{\mu}$ is in $C^1(\mathbf{R})$. If for μ in $\mathbf{M}_1(\mathbf{R})$ we define μ' such that $d\mu'(x) = ix\,d\mu(x)$, then $(\mu')^\wedge = \hat{\mu}'$. For μ invertible in $\mathbf{M}_1(\mathbf{R})$ define $\ell(\mu) = \mu' * \mu^{-1}$. With a little work one can prove

Theorem 4. *If* μ *is an invertible measure in* $M_1(\mathbf{R})$, $\mu = \varrho^n * e^{\omega} * \delta_c$, *and* $\nu = \varrho^n * e^{\omega}$, *then* $c = i\ell(\mu)(\{0\})$, $n = i \lim_{n\to\infty} \frac{1}{2\nu} \ell(\nu)([-r, r])$, *and* $d\mu(x) = \frac{i}{x} d\ell(e^{\omega})(x) + k\, d\delta_0(x)$ *for some complex* k .

With this result it is possible, knowing both μ' and μ^{-1} , to compute n, c and ω .

We conclude with a few comments on some questions raised by our results.

Firstly, we want to exhibit an example of a non-invertible measure μ for which W_μ is invertible on $L^2(\mathbf{R}^+)$. Since the closure of $\mathbf{R}$ does not contain the Šilov boundary for $M(\mathbf{R})$ (cf. [9]), it is possible to choose a measure ν supported on $[1,\infty)$ such that $\|\hat{\nu}\|_\infty$ is less than the spectral radius of ν . The Laplace transform extends $\hat{\nu}$ analytically to the upper half plane and $\lim_{\operatorname{Im} z\to\infty} \hat{\nu}(z) = 0$. Hence, we have $|\hat{\nu}(z)| \leq \|\hat{\nu}\|_\infty$ for z in the closed upper half plane by the maximum modulus principle. If λ is now chosen in the spectrum of ν such that $|\lambda| > \|\hat{\nu}\|_\infty$, then the measure $\mu = \lambda\delta_0 - \nu$ fails to be invertible in $M(\mathbf{R})$ but W_μ is invertible on $L^2(\mathbf{R}^+)$ since both $\hat{\mu}(z)$, $1/\hat{\mu}(z)$ are bounded and analytic in the upper half plane.

Secondly, if $B(\mathbf{R})$ denotes the closure in $L^\infty(\mathbf{R})$ of the Fourier-Stieltjes transforms of the measures in $M(\mathbf{R})$ and μ is a measure in $M(\mathbf{R})$ for which $\hat{\mu} = e^f$ for some f in $B(\mathbf{R})$, then W_μ is invertible on $L^2(\mathbf{R}^+)$. If $f = \nu_1 + \nu_2 + g$, where ν_1 and ν_2 are measures supported on $[0,\infty)$ and $(-\infty, 0]$, respectively, and g is a function in $B(\mathbf{R})$ for which $\|g\| < 1/2$, then $W_\mu = W_{e^{\nu_2}} W_{e^g} W_{e^{\nu_1}}$ with all three factors invertible. Whether this condition is also necessary is unknown and a proof probably depends on having a concrete representation for the abstract index group of $B(\mathbf{R})$ such as given for $M(\mathbf{R})$ in Theorem A. A similar sufficient condition could be given for the invertibility of W_μ on $L^p(\mathbf{R}^+)$.

Thirdly, we observe that the operators W_μ fit into the context explored by Douglas and Sarason in [5]. In particular, if μ is supported on $[a,\infty)$ for $a<0$, then $\mu(t) = e^{iat}(e^{iat}\mu(t))$ and the second factor has a bounded analytic extension to the upper half plane.

Fourthly, the analogue of the theorem due to Coburn [1], that a Wiener-Hopf operator can not have both a nontrivial kernel and cokernel, is undoubtedly true for the operators which we are investigating; we have, however, not checked the details of the proof which would probably involve the distribution Fourier transform.

Lastly, we have not considered the question of extending the index theorem of [3] to cover this class of operators. In fact, it seems quite hard to extend these results to the class of operators on $L^2(\mathbf{R}^+)$ arising from measures having no continuous singular part.

REFERENCES

[1] L.A. Coburn, Weyl's Theorem for non normal operators, *Michigan Math. J.*, 13 (1966), 285-286.

[2] L.A. Coburn and R.G. Douglas, Translation operators on the half-line, *Proc. Nat. Acad. Sci. U.S.A.*, 62 (1969), 1010-1013.

[3] L.A. Coburn, R.G. Douglas, D.G. Schaeffer and I.M. Singer, On C^*-algebras of operators on a half-space. II: Index Theory.

[4] R.G. Douglas, On the spectrum of Toeplitz and Wiener-Hopf Operators, ISNM vol. 10 (Basel, 1969).

[5] R.G. Douglas and D.E. Sarason, A Class of Toeplitz Operators, *Indiana J. Math.*, (1971).

[6] I.C. Gohberg and I.A. Feldman, On Wiener-Hopf integral-difference equations, *Dokl. Akad. Nauk USSR*, 183 (1968), 25-28; *Soviet Math. Dokl.*, 9 (1968), 1312-1316.

[7] M.G. Kreĭn, Integral equations on half line with kernel depending upon the difference of the argument, *Uspekhi Mat. Nauk*, 13 (1958), 1-120.

[8] W. Rudin, *Fourier Analysis on Groups* (New York, 1962).

[9] J.L. Taylor, The Šilov boundary of the algebra of measures on a group, *Proc. Amer. Math. Soc.*, 16 (1965), 941-945.

[10] J.L. Taylor, The cohomology of the spectrum of a measure algebra, to appear.

[11] N. Wiener und E. Hopf, Über eine Klasse singulärer Integral-gleichungen, *S.–B. Preuss Akad. Wiss. Berlin.* Phys.–Math. Kl., 30/32 (1931), 696-706.

COLLOQUIA MATHEMATICA SOCIETATIS JÁNOS BOLYAI
5. HILBERT SPACE OPERATORS, TIHANY (HUNGARY), 1970

A generalization of Schäffer's matrix

E. DURSZT

1. Let T be a linear operator on a Hilbert space H and let $\varrho > 0$. If U is a unitary operator on a Hilbert space $K(\supset H)$ such that

(1) $$T^n x = \varrho P U^n x \qquad (x \in H;\ n = 1, 2, \dots),$$

then U is called a unitary ϱ-dilation of T (cf. [2], ch. I). P always denotes the orthoprojection of K onto H. U is *minimal*, if

(2) $$K = \bigvee_{n=-\infty}^{\infty} U^n H .$$

The minimal U is uniquely determined up to isomorphism. The proof is similar to that given for $\varrho = 1$ in ch. I of [2].

In the special case $\varrho = 1$, J.J. Schäffer [1] has given a matrix of operator entries, which is a representation of a not necessarily minimal U. For a representation of a minimal U see e.g. [2], ch. I.

Our purpose is to generalise this result for $\varrho > 0$.

It is a pleasure to thank C. Foiaş and A. Rácz for the conversations which inspired this investigation.

2. As the case $\varrho = 1$ is well known, we shall assume that $\varrho \neq 1$; r will stand for $\frac{1}{\varrho}$. (1) implies that $\|T\| \leqq \varrho$, consequently $0 < \min\{1, (\varrho-1)^2\} I$ is a lower bound for both $I+(1-2r)T^*T$ and $I+(1-2r)TT^*$. Thus the self-adjoint operators

$$D_T = (I+(1-2r)T^*T)^{\frac{1}{2}} \quad \text{and} \quad D_{T*} = (I+(1-2r)TT^*)^{\frac{1}{2}}$$

exist and have bounded inverses.

We define

$$K' = \bigoplus_{n=-\infty}^{\infty} H_n ,$$

where every H_n is a copy of H. The (i,j)-th entry of each operator-matrix defined below maps the j-th direct summand in K' into the i-th one. The elements of K' will be denoted by $\oplus x_n$, where x_n is an element of the n-th direct summand. Using the notations

$$A = (r-1)D_T^{-1}TD_T^{-1}, \qquad B = (r-1)D_{T*}^{-1}TD_{T*}^{-1}, \qquad C = (1-r)D_{T*}^{-1}T^2D_T^{-1}$$

we introduce the matrix

$$M = \begin{pmatrix} \ddots & \ddots & \ddots & & & & & \\ \ddots & I & A^* & 0 & & & & \\ \ddots & A & I & A^* & 0 & & & \\ \ddots & 0 & A & I & A^*D_T & C^* & & \\ & & 0 & D_TA & \boxed{I} & D_{T*}B^* & 0 & \\ & & & C & BD_{T*} & I & B^* & 0 \; \ddots \\ & & & & 0 & B & I & B^* \; \ddots \\ & & & & & 0 & B & I \; \ddots \\ & & & & & & \ddots & \ddots \end{pmatrix}$$

Clearly M is a bounded operator on K'.

We shall prove that M is (not necessarily strictly) positive (Th. 1). Then we define K as the factor space of K' with respect to the positive semi-definite bilinear form

$$[x,y] = (Mx,y),$$

and denote by $\dot{\oplus} x_n$ the element of K represented by $\oplus x_n$. We identify the element x of H with $\dot{\oplus} x_n$, where $x_0 = x$ and $x_n = 0$ if $n \neq 0$. This identification obviously preserves linearity and metric. $\dot{\oplus} x_n$ is orthogonal to H if and only if for every $y_0 \in H$ and $y_n = 0 \quad (n \neq 0)$.

$$0 = [\oplus x_n, \oplus y_n] = (M(\oplus x_n), \oplus y_n) =$$

$$= (r-1)(TD_T^{-1} x_{-1} + x_0 + (r-1)T^*D_{T^*}^{-1} x_1, y_0),$$

i.e. if $\quad x_0 = (1-r)(TD_T^{-1} x_{-1} + T^*D_{T^*}^{-1} x_1)$. Thus

$$(I-P)(\dot{\oplus} x_n) = \dot{\oplus} z_n,$$

where

$z_0 = (1-r)(TD_T^{-1} x_{-1} + T^*D_{T^*}^{-1} x_1), \quad z_n = x_n$ if $n \neq 0$ and

$$(3) \quad \begin{cases} P(\dot{\oplus} x_n) = \dot{\oplus} y_n \\ y_0 = x_0 + (r-1)TD_T^{-1} x_{-1} + T^*D_{T^*}^{-1} x_1, \quad y_n = 0 \quad n \neq 0 . \end{cases}$$

We define a matrix of operator entries as follows

$$V = \begin{pmatrix} \ddots & \ddots & \ddots & & & \\ 0 & I & 0 & & & \\ & 0 & D_T & -T^* & & \\ & & \boxed{T} & (I-TT^*)D_{T^*}^{-1} & 0 & \\ & & & 0 & I & \ddots \\ & & & & 0 & \ddots \end{pmatrix}$$

Clearly V maps K' into itself; moreover,

$$V' = \begin{pmatrix} \ddots & & & & & \\ \ddots & 0 & & & & \\ \ddots & I & 0 & & & \\ \ddots & 0 & D_T^{-1}(I-T^*T) & \boxed{T^*} & & \\ & & -T & D_{T^*} & 0 & \\ & & & 0 & I & 0 \\ & & & \ddots & \ddots & \ddots \end{pmatrix}$$

is the inverse of V on K'. In order to verify this statement, we use the relations

$$(4) \quad \begin{cases} TD_T = D_{T^*}T, & TD_T^{-1} = D_{T^*}^{-1}T, \\ T^*D_{T^*} = D_T T^*, & T^*D_{T^*}^{-1} = D_T^{-1}T^*, \end{cases}$$

which are valid because both D_T and D_T^{-1} are limits of polynomials of T^*T, and similarly, D_{T^*} and $D_{T^*}^{-1}$ are limits of polynomials of TT^*. So V induces an invertible operator U on K. We shall prove that U is isometric and satisfies (1) and (2), i.e. U is a minimal unitary ϱ-dilation of T. (Th. 2.)

Using the notation $K_1 = \bigoplus_{n=-\infty}^{-1} \overline{D_T H_n} \oplus H_0 \bigoplus_{n=1}^{\infty} \overline{D_{T^*} H_n}$, we have $K' = K_1$ if $\varrho \neq 1$. In case $\varrho = 1$, K_1 is the space of the minimal unitary dilation (cf. [2], Ch. I). Moreover, if $\varrho = 1$ then D_{T^*} has a not necessarily unique and bounded inverse on $D_{T^*} H$, and hence the closure of $(I-TT^*)D_{T^*}^{-1}$ equals equals D_{T^*} on $\overline{D_{T^*}H}$. Thus V equals the minimal Schaffer matrix on K_1.

3. No we are going to prove the theorems mentioned above.

Theorem 1. *If* $\varrho \neq 1$ *and* T *has any unitary* ϱ*-dilation, then* $M \geqq 0$ *on* K'.

Proof. Let U be an arbitrary unitary ϱ-dilation of T, let $x, y \in H$; and let n and k be two integers. If $n \geqq k+2$, then

$$(U^n(U-T)x,\ U^k(U-T)y) =$$

$$= (PU^{n-k}x,y) - (PU^{n-k-1}Tx,y) - (PU^{n-k+1}x,Ty) + (PU^{n-k}Tx,Ty) =$$

$$= r(T^{n-k}x,y) - r(T^{n-k}x,y) - r(T^{n-k+1}x,Ty) + r(T^{n-k+1}x,Ty) = 0.$$

Similar computation can be done if $k \geqq n+2$. So we have:

$$(U^n(U-T)x,\ U^k(U-T)y) = 0 \quad \text{if} \quad |n-k| \geqq 2.$$

Using the fact, that (1) implies

$$T^{*n}x = \varrho PU^{*n}x \qquad (x \in H, \quad n = 1,2,\dots),$$

we get:

$$(U^n(U^*-T^*)x,\ U^k(U^*-T^*)y) = 0 \qquad \text{if} \qquad |n-k| \geqq 2,$$

$$(U^n(U-T)x,\ U^k(U^*-T^*)y) = 0 \qquad \text{if} \qquad |n-k| \geqq 1,$$

$$(U^n(U-T)x, y) = 0 \qquad \text{if} \qquad n \neq 0,$$

$$(U^n(U^*-T^*)x, y) = 0 \qquad \text{if} \qquad n \neq 0,$$

$$(U(U-T)x, (U-T)y) = (r-1)(Tx, y),$$

$$((U-T)x, y) = (r-1)(Tx, y),$$

$$((U-T)x, (U-T)y) = (D_T^2 x, y),$$

$$(U^{-1}(U^*-T^*)x, (U^*-T^*)y) = (r-1)(T^*x, y),$$

$$((U^*-T^*)x, y) = (r-1)(T^*x, y),$$

$$((U^*-T^*)x, (U^*-T^*)y) = (D_{T^*}^2 x, y), \text{ and}$$

$$((U-T)x, (U^*-T^*)y) = (1-r)(T^2x, y),$$

Let $\oplus\, x_n$ be an arbitrary element of K_r' and let us introduce the notations:

$$y_n = D_T x_n \quad (n<0), \qquad y_0 = x_0, \qquad y_n = D_{T^*} x_n \quad (n>0).$$

Now we have

$$0 \leqq \left\| \sum_{n=\infty}^{1} U^{n-1}(U-T)y_{-n} + y_0 + \sum_{n=1}^{\infty} U^{*n-1}(U^*-T^*)y_n \right\|^2 =$$

$$= \sum_{n=-\infty}^{-2} \left(U(U-T)y_{n-1} + (U-T)y_n + U^{-1}(U-T)y_{n+1}, (U-T)y_n\right) +$$

$$+ \left(U(U-T)y_{-2} + (U-T)y_{-1} + y_0 + (U^*-T^*)y_1, (U-T)y_{-1}\right) +$$

$$+ \left((U-T)y_{-1} + y_0 + (U^*-T^*)y_1, y_0\right) +$$

$$+ \left((U-T)y_{-1} + y_0 + (U^*-T^*)y_1 + U^{-1}(U^*-T^*)y_2, (U^*-T^*)y_1\right) +$$

$$+ \sum_{n=2}^{\infty} \left(U(U^*-T^*)y_{n-1} + (U^*-T^*)y_n + U^{-1}(U^*-T^*)y_{n+1}, (U^*-T^*)y_n\right) =$$

$$= \sum_{n=-\infty}^{-2} \left((r-1)Ty_{n-1} + D_T^2 y_n + (r-1)T^* y_{n+1}, y_n\right) +$$

$$+ \left((r-1)Ty_{-2} + D_T^2 y_{-1} + (r-1)Tx_0 + (1-r)T^{*2} y_1, y_{-1}\right) +$$

$$+ \left((r-1)Ty_{-1} + y_0 + (r-1)T^* y_1, y_0\right) +$$

$$+ \left((1-r)Ty_{-1} + (r-1)Ty_0 + D_{T^*}^2 y_1 + (r-1)T^* y_2, y_1\right) +$$

$$+ \sum_{n=2}^{\infty} \left((r-1)Ty_{n-1} + D_{T^*}^2 y_n + (r-1)T^* y_{n+1}, y_n\right) =$$

$$= \sum_{n=-\infty}^{-2} ((r-1)D_T^{-1}TD_T^{-1}x_{n-1} + x_n + (r-1)D_T^{-1}T^*D_T^{-1}x_{n+1}, x_n) +$$

$$+ ((r-1)D_T^{-1}TD_T^{-1}x_{-2} + x_{-1} + (r-1)D_T^{-1}T^*x_0 + (1-r)D_T^{-1}T^{*2}D_T^{-1}x_1, x_{-1}) +$$

$$+ ((r-1)TD_T^{-1}x_{-1} + x_0 + (r-1)T^*D_{T^*}^{-1}x_1, x_0) +$$

$$+ ((1-r)D_{T^*}^{-1}T^2D_T^{-1}x_{-1} + (r-1)D_{T^*}^{-1}Tx_0 + x_1 +$$

$$+ (r-1)D_{T^*}^{-1}T^*D_{T^*}^{-1}T^*D_{T^*}^{-1}x_0, x_1) +$$

$$+ \sum_{n=2}^{\infty} ((r-1)D_{T^*}^{-1}TD_{T^*}^{-1}x_{n-1} + x_n + (r-1)D_{T^*}^{-1}T^*D_{T^*}^{-1}x_{n+1}, x_n) =$$

$$= (M(\oplus x_n), \oplus x_n).$$

So we have $M \geqq 0$ on K'.

We prove by an example that M is not necessarily strictly positive on K'.

Let us consider a Hilbert space spanned by the orthonormal basis: $\{z_n\}_{n=-\infty}^{\infty}$ and let H be the subspace spanned by z_{-1} and z_0. The unitary operator defined by $Uz_n = z_{n-1}$ is a unitary ϱ-dilation of the operator T defined on H by $Tz_{-1} = 0$ and $Tz_0 = \varrho z_{-1}$. Clearly $Tz_{-1} = \varrho z_0$ and $D_T z_0 = |\varrho - 1| z_0$. Define y_n as follows:

$$y_0 = (\varrho - 1) z_{-1}, \quad y_{-1} = D_T z_0 \qquad \text{and} \quad y_n = 0 \quad \text{if} \quad n \neq 0, -1.$$

Now $\oplus y_n \in K'$ and we have:

$$(M(\oplus y_n), \oplus y_n) = (y_{-1} + (r-1)D_T^{-1}T^*y_0, y_{-1}) + ((r-1)TD^{-1}y_{-1} + y_0, y_0) =$$

$$= \| y_{-1} \|^2 + 2(r-1) \operatorname{Re} (D_T^{-1} y_{-1}, T^* y_0) + \| y_0 \|^2 =$$

$$= (\varrho-1)^2 \| z_0 \|^2 + 2(r-1) \operatorname{Re} (z_0, (\varrho-1) z_0) + (\varrho-1)^2 \| z_0 \|^2 = 0 .$$

Theorem 2. *If* $M \geqq 0$ *on* K', *then the operator* U *induced by* V *on* K *is a minimal unitary* ϱ*-dilation of* T.

Proof. Let x be arbitrary element of H. We have already identified x with the element $\dot{\oplus} x_k$ of K, where $x_0 = x$ and $x_k = 0$ if $k \neq 0$. The definition of U shows that, for n 1,2,... , $U^n \dot{\oplus} x_k = \dot{\oplus} y_k$, where

$$y_k = \begin{cases} 0 & \text{if} \quad k > 0, \\ T^n x & \text{if} \quad k = 0, \\ D_T T^{n+k} x & \text{if} \quad 0 > k \geqq -n, \\ 0 & \text{if} \quad k < -n . \end{cases}$$

This fact and (3) imply that

$$P U^n x = r T^n x \qquad (x \in H, \quad n = 1, 2, \dots),$$

i.e. (1) is fulfilled.

Now let us consider $\dot{\oplus} x_n$, where $x_\nu = x \in H$ for a given $\nu \neq 0$ and $x_n = 0$ otherwise. Using the fact that V induces U and V' induces U^{-1} on K, an easy computation shows that

$$\dot{\oplus} x_n = \begin{cases} U^{-\nu-1} (U - T) D_T^{-1} x & \text{if} \quad \nu < 0, \\ U^{-\nu+1} (U^{-1} - T^*) D_{T^*}^{-1} x & \text{if} \quad \nu > 0 . \end{cases}$$

Thus (2) is fulfilled.

Finally we have to verify that U is an isometric operator on K. For this purpose it suffices to prove that V is an M-isometry on K'. Now let $\oplus x_n$ be an arbitrary element of K', then $V(\oplus x_n) = \oplus y_n$, where

$$y_{-1} = D_T x_0 - T^* x_1, \quad y_0 = Tx_0 + (I - TT^*) D_{T^*}^{-1} x_1, \quad y_n = x_{n+1}$$

if $n \neq -1, 0$.

Thus using (4) we get:

$$(M(\oplus y_n), \oplus y_n) =$$

$$= \sum_{n=-\infty}^{-3} ((r-1) D_T^{-1} T D_T^{-1} x_n + x_{n+1} + (r-1) D_T^{-1} T^* D_T^{-1} x_{n+2}, x_{n+1}) +$$

$$+ ((r-1) D_T^{-1} T D_T^{-1} x_{-2} + x_{-1} + (r-1) D_T^{-1} T^* D_T^{-1} (D_T x_0 - T^* x_1, x_{-1}) +$$

$$+ ((r-1) D_T^{-1} T D_T^{-1} x_{-1} + D_T x_0 - T^* x_1 + (r-1) D_T T^* (Tx_0 + I - TT^*) D_{T^*}^{-1} x_1 +$$

$$+ (1-r) D_T^{-1} T^{*2} D_{T^*}^{-1} x_2, D_T x_0 - T^* x_1) +$$

$$+ ((r-1) T D_T^{-1} (D_T x_0 - T^* x_1) + Tx_0 + (I - TT^*) D_{T^*}^{-1} x_1 +$$

$$+ (r-1) T^* D_{T^*}^{-1} x_2, Tx_0 + (I - TT^*) D_{T^*}^{-1} x_1) +$$

$$+ ((1-r) D_{T^*}^{-1} T^2 D_T^{-1} (D_T x - T^* x_1) + (r-1) D_{T^*}^{-1} T (Tx_0 + (I - TT^*) D_{T^*}^{-1} x_1 +$$

$$+ x_2 + (r-1) D_{T^*}^{-1} T^* D_{T^*}^{-1} x_3, x_2) +$$

$$+ \sum_{n=2}^{\infty} ((r-1) D_{T^*}^{-1} T D_{T^*}^{-1} x_n + x_{n+1} + (r-1) D_{T^*}^{-1} T^* D_{T^*}^{-1} x_{n+2}, x_{n+1}) =$$

$$= \sum_{m=-\infty}^{-2} ((r-1)D_T^{-1}TD_T^{-1}x_{m-1}+x_m+(r-1)D_T^{-1}T^*D_T^{-1}x_{m+1}, x_m)+$$

$$+((r-1)D_T^{-1}TD_T^{-1}x_{-2}+x_{-1}+(r-1)D_T^{-1}T^*x_0+(1-r)D_T^{-1}T^{*2}D_{T*}^{-1}x_1, x_{-1})+$$

$$+((r-1)TD_T^{-1}x_{-1}+x_0+(r-1)T^*D_{T*}^{-1}x_1, x_0)+$$

$$+((1-r)D_{T*}^{-1}T^2D_T^{-1}x_{-1}+(r-1)D_{T*}^{-1}Tx_0+x_1+(r-1)D_{T*}^{-1}TD_{T*}^{-1}x_2, x_1)+$$

$$+((r-1)D_{T*}^{-1}TD_{T*}^{-1}x_1+x_2+(r-1)D_{T*}^{-1}T^*D_{T*}^{-1}x_3, x_2)+$$

$$+\sum_{m=3}^{\infty}((r-1)D_{T*}^{-1}TD_{T*}^{-1}x_{m-1}+x_m+(r-1)D_{T*}^{-1}T^*D_{T*}^{-1}x_{m+1}, x_m) =$$

$$= (M(\oplus x_m), \oplus x_m).$$

So we have that V is M-isometric on K', and this completes the proof.

REFERENCES

[1] J.J. Schäffer, On unitary dilations of contractions, *Proc. Amer. Math. Soc.*, 6 (1955), 322.

[2] B. Sz.-Nagy – C. Foiaş, *Analyse harmonique des opérateurs de l'espace de Hilbert* (Budapest, 1967).

COLLOQUIA MATHEMATICA SOCIETATIS JÁNOS BOLYAI
5. HILBERT SPACE OPERATORS, TIHANY (HUNGARY), 1970

A classification of doubly cyclic operators

C. FOIAŞ

1. Let $\mathfrak{H}$ be a complex Hilbert space and let T be an operator on $\mathfrak{H}$.* A vector $f \in \mathfrak{H}$ is called cyclic for T if $T^n f$ $(n = 0,1,2,...)$ span $\mathfrak{H}$; it is called *doubly cyclic for* T if it is cyclic for both T and T^*. The operator T is called *doubly cyclic* if there exists a doubly cyclic vector for T. The interest of studying doubly cyclic operators lies in the fact that any unicellular operator is doubly cyclic (see [5], Cor. III. 7.8). Note also that the study of doubly cyclic operators can be restricted to that of the system $\{T, f\}$ formed by a completely non unitary (c.n.u.) contraction T and a vector f of norm 1 which is doubly cyclic for T. In the family $\mathbf{C}$ of these systems we define the following relation: Two systems $\{T, f\}$, $\{T_1, f_1\} \in \mathbf{C}$ will be called $\mathbf{C}$*-isomorphic* (notation: $\{T, f\} \approx \{T_1, f_1\}$) if there exists a linear injective closed transformation X with dense domain satisfying

* By an operator we will always mean a linear bounded everywhere defined transformation; the Hilbert spaces considered will be always complex.

(1) $T_1 X \subset X T$,

(2) $X f = f_1$, $X^* f_1 = f$.*

Since $\{T, f\} \approx \{T_1, f_1\}$ *if and only if*

(3) $(T^n f, f) = (T_1^n f_1, f_1)$ *for all integers* $n \geq 0$,

$\mathbf{C}$ *-isomorphism is an equivalence relation in* $\mathbf{C}$. Our main purpose is to give a concrete $\mathbf{C}$ -isomorphic model for any system of $\mathbf{C}$.**

2. Our models will involve the unilateral shift of multiplicity one in its (unitary) realization as multiplication operator with the independent variable in the numerical Hardy space H^2 (see [5], chap. II and III, or [3], Chap.3). Let us recall the reader that

$$f : z \longrightarrow f(z) = a_0 + a_1 z + a_2 z^2 + \dots$$

belongs to H^2 if and only if

$$\| f \|^2 = |a_0|^2 + |a_1|^2 + |a_2|^2 + \dots < \infty ,$$

and that the multiplication operator is given by $Sf(z) = zf(z)$ for $|z| < 1$ and $f \in H^2$. The functions $\varphi \in H^2$ which are cyclic for S have been characterized long time ago by A. Beurling [1] in the following way: $\varphi \in H^2$ is cyclic for S if and only if it is *outer*, that is if

$$\varphi(z) = k \cdot \exp\left(\int_0^{2\pi} \frac{e^{it} + z}{e^{it} - z} \log |\varphi(e^{it})| \frac{dt}{2\pi} \right) \qquad \text{for} \quad |z| < 1 ,$$

where k is a constant, $|k| = 1$. On the other hand, the characterization of those functions $\psi \in H^2$ which are cyclic for S^* is recent [2]: $\psi \in H^2$ is cyclic for S^* if and only if its boundary values on $\{e^{it} : 0 \leq t < 2\pi\}$ do not agree almost everywhere with those of any functions on u/v , where u and v are analytic bounded functions on $\{ z : |z| > 1 \}$. If ψ is not cyclic for S^* then

* Obviously the conditions are redundant but we preferred to list all the useful ones.

** This Note was prepared during the stay of the author at Indiana University in the summer quarter 1970; expressing his thanks to this University he also thanks his colleagues A. Brown and R.G. Douglas for fruitful discussions on the subject.

$$\bigvee_{n \geqq 0} (S^*)^n \psi = H^2 \ominus mH^2$$

with a suitable inner function m (i.e. $|m(e^{it})| = 1$ almost everywhere.* We shall denote by $\mathfrak{H}(m)$ this space and by $S(m)$ the compression of S to $\mathfrak{H}(m)$ i.e. $S(m) = (S^* | \mathfrak{H}(m))^*$. Any c.n.u. contraction T on a Hilbert space $\mathfrak{H}$ of class C_{00} (i.e. such that $T^n \longrightarrow 0$, $(T^*)^n \longrightarrow 0$ strongly) and of defect indices $= 1$ (i.e. $\dim(I-T^*T)\mathfrak{H} = 1 = \dim(I-T^*T)\mathfrak{H}$) is unitarily equivalent with a uniquely determined $S(m)$ (see [5], Chap. VI). It is easy to see that a function $\varphi \in \mathfrak{H}(m)$ is cyclic for $S(m)$ if and only if φ and m have no non constant inner common divisor; in particular any outer function $\varphi \in \mathfrak{H}(m)$ is cyclic for $S(m)$. Let us also remark that *no system* $\{S, \varphi\} \in \mathbf{C}$ *can be* $\mathbf{C}$*-isomorphic with some system* $\{S(m), \varphi\} \in \mathbf{C}$. Indeed, if this were the case we would deduce that $m(S) = 0$ (since $m(S(m)) = 0$; see [5], Sec. III. 4), which is impossible because $m(S)$ is isometric.

We shall conclude these preliminaries with an improved particular case of the main theorem of [6]:

Proposition 1. *For any* $\{T, f\} \in \mathbf{C}$ *there exists an operator* A *from* H^2 *into* $\mathfrak{H}$ (the Hilbert space on which T operates) *and an outer function* $\varphi \in H^2$ *of norm* 1, *such that*

(4) $AS = TA$,

(5) $\|A\| \leq 1$, $A\varphi = f$ *and* $\overline{AH^2} = \mathfrak{H}$.

We omit the proof of this proposition since it differs from that given in [6] only in the use of the structure of isometric operatore with cyclic vectors, well-known by specialists.

3. Our models determining the system of $\mathcal{L}$ up to a $\mathcal{L}$-isomorphism are given by the following theorem, which constitutes the main result of this Note.

Theorem. *Every system* $\{T, f\} \in \mathcal{L}$ *is* $\mathcal{L}$*-isomorphic to a system of* $\mathcal{L}$ *either of the form* $\{S, \varphi\}$ *or of the form* $\{S(m), \varphi\}$; *in both cases* φ *is a suitable outer function. For such systems*

* "Almost everywhere" will always be considered with respect to Lebesgue measure.

$$\{S,\varphi\} \approx \{S,\varphi_1\}, \quad \textit{resp.} \quad \{S(m),\varphi\} \approx \{S(m_1),\varphi_1\},$$

if and only if

$$\varphi = a\varphi_1, \quad \textit{resp.}\ \varphi = a\varphi_1 \ \textit{and} \quad m = bm_1,$$

where a *, resp.* a *and* b *, are constants of modulus* 1 .

Proof. If the operator A given by Proposition 1 is injective then to verify that $\{S,\varphi\} \approx \{T,f\}$ it remains only to determine whether $X = A$ satisfies the second condition (2) with φ and f in the role of f and f_1 ; indeed this relation plainly implies that φ is cyclic for S^* , thus doubly cyclic for S . To prove $A^*f = \varphi$ note first that $A\varphi = f$ and $\|\varphi\| = 1 = \|f\|$ imply $((I - A^*A)\varphi,\varphi) = 0$. Since $\|A\| = 1$ it results $(I - A^*A)\varphi = 0$, that is $A^*f = \varphi$. If the operator A is not injective then its kernel $\operatorname{Ker} A$, being invariant for S , is of the form mH^2 with an appropriate inner function m . Put $X = A|\mathfrak{H}(m)$. Obviously $\|X\| \leq 1$, X is injective and, for any $g \in \mathfrak{H}(m)$.

$$TXg = TAg = ASg = AS(m)g + A(S - S(m))g = AS(m)g$$

(since $Sg - S(m)g \in \operatorname{Ker} A$). Let now $\varphi = h + g$, where $h \in \mathfrak{H}(m)$, $g \in mH^2 = \operatorname{Ker} A$; then

$$1 = \|\varphi\| \geq \|h\| \geqq \|Xh\| = \|Ah\| = \|A\varphi\| = \|f\| = 1,$$

whence $\|\varphi\| = \|h\|$, therefore $\varphi = h \in \mathfrak{H}(m)$. The relation $X^*f = \varphi$ follows as the analogous one for A ; in this way we obtain that φ is also cyclic for $S(m)^*$ and that X realizes the **C**-isomorphism $\{S(m),\varphi\} \approx \{T,f\}$.

Let now $\{S,\varphi\} \approx \{S,\varphi_1\}$; then in virtue of (3) we have

$$(S^n\varphi,\varphi) = (S^n\varphi_1,\varphi_1) \quad \text{for all} \quad n = 0,1,2,\ldots,$$

so that

$$\int_0^{2\pi} (|\varphi(e^{it})|^2 - |\varphi_1(e^{it})|^2)\, e^{-int}\, \frac{dt}{2\pi} = 0 \quad \text{for} \quad n = 0,1,2,\ldots,$$

Clearly this implies

$$|\varphi(e^{it})| = |\varphi_1(e^{it})| \qquad \text{almost everywhere,}$$

so that, φ_1 and φ being outer, we must have $\varphi = a\varphi_1$ with a suitable constant a, $|a| = 1$. Let now $\{S(m), \varphi\} \approx \{S(m_1), \varphi_1\}$; then taking into account that for all $n = 0, 1, 2, \ldots$

$$(S(m)^n \varphi, \varphi) = (S^n \varphi, \varphi) \text{ and } (S(m_1)^n \varphi_1, \varphi_1) = (S^n \varphi_1, \varphi_1),$$

we can deduce, as above, that $\varphi = a\varphi_1$. On the other hand since

$$\mathfrak{H}(m) = \bigvee_{n \geqq 0} (S(m)^*)^n \varphi = \bigvee_{n \geqq 0} (S^*)^n \varphi =$$

$$= \bigvee_{n \geqq 0} (S^*)^n \varphi_1 = \bigvee_{n \geqq 0} (S(m_1)^*)^n \varphi_1 = \mathfrak{H}(m_1),$$

we have $mH^2 = m_1H^2$, whence $m = bm_1$ with a suitable constant b, $|b| = 1$. This finishes the proof.

Remarks. 1^0. An operator $S(m)$ can occur in the theorem if and only if T is an operator of class C_0 (for definition see [5], Sec. III. 4). In particular, if $\|T\| < 1$ and $\mathfrak{H}$ is of infinite dimension, then in the theorem will occur the operator S since T cannot be of class C_0 (see [6], Prop. 4).

2^0. The property of S and $S(m)$ given by the second statement of the theorem is not valid for all c.n.u. contractions.

Indeed, if $T = \frac{\pi}{2} V$, where V is the simplest Volterra operator, i.e. the operator defined on $L^2(0,1)$* by

$$Vf(x) = \int_0^{2\pi} f(t)\,dt \quad \text{for } x \in (0,1),$$

and

* This is the usual L^2 -space of all square Lebesgue integrable functions defined on (0,1).

$$f_0(x) = \sqrt{2}\cos\frac{\pi}{2}x, \quad f_1(x) = \sqrt{2}\sin\frac{\pi}{2}x \quad \text{for} \quad x \in (0,1),$$

then $\{T, f_0\}, \{T, f_1\} \in \mathbf{C}$ and $\{T, f_0\} \approx \{T, f_1\}$, the $\mathbf{C}$-isomorphism being realized by $X = \frac{\pi}{2} V$. Howerer f_0 and f_1 are not collinear.

3^0. It may seem that a more natural relation between operators T and T_1 would be the following: *There exists a linear closed injective transformation* X *with dense domain and dense range such that* $T_1 X \subset X T$. Obviously this relation is reflexive and symmetric; howerer *this relation is not transitive* as the following example shows:

Let

$$T_j = \frac{1}{3} S + (-1)^j \frac{1}{2} \qquad \text{for} \qquad j = 1, 2,$$

and let φ be any doubly cyclic vector (in fact function) for S, of norm 1. Then $\{T_j, \varphi\} \in \mathbf{C}$ for $j = 1, 2$, and since $\|T_j\| < 1$, the model given by the theorem for $\{T_j, \varphi\}$ will involve the operator S, in both cases $j = 1, 2$. In this manner, T_1 and T_2 are "equivalent" (in the above underlined sense) to S. If this "equivalence" were transitive then there would exist a closed injective transformation X with dense domain $\mathcal{D}_X$ and dense range $X\mathcal{D}_X$ such that $XT_1 \subset T_2 X$. Consider $T = T_1 \oplus T_2$. The spectrum $\sigma(T)$ is the union of the disjoint disks $\{z : |z + (-1)^j \frac{1}{2}| \leq \frac{1}{3}\}$, $j = 1, 2$. The graph $\mathcal{G}_X = \{h \oplus Xh : h \in \mathcal{D}_X\}$ is invariant to T and the resolvent set $\varrho(T)$ of T is connected and containing the circle $\Gamma = \{z : |z + \frac{1}{2}| = \frac{1}{2}\}$, we deduce

$$\left(\frac{1}{2\pi i} \int_\Gamma (z - T)^{-1} dz\right) \mathcal{G}_X \subset \mathcal{G}_X. \tag{6}$$

But the definition of T yields

$$\left(\frac{1}{2\pi i} \int_\Gamma (z - T)^{-1} dz\right)(h_1 \oplus h_2) = h_1 \oplus 0,$$

so that, by (6), we infer that

$$\{h \oplus 0 : h \in \mathcal{D}_X\} \subset \mathcal{G}_X$$

contradicting the injectivity of X .*

REFERENCES

[1] A. Beurling, On two problems concerning linear transformations in Hilbert space, *Acta Math.*, 81 (1949), 239-255.

[2] R.G. Douglas – H.S. Shapiro – A.S. Shields, Cyclic vectors and invariant subspaces for the backward shift operator, *Ann. Inst. Fourier,* 20 (1970), 37-76.

[3] P.R. Halmos, *A Hilbert Space Problem Book* (Princeton, 1967).

[4] G.W. Mackey, Infinite-dimensional group representations, *Bull. Amer. Math. Soc.,* 69 (1963), 628-686.

[5] B. Sz.-Nagy – C. Foiaş, *Analyse harmonique des opérateurs de l'espace de Hilbert* (Budapest, 1967).

[6] B. Sz.-Nagy – C. Foiaş, Vecteurs cycliques et quasi-affinités, *Studia Math.,* 31 (1968), 35-42.

*Considering the group representations $\{\exp(tT_j : -\infty < t < \infty\}$, $j = 1,2$, one can connect Remark 3 to a question raised in [4], Sec. 8.

COLLOQUIA MATHEMATICA SOCIETATIS JÁNOS BOLYAI
5. HILBERT SPACE OPERATORS. TIHANY (HUNGARY), 1970

Problems of linear algebra and classification of quadruples of subspaces in a finite-dimensional vector space

I. M. GELFAND and V. A. PONOMAREV

§ 1. BASIC DEFINITIONS AND STATEMENT OF RESULT

One of the natural problems of linear algebra is the classification of systems of subspaces in a finite-dimensional vector space. Many problems of linear algebra can be reduced to the classification of systems. We shall see that problems as bringing a linear transformation to a canonical form, finding a complete system of invariants as well as the canonical form of a pair of linear mappings of a space E into F, the classification of linear relations, and a number of others are special cases of the theorem to be proved here.

1. Let P be a finite-dimensional vector space over an algebraically closed field K of characteristic 0. Further let $E_1, E_2, \dots, E_r$ be a finite set of its subspaces. For systems of this kind we shall write $S = \{P; E_1, \dots, E_r\}$. Let us consider two systems $S = \{P; E_1, \dots, E_r\}$ and $T = \{Q, F_1, \dots, F_r\}$. By a morphism α : $S \longrightarrow T$ of the system S into the system T we shall mean a linear mapping

$\alpha: P \longrightarrow Q$ such that $\alpha E_i \subseteq F_i$ for every i. Denote by C^r the category of all systems $S = \{P; E_1, \ldots, E_r\}$ with the same number of subspaces E_i. For any pair S, T of systems $\mathrm{Hom}(S,T)$ stands for the set of all morphisms $\alpha: S \longrightarrow T$. The category C^r is additive since for every S and T the set $\mathrm{Hom}(S,T)$ is an abelian group (more accurately, a finite-dimensional vector space). The category, however, is not abelian.

We shall say that the system S is decomposable if there is a family of subspaces P_t in the space P $(t = 1, 2, \ldots, k;\ k > 1,\ P_t \neq 0)$ such that $\sum_t P_t = P$, where the sum is a direct sum and, moreover, each subspace E_i is decomposable on the family $\{P_t\}$:

$$E_i = \sum_t (E_i \cap P_t).$$

In the case $r = 1$ two systems $S = \{P; E_1\}$ and $T = \{Q; F_1\}$ are isomorphic if $\dim P = \dim Q$ and $\dim E_1 = \dim F_1$. In C^1 there are only two mutually non-isomorphic non-decomposable systems: $S_0 = \{P; 0\}$ where $\dim P = 1$, and $S_1 = \{P; E\}$ where $\dim P = 1$, $E = P$. Every system $S \in C^1$ is isomorphic to the direct sum of a finite number of non-decomposable subsystems S_0 and S_1.

Likewise trivial is the classification of systems belonging to C^2.

Every non-decomposable system in C^2 is isomorphic to one of the 4 systems: $S_0 = \{P; 0, 0\}$, where $\dim P = 1$; $S_1 = \{P; E_1, 0\}$, where $\dim P = 1$, $E_1 = P$; $S_2 = \{P; 0, E_2\}$, where $\dim P = 1$, $E_2 = P$; and $S_{1,2} = \{P; E_1, E_2\}$, where $\dim P = 1$, $E_1 = E_2 = P$. Any system $S \in C^2$ is isomorphic to the direct sum of a finite number of non-decomposable systems $S_0, S_1, S_2, S_{1,2}$. The system S has 4 numbers as invariants: $n = \dim P$, $m_1 = \dim E_1$, $m_2 = \dim E_2$ and $m_{1,2} = \dim(E_1 \cap E_2)$. Two systems, S and S' are isomorphic if and only if their invariants coincide.

The systems $S \in C^3$ have 9 numbers as invariants: $m_{1,2,3} = \dim(E_1 \cap E_2 \cap E_3)$; $m_{i,j} = \dim(E_i \cap E_j)$ $(i < j)$; $\ell = \dim(E_1 \cap (E_2 + E_3))$; $m_i = \dim E_i$ and $n = \dim P$. In this case there are 9 non-isomorphic non-decomposable systems; in 8 of them $\dim P = 1$, whereas E_i equals either P or 0. Besides, there is one non-decomposable system $S = \{P; E_1, E_2, E_3\}$ with $\dim P = 2$, $\dim E_i = 1$, where the E_i are three non-intersecting subspaces.

It is interesting to note that the Dedekind lattice generated by a triple E_1, E_2, E_3 of subspaces is a free Dedekind lattice with 3 generators. This lattice is finite and consists of 28 elements, see Birkhoff [2].

Everything tends to show that the central problem is the classification of quadruples i.e. systems of the form $S=\{P; E_1, \ldots, E_4\}$. The main technical device for its solution is the introduction of certain functors Φ^+ and Φ^- which enable us to reduce the additive category C^4 to an abelian category. These functors are defined in the category C^r with any finite number of subspaces as well. Why the case $r = 4$ plays an exceptional role will turn out in course of the proofs (see p. 191) and also by considering the systems presented just below as examples.

2. Let us mention now some examples of systems belonging to C^4.

1) In the finite-dimensional space E there is given a linear transformation $A: E \longrightarrow E$. With the pair E, A we associate a system $S = \{P; E_1, \ldots, E_4\}$ in the following way. Put $P = E \oplus E$; $E_1 = E \oplus \{0\}$, i.e. E_1 consists of all pairs $(x, 0)$, $x \in E$; $E_2 = \{0\} \oplus E$, the subspace of all pairs $(0, y)$, $y \in E$; further let E_3 be the graph of the transformation A, i.e. the subspace of all pairs (x, Ax), $x \in E$; finally let E_4 be the diagonal of $E \oplus E$, i.e. $E_4 = \{(x, x) \mid x \in E\}$.

2) Let E and F be two finite-dimensional vector spaces. Consider two linear mappings $A: E \longrightarrow F$ and $B: F \longrightarrow E$. The solution of this problem of linear algebra does not involve substantial difficulties in comparison with the classification problem for one linear transformation. The solution can be found e.g. in [4]. The corresponding system is $S = \{P; E_1, \ldots, E_4\}$, where $P = E \oplus F$; $E_1 = E \oplus \{0\}$; $E_2 = \{0\} \oplus F$; $E_3 = \{(x, Ax) \mid x \in E\}$, the graph of the mapping A; $E_4 = \{(By, y) \mid y \in F\}$. More generally: $P = E \oplus F$, $E_i = E \oplus \{0\}$, $E_j = \{0\} \oplus F$, $E_k = \{(x, Ax) \mid x \in E\}$, $E_\ell = \{(By, y) \mid y \in F\}$ with $\{i, j, k, \ell\} = \{1, 2, 3, 4\}$.

3) E and F are finite-dimensional spaces, $A_1: E \longrightarrow F$ and $A_2: E \longrightarrow F$ two linear mappings. More frequently in the literature this problem is stated in a different form, namely as the problem of classification of matrix pencils. Its solution was obtained by Kronecker [1]. For the corresponding system S we have: $P = E \oplus F$; $E_1 = E \oplus \{0\}$; $E_2 = \{0\} \oplus F$; E_3 is the graph of A_1, i.e. $E_3 = \{(x, A_1 x) \mid x \in E)$; E_4 equals the graph of A_2.

4) On the space E there is given an additive relation A; in other words, A is an arbitrary subspace of the direct sum $E \oplus E$. The corresponding system is S: $P = E \oplus E$; $E_1 = E \oplus \{0\}$; $E_2 = \{0\} \oplus E$; $E_3 = A$; E_4 the diagonal of

$E \oplus E$. The concept of additive relation was introduced by MacLane [5]. Relations seem to be a natural tool in a number of algebraic problems. For an example of such applications we refer to the paper [3] of the authors, where using the technique of relations the classification of a pair of commuting linear transformations a, b satisfying $ab = ba = 0$ is given.

5) On a pair of space E, F there are given two additive relations $A_1 \subseteq E \oplus F$ and $A_2 \subseteq E \oplus F$. The corresponding quadruple S: $P = E \oplus F$; $E_i = E \oplus \{0\}$; $E_j = \{0\} \oplus F$; $E_k = A_1$; $E_\ell = A_2$; where $\{i, j, k, \ell\} = \{1, 2, 3, 4\}$. Note that this problem includes all the foregoing examples as special cases. The solution can be found in [6].

It is interesting to remark that there are quadruples $S = \{P; E_1, \ldots, E_4\}$ which cannot be reduced to a pair of relations (Example 5). Thus the classification problem for quadruples $S \in C^4$ generalizes many problems of linear algebra and is distinguished at the same time by the fact that it is more natural.

3. Now we are going to present the basic result of this paper: the canonical description of all non-decomposable quadruples. Important invariants of a non-decomposable quadruple $S = \{P; E_1, \ldots, E_4\}$ are the two numbers $n(S) = \dim P$ and the defect $\rho(S) = \sum_{i=1}^{4} \dim E_i - 2 \dim P$. In §3 we prove that if the quadruple S is non-decomposable then its defect $\rho(S)$ can assume only one of the 5 values $-2, -1, 0, 1, 2$. The defect is an essential characteristic of the system. While systems with defect 0 correspond to a pair of linear transformations $A: E \longrightarrow F$, $B: F \longrightarrow E$, (see Example 2), systems with defect ± 1 describe relations $A \subseteq E \oplus F$, $B \subseteq E \oplus F$ (see Example 5) that do not reduce to the type 2). As to the systems with defect ± 2, they cannot be described in classical terms.

First we write down the canonical form of non-decomposable quadruples for which $n(S) = \dim P$ is even: $n(S) = 2k$. These will be quadruples $S_{i,j}(2k, 0)$, $i, j \in \{1, 2, 3, 4\}$; $S(2k, 0; \lambda)$, $\lambda \in K$ with defect 0 ; and $S_i(2k, \rho)$, $i \in \{1, 2, 3, 4\}$, $\rho = \pm 1$ with defect ± 1 (the number $2k$ always denotes $n(S) = \dim P$, while the next one represents the defect $\rho(S)$ of the system S).

Let us define the systems $S_3(2k, -1)$, $S_3(2k, 1)$, $S_{1,3}(2k, 0)$ and $S(2k, 0; \lambda)$. In each of them P is a space with basis $\{e_1, \ldots, e_k, f_1, \ldots, f_k\}$, and the subspaces E_i are spanned, respectively, by the sets of vectors indicated below.

$S_3(2k,-1)$:

$P=\{e_1,\dots,e_k,f_1,\dots,f_k\}$,

$E_1=\{e_1,\dots,e_k\}$,

$E_2=\{f_1,\dots,f_k\}$,

$E_3=\{(e_2+f_1),\dots,(e_k+f_{k-1})\}$,

$E_4=\{(e_1+f_1),\dots,(e_k+f_k)\}$;

$S_3(2k,1)$:

$P=\{e_1,\dots,e_k,f_1,\dots,f_k\}$,

$E_1=\{e_1,\dots,e_k\}$,

$E_2=\{f_1,\dots,f_k\}$,

$E_3=\{e_1,(e_2+f_1),\dots,(e_k+f_{k-1}),f_k\}$,

$E_4=\{(e_1+f_1),\dots,(e_k+f_k)\}$;

$S_{1,3}(2k,0)$:

$P=\{e_1,\dots,e_k,f_1,\dots,f_k\}$,

$E_1=\{e_1,\dots,e_k\}$,

$E_2=\{f_1,\dots,f_k\}$,

$E_3=\{e_1,(e_2+f_1),\dots,(e_k+f_{k-1})\}$,

$E_4=\{(e_1+f_1),\dots,(e_k+f_k)\}$;

$S(2k,0;\lambda)$:

$P=\{e_1,\dots,e_k,f_1,\dots,f_k\}$,

$E_1=\{e_1,\dots,e_k\}$,

$E_2=\{f_1,\dots,f_k\}$,

$E_3=\{(e_1+\lambda f_1),(e_2+f_1+\lambda f_2),\dots,(e_k+f_{k-1}+\lambda f_k)\}$,

$E_4=\{(e_1+f_1),\dots,(e_k+f_k)\}$.

Every other system $S_i(2k,\varrho)$, $S_{i,j}(2k,0)$ can be obtained from the systems $S_3(2k,\varrho)$, $S_{1,3}(2k,0)$ by a suitable permutation of the indices of the subspaces E_i. Let σ be a permutation of the indices $1,2,3,4$. We shall denote by σS the system $\{P;E_1',\dots,E_4'\}$ obtained from $S=\{P;E_1,\dots,E_4\}$ by the following change in the indices of subspaces: $E_i'=E_{\sigma^{-1}(i)}$.

We put

$$S_i(2k,\varrho)=\sigma_{3,i}S_3(2k,\varrho),\qquad \varrho=-1,1,$$

$$S_{i,j}(2k,0)=\sigma_{1,i}\sigma_{3,j}S_{1,3}(2k,0),\qquad i,j\in\{1,2,3,4\},$$

where $\sigma_{i,j}$ denotes the permutation exchanging the indices i, j and leaving all other indices invariant.

We shall show that if P is a space over an algebraically closed field K, the system $S=\{P;E_1,\dots,E_4\}$ is non-decomposable and $\dim P=2k$, then S is isomorphic to one and only one of the systems $S_i(2k,-1)$, $S_i(2k,1)$,

$i = 1,2,3,4$; $S_{ij}(2k,0)$, where $i<j$, $i,j \in \{1,2,3,4\}$; $S(2k,0;\lambda)$, where $\lambda \in K$, $\lambda \neq 0$, $\lambda \neq 1$.

Hence, in particular, it will follow that for any permutation σ we have the equalities: 1) $\sigma S_i(2k,\varrho) \cong S_{\sigma(i)}(2k,\varrho)$; 2) $S_{i,j}(2k,0) \cong S_{j,i}(2k,0)$; 3) $\sigma S_{i,j}(2k,0) \cong S_{\sigma(i),\sigma(j)}(2k,0)$.

Next we give the canonical form of the non-decomposable quadruples $S_{i,j}(2k+1,0)$; $S_i(2k+1,\varrho)$, $\varrho = -1,1$; $S(2k+1,\varrho)$, $\varrho = -2,2$, for which $n(S) = \dim P$ is odd: $n(S) = 2k+1$.

Let us first define the systems $S(2k+1,-2)$, $S_1(2k+1,-1)$, $S_{1,3}(2k+1,0)$, $S_2(2k+1,1)$, $S(2k+1,2)$. In each of these systems the space P has the basis $\{e_1,\dots,e_{k+1}, f_1,\dots,f_k\}$, while the subspaces E_i are spanned, respectively, by the following sets of vectors in P.

$S_1(2k+1,-1)$:

$P = \{e_1,\dots,e_k,e_{k+1},f_1,\dots,f_k\}$,

$E_1 = \{e_1,\dots,e_k,e_{k+1}\}$,

$E_2 = \{f_1,\dots,f_k\}$,

$E_3 = \{(e_2+f_1),\dots,(e_{k+1}+f_k)\}$,

$E_4 = \{(e_1+f_1),\dots,(e_k+f_k)\}$;

$S_2(2k+1,1)$:

$P = \{e_1,\dots,e_k,e_{k+1},f_1,\dots,f_k\}$,

$E_1 = \{e_1,\dots,e_k,e_{k+1}\}$,

$E_2 = \{f_1,\dots,f_k\}$,

$E_3 = \{e_1,(e_2+f_1),\dots,(e_{k+1}+f_k)\}$,

$E_4 = \{(e_1+f_1),\dots,(e_k+f_k),e_{k+1}\}$;

$S_{1,3}(2k+1,0)$:

$P = \{e_1,\dots,e_k,e_{k+1},f_1,\dots,f_k\}$,

$E_1 = \{e_1,\dots,e_k,e_{k+1}\}$,

$E_2 = \{f_1,\dots,f_k\}$,

$E_3 = \{e_1(e_2+f_1),\dots,(e_{k+1}+f_k)\}$,

$E_4 = \{(e_1+f_1),\dots,(e_k+f_k)\}$;

$S(2k+1,-2)$:

$P = \{e_1,\dots,e_k,e_{k+1},f_1,\dots,f_k\}$,

$E_1 = \{e_1,\dots,e_k\}$,

$E_2 = \{f_1,\dots,f_k\}$,

$E_3 = \{(e_2+f_1),\dots,(e_{k+1}+f_k)\}$,

$E_4 = \{(e_1+f_2),\dots,(e_{k-1}+f_k),(e_k+e_{k+1})\}$;

$S(2k+1,2)$:

$$P = \{e_1, \dots, e_k, e_{k+1}, f_1, \dots, f_k\},$$
$$E_1 = \{e_1, \dots, e_k, e_{k+1}\},$$
$$E_2 = \{f_1, \dots, f_k, e_{k+1}\},$$
$$E_3 = \{e_1, (e_2+f_1), \dots, (e_{k+1}+f_k)\},$$
$$E_4 = \{f_1, (e_1+f_2), \dots, (e_{k-1}+f_k), (e_k+e_{k+1})\}.$$

We notice that in the system $S(2k+1,-2)$ the subspaces E_i do not intersect each other and have one and the same dimension k. In the system $S(2k+1,2)$ every subspace has dimension $k+1$ and for any pair (i,j) we have $\dim(E_i \cap E_j) = 1$.

Put

$$S_i(2k+1,-1) = \sigma_{1,i} S_1(2k+1,-1),$$
$$S_i(2k+1,1) = \sigma_{2,i} S_2(2k+1,1),$$
$$S_{i,j}(2k+1,0) = \sigma_{1,i}\sigma_{3,j} S_{1,3}(2k+1,0),$$
$$i,j \in \{1,2,3,4\}.$$

We shall show that if the system $S = \{P; E_1, \dots, E_4\}$ is non-decomposable and $\dim P = 2k+1$, then S is isomorphic to one and only one of the systems $S(2k+1,-2)$; $S_i(2k+1,-1)$; $S_{i,j}(2k+1,0)$, $i<j$; $S_i(2k+1,1)$; $S(2k+1,2)$, where $i,j \in \{1,2,3,4\}$.

Hence it follows in particular that for any permutation σ we have the equalities: 1) $\sigma S(2k+1,-2) \cong S(2k+1,-2)$; 2) $\sigma S_i(2k+1,\rho) \cong S_{\sigma(i)}(2k+1,\rho)$, $\rho = -1,1$; 3) $\sigma S(2k+1,2) \cong S(2k+1,2)$; 4) $\sigma S_{i,j}(2k+1,0) \cong S_{\sigma(i),\sigma(j)}(2k+1,0)$.

Main Theorem. *Let* $S = \{P; E_1, \dots, E_4\}$ *be a system consisting of a finite-dimensional space* P *over the algebraically closed field* K *and a quadruple of subspaces* $E_i \subseteq P$.

Then S *can be decomposed into the direct sum of non-decomposable subsystems of the following types:* $S_{i,j}(m,0)$, $i<j$, $i,j \in \{1,2,3,4\}$, $m = 1,2,\dots$; $S(2m,0,\lambda)$, $\lambda \in K$, $\lambda \neq 0$, $\lambda \neq 1$, $m = 1,2,\dots$; $S_i(m,-1)$, $S_i(m,1)$, $i \in \{1,2,3,4\}$, $m = 1,2,\dots$; $S(2k+1,-2)$, $S(2k+1,2)$, $k = 0,1,\dots$.

4. Let us return now to the examples of Section 2 and describe into what kind of non-decomposable quadruples the respective systems split up.

1) Let the quadruple S correspond to the pair E, A, where A is a linear transformation, $A: E \longrightarrow E$, so that $P = E \oplus E$, $E_1 = E \oplus \{0\}$, $E_2 = \{0\} \oplus E$, E_3 is the graph of A, and E_4 is the diagonal. Then the quadruple $S = \{P; E_1, \ldots, E_4\}$ can be decomposed into the direct sum of non-decomposable quadruples of the types $S_{1,3}(2k,0)$, $S_{3,4}(2k,0)$ and $S(2k,0;\lambda)$, where $\lambda \in K$, $\lambda \neq 0$, $\lambda \neq 1$.

2) Let the quadruple S be constructed from two spaces E, F and two linear mappings $A: E \longrightarrow F$, $B: F \longrightarrow E$ so that $P = E \oplus F$, $E_1 = E \oplus \{0\}$, $E_2 = \{0\} \oplus F$, while E_3 is the graph of A, and $E_4 = \{(by, y) \mid y \in F\}$.

We remark that in quadruples of this kind the equalities $E_i \cap E_j = 0$, $E_i + E_j = P$ hold for every pair (i,j) with the possible exception of the pairs $(1,3)$, $(2,4)$, $(3,4)$. Moreover, $\dim E_1 = \dim E_3$ and $\dim E_2 = \dim E_4$. These equalities, naturally, are preserved when decomposing the quadruple into a direct sum. Reasoning along these lines it can be shown that the quadruple S may be decomposed into the direct sum of quadruples of the types $S(2k,0;\lambda)$, $\lambda \neq 0$, $\lambda \neq 1$; $S_{3,4}(2k,0)$, $k = 1,2,\ldots$; $S_{1,3}(m,0)$, $S_{2,4}(m,0)$, $m = 1,2,\ldots$.

3) Let S be constructed from two spaces E, F and two mappings $A_i: E \longrightarrow F$ so that $P = E \oplus F$; $E_1 = E \oplus \{0\}$; $E_2 = \{0\} \oplus F$; E_3 is the graph of A_1, and E_4 is that of A_2. In a quadruple of this type the equalities $E_i \cap E_j = 0$, $E_i + E_j = P$ are satisfied for the following pairs (i,j): $(1,2)$, $(2,3)$, $(2,4)$. Moreover, $\dim E_1 = \dim E_3 = \dim E_4$ and $\dim E_1 + \dim E_2 = \dim P$. These equalities are preserved when decomposing S into a direct sum. Therefore the quadruple S can be decomposed into the direct sum of quadruples of the following types: a) $S(2k,0;\lambda)$, $\lambda \neq 0$, $\lambda \neq 1$; b) $S_{3,4}(2k,0)$; c) $S_{1,3}(2k,0)$; d) $S_{1,4}(2k,0)$; e) $S_2(2k+1,-1)$ $k = 0,1,2,\ldots$; f) $S_2(2k+1,1)$, $k = 0,1,2,\ldots$.

Decomposing in each of the non-decomposable quadruples the space P into a direct sum $P \cong E_1 \oplus E_2$ and considering the subspaces E_3, E_4 as the graphs of the corresponding operators, we obtain the following proposition, which corresponds to a result of Kronecker [1]:

Proposition. *Let E, F be finite-dimensional vector spaces and let A_1, A_2 be to linear mappings of E into F. Then the spaces E, F can be decomposed into* *direct sums* $E = \bigoplus_t E^{(t)}$, $F = \bigoplus_t F^{(t)}$ *of subspaces satisfying the relations* $A_i E^{(t)} \subseteq F^{(t)}$ $(i=1,2)$. *Furthermore, in the subspaces* $E^{(t)}$, $F^{(t)}$ *the bases* $\{e_1,\dots,e_k\}$ *resp.* $\{f_1,\dots,f_{k'}\}$ *can be chosen so that the effect of the operators* A_1, A_2 *is described by one of the following schemes:*

a) $\dim E^{(t)} = \dim F^{(t)} = k$;

$A_1e_1 = \lambda f_1,\ A_1e_2 = f_1+\lambda f_2, \dots, A_1e_k = f_{k-1}+\lambda f_k \qquad (\lambda \neq 0,\ \lambda \neq 1)$;

$A_2e_1 = f_1,\ A_2e_2 = f_2, \dots, A_2e_k = f_k$.

(The corresponding quadruple is $S(2k,0;\lambda)$, $\lambda \neq 0$, $\lambda \neq 1$ *.)*

b) $\dim E^{(t)} = \dim F^{(t)} = k$;

$A_1e_1 = f_1,\ A_1e_2 = f_1+f_2, \dots, A_1e_k = f_{k-1}+f_k$;

$A_2e_1 = f_1,\ A_2e_2 = f_2, \dots, A_2e_k = f_k$.

(The corresponding quadruple is $S(2k,0;1) \cong S_{3,4}(2k,0)$*.)*

c) $\dim E^{(t)} = \dim F^{(t)} = k$;

$A_1e_1 = 0,\ A_1e_2 = f_1, \dots, A_1e_k = f_{k-1}$;

$A_2e_1 = f_1,\ A_2e_2 = f_2, \dots, A_2e_k = f_k$.

(The corresponding quadruple is $S_{1,3}(2k,0)$ *.)*

d) $\dim E^{(t)} = \dim F^{(t)} = k$;

$A_1e_1 = f_1,\ A_1e_2 = f_2, \dots, A_1e_k = f_k$;

$A_2e_1 = 0,\ A_2e_2 = f_1, \dots, A_2e_k = f_{k-1}$.

(The corresponding quadruple is $S_{1,4}(2k,0)$ *.)*

e) $\dim E^{(t)} = k$, $\dim F^{(t)} = k+1 \qquad (k = 0,1,2,\dots)$;

$A_1e_1 = f_1,\ A_1e_2 = f_2, \dots, A_1e_k = f_k$;

$A_2e_1 = f_2,\ A_2e_2 = f_3, \dots, A_2e_k = f_{k+1}$.

(The corresponding quadruple is $S_2(2k+1,-1)$ *.)*

f) $\dim E^{(t)} = k+1$, $\dim F^{(t)} = k$ $(k = 0,1,2,\dots)$;

$A_1e_1 = 0$, $A_1e_2 = f_1, \dots, A_1e_{k+1} = f_k$;

$A_2e_1 = f_1$, $A_2e_2 = f_2, \dots, A_2e_k = f_k$, $A_2e_{k+1} = 0$.

(The corresponding quadruple is $S_2(2k+1,1)$.*)*

4) Let S be the quadruple that corresponds to the additive relation $A \subseteq E \oplus E$, i.e. $P = E \oplus E$, $E_1 = E \oplus \{0\}$, $E_2 = \{0\} \oplus E$, $E_3 = A$, and E_4 is the diagonal in $E \oplus E$. Then S can be decomposed into the direct sum of quadruples of the following types: $S(2k,0;\lambda)$, $\lambda \neq 0$, $\lambda \neq 1$; $S_{1,3}(2k,0)$; $S_{2,3}(2k,0)$; $S_{3,4}(2k,0)$; $S_4(2k,-1)$; $S_4(2k,+1)$; $k = 1,2,3,\dots$.

5) Let S be the quadruple which corresponds to the pair of additive relations $A_1 \subseteq E \oplus F$, $A_2 \subseteq E \oplus F$. That is, $P = E \oplus F$, $E_1 = E \oplus \{0\}$, $E_2 = \{0\} \oplus F$, $E_3 = A_1$, $E_4 = A_2$. Then S can be decomposed into the direct sum of quadruples of the following types: $S(2k,0;\lambda)$, $\lambda \neq 0$, $\lambda \neq 1$; $S_{3,4}(2k,0)$; $S_{i,j}(k,0)$, $(i,j) \neq (1,2)$, $(i,j) \neq (3,4)$; $S_1(2k-1,\rho)$; $S_2(2k-1,\rho)$; $S_3(2k,\rho)$; $S_4(2k,\rho)$; $\rho = -1,+1$; $k = 1,2,3,\dots$.

5. Let us give a short summary of the paper.

In § 2 the category C^r of systems $S = \{P; E_1, \dots, E_r\}$ is considered. As a main tool we introduce the functors Φ^+, Φ^-, which carry the system S into systems $S^+ = \Phi^+(S)$, $S^- = \Phi^-(S)$. It seems likely that these functors are interesting on their own right, too. The fundamental properties of Φ^+ and Φ^- are the following: 1) if the system S is non-decomposable, then also S^+, S^- are non-decomposable; 2) if S non-decomposable, $S \neq 0$ and $\Phi^+(S) \neq 0$ (resp. $\Phi^-(S) \neq 0$), then the system $\Phi^-\Phi^+(S)$ (resp. $\Phi^+\Phi^-(S)$) is isomorphic to S; 3) if $S \in C^4$ (i.e. $S = \{P; E_1, \dots, E_4\}$), S is non-decomposable, $S \neq 0$ and $\Phi^+(S) \neq 0$ (resp. $\Phi^-(S) \neq 0$), then the system $S^+ = \Phi^+(S)$ $(S^- = \Phi^-(S))$ has the same defect $\rho(S^+)$ $(\rho(S^-))$ as the system S.

In § 3 with the use of the functors Φ^+, Φ^- a complete classification for non-decomposable quadruples with defect $\rho(S) \neq 0$ is given. It is shown that if the quadruple S is non-decomposable and its defect $\rho(S)$ is negative, then there exists a number $\ell \geq 1$ such that $(\Phi^+)^{\ell-1}S \neq 0$, but $(\Phi^+)^{\ell}S = 0$. Moreover, along with S also the system $S^{+(\ell-1)} = (\Phi^+)^{\ell-1}S$ is non-decomposable and has the same defect as S. Note that $S^{+(\ell-1)}$ satisfies the relation $\Phi^+S^{+(\ell-1)} = 0$. Systems for which the condition $\Phi^+S = 0$ is fulfilled are termed simplest relative to Φ^+. Non-decomposable simplext systems can be classified easily.

If, on the other hand, the non-decomposable system S $(\rho(S)<0)$ satisfies the relations $(\Phi^+)^{\ell}S = 0$, $S^{+(\ell-1)} = (\Phi^+)^{\ell-1}S \neq 0$, then S is uniquely determined by the corresponding simplest system $S^{+(\ell-1)}$ and the number ℓ ; namely the following isomorphism is valid:

$$S \cong (\Phi^-)^{\ell-1}S^{+(\ell-1)} .$$

A similar statement holds for systems S with defect $\rho(S)>0$. To every system of this kind there corresponds a number $\ell \geq 1$ such that $(\Phi^-)^{\ell-1}S \neq 0$, $(\Phi^-)^{\ell}S = 0$. The system $S_{\ell-1} = (\Phi^-)^{\ell-1}S$ is simplest relative to Φ^-, i.e. $\Phi^- S_{\ell-1} = 0$. The system S is uniquely determined by the corresponding simplest system $S_{\ell-1}$ and the number ℓ: $S \cong (\Phi^+)^{\ell-1}S_{\ell-1}$.

§§4,5,6 are devoted to the classification of non-decomposable systems with defect 0. The direct sum of any number of these systems is said to be a regular system. The functors Φ^+, Φ^- do not affect regular systems. Namely if $S = \{P; E_1, \ldots, E_4\}$ is a regular system, then for any k the quadruples $S^{(k)} = (\Phi^+)^k S = \{P^{(k)}; E_1^{(k)}, \ldots, E_4^{(k)}\}$, $S^{(-k)} = (\Phi^-)^k S = \{P^{(-k)}; E_1^{(-k)}, \ldots, E_4^{(-k)}\}$ satisfy the relations $\dim P^{(k)} = \dim P^{(-k)} = \dim P$.

In § 4 the notion of operator quadruple is introduced. The regular quadruple S is called an operator quadruple provided its constituents satisfy the equalities $E_i \cap E_j = 0$, $E_i + E_j = P$ for each of the four pairs (i,j), where $i<j$; $i,j \in \{1,2,3,4\}$. If these equalities are fulfilled for a pair (k,ℓ), then they are fulfilled also for the pair (r,s) with $\{k,\ell,r,s\} = \{1,2,3,4\}$. It turns out that with every operator quadruple one can associate two spaces E, F and two linear mappings $A: E \longrightarrow F$, $B: F \longrightarrow E$ (see Example 2) of Section 2). In § 4 it is proved that every non-decomposable regular quadruple is an operator quadruple.

In § 5 a classification of non-degenerate operator quadruples is given. A quadruple is said to be non-degenerate if the product BA is a non-degenerate linear transformation (A and B stand for the transformations corresponding to the quadruple).

In § 6 the classification of nilpotent operator quadruples is given. An operator quadruple is called nilpotent if the product BA is nilpotent.

We think that the proofs contained in §§ 4-6, in contrast to those of §§ 2-3, can be simplified.

The classification problem of quadruples includes, roughly speaking, about one half of the problems of linear algebra.

In a forthcoming paper the authors hope to examine another class of problems in linear algebra.

§ 2. THE FUNCTORS Φ^+ AND Φ^-

In this § we introduce two functors Φ^+, Φ^-, in the category C^r and study their properties. These functors play a major role in the classification of quadruples.

1. Let us first describe some properties of the category C^r. Let $S = \{P; E_1, \dots, E_r\}$ and $T = \{Q; F_1, \dots, F_r\}$ be any pair of systems, and $\alpha: P \longrightarrow Q$ a linear mapping. We shall say that the mapping α induces (or: defines) a morphism $\alpha: S \longrightarrow T$ is for every i the inclusion $\alpha E_i \subset F_i$ is valid. Thus $\mathrm{Hom}(S,T)$ is a part of $\mathrm{Hom}(P,Q)$.

Monomorphisms μ and epimorphisms τ in the category C^r can be defined as follows. A monomorphism $\mu: S \longrightarrow T$ is a morphism induced by a mapping $\mu: P \longrightarrow Q$ without kernel. An epimorphism $\tau: S \to S'$, where $S' = \{P'; E'_1, \dots, E'_r\}$, is a morphism induced by a mapping $\tau: P \longrightarrow P'$ such that $\tau P = P'$.

We shall say that $S^\circ = \{P^\circ; E^\circ_1, \dots, E^\circ_r\}$ is a subsystem of the system $S = \{P, E_1, \dots, E_r\}$ if there exists a monomorphism $\mu: S^\circ \longrightarrow S$, i.e., if P° is a subspace of P, and $E^\circ_i \subset E_i$. Analogously, we say that $S' = \{P'; E'_1, \dots, E'_r\}$ is a quotient system of the system $S = \{P; E_1, \dots, E_r\}$, if there exists an epimorphism $\tau: S \longrightarrow S'$, i.e., if P' is a quotient space of P, and $\tau E_i \subset E'_i$.

The system $S^* = \{P^*; \bar{E}_1, \dots, \bar{E}_r\} \in C^r$ will be called the conjugate of the system $S = \{P, E_1, \dots, E_r\}$, if the space P^* is the conjugate of P, whereas $\bar{E}_i$ consists of all functionals in P^* vanishing on E_i (i.e. the subspace $\bar{E}_i$ is orthogonal to E_i).

The morphism $\alpha^*: S_1^* \longrightarrow S^*$ will be called the conjugate of the morphism $\alpha: S \to S_1$, provided the mapping $\alpha^*: P_1^* \longrightarrow P^*$ that induces it is the conjugate of $\alpha: P \longrightarrow P_1$. It is not difficult to show that the function which carries over the system S into its conjugate S^* and associates with every morphism $\alpha: S \longrightarrow S_1$ the conjugate morphism $\alpha^*: S_1^* \longrightarrow S^*$ is a contravariant functor from C^r to C^r.

2. The functor Φ^+. Let $S = \{P; E_1, \ldots, E_r\}$ be an object in C^r. We associate with it an object $\Phi^+(S) = S^+ = \{P^+; E_1^+, \ldots, E_r^+\}$ in the following manner. We consider the space $R = \bigoplus_{t=1}^{r} E_t$, the direct sum of the subspaces E_t.

Let ξ be a vector belonging to R, i.e. $\xi = (x_1, \ldots, x_r)$, where $x_t \in E_t$. Define a mapping $\tau: R \longrightarrow P$ with the aid of the relation

$$\tau\xi = \sum_{t=1}^{r} x_t .$$

Denote the kernel of the mapping τ by P^+. Thus P^+ consists of all vectors satisfying the equality $\sum_t x_t^+ = 0$. Further denote by E_t^+ the subspace of P^+ consisting of all vectors of the form $\xi^0 = (x_1, \ldots, x_{t-1}, 0, x_{t+1}, \ldots, x_r)$ with $\sum_{j \neq t} x_j = 0$.

The system $\{P^+; E_1^+, \ldots, E_r^+\}$ will be denoted by S^+. So with every object $S \in C^r$ we have associated an object $S^+ = \Phi^+(S) \in C^r$.

If α is a morphism, $\alpha: S \longrightarrow S'$, then we can define the morphism $\alpha^+ = \Phi^+(\alpha)$, $\alpha^+: S^+ \longrightarrow (S')^+$ in a natural way. Let namely $R = \bigoplus_{t=1}^{r} E_r$, $R' = \bigoplus_{t=1}^{r} E_r'$. For any vector $\xi \in R$, $\xi = (x_1, \ldots, x_r)$ we set $A\xi = (\alpha x_1, \ldots, \alpha x_r)$. By definition, the mapping α satisfies the relations $\alpha E_i \subset E_i$. Consequently, $\xi = A\xi \in R'$. Further let $\xi \in P^+$, i.e. $\sum_t x_t = 0$. Then $\sum \alpha x_t = 0$, i.e. $A\xi = (\alpha x_1, \ldots, \alpha x_r) \in (P')^+$. Therefore $A\xi$ belongs to the subspace $(P')^+$. Let us denote by α^+ the restriction of the mapping A to the subspace P^+. We have shown that $\alpha^+ P^+ \subseteq (P')^+$. It is not difficult to verify that $\alpha^+ E_i^+ \subseteq (E_i')^+$. Thus we have constructed a morphism $\alpha^+ = \Phi^+(\alpha): S^+ \longrightarrow (S')^+$. It can be easily seen that Φ^+ is a covariant functor from C^r to C^r.

3. The functor Φ^-. Now for every object $S \in C^r$ we define an object $\Phi^-(S) = S^- = \{P^-; E_1^-, \ldots, E_r^-\}$. Consider the space $Q = \bigoplus_{t=1}^{r} P/E_t$, the direct sum of the quotient spaces P/E_t. Let $\beta_t: P \longrightarrow P/E_t$ be the natural mapping. The mapping

$$x \longrightarrow (\beta_1 x, \beta_2 x, \ldots, \beta_r x)$$

of the space P into the space Q will be denoted by μ. Put further $P^- = Q/\mathrm{Im}\,\mu$

and $\tau: Q \longrightarrow P^-$, where τ is the natural epimorphism. The subspace of Q consisting of the vectors $(0,\ldots,0,\beta_t x,0,\ldots,0)$ will be denoted by L_t, and the subspace τL_t by E_t^-.

Let us denote the system $\{P^-; E_1^-,\ldots,E_r^-\}$ by S^-. Thus with each object $S \in C^r$ we have associated an object $S^- = \Phi^-(S) \in C^r$. The morphism $\alpha^- = \Phi^-(\alpha)$: $S^- \longrightarrow (S')^-$ can be built up from the morphism $\alpha: S \longrightarrow S'$ in a natural way. It is easy to verify that Φ^- is a covariant functor from C^r to C^r.

4. In this section we prove two technical lemmas, which will be utilized in the next section when studying the properties of the functors Φ^+, Φ^-.

It should be noted that in the course of defining the functors Φ^+ and Φ^- we constructed an exact sequence of spaces, but not of systems:

$$0 \longrightarrow P^+ \xrightarrow{\mu} R \xrightarrow{\tau} P.$$

Here $R = \bigoplus_{t=1}^{r} E_r$. Let us denote by π_t the projector on R given by the relation

$$\pi_t(x_1,\ldots,x_t,\ldots,x_r) = (0,\ldots,0,x_t,0,\ldots,0).$$

It can be shown that

$$E_t = \mathrm{Im}(\tau\pi_t), \qquad E_t^+ = \mathrm{Ker}(\pi_t\mu).$$

Now let $\Gamma: 0 \longrightarrow \bar{P} \xrightarrow{\mu} R \xrightarrow{\tau} P$ be any exact sequence of spaces, and assume that on the space R a family $\{\pi_t\}_{t=1,\ldots,r}$ of projectors is given so that

$$\pi_i\pi_j = 0 \quad \text{if} \quad i \neq j; \qquad (\pi_i)^2 = \pi_i,$$

$$\sum_{t=1}^{r} \pi_t = 1.$$

Put $E_t = \mathrm{Im}(\tau\pi_t)$, $\bar{E}_t = \mathrm{Ker}(\pi_t\mu)$, $S = \{P; E_1,\ldots,E_r\}$, $\bar{S} = \{\bar{P}; \bar{E}_1,\ldots,\bar{E}_r\}$. We shall say that Γ defines the systems $S, \bar{S}$. The next lemma exhibits conditions for the validity of the isomorphism $\bar{S} \cong \Phi^+(S)$.

Lemma 2.1. *Let* $\Gamma: 0 \longrightarrow \bar{P} \xrightarrow{\mu} R \xrightarrow{\tau} P$ *be an exact sequence of finite-dimensional spaces, and* $\{\pi_t\}_{t=1,\ldots,r}$ *a family of projectors on the space* R *such that* $\sum_t \pi_t = 1$; $\pi_i\pi_j = 0$ *if* $i \neq j$. *Further let* $S = \{P; \mathrm{Im}\,\tau\pi_t\}_{t=1,\ldots,r}$ *and*

$\bar{S} = \{\bar{P};\ \mathrm{Ker}\,\pi_t\mu\}_{t=1,\dots,r}$ *be the systems defined by* Γ . *If, moreover, for every* t *the relation*

(2.1) $$\mathrm{Ker}\,\pi_t = \mathrm{Ker}\,\tau\pi_t$$

is satisfied, then

$$\bar{S} \cong \Phi^{+}(S) .$$

Proof. Let us prove that the subspace $\mathrm{Jm}\,\pi_t \subset R$ and the subspace $E_t = \mathrm{Jm}\,\tau\pi_t$ are isomorphic. By definition, the subspace $\mathrm{Jm}\,\pi_t$ is complementary in R to $\mathrm{Ker}\,\pi_t = \mathrm{Ker}\,\tau\pi_t$, i.e. $\mathrm{Jm}\,\pi_t \cap \mathrm{Ker}\,\tau\pi_t = 0$, $\mathrm{Jm}\,\pi_t + \mathrm{Ker}\,\tau\pi_t = R$. Consequently, the restriction $\tau\pi_t \,|\, \mathrm{Jm}\,\pi_t :\ \mathrm{Jm}\,\pi_t \longrightarrow \mathrm{Jm}\,\tau\pi_t$ is an isomorphism. Hence we obtain easily that $R \cong \bigoplus_t \mathrm{Jm}\,\tau\pi_t$.

Now it is no longer difficult to verify that the system $\bar{S} = \{P;\ \mathrm{Ker}\,\pi_t\mu\}_{t=1,\dots,r}$ is isomorphic to the system S^{+} , where $S = \{P;\ \mathrm{Jm}\,\tau\pi_t\}_{t=1,2,\dots,r}$.

Similarly, when defining the functor Φ^{-}, we constructed an exact sequence of finite-dimensional spaces,

$$P \xrightarrow{\ \mu\ } Q \xrightarrow{\ \tau\ } P' \longrightarrow 0 ,$$

where $Q = \bigoplus_{t=1}^{r} P/E_t$. Clearly, also on Q it is possible to define a family $\{\pi_t\}_{t=1,\dots,r}$ of projectors, so that the subspaces $E_t \subset P$, $E_t^{-} \subset P^{-}$ can be defined in the following way:

$$E_t = \mathrm{Ker}\,\pi_t\mu , \qquad E_t^{-} = \mathrm{Jm}\,\tau\pi_t .$$

Lemma 2.2. *Let* $\Gamma :\ P \xrightarrow{\ \mu\ } Q \xrightarrow{\ \tau\ } \bar{P} \longrightarrow 0$ *be an exact sequence of spaces, and* $\{\pi_t\}_{t=1,\dots,r}$ *a family of projectors on* Q *such that* $\sum \pi_t = 1$ *and* $\pi_i\pi_j = 0$ *if* $i \neq j$. *Furthermore, let* $S = \{P;\ \mathrm{Ker}\,\pi_t\mu\}_{t=1,\dots,r}$, $\bar{S} = \{\bar{P};\ \mathrm{Jm}\,\tau\pi_t\}_{t=1,\dots,r}$ *be the systems defined with the help of* Γ . *If for every* t

(2.2) $$\mathrm{Jm}\,\pi_t\mu = \mathrm{Jm}\,\pi_t ,$$

then

$$\bar{S} \cong \Phi^{-}(S) .$$

The proof of this lemma is dual to that of Lemma 2.1.

5. Basic properties of the functors Φ^+, Φ^-.

Theorem 2.3. *Let* $S = \{P; E_1, \dots, E_r\}$ *be any system. Then:*

1) $(\Phi^+(S))^* \cong \Phi^-(S^*)$, *where* S^* *is the system conjugate to* S.

2) $(\Phi^-(S))^* \cong \Phi^+(S^*)$.

3) *If* S *is isomorphic to the direct sum of some systems* S_i, *i.e.* $S \cong \bigoplus_{i=1}^{k} S_i$, *then* $\Phi^+(S) \cong \bigoplus_{i=1}^{k} (\Phi^+(S_i))$, $\Phi^-(S) = \bigoplus_{i=1}^{k} (\Phi^-(S_i))$.

4) *There exists a natural monomorphism* $\alpha: \Phi^-\Phi^+(S) \longrightarrow S$.

5) *There exists a natural epimorphism* $\tau: S \longrightarrow \Phi^+\Phi^-(S)$.

Proof of 4) Denote by E'_t the subspace of E_t that is equal to the intersection $E_t \cap (\sum_{j \neq t} E_j)$. Put

$$P' = \sum_{t=1}^{r} E'_t .$$

Denote the system $\{P'; E'_1, \dots, E'_r\}$ by S'. We are going to prove that there exists a natural isomorphism

$$\Phi^-\Phi^+(S) \cong S' .$$

Starting with the system $S' = \{P'; E'_1, \dots, E'_r\}$ one can build up the system $\Phi^+(S') = \{(P')^+; (E'_1)^+, \dots, (E'_r)^+\}$ and the exact sequence of spaces

$$0 \longrightarrow (P')^+ \xrightarrow{\mu'} R' \xrightarrow{\tau} P', \tag{2.3}$$

where $R' = \bigoplus_{t=1}^{r} E'_t$.

Denote the imbedding $P' \longrightarrow P$ by α. From the mapping α we can construct, in a natural way, two imbeddings $A: R' \longrightarrow R$, $\alpha^+: (P')^+ \longrightarrow P^+$ such that the following diagram is commutative:

$$\begin{array}{ccccccc} 0 & \longrightarrow & P^+ & \xrightarrow{\mu} & R & \xrightarrow{\tau} & P \\ & & \uparrow{\scriptstyle \alpha^+} & & \uparrow{\scriptstyle A} & & \uparrow{\scriptstyle \alpha} \\ 0 & \longrightarrow & (P')^+ & \xrightarrow{\mu'} & R' & \xrightarrow{\tau'} & P' \end{array} \tag{2.4}$$

a) Let us show that the linear mapping α^+ which appears in this diagram is an isomorphism, i.e. the subspaces $\mathrm{Ker}\,\tau$, $\mathrm{Ker}\,\tau'$ are isomorphic. Let ξ^0 be any vector of $\mathrm{Ker}\,\tau$, i.e. $\xi^0 = (x_1^0, \ldots, x_r^0)$, where $x_t^0 \in E_t$, $\sum_t x_t^0 = 0$. Then for any t we have $x_t^0 = -\sum_{j \neq t} x_j^0$, i.e. $x_t^0 \in E_t \cap (\sum_{j \neq t} E_j) = E_t'$. Thus $\xi^0 \in \in \bigoplus_{t=1}^{r} E_t' = R'$. We have proved that $A\,\mathrm{Ker}\,\tau' = \mathrm{Ker}\,\tau$. As A is a monomorphism, the mapping $\alpha^+ = A / \mathrm{Ker}\,\tau$ is an isomorphism.

b) It can be proved in an analogous manner that $\alpha^+(E_t')^+ = E_t^+$. This means that the isomorphism α^+ induces a system isomorphism $\alpha^+ : \Phi^+(S') \longrightarrow \Phi^+(S)$.

c) Let us show that the mapping $\tau' : R' \longrightarrow P'$ is an epimorhpism, i.e. $\mathrm{Im}\,\tau' = P'$. By definition, $P' = \sum_{t=1}^{r} E_t$. Therefore any vector $x' \in P'$ can be represented as a sum $x' = \sum_{t=1}^{r} x_t'$, $x_t' \in E_t'$. From the definition of the mapping τ' it follows that $\tau(x_1', \ldots, x_r') = \sum_t x_t' = x'$, i.e. the vector x' admits an inverse image, the statement to be proved.

d) Let us prove now that for any t

(2.5) $$\mathrm{Im}\,\pi_t' \mu' = \mathrm{Im}\,\pi_t' .$$

The inclusion $\mathrm{Im}\,\pi_t' \mu' \subset \mathrm{Im}\,\pi_t'$ is evident, so it remains to prove that $\mathrm{Im}\,\pi_t' \subset \subset \mathrm{Im}\,\pi_t' \mu'$. In the course of the proof we shall make use of the following simple lemma, stated here without proof:

Lemma 2.4. *Let $E_1, \ldots, E_r$ be subspaces of the space P, and let $E_t' = = E_t \cap (\sum_{j \neq t} E_j)$. Then for every t we have the relation*

$$E_t' \subset (\sum_{j \neq t} E_j') .$$

From this lemma it follows that any vector $x_t' \in E_t'$ can be represented as $x_t' = \sum_{j \neq t} x_j'$. The definition of τ implies that the vector

$$\xi^0 = (-x_1', \ldots, -x_{t-1}', x_t' - x_{t+1}', \ldots, -x_r')$$

belongs to the subspace $\mathrm{Ker}\,\tau' = \mathrm{Im}\,\mu'$. Set

$$\xi_t^\circ = \pi_t \xi^\circ = (0,\dots,0,\dots,x_t,0,\dots,0) .$$

Since $\xi^\circ \in \operatorname{Im} \mu'$, we have $\xi_t^\circ \in \operatorname{Im} \pi_t' \mu'$. The vector x_t' is, by construction, an arbitrary element of E_t'. Thus ξ_t° is an arbitrary element of the subspace $\operatorname{Im} \pi_t' \cong$ $\cong E_t'$. Equality (2.5) is proved.

Comparing the results c), d) with Lemma 2.2 we obtain:

$$S' \cong \Phi^- \Phi^+(S') .$$

Under b) we have proved that $\Phi^+(S') \cong \Phi(S)$. Consequently, $S' \cong \Phi^- \Phi^+(S)$.

Thus we have proved that $\Phi^- \Phi^+(S)$ is isomorphic to the subsystem S' of S. Property 4 is established. We omit the proofs of the assertions 1), 2), 3), 5), of Theorem 2.3 which are not more complicated than the one carried out above.

6. Simplest systems. We shall say that the system S is simplest with respect to the functor $\Phi^+ (\Phi^-)$, if $\Phi^+(S) = 0$ $(\Phi^-(S) = 0)$.

Proposition 2.5. *The system* $S = \{P; E_1,\dots,E_r\}$ *satisfies the equality* $\Phi^+(S) =$ $= 0$ *if and only if for every* t *we have*

$$E_t' = (E_t \cap (\sum_{j \neq t} E_j)) = 0 . \tag{2.6}$$

Proof. The condition is necessary. Put $S' = \{P'; E_1',\dots,E_r'\}$, where $P' =$ $= \sum_t E_t$. In Theorem 2.1 we proved that $S' \cong \Phi^- \Phi^+(S)$. If $\Phi^+(S) = 0$, then $S' =$ $\cong \Phi^-(0) = 0$. In particular, each of the subspaces E_t' is equal to zero.

The condition is sufficient. The equalities (2.6) say that $\sum_t E_t \cong \bigoplus_t E_t = R$. Therefore the mapping $\tau : R \longrightarrow P$, defined by the formula $\tau(x_1,\dots,x_r) = \sum_t x_t$ has kernel zero. So $P^+(0)$, $P^+(S) = 0$, what was to be proved.

The next result is the dual of Proposition 2.5.

Proposition 2.6. *The system* $S = \{P; E_1,\dots, E_r\}$ *satisfies the equality* $\Phi^-(S) = 0$ *if and only if for every* t *we have*

$$(E_t + (\bigcap_{j \neq t} E_j)) = P .$$

Now it is easy to give a complete classification of simplest systems.

Let S be a simplest system with respect to Φ^+, i.e. $\Phi^+(S) = 0$. As P is a space over the field K, we can choose a basis $e(i,j)$ in P so that the vectors $e(t,j)$ $(j = 1,2,\dots,k_t)$ form a basis of the subspace E_t $(t = 1,2,\dots,r)$, and the subspace P_0 spanned by the vectors $e(0,j)$ $(j = 1,2,\dots,k_0)$ is complementary to the subspace $(\sum_{t=1}^{r} E_t)$ (i.e. $P_0 \cap (\sum_{t=1}^{r} E_t = 0$, $P_0 + (\sum_{t=1}^{r} E_t) = P$). The subspace spanned by the vector $e(i,j)$ will be denoted by $P(i,j)$. It is easy to see that the system S is decomposable on the subspaces $P(i,j)$ into the direct sum of some subsystems $S(i,j)$. These subsystems are the following:

$$S(0,j) = \{P(0,j);\ E(0,j,t)\}_{t=1,\dots,r}\ ,$$

where $E(0,j,t) = 0$ for every t, while for $i \neq 0$

$$S(i,j) = \{P(i,j);\ E(i,j,t)\}_{t=1,\dots,r}\ ,$$

where $E(i,j,i) = P(i,j)$, and $E(i,j,t) = 0$ if $i \neq t$. It is easy to see that for every j_1, j_2 we have

$$S(i,j_1) \cong S(i,j_2) \qquad (i = 0,1,\dots,r)\ ,$$

and that each of the systems $S(i,j)$ is non-decomposable. Put $S(i) \cong S(i,j)$ $(i = 0,1,\dots,r)$. We have proved the following

Proposition 2.7. *Let* S *be a simplest system relative to the functor* Φ^+ *i.e.* $\Phi^-(S) = 0$. *Then* S *can be decomposed into the direct sum of subsystems, each subsystem being isomorphic to one of the systems* $S(i)$ $(i = 0,1,\dots,r)$. *Here* $S(0) = \{P; 0,\dots,0\}$, $\dim P = 1$, *and for* $i \neq 0$, $S(i) = \{P; 0,\dots,0, E_i, 0,\dots,0\}$ *with* $E_i = P$, $\dim P = 1$.

The proof of the following result is also elementary.

Proposition 2.8. *Let* S *be a simplest system relative to the functor* Φ^-, *i.e.* $\Phi^-(S) = 0$. *Then* S *can be decomposed into the direct sum of subsystems, each subsystem being isomorphic to one of the systems* $S(i)$ $(i = 0,1,\dots,r)$. *Here* $S(0) = \{P; E_1,\dots,E_r\}$ *with* $E_t = P$ *for every* t *and* $\dim P = 1$ *; for* $i \neq 0$ *the system* $S(i)$ *is given by the formula*

$$S(i) = \{P;\ E_1,\dots,E_{i-1}, 0, E_{i+1},\dots,E_r\}$$

with $E_t = P$ *if* $t \neq i$ *and* $\dim P = 1$.

7. In this section we describe the so-called reduced systems i.e. systems complementary to simplest ones.

Definition 2.9. The system $S = \{P; E_1, \ldots, E_r\}$ is said to be *reduced from above* if for every t we have

$$(2.7) \qquad (\sum_{j \neq t} E_j) = P.$$

Definition 2.10. The system S is said to be *reduced from below* if for every t we have

$$(2.8) \qquad (\bigcap_{j \neq t} E_j) = 0.$$

Proposition 2.11. a) *The system* S *in reduced from above if and only if*

$$\Phi^- \Phi^+(S) \cong S.$$

b) *The system* S *is reduced from below if and only if*

$$\Phi^+ \Phi^-(S) \cong S.$$

Proof of property a). In the course of proving assertion 4 of Theorem 2.3 we established that the system $\Phi^- \Phi^+(S)$ is isomorphic to the subsystem $S' = \{P'; E'_1, \ldots, E'_r\}$, where $E'_t = E_t \cap (\sum_{j \neq t} E_j)$, $P' = \sum_t E'_t$. Let S be reduced from above. Then from the equality (2.7) it follows that $E'_t = E_t$ for every t, so that $S' = S$. Consequently, according to Theorem 2.1, $\Phi^- \Phi^+(S) \cong S$.

Let, conversely, $\Phi^- \Phi^+(S) = S$. Then $S' = S$ i.e.

$$(2.9) \qquad E'_t = E_t$$

for every t and

$$(2.10) \qquad P' = P.$$

Equality (2.9) can be rewritten in the form $E_t \cap (\sum_{j \neq t} E_j) = E_t$. Therefore

$$(2.11) \qquad E_t \subseteq (\sum_{j \neq t} E_j)$$

for every t .

From (2.9) and (2.10) it follows that $P = \sum_t E_t = \sum E_t$. Hence, making use of (2.11), we find: $P = E_t + (\sum_{j \neq t} E_j) = \sum_{j \neq t} E_j$. Qu.e.d.

The proof of the assertion 2.11 b) is dual to the one performed just now.

Proposition 2.12. a) *If the system* S *is reduced from above and decomposable,* $S = \bigoplus_{i=1}^{k} S_i$, $k > 1$, $S_i \neq 0$*; then the subsystems* S_i *are reduced from above, too.*

b) *If the system* S *is reduced from below and decomposable,* $S = \bigoplus_{i=1}^{k} S_i$, $k > 1$, $S_i \neq 0$, *then also the subsystems* S_i *are reduced from below.*

Proof. Let S be reduced from above. Then by Theorem 2.3 $\phi^-\phi^+(S) \cong \bigoplus_{i=1}^{k} \phi^-\phi^+(S_i)$. Assume that for some of the systems S_i we have $\phi^-\phi^+(S_i') \cong S_i' \subset S_i$, where $S_i' \neq S_i$. Then

$$\phi^-\phi^+(S) \cong \bigoplus_{i=1}^{k} \phi^-\phi^+(S_i) \cong \bigoplus_{i=1}^{k} S_i' \subset \bigoplus_{i=1}^{k} S_i = S ,$$

where $\bigoplus_{i=1}^{k} S_i' \neq \bigoplus_{i=1}^{k} S_i$. We obtained that $\phi^-\phi^+(S) \not\cong S$, contrary to the assumption. Hence $\phi^-\phi^+(S_i) \cong S_i$ for every i .

For systems reduced from below the proof is similar.

Proposition 2.13. *For any system* S *we have the relations*

(2.12) $$\phi^+\phi^-\phi^+(S) \cong \phi^+(S),$$

(2.13) $$\phi^-\phi^+\phi^-(S) \cong \phi^-(S).$$

In other words, $\phi^+(S)$ *is reduced from below, while* $\phi^-(S)$ *is reduced from above.*

Let us prove the isomorphism (2.12). Put, in the same way as before, $\phi^+(S) = \{P^+; E_1^+, \dots, E_r^+\}$. It is sufficient to show that

(2.14) $$(\bigcap_{j \neq t} E_j^+) = 0$$

for every t.

By definition, the subspace E_j^+ consists of vectors

$$\xi_j^0 = (x_1^0, \dots, x_{j-1}^0, 0, x_{j+1}^0, \dots, x_r^0)$$

with $\sum_{i \neq j} x_i^0 = 0$. Consequently, the subspace $(\bigcap_{j \neq t} E_j^+)$ consists of vectors

$$(0, \dots, 0, x_t^0, 0, \dots, 0)$$

such that $x_t^0 = 0$. Equality (2.14), and at the same time the isomorphism (2.12), are proved.

The isomorphism (2.13) is dual to the isomorphism (2.12).

To conclude this section we state two interesting assertions without proof.

Proposition 2.14. *Let S be any system.*

a) *Then there exist two subsystems $S^{(1)}$, $S^{(0)}$ such that*

$$S \cong S^{(1)} \oplus S^{(0)} .$$

The subsystem $S^{(1)}$ in uniquely determined and $S^{(1)} \cong \Phi^-\Phi^+(S) = \Phi^-\Phi^+(S^{(1)})$, i.e. $S^{(1)}$ is reduced from above; the subsystem $S^{(0)}$, in turn, is simplest relative to the functor Φ^+, i.e. $\Phi^+(S^{(0)}) = 0$.

b) *There exist two subsystems $S(1), S(0)$ such that*

$$S \cong S(1) \oplus S(0) .$$

The subsystem $S(0)$ is uniquely determined and $\Phi^-(S^{(0)}) = 0$; i.e. $S(0)$ is a simplest system relative to the functor Φ^-, whereas the subsystem $S(1)$ satisfies the relation $S(1) \cong \Phi^+\Phi^-(S) \cong \Phi^+\Phi^-(S^{(1)})$, i.e. $S(1)$ is reduced from below.

Theorem 2.15 of the next section follows at once from these assertions. In order to save space we shall prove Theorem 2.15 directly.

8. The action of the functors Φ^+, Φ^- on non-decomposable systems. In this section we prove some rather important properties of the functors Φ^+, Φ^-.

Theorem 2.15. *Let S be a non-decomposable system, different from zero. Then there are two mutually exclusive possibilities:*

1) $S^+ = 0$ *, which is, in the case of non-decomposable systems, equivalent*

to the inequality $\Phi^-\Phi^+(S) \not\cong S$.

2) $S^+ \neq 0$, *which is, for non-decomposable systems, equivalent to the equality* $\Phi^-\Phi^+(S) \cong S$.

Analogously, for the system S^- *we have: either*

3) $S^- = 0$ *i.e.* $\Phi^+\Phi^-(S) \not\cong S$, *or*

4) $S^- \neq 0$ *i.e.* $\Phi^+\Phi^-(S) \cong S$.

Proof. Let $S^+ = 0$. Then $\Phi^-\Phi^+(S) = \Phi^-(S^+) = \Phi^-(0) = 0$, i.e. $\Phi^-\Phi^+(S) \not\cong S$.

Assume, conversely, that the system S is non-decomposable and $\Phi^-\Phi^+(S) \not\cong S$. From Proposition 2.11 it follows that for some t

$$\sum_{j \neq t} E_j \neq P.$$

In this connection two cases are possible: 1) $\sum_j E_j \neq P$; 2) $\sum_j E_j = P$. Let us treat them separately.

1) Let $\sum_j E_j \neq P$. Denote the subspace $\sum_j E_j$ by P_0. Since P is a space over the field K, there exists a subspace $\bar{P}_0$ complementary to P_0 i.e. such that $P_0 \cap \bar{P}_0 = 0$, $P_0 + \bar{P}_0 = P$. Put $S_0 = \{P_0; E_1, \dots, E_r\}$, $\bar{S}_0 = \{\bar{P}_0; 0, \dots, 0\}$. It is easy to see that $S = S_0 \oplus \bar{S}_0$. By assumption the system S is non-decomposable and $S_0 \subset S$, $S_0 \neq S$. Therefore $S_0 = 0$, $S = \bar{S}_0$. It is not difficult to recognize that $\bar{S}_0$ is a simplest system.

Consequently, if $\sum_j E_j \neq P$, then $\Phi^+(S) = 0$.

2) Let $\sum_j E_j = P$, $\sum_{j \neq t} E_j \neq P$. Introduce the notations $P_t = \sum_{j \neq t} E_j$, $E'_t = E_t \cap P_t$,

$$S'_t = \{P_t; E_1, \dots, E_{t-1}, E'_t, E_{t+1}, \dots, E_r\}.$$

Choose a subspace $\bar{E}_t$ which is complementary to E'_t in E_t, i.e. $\bar{E}_t \oplus E'_t = E_t$. Further put $\bar{P}_t = \bar{E}_t$ and

$$\bar{S}_t = \{\bar{P}_t; 0, \dots, 0, \bar{E}_t, 0, \dots, 0\}.$$

It is easy to see that $S = S'_t \oplus \bar{S}_t$. By assumption the system S is non-decomposable and $P_t \neq P$. Thus $P_t = 0$, $S_t' = 0$. Hence $S = \bar{S}_t$. It is easy to recognize that $\bar{S}_t$ is a simplest system: $\Phi^+(\bar{S}_t) = 0$. Therefore also in this case $\Phi^+(S) = 0$.

We have proved that the assertions $S^+ = 0$ and $\Phi^-\Phi^+(S) \not\cong S$ are equivalent provided the system $S \neq 0$ is non-decomposable. Consequently, the assertions $S^+ \neq 0$ and $\Phi^-\Phi^+(S) \cong S$ are also equivalent.

The second part of the theorem can be proved in a dual way.

Theorem 2.16. *If the system* S *is non-decomposable, then also* S^+ *and* S^- *are non-decomposable.*

Proof. From Theorem 2.4 it follows that there are two possibilities: either $S^+ = 0$ or $S^+ \neq 0$. If $S^+ = 0$, then the theorem is proved.

Let us now consider the second possibility: $S^+ \neq 0$. This is equivalent to S being reduced from above, i.e. $\Phi^-\Phi^+(S) \cong S$.

Suppose that the system $S^+ = \Phi^+(S)$ can be decomposed into the direct sum of non-decomposable systems: $S^+ = \bigoplus_{i=1}^{k} S_i^+$, where $S_i^+ \neq 0$ and $k > 1$. According to Proposition 2.13 the system $S^+ = \Phi^+(S)$ is reduced from below, i.e. $\Phi^+\Phi^-(S^+) \cong S^+$. Therefore, on account of Proposition 2.12, the same can be stated for every subsystem $S_i^+ \subset S^+$, i.e. $\Phi^+\Phi^-(S_i^+) \cong S_i^+$. Since the subsystem S_i^+ is non-decomposable, from Theorem 2.15 it follows that $\Phi^-(S_i^+) \neq 0$.

Making use of Theorem 2.3 we find: $\Phi^-(S^+) \cong \bigoplus_{i=1}^{k} (\Phi^-(S_i^+))$, where $k > 1$ and, as we have shown, $\Phi^-(S_i^+) \neq 0$. Thus we obtain that the system $\Phi^-(S^+) = \Phi^-\Phi^+(S)$ is decomposable. Moreover, it has been previously proved that $\Phi^-\Phi^+(S) \cong S$. Hence S is decomposable, in contradiction with the basic assumption. That S^- is non-decomposable can be proved in a similar fashion.

9. Families of non-decomposable systems. Let $\{P; E_1, \ldots, E_r\}$ be a non-decomposable system. Put $S^{(1)} = \Phi^+(S), \ldots, S^{(i+1)} = \Phi^+(S^{(i)}), \ldots, S^{(-1)} = \Phi^-(S), \ldots, S^{(-i-1)} = \Phi^-(S^{(-i)}), \ldots$. We obtain a sequence of systems:

$$\ldots, S^{(-i)}, \ldots, S^{(-2)}, S^{(-1)}, S, S^{(1)}, S^{(2)}, \ldots, S^{(i)}, \ldots \quad (2.15)$$

An arbitrary member of this sequence will be denoted by $S^{(i)}$. From Theorem 2.16 it follows that every $S^{(i)}$ is non-decomposable.

We define the family $A(S)$ corresponding to the system S as the class of all systems isomorphic to one of the systems $S^{(i)}$ appearing in the sequence (2.15). In other words, S_1 belongs to $A(S)$ if one of the following three conditions is fulfilled: either $S_1 \cong S$, or $S_1 \cong (\Phi^+)^i(S)$ for some $i > 0$, or $S_1 \cong (\Phi^-)^j(S)$ for some $j > 0$.

For any non-decomposable system S there are three possibilities available:

1) $S^{(k)} = 0$ for some $k > 0$, i.e. $(\Phi^+)^k(S) = 0$;

2) $S^{(k)} = 0$ for some $\ell < 0$, i.e. $(\Phi^-)^\ell(S) = 0$;

3) every member of the family $A(S)$ is different from zero.

Note that in the case 1) $S^{(i)} = 0$ for every $i \geq k$, while in the case 2) $S^{(j)} = 0$ for every $j \leq \ell$. In the cases 1) and 2) we shall say that the family $A(S)$ is elementary.

Proposition 2.17. *Let* $S^{(i_1)}, S^{(i_2)}$ *be two non-zero systems belonging to* $A(S)$. *Let* $i_1 > i_2$. *Then*

$$S^{(i_2)} \cong (\Phi^+)^{i_2 - i_1}(S^{(i_1)}), \tag{2.16}$$

$$S^{(i_1)} \cong (\Phi^-)^{i_2 - i_1}(S^{(i_2)}). \tag{2.17}$$

We first carry out the proof for the case $i_2 = i_1 + 1$.

We shall consider two possibilities separately: 1) $i_1 \geq 0$; 2) $i_1 < 0$.

1) Let $i_1 \geq 0$, $i_2 = i_1 + 1$. Then, by definition,

$$S^{(i_2)} = \Phi^+(S^{(i_1)}).$$

Since $\Phi^+(S^{(i_1)}) \neq 0$, therefore, in view of Theorem 2.15, $\Phi^-\Phi^+(S^{(i_1)}) \cong S^{(i_1)}$. Substituting $\Phi^+(S^{(i_1)})$ by the isomorphic system $S^{(i_2)}$ we obtain: $\Phi^-(S^{(i_2)}) \cong S^{(i_1)}$.

2) Let $i_1 < 0$, $i_2 = i_1 + 1$. Then, by definition,

$$S^{(i_1)} = \Phi^-(S^{(i_2)}).$$

Since $\Phi^-(S^{(i_2)}) \neq 0$, therefore, in view of Theorem 2.15, $\Phi^+\Phi^-(S^{(i_2)}) \cong S^{(i_2)}$. Substituting in this relation $\Phi^-(S^{(i_2)}) = S^{(i_1)}$ we find:

$$\Phi(S^{(i_1)}) \cong S^{(i_2)} .$$

We have proved that if the systems $S^{(i)}$, $S^{(i+1)}$ are different from zero, then

$$S^{(i+1)} \cong \Phi^+(S^{(i)}),$$

$$S^{(i)} \cong \Phi^-(S^{(i+1)}) .$$

Let us deal now with the general case. If the systems $S^{(i_1)}, S^{(i_2)}$ $(i_1 < i_2)$ are different from zero, then it can be easily shown that every $S^{(i)}$ with $i_1 \le i \le i_2$ is also different from zero. Therefore, making use of the above argument, we may write: $S^{(i_1+1)} \cong \Phi^+(S^{(i_1)})$, $S^{(i_1+2)} \cong \Phi^+(S^{(i_1+1)}) \cong (\Phi^+)^2(S^{(i_1)}), \ldots, S^{(i_2)} \cong$ $\cong (\Phi^+)^{i_2-i_1}(S^{(i_1)})$.

The isomorphism (2.17) can be proved similarly.

Corollary 2.18. *Let* S_1 *be a non-zero system belonging to the family* $A(S)$ *Then the families* $A(S)$ *and* $A(S_1)$ *coincide.*

Proof. From Proposition 2.17 it follows that if $S_1 \in A(S)$, $S_1 \neq 0$, then for every $k > 0$ the system $S_2 = (\Phi^+)^{\ell}(S_1)$ and for every $\ell > 0$ the system $S_3 = (\Phi^-)^{\ell} S_1$ belongs to $A(S)$. The corollary is proved.

Proposition 2.19. a) *Let* A *be an elementary family of non-decomposable systems containing a simplest system* $S_0 \neq 0$, $\Phi^+(S_0) = 0$. *Then for every non-zero system* $S \in A$ *there exists a number* $k > 0$ *such that*

$$S \cong (\Phi^-)^k(S_0) .$$

b) *Let* A *be an elementary family of non-decomposable systems containing a simplest system* $S^0 \neq 0$, $\Phi^+(S_0) = 0$. *Then for every non-zero system* $S \in A$ *there exists a number* $k > 0$ *such that*

$$S \cong (\Phi^+)^{\ell}(S^0) .$$

Proof. Since $S_0 \in A$ and $S_0 \neq 0$, therefore, in view of Corollary 2.18, $A = A(S_0)$. By definition $\Phi^+(S_0) = 0$. Hence also $(\Phi^+)^k(S_0) = 0$ for every $k > 0$. All other systems in $A(S_0)$ can be represented in the form $S = (\Phi^-)^k(S_0)$. Proposition 2.19 a) is proved. Proposition 2.19 b) can be established in a similar way.

10. Let $S = \{P; E_1, \ldots, E_r\}$ be a non-decomposable system and $A(S)$ the corresponding family. Introduce the notations $\dim P = n$, $\dim E_t = m_t$. If $S_1 = S^{(i)} = \{P^{(i)}; E_1^{(i)}, \dots, E_r^{(i)}\}$ is another system belonging to $A(S)$, we set $n^{(i)} = \dim P^{(i)}$, $m_t^{(i)} = \dim E_t^{(i)}$.

Let us express $n^{(i)}$ and $m_t^{(i)}$ as functions of the numbers n and $\{m_t\}_{t=1,\dots,r}$.

Proposition 2.20. *Let* $S = \{P; E_1, \dots, E_r\}$ *be a non-decomposable system. Let* $S^{(1)} = \Phi^+(S) = \{P^{(1)}; E_1^{(1)}, \dots, E_r^{(1)}\}$, $S^{(1)} \neq 0$. *Then*

(2.18)
$$n^{(1)} = \sum_{t=1}^{r} m_t - n ,$$

(2.19)
$$m_t^{(1)} = \sum_{j \neq t} m_j - n .$$

Proof. a) Recall that the space $P^{(1)}$ appearing in the system $S^{(1)} = \Phi^+(S)$ is equal to the kernel of the mapping $\tau: R \longrightarrow P$, where $R = \bigoplus_{t=1}^{r} E_t$. Thus

(2.20)
$$n^{(1)} = \dim P^{(1)} = \dim(\operatorname{Ker}\tau) = \dim R - \dim(\operatorname{Im}\tau) = \\ = \sum_{t=1}^{r} m_t - \dim(\operatorname{Im}\tau).$$

From the definition of τ it follows that $\operatorname{Im}\tau = (\sum_{t=1}^{r} E_t)$. The system S is, according to the assumption, non-decomposable and we have $\Phi^+(S) \neq 0$. Therefore, by Theorem 2.15, S is reduced from above, i.e. $\Phi^-\Phi^+(S) \cong S$. In other words, $\sum_{j \neq t} E_j = P$ for every t. Hence $\operatorname{Im}\tau = P$, $\dim(\operatorname{Im}\tau) = n$. Substituting this expression into (2.20) we find

$$n^{(1)} = \sum_{t} m_t - n .$$

b) Now let us prove (2.19). By definition, the subspace $E_t^{(1)}$ consists of all vectors

$$(x_1, \dots, x_{t-1}, 0, x_{t+1}, \dots, x_r)$$

such that $\sum_{j \neq t} x_j = 0$. Therefore $E_t^{(1)}$ is isomorphic to the kernel of the mapping

$\tau_t : R_t \longrightarrow P$, where $R_t = \bigoplus_{j \neq t} E_j$. Consequently,

$$m_t^{(1)} = \dim E_t^{(1)} = \dim(\mathrm{Ker}\,\tau_t) = \dim R_t - \dim(\mathrm{Jm}\,\tau_t) =$$

$$= \sum_{j \neq t} m_j - \dim(\mathrm{Jm}\,\tau_t) .$$

It is clear that $\mathrm{Jm}\,\tau_t = \sum_{j \neq t} E_j$. As S is reduced from above, $\mathrm{Jm}\,\tau_t = P$. Thus

$$m_t^{(1)} = \sum_{j \neq t} m_j - \dim(\mathrm{Jm}\,\tau_t) = \sum_{j \neq t} m_j - n .$$

The proposition is proved.

Proposition 2.21. *Let* $A(S)$ *be a family of non-decomposable systems, and let* $S^{(k)} = \{P^{(k)}; E_1^{(k)}, \dots, E_r^{(k)}\}$, $S^{(k+1)} = \{P^{(k+1)}; E_1^{(k+1)}, \dots, E_r^{(k+1)}\}$ *be two non-zero systems belonging to this family. Then*

(2.21) $$n^{(k+1)} = \sum_t m_t^{(k)} - n^{(k)} ,$$

(2.22) $$m_t^{(k+1)} = \sum_{j \neq t} m_j^{(k)} - n^{(k)} .$$

Proof. From Proposition 2.17 it follows that $S^{(k+1)} \cong \Phi^+(S^{(k)})$. Consequently, Proposition 2.20 may be applied to the systems $S^{(k)}, S^{(k+1)}$, which at once yields the validity of (2.21) and (2.22).

We are going to utilize the system (2.21), (2.22) of recursion formulas for finding the interdependence between the numbers $n^{(k)}$ with different k. For this sake we calculate the value $n^{(k+2)}$:

$$n^{(k+2)} = \sum_{t=1}^{r} m_t^{(k+1)} - n^{(k+1)} = \sum_{t=1}^{r} \Big(\sum_{j \neq t} m_j^{(k)} - n^{(k)} \Big) - n^{(k+1)} =$$

$$= (r-1) \sum_{t=1}^{r} m_t^{(k)} - r n^{(k)} - n^{(k+1)} .$$

Equality (2.21) implies that $\sum_{t=1}^{r} m_t^{(k)} = n^{(k+1)} + n^{(k)}$. Substituting this expression into the preceding formula we obtain:

$$n^{(k+2)} = (r-1)(n^{(k+1)} + n^{(k)}) - rn^{(k)} - n^{(k+1)} = (r-2)n^{(k+1)} - n^{(k)} .$$

Thus we have the following result:

Proposition 2.22. *Let* $S^{(k)}, S^{(k+1)}, S^{(k+2)}$ *be non-zero, non-decomposable systems belonging to the family* $A(S)$. *Then the dimensions of the spaces* $P^{(k)}, P^{(k+1)}$, $P^{(k+2)}$ *are related by the equation*

$$n^{(k+2)} - (r-2)n^{(k+1)} + n^{(k)} = 0 . \tag{2.23}$$

The first thing that becomes apparent when solving this system of equations is the exceptional behaviour of systems with $r = 4$. We are going to treat the cases $r = 4$ and $r > 4$ separately.

For $r = 4$ equation (2.23) can be brought to the form

$$n^{(k+2)} - n^{(k+1)} = n^{(k+1)} - n^{(k)} . \tag{2.24}$$

Recall that the number $\sum_{t=1}^{4} m_t - 2n$ is termed the defect $\rho(S)$ of the system $S = \{P; E_1, \dots, E_4\}$. From (2.21) it follows that

$$n^{(k+1)} - n^{(k)} = \sum_{t=1}^{4} m_t^{(k)} - 2n^{(k)} = \rho(S^{(k)}) .$$

Therefore the equality (2.24) can be written so:

$$\rho(S^{(k+1)}) = \rho(S^{(k)}) . \tag{2.25}$$

By the way, equality (2.25) can also be obtained directly from (2.21), (2.22). Hence (2.25) holds for arbitrary non-zero systems $S^{(k)}, S^{(k+1)} \in A(S)$.

Proposition 2.22. a) *If the systems* $S, S^{(1)}, \dots, S^{(k)}$ $(k \geq 1)$ *appearing in the family* $A(S)$ *are different from zero, then each of then has the same defect* $\rho = \rho(S)$. *Moreover,*

$$n^{(i)} = n^{(0)} + i\rho(S); \qquad i = 1, 2, \dots, k .$$

b) *If the systems* $S^{(-k)}, \dots, S^{(-1)}, S$ $(k \geq 1)$ *appearing in the family* $A(S)$ *are different from zero, then each of them has the same defect* $\rho = \rho(S)$. *Moreover,*

$$n^{(i)} = n^{(0)} + i\rho(S); \qquad i = -1, -2, \dots, -k .$$

The proof of this proposition evidently follows from the equalities (2.24), (2.25).

In the next § we study more closely those families $A(S)$ of quadruples satisfying the relation $\rho(S) \neq 0$.

In the case $r > 4$ the solution of the system of equations (2.23) takes a fundamentally different form:

$$n^{(k)} = C\mu_1^k + D\mu_2^k ,$$

where μ_1, μ_2 are the roots of the equation $x^2 - (r-2)x + 1 = 0$. These roots are given, respectively, by the formula $\mu_{1,2} = \frac{1}{2}(r-2 \pm \sqrt{r(r-4)})$. Here $\mu_1\mu_2 = 1$. Setting $\mu = \frac{1}{2}(r-2+\sqrt{r(r-4)})$ we obtain:

$$n^{(k)} = C\mu^k + D\mu^{-k} .$$

If $k > 0$, then the constants C, D can be determined from the system of equalities

$$n = n^{(0)} = C + D ,$$

$$n^{(1)} = C\mu + D\mu^{-1} .$$

On account of Proposition 2.20 we have $n^{(1)} = \sum_{j=1}^{r} m_j - n$. Substitute $\sum_{j=1}^{r} m_j = r\bar{m}$. Then, after simple calculations, we obtain:

$$C = \frac{2\sqrt{r}\,\bar{m} - n(\sqrt{r} - \sqrt{r-4})}{2\sqrt{r-4}} ,$$

$$D = \frac{n(\sqrt{r} + \sqrt{r-4}) - 2\sqrt{r}\,\bar{m}}{2\sqrt{r-4}} .$$

§3. CLASSIFICATION OF NON-DECOMPOSABLE QUADRUPLES WITH NON-ZERO DEFECT

In this § we give a complete classification of elementary quadruples and describe their canonical form.

Recall that there exist 5 mutually non-isomorphic non-decomposable quadruples satisfying the condition $\Phi^{+}(S)=0$ and 5 mutually non-isomorphic non-decomposable quadruples for which $\Phi^{-}(S)=0$. We now introduce new notations. Set

$$S(1,-2)=\{P;0,0,0,0\};$$

$$S(1,2)=\{P;E_1,E_2,E_3,E_4\}, \text{ where every } E_j=P;$$

$$S_i(1,-1)=\{P;E_1,E_2,E_3,E_4\}, \text{ where } E_j|_{j\neq i}=0, \quad E_i=P;$$

$$S_i(1,1)=\{P;E_1,E_2,E_3,E_4\}, \text{ where } E_j|_{j\neq i}=P, \quad E_i=0.$$

In the notation $S(1,-2)$ the first number equals $\dim P$, whereas the second indicates the defect $\rho(S)$. Consequently, non-decomposable quadruples satisfying the condition $\Phi^{+}(S)=1$ have defect -2 or -1, while those satisfying $\Phi^{-}(S)=0$ have defect 1 or 2. We shall see that all non-decomposable quadruples with non-zero defect can be obtained from simplest quadruples under the action of the functors Φ^{+} or Φ^{-}.

Theorem 3.1. *Let S be a non-decomposable quadruple with non-zero defect. Then the defect $\rho(S)$ is equal to one of the numbers $-2,-1,1,2$. Moreover:*

1) *If $\rho(S)=-2$, then $\dim P=n(S)=2k+1$ and*

$$S\cong(\Phi^{-})^{k}S(1,-2); \tag{3.1}$$

2) *if $\rho(S)=-1$, then $n(S)=n$ can be any positive integer, and for some i we have*

$$S\cong(\Phi^{-})^{n-1}S_i(1,-1); \tag{3.2}$$

3) *if $\rho(S)=1$, then $n(S)=n$ can be any positive integer, and for some i we have*

$$S\cong(\Phi^{+})^{n-1}S_i(1,1); \tag{3.3}$$

4) *if $\rho(S)=2$, then $n(S)=2k+1$ and*

(3.4) $$S \cong (\Phi^+)^k S(1,2).$$

Proof. Let the system S be such that $\rho(S) < 0$. Let us first prove that the family $A(S)$ is elementary and contains a system $\tilde{S}$ with $\Phi^+(\tilde{S}) = 0$. Suppose the contrary: let each of the systems $S, S^{(1)}, S^{(2)}, \dots, S^{(i)}, \dots$ be different from zero, i.e. $(\Phi^+)^i S \neq 0$ for every i. Proposition 2.22 implies that in this case all systems $S^{(i)}$ have the same defect $\rho(S)$ and the following relation holds:

(3.5) $$n^{(i)} = n^{(0)} + i\rho(S), \quad \text{where } n^{(0)} = n(S).$$

By assumption, $\rho(S) < 0$. Hence the numbers $n^{(i)}$ $(i = 0,1,2,\dots)$ form a decreasing arithmetical progression. Therefore one can find an i such that $n^{(i)} < 0$. But this is impossible, since $n^{(i)} = \dim P^{(i)} \geq 0$.

Consequently, there exists a $k > 0$ such that $S^{(k-1)} \neq 0$, $S^{(k)} = \Phi^+(S^{(k-1)}) = 0$. In other words, the family $A(S)$ is elementary and contains a **simplest system** $S_0 = S^{(k-1)}$ relative to the functor Φ^+.

According to Proposition 2.19 the family $A(S)$ coincides with $A(S_0)$. Note that the simplest system S_0 $(\Phi^+(S_0) = 0)$ is isomorphic to one of the 5 systems $S(1,-2)$; $S_i(1,-1)$, $i = 1,2,3,4$. Thus the defect $\rho(S_0)$ of the system S_0 is equal to -2 or -1.

Let us now show that in the family $A(S_0) = A(S)$ all systems of the form $S_0^{(-i)} = (\Phi^-)^i S_0$, $i > 0$, are different from zero. Suppose the contrary: let the systems $S_0^{(-k)}, \dots, S_0^{(-1)}, S_0$ be different from zero, but $S_0^{(-k-1)} = 0$. Then $\Phi^- S_0^{(-k)} = 0$, i.e. the system $S_0^{(-k)}$ is elementary with respect to the functor Φ^-; so its defect is 1 or 2. On the other hand, on account of Proposition 2.22, each of the systems $S_0^{(-k)}, \dots, S_0^{(-1)}, S_0$ has the same defect $\rho = \rho(S_0)$, which is less than zero as we have just shown. We got a contradiction.

Thus all systems $S_0^{(-i)} = (\Phi^-)^i S_0$, $i = 1,2,\dots$ are different from zero. From Proposition 2.22 it follows that they have each the same defect $\rho = \rho(S_0)$ and

(3.6) $$n(S_0^{(-i)}) = n(S_0) - i\rho(S_0).$$

The system S with which we began our construction belongs, by Proposition 2.19, to the family $A(S_0) = A(S)$. Hence there exists a number $k > 0$ such that $S \cong (\Phi^-)^k(S_0)$. We know that S_0 is a simplest system with respect to Φ^+

and equals one of the 5 systems $S(1,-2)$; $S_i(1,-1)$, $i=1,2,3,4$.

Thus: 1) if $S_0 = S(1,-2)$, then $\varrho(S) = \varrho(S_0) = -2$,

$$S \cong (\Phi^-)^k S(1,-2)$$

and, in view of (3.6),

$$n(S) = 1+2k\,;$$

2) if $S_0 = S_i(1,-1)$, where $i \in \{1,2,3,4\}$, then $\varrho(S) = \varrho(S_0) = -1$,

$$S \cong (\Phi^-)^k S_i(1,-1)$$

and

$$n(S) = 1+k\,.$$

Consequently, in the case $\varrho(S) < 0$ the theorem is proved.

The proof of the theorem for the case $\varrho(S) > 0$ can be performed in a similar way. The difference is that in the case $\varrho(S) > 0$ the functors Φ^+, Φ^- must be interchanged. The theorem is proved.

This theorem yields, in essence, a complete description of all non-decomposable quadruples with defect $\varrho(S) \neq 0$.

2. Let us now show that the canonical systems $S_i(n, \varrho)$, $\varrho = \pm 1$, $i=1,2,3,4$; $S(2k+1,\varrho)$, $\varrho = \pm 2$ described in Section 3 of § 1 are non-decomposable and they exhaust, up to equivalence, all non-decomposable systems.

Theorem 3.2. *The systems* $S_i(n,\varrho)$, $\varrho = \pm 1$, $i = 1,2,3,4$; $S(2k+1,\varrho)$, $\varrho = \pm 2$ *are elementary and non-decomposable. Moreover,*

(3.7) $$S_i(n,-1) \cong (\Phi^-)^{n-1} S_i(1,-1)\,,$$

(3.8) $$S_i(n,1) \cong (\Phi^+)^{n-1} S_i(1,1)\,,$$

(3.9) $$S(2k+1,-2) \cong (\Phi^-)^k S(1,-2)\,,$$

(3.10) $$S(2k+1,2) \cong (\Phi^+)^k S(1,2)\,.$$

The proof will be carried out in the case of the isomorphism (3.8). We first remark that the system $S_i(1,1)$ given by the expressions on pp. 168-169 is a simplest non-decomposable system satisfying $\Phi^-(S_i(1,1)) = 0$. It is also easy to see that for every permutation σ we have

$$\sigma S_i(1,1) \cong S_{\sigma(i)}(1,1).$$

We are going to prove the isomorphism (3.8) by induction. Let $S_i(2k,1) \cong (\Phi^+)^{2k-1} S_i(1,1)$ and $\sigma S_i(2k,1) \cong S_{\sigma(i)}(2k,1)$.

a) Let us prove that $S_i(2k+1,1) \cong \Phi^+(S_i(2k,1))$. Put $i = 3$. Recall that $S_3(2k,1)$ has the following canonical form:

$$
\begin{aligned}
P &= \{e_1, \dots, e_k, f_1, \dots, f_k\}, \\
E_1 &= \{e_1, \dots, e_k\}, \\
E_2 &= \{f_1, \dots, f_k\}, \\
E_3 &= \{e_1, e_2 + f_1, \dots, e_k + f_{k-1}, f_k\}, \\
E_4 &= \{e_1 + f_1, \dots, e_k + f_k\}.
\end{aligned} \tag{3.11}
$$

Denote the basis vectors of E_3 and E_4 by $g_1, \dots, g_{k+1}$ and $h_1, \dots, h_k$, respectively. Then the following equalities are valid:

$$
\begin{array}{ll}
e_1 - g_1 = 0, & e_1 + f_1 - h_1 = 0, \\
e_2 + f_1 - g_2 = 0, & e_2 + f_2 - h_2 = 0, \\
\dots\dots\dots & \dots\dots\dots \\
e_k + f_{k-1} - g_k = 0, & e_k + f_k - h_k = 0. \\
f_k - g_{k+1} = 0, &
\end{array} \tag{3.12}
$$

Recall that the space P^+ occuring in the system $\Phi^+(S)$ consists of all vectors (x_1, x_2, x_3, x_4) with $x_i \in E_i$, $\sum_i^4 x_i = 0$. The subspace E_1^+ consists of all vectors $\xi = (0, x_2, x_3, x_4)$ with $\sum_{i=2}^4 x_i = 0$, etc.

Introduce the following notations:

$$(-e_1,-f_1,0,h_1) = g_1^+, \qquad (e_1,0,-g_1,0) = h_1^+,$$

$$(-e_2,-f_2,0,h_2) = g_2^+, \qquad (e_2,f_1,-g_2,0) = h_2^+,$$

$$\cdots\cdots \qquad \cdots\cdots$$

$$(-e_k,-f_k,0,h_k) = g_k^+, \qquad (e_k,f_{k-1},-g_k,0) = h_k^+,$$

$$(0,f_k,-g_{k+1},0) = h_{k+1}^+ .$$

It is not difficult to show that the subspace of $R = \oplus E_i$ spanned by the vectors $g_1^+,\dots,g_k^+, h_1^+,\dots,h_{k+1}^+$ is the desired subspace P^{+^i}; in particular, $\dim P^+ = 2k+1$. Moreover, the subspace spanned by the vectors $g_1^+,\dots,g_k^+$ is E_3^+ so that $\dim E_3^+ = k$, and the subspace spanned by the vectors $h_1^+,\dots,h_{k+1}^+$ is E_4^+ so that $\dim E_4^+ = k+1$.

Further set

$$e_1^+ = g_1^+ + h_1^+ = (0,-f_1,-g_1,h_1),$$

$$e_2^+ = g_2^+ + h_2^+ = (0,f_1-f_2,-g_2,h_2),$$

$$\cdots\cdots$$

$$e_k^+ = g_k^+ + h_k^+ = (0,f_{k-1}-f_k,-g_k,h_k),$$

$$e_{k+1}^+ = h_{k+1}^+ = (0,f_k,-g_{k+1},0);$$

$$f_1^+ = h_1^+ = (e_1,0,-g_1,0),$$

$$f_2^+ = g_1^+ + h_2^+ = (e_2-e_1,0,-g_2,h_1),$$

$$\cdots\cdots$$

$$f_k^+ = g_{k-1}^+ + h_k^+ = (e_k-e_{k-1},0,-g_k,h_{k-1}),$$

$$f_{k+1}^+ = g_k^+ + h_{k+1}^+ = (-e_k,0,-g_{k+1},h_k).$$

It is not difficult to show that

$$E_1^+ = \{e_1^+, \dots, e_{k+1}^+\}, \qquad E_2^+ = \{f_1^+, \dots, f_{k+1}^+\} \quad .$$

and $\dim E_1^+ = \dim E_2^+ = k+1$.

Thus we have constructed the system $\Phi^+(S_3(2k,1)) = \{P^+; E_1^+, \dots, E_4^+\}$, where

$$P^+ = \{g_1^+, \dots, g_k^+, h_1^+, \dots, h_{k+1}^+\},$$
$$E_1^+ = \{g_1^+ + h_1^+, g_2^+ + h_2^+, \dots, g_k^+ + h_k^+, h_{k+1}^+\},$$
$$E_2^+ = \{h_1^+, g_1^+ + h_2^+, \dots, g_{k-1}^+ + h_k^+, g_k^+ + h_{k+1}^+\},$$
$$E_3^+ = \{g_1^+, g_2^+, \dots, g_k^+\},$$
$$E_4^+ = \{h_1^+, h_2^+, \dots, h_k^+, h_{k+1}^+\}.$$

This system can be obtained from the canonical system $S_2(2k+1,1)$ (see p. 168) by means of the permutation $\sigma = \begin{pmatrix} 1 & 2 & 3 & 4 \\ 4 & 3 & 2 & 1 \end{pmatrix}$, i.e.

$$\Phi^+(S_3(2k,1)) \cong \sigma_{1,4}\, \sigma_{2,3}\, S_2(2k+1,1) .$$

By definition (see p. 169), $\sigma_{2,3} S_2(2k+1,1) = S_3(2k+1,1)$. Thus we have proved the isomorphism

$$\Phi^+(S_3(2k,1) \cong \sigma_{1,4}\, S_3(2k+1,1) .$$

According to the induction hypothesis $S_3(2k,1)$ is non-decomposable and we have $S_3(2k,1) \cong (\Phi^+)^{2k-1} S_3(1,1)$. Consequently, $\sigma_{1,4} S_3(2k+1,1) \cong$ $\cong (\Phi^+)^{2k} S_3(1,1)$. Therefore the system $\sigma_{1,4} S_3(2k+1,1)$ is elementary and non-decomposable.

It is obvious that rearranging the indices of the subspaces in a non-decomposable system S, $\rho(S)=1$, we get a non-decomposable system with defect $\rho(S)=1$ and dimension $\dim P$ unchanged. For instance, the system $\sigma_{1,4} S = \sigma_{1,4}\sigma_{1,4} S_3(2k+1,1) =$ $= S_3(2k+1,1)$ is non-decomposable.

From Theorem 3.1 it follows easily that for every pair of numbers $(n(S), \rho(S))$ of the form $(n,1)$ there exist 4 mutually non-isomorphic, non-decomposable systems: $S_i(n,1) = (\Phi^+)^{n-1} S_i(1,1), \quad i = 1,2,3,4$.

Note that for the system $S_i(2k+1,1)$ we have the relations $\dim E_i = k$,

$\dim E_j|_{j\neq i} = k+1$. Thus the four systems $S_i(2k+1,1)$, $i = 1,2,3,4$, are pairwise non-isomorphic and they exhaust the class of non-decomposable systems with $\rho(S)=1$, $n(S) = 2k+1$.

Hence it follows that

$$\sigma S_i(2k+1,1) \cong S_{\sigma(i)}(2k+1,1)$$

for every permutation σ . So we have proved that

$$S_3(2k+1,1) \cong (\Phi^+)^{2k} S_3(1,1) .$$

Performing in this relation the necessary change of the indices of the subspaces E_i entering into the systems $S_3(1,1)$, $S_3(2k+1,1)$ we obtain:

$$S_i(2k+1,1) \cong (\Phi^+)^{2k} S_i(1,1) .$$

For completing the inductive proof of the isomorphism (3.8) it remains to prove that

$$\Phi^+(S_i(2k+1,1)) \cong S_i(2k+2,2) .$$

This happens in an analogous manner.

The isomorphism (3.10), that reads $S(2k+1,2) \cong (\Phi^+)^k S(1,2)$, can be established similarly. As to the isomorphisms (3.7) and (3.9), they are dual to (3.8) and (3.10).

The theorem is proved.

As an end of this § we announce the following generalization:

Theorem 3.3. *Let* S *be a non-decomposable system with defect* $\rho(S)\neq 0$ *. Then* S *is isomorphic to one and only one of the canonical non-decomposable systems* $S_i(n,\rho)$, $i=1,2,3,4$, $\rho=\pm 1$, $n=1,2,\dots,i$; $S(2k+1,\rho)$, $\rho=\pm 2$.

§4. REGULAR AND OPERATOR SYSTEMS

1. Systems that are equal to the direct sum of non-decomposable systems with defect zero will be called *regular* systems.

Theorem 4.1. *The system* S *is regular, i.e.* S *can be decomposed into the direct sum* $\bigoplus_r S_r$ *of non-decomposable systems* S_r *with defect* $\rho(S_r)$ *equal to zero if and only if* S *has defect zero and any subsystem* S_1 *of* S *has defect* $\rho(S_1)\leq 0$.

The proof of the theorem is based on the following simple lemma, which we state without proof.

Lemma 4.2. *If* $\mu: S \longrightarrow T$ *is a monomorphism, then* $\mu^+ = \Phi^+(\mu):$ $\Phi^+(S) \longrightarrow \Phi^+(T)$ *is also a monomorphism.*

The condition is necessary. Suppose that the system S contains a subsystem S_1, with defect $\varrho(S_1) > 0$. Then S_1 contains a non-decomposable subsystem S_2 with defect $\varrho(S_2) > 0$. Let $\mu: S_2 \longrightarrow S$ be the corresponding monomorphism. Owing to Lemma 4.2 the morphism $\mu^+ = \Phi^+(\mu): \Phi^+(S_2) \longrightarrow \Phi^+(S)$ will also be a monomorphism. Moreover, all morphisms $\mu^{(k)} = (\Phi^+)^k(\mu): S_2^{(k)} \longrightarrow S^{(k)}$, where $S^{(k)} = (\Phi^+)^k(S)$ are monomorphisms.

By definition, $\varrho(S_2) > 0$. Consequently (see Proposition 2.22), the dimensions $n^{(k)} = n(S_2^{(k)})$ of the systems $S_2^{(k)} = (\Phi^+)^k(S_2)$ form an arithmetical progression:

$$n^{(k)} = n(S_2) + \varrho(S_2)k.$$

Therefore the family

$$A(S_2) = \{\dots, S_2^{(-1)}, S_2, S_2^{(1)}, S_2^{(2)}, \dots\}$$

contains systems with arbitrarily high dimension. Let us now turn our attention to the family $A(S)$ of the system S. According to the hypothesis, the system S is regular, i.e. S can be decomposed into the direct sum $\bigoplus_r S_r$ of non-decomposable subsystems S_r with defect 0. Hence the system $S^{(k)} = (\Phi^+)^k(S) \in A(S)$ is also decomposable: $S^{(k)} \cong \bigoplus_r (\Phi^+)^k(S_2) = \bigoplus_r S_r^{(k)}$. Since $\varrho(S_2) = 0$, Proposition 2.22 yields

$$n(S_2^{(k)}) = n(S_2)$$

so that

$$n(S^{(k)}) = n(S).$$

Thus, on the one hand, for every k the mapping $\mu^{(k)}: S_2^{(k)} \longrightarrow S^{(k)}$ is a monomorphism while, on the other hand, there exists a k satisfying $n(S_2^{(k)}) > n(S^{(k)})$. We have got a contradiction.

The condition is sufficient. Let the system S have the property that

$\varrho(S)=0$ and that $\varrho(S')\leq 0$ for every subsystem $S'\subset S$. If S is non-decomposable, then $\varrho(S)=0$; hence S is regular.

If the system S is decomposable, $S=\bigoplus_{r=1}^{k} S_r$ $(k>1)$, then $\varrho(S)=$ $=\sum_r \varrho(S_r)$. By the assumption, $\varrho(S)=0$ and $\varrho(S_r)\leq 0$. Therefore $\varrho(S_r)=0$, i.e. the system S is regular. The theorem is proved.

2. Invariant subspaces. We shall say that the subspace F of the space P appearing in the regular system $S=\{P;\ E_1,\dots,E_4\}$ is invariant, if the corresponding subsystem $S_F=\{F;\ E_i\cap F\}_{i=1,2,3,4}$ has defect $\varrho(S_F)=0$.

Proposition 4.3. *Let* $S=\{P;\ E_1,\dots,E_4\}$ *and* $T=\{Q;\ F_1,\dots,F_4\}$ *be regular quadruples and let* $\alpha:\ S\longrightarrow T$ *be a morphism. Denote by* P^0 *and* Q' *the kernel and the image, respectively, of the linear mapping* $\alpha:\ P\longrightarrow Q$ *inducing the morphism* α. *Then* P^0 *and* Q *are invariant subspaces in* P *resp.* Q'.

Proof. Set $E_i^0=E_i\cap P^0$ and $S^0=\{P^0;\ E_1^0,\dots,E_4^0\}$. The system S being regular, $\varrho(S^0)\leq 0$.

Denote by S' the subsystem of T equal to $\{Q';\ \alpha E_1,\dots,\alpha E_4\}$, where $Q'=\alpha P$. We are looking for the defect $\varrho(S')$. By definition, P^0 is the kernel of the linear mapping $\alpha:\ P\longrightarrow Q$. Therefore

$$\dim Q'=\dim\alpha P=\dim P-\dim P^0,$$
$$\dim\alpha E_i=\dim E_i=\dim(E_i\cap P^0)=\dim E_i-\dim E_i^0.$$

Hence we obtain:

$$\varrho(S')=\sum_{i=1}^{4}\dim(\alpha E_i)-2\dim Q'=$$

$$=\sum_{i=1}^{4}\dim E_i-2\dim P-\Big(\sum_{i=1}^{4}\dim E_i^0-2\dim P^0\Big)=\varrho(S)-\varrho(S^0).$$

As the system S is regular, $\varrho(S)=0$, so that

(4.1) $$\varrho(S')=-\varrho(S^0).$$

Previously we have proved that $\varrho(S^0)\leq 0$. Now observe that S' being a subsystem of the regular system T we have $\rho(S') \leq 0$. Comparing these inequalities with (4.1) we obtain that $\rho(S^0) = \rho(S') = 0$. So the subspaces P^0 and $Q' = \alpha P$ are invariant. The proposition is proved.

The quotient system $S' = \{Q'; \alpha E_1, \dots, \alpha E_4\}$ is the co-image of the morphism α (in symbols: $S' = \mathrm{Coim}\, \alpha$). By T' we denote $\mathrm{Im}\,\alpha$, the image of the morphism α. Note that $T' = \{Q'; (F_1 \cap Q'), \dots, (F_4 \cap Q')\}$, where $Q' = \alpha P$. It is evident that $\alpha E_i \subseteq F_i \cap Q'$. Therefore

(4.2) $$\rho(S') \leq \rho(T').$$

As we have already proved, $\rho(S') = 0$. Moreover, T' is a subsystem of the regular system T, hence $\rho(T') \leq 0$. Comparing these relations with the inequality (4.2) we obtain: $\rho(S') = \rho(T')$. This means that $\alpha E_i = F_i \cap Q'$. Thus the systems

$$S' = \{Q'; \alpha E_1, \dots, \alpha E_4\}, \qquad T' = \{Q'; (F_1 \cap Q), \dots, (F_4 \cap Q')\}$$

are equal to each other, i.e.

$$\mathrm{Coim}\,\alpha = \mathrm{Im}\,\alpha.$$

Denote by C_0^4 the full subcategory of regular systems. From the equality (4.2) it follows that the subcategory C_0^4 is abelian.

If $S = \{P; E_1, \dots, E_4\}$ is a regular system, i.e. $S \in C_0^4$, then to the invariant subspace $F \subseteq P$ there corresponds the subsystem $S_F = \{F; F \cap E_1, \dots, F \cap E_4\}$, and $S_F \in C_0^4$, i.e. S_F is regular.

In the abelian subcategory C_0^4 the notion of quotient system can be defined in a natural way, so that we have:

Proposition 4.4. *Let* $S = \{P; E_1, \dots, E_4\}$ *be a regular system, and* F *an invariant subspace in* P. *Then the quotient system* $S' = \{P/F; \tau E_1, \dots, \tau E_4\}$, *where* $\tau: P \longrightarrow P/F$ *denotes the natural mapping, is also regular.*

Similarly, in the abelian subcategory C_0^4 the following result can be easily proved:

Proposition 4.5. *Let* $S = \{P; E_1, \dots, E_4\}, T = \{Q; F_1, \dots, F_4\}$ *be two regular systems, and let* $\alpha: S \longrightarrow T$ *be a morphism. If* G *is an invariant subspace in* Q, *then the inverse image* $\alpha^{-1}(G)$ *of* G *is an invariant subspace in* P.

We also omit the likewise not complicated proof of the following assertion:

Proposition 4.6. *The set of invariant subspaces in the regular system* S *is a lattice, i.e. if two subspaces* $F_1, F_2 \subseteq P$ *are invariant, then the subspaces* $F_1 \cap F_2$, $F_1 + F_2$ *are also invariant.*

Let us now show that in a regular system S the subspaces $E_i \cap E_j$ are invariant. Consider, as an example, the subspace $F = E_1 \cap E_2$. We have the relations $F \cap E_1 = F$, $F \cap E_2 = F$, $F \cap E_3 = E_1 \cap E_2 \cap E_3$, $F \cap E_4 = E_1 \cap E_2 \cap E_3$. Therefore $\varrho(S_F) = \sum_i \dim(F \cap E_i) - 2\dim F = \dim(E_1 \cap E_2 \cap E_3) + \dim(E_1 \cap E_2 \cap E_4)$. In view of Theorem 4.1 $\varrho(S_F) \leq 0$. Thus in the present case $\varrho(S_F) = 0$, i.e. the subspace $F = E_1 \cap E_2$ is invariant.

We have proved the following

Proposition 4.7. *In a regular system* $S = \{P; E_1, \dots, E_4\}$ *the subspaces* $E_i \cap E_j$ *are invariant.*

Corollary 4.8. *In a regular system* S *each of the subspaces* $E_i \cap E_j \cap E_k$ $(i \neq j,\ i \neq k,\ j \neq k)$ *is zero.*

3. Operator quadruples. We shall say that the system $S = \{P; E_1, \dots, E_4\}$ is an *operator system* if the following condition is fulfilled:

The subspaces $E_1, \dots, E_4$ are such that for four pairs of indices (i, j), $i < j$, $i, j \in \{1, 2, 3, 4\}$ the equalities

$$E_i \cap E_j = 0, \qquad E_i + E_j = P \tag{4.4}$$

are valid, and if these equalities hold for a pair (k, ℓ), then they hold also for the pair (r, s) satisfying $\{k, \ell, r, s\} = \{1, 2, 3, 4\}$.

We shall distinguish 4 types of non-decomposable operator quadruples.

1^0. The equalities

$$E_i \cap E_j = 0, \qquad E_i + E_j = P$$

are valid for every pair (i, j), $i, j \in \{1, 2, 3, 4\}$.

2^0. The equalities (4.4) are valid for every pair $(i, j) \in \{(1,3), (2,4), (1,4), (2,3)\}$ (this set of pairs will be denoted by $M(1,2)$), and

$$E_1 \cap E_2 \neq 0 \quad \text{or} \quad E_3 \cap E_4 \neq 0 .$$

3^0. The equalities (4.4) are valid for every pair $(i,j) \in M(1,3) = \{(1,2), (3,4), (1,4), (2,3)\}$ and

$$E_1 \cap E_3 \neq 0 \quad \text{or} \quad E_2 \cap E_4 \neq 0 .$$

4^0. The equalities (4.4) are valid for every pair $(i,j) \in M(1,4) = \{(1,2), (3,4), (1,3), (2,4)\}$ and

$$E_1 \cap E_4 \neq 0 \quad \text{or} \quad E_2 \cap E_3 \neq 0 .$$

We shall say that an operator quadruple is a quadruple of type 1^0, K^0 provided it is the direct sum of a finite number of non-decomposable quadruples of type 1^0 and K^0, where $K = \{1,2,3,4\}$.

Proposition 4.9. *The quadruple* S *is an operator quadruple of type* 1^0, K^0 *if and only if the equalities* $E_i \cap E_j = 0$, $E_i + E_j = P$ *are satisfied for all pairs* $i,j \in M(1,k)$.

The proof will be carried out for the case $k=3$. The condition is obviously necessary. Let us prove that it is sufficient, too. Let the quadruple S be such that the equalities (4.4) are satisfied in it for every pair $(i,j) \in M(1,3) = \{(1,2), (3,4), (1,4), (2,3)\}$. If S is non-decomposable, then there are two possibilities.

a) The equalities (4.4) are valid also for those pairs (i,j) not belonging to $M(1,3)$. This means that the equalities (4.4) are satisfied for all pairs (i,j) $i,j \in \{1,2,3,4\}$, i.e. the quadruple S is a non-decomposable operator quadruple of type 1^0.

b) If the equalities (4.4) hold for the pairs $(i,j) \in M(1,3)$, then

$$\dim E_i + \dim E_j = \dim P \tag{4.5}$$

for every pair $(i,j) \in M(1,3)$. The solution of this system of equations is elementary:

$$\begin{aligned} \dim E_1 &= \dim E_3 , \\ \dim E_2 &= \dim E_4 , \\ \dim E_1 + \dim E_2 &= \dim P . \end{aligned} \tag{4.6}$$

Set $\dim E_i = m_i$, $\dim P = n$. Then (4.6) can be rewritten as follows:

$$m_1 = m_3,$$
(4.7) $$m_2 = m_4,$$
$$m_1 + m_2 = n.$$

Let the quadruple S be such that the equalities $E_i \cap E_j = 0$, $E_i + E_j = P$ are not satisfied in it for some of the pairs $(1,3)$, $(2,4)$ which do not belong to $M(1,3)$. At this stage it is convenient to distinguish the following, mutually exclusive possibilities: 1) $m_1 + m_3 > n$ or $m_2 + m_4 > n$; 2) the number n is even, and $m_i = m = \frac{n}{2}$ for every i. Let us consider them separately. 1) If $m_1 + m_3 > n$ or $m_2 + m_4 > n$, then $E_1 \cap E_3 \neq 0$ or $E_2 \cap E_4 \neq 0$; hence in this case the non-decomposable quadruple S is of type 3°. 2) If $m_i = m = \frac{n}{2}$ for every i, then the failure of the equalities (4.4) for the pairs $(i,j) = (1,3)$ and $(2,4)$ also means that $E_1 \cap E_3 \neq 0$ or $E_2 \cap E_4 \neq 0$. Therefore also in this case the quadruple is of type 3°.

Consequently, if the quadruple S is non-decomposable and the equalities (4.4) are satisfied in it for all pairs $(i,j) \in M(1,3)$, then S is either of type 1° or of type 3°.

Let us now examine the case where the quadruple S can be decomposed into the direct sum $S = \bigoplus_{t=1}^{q} S^{(t)}$, $q > 1$, $S^{(t)} \neq 0$, of non-decomposable subsystems $S^{(t)} = \{P^{(t)}; E_1^{(t)}, \dots, E_4^{(t)}\}$. Let us show that in each subsystem $S^{(t)}$ the equalities $E_i^{(t)} \cap E_j^{(t)} = 0$, $E_i^{(t)} + E_j^{(t)} = P^{(t)}$ are valid for every pair $(i,j) \in M(1,3)$. The equality $E_i^{(t)} \cap E_j^{(t)} = 0$ follows obviously from the relation $E_i \cap E_j = 0$, since $E_i^{(t)} \subset E_i$, $E_j^{(t)} \subset E_j$. Suppose that for some pair $(i,j) \in M(1,3)$ there exists a t with

$$E_i^{(t)} + E_j^{(t)} \subset P^{(t)}.$$

Summing the inclusions $E_i^{(t)} + E_j^{(t)} \subseteq P^{(t)}$ over t we obtain: $\sum_t E_i^{(t)} + \sum_t E_j^{(t)} \subset \sum_t P^{(t)}$. Hence (by the definition of decomposable systems) $E_i + E_j \subset P$. This is a contradiction. Thus for every pair $(i,j) \in M(1,3)$ and every t the equalities $E_i^{(t)} \cap E_j^{(t)} = 0$, $E_i^{(t)} + E_j^{(t)} = P^{(t)}$ are satisfied. The subsystems are, by definition, non-decomposable. Therefore, as we have proved above, each subsystem $S^{(t)}$ is a non-decomposable subsystem of type 1^0 or 3^0. Consequently, the whole system S is an operator quadruple of type $1^0, 3^0$. The proposition is proved.

Corollary 4.10. *Let* $S = \{P; E_1, \ldots, E_4\}$ *be an operator system of type* $1^0, K^0, K \in \{1,2,3,4\}$. *Then*

$$\dim E_1 = \dim E_k$$

$$\dim E_r = \dim E_s$$

$$\dim E_1 + \dim E_r = \dim P,$$

where $\{k, r, s\} = \{2,3,4\}$.

Proposition 4.11. *An operator quadruple* S *is regular.*

Proof. Let, for instance, S be an operator quadruple of type 1^0, 3^0, so that the equalities $E_i \cap E_j = 0$, $E_i + E_j = P$ are satisfied in S for every pair $(i,j) \in M(1,3) = \{(1,2),(3,4),(1,4),(2,3)\}$. In Corollary 4.10 we proved that in this case $\dim E_1 = \dim E_3$ i.e. $m_1 = m_3$, and $\dim E_2 = \dim E_4$ i.e. $m_2 = m_4$. Moreover, $m_1 + m_2 = n$. Consequently, $\varrho(S) = \sum_{i=1}^{4} m_i - 2n = 0$.

Let now $\tilde{S} = \{\tilde{P}; \tilde{E}_1, \ldots, \tilde{E}_4\}$ be a subsystem of S. As $E_1 \cap E_2 = 0$, $E_3 \cap E_4 = 0$, it is clear that

$$\tilde{E}_1 \cap \tilde{E}_2 = \tilde{E}_3 \cap \tilde{E}_4 = 0. \tag{4.8}$$

Set $\dim \tilde{P} = \tilde{n}$, $\dim \tilde{E}_i = \tilde{m}_i$. The equalities (4.8) say that $\tilde{m}_1 + \tilde{m}_2 \leq \tilde{n}$, $\tilde{m}_3 + \tilde{m}_4 \leq \tilde{n}$ So we find: $\varrho(\tilde{S}) = \sum_i \tilde{m}_i - 2\tilde{n} \leq 0$.

We have proved that $\varrho(S) = 0$, and that any subsystem $\tilde{S}$ of S has defect $\varrho(\tilde{S}) \leq 0$, which means (Theorem 4.1) that the system S is regular.

Let us now make comment concerning the term "operator" quadruple.

Let S be an operator quadruple of type 1^0, 3^0. Then

$$P \cong E_1 \oplus E_2.$$

Denote by $\tilde{E}_3$ and $\tilde{E}_4$ the subspaces of $E_1 \oplus E_2$ isomorphic to E_3

and E_4 . Furthermore, set $\tilde{E}_1 = E_1 \oplus \{0\}$, $\tilde{E}_2 = \{0\} \oplus E_2$. Obviously, the subspaces $\tilde{E}_i$ possess the same properties as the subspaces E_i . It is not hard to verify that the conditions $\tilde{E}_3 \cap \tilde{E}_2 = 0$, $\tilde{E}_3 + \tilde{E}_2 = \tilde{P}$ mean that the subspace $\tilde{E}_3$ of the direct sum $E_1 \oplus E_2$ is the graph of a linear mapping $A: E_1 \longrightarrow E_2$. Similarly, the conditions $\tilde{E}_1 \cap \tilde{E}_4 = 0$, $\tilde{E}_1 + \tilde{E}_4 = P$ mean that there exists a linear mapping B: $E_2 \longrightarrow E_1$ such that $\tilde{E}_4$ consists of all pairs (By, y), $y \in E_2$. Thus to an operator quadruple S of type 1^0, 3^0 there correspond two spaces E and F, $E \cong E_1$, $F \cong E_2$, with two linear mappings $A: E \longrightarrow F$ and $B: F \longrightarrow E$.

4. In this section we prove that the classification of regular quadruples can be reduced to the classification of operator quadruples.

We shall say that the operator quadruple is non-degenerate, if the equalities $E_i \cap E_j = 0$, $E_i + E_j = P$ are satisfied in it for every pair (i, j), $i, j \in \{1,2,3,4\}$.

Theorem 4.12. *If the system* S *is regular and for every pair* (i, j) *we have*

$$E_i \cap E_j = 0$$

then S *is a non-degenerate operator quadruple, i.e.*

$$E_i + E_j = P$$

for every pair (i, j), *and the dimension of each* E_i *is equal to* m, *where* $\dim P = 2m$.

Proof. Set, as usually, $\dim E_i = m_i$, $\dim P = n$. From the conditions $E_1 \cap E_2 = 0$, $E_3 \cap E_4 = 0$ it follows that $\dim(E_1 + E_2) = m_1 + m_2$, $\dim(E_3 + E_4) = m_3 + m_4$. Therefore $m_1 + m_2 \leq n$, $m_3 + m_4 \leq n$. Consequently, $\sum_i m_i \leq 2n$, i.e. $\rho(S) = \sum_{i=1}^{4} m_i - 2n \leq 0$. By assumption $\rho(S) = 0$. This equality is possible only if $m_1 + m_2 = n$, $m_3 + m_4 = n$. We have proved that

$$E_1 \oplus E_2 \cong P, \qquad E_3 \oplus E_4 \cong P .$$

In a similar way it can be proved that $m_i + m_j = n$, i.e.

$$E_i \oplus E_j \cong P$$

for every pair of indices.

Now it is no longer difficult to see that the dimension of each subspace E_i is equal to m and $n = 2m$. The theorem is proved.

Theorem 4.13. *If the quadruple* S *is non-decomposable and* $\rho(S) = 0$, *then* S *is an operator quadruple belonging to one of the four types* $1^0, 2^0, 3^0, 4^0$.

Proof. If the quadruple S is non-decomposable, then there are two cases: a) all intersections $E_i \cap E_j$ are zero; b) there is a pair (i,j) with $E_i \cap E_j \neq 0$. Case a) was discussed in Theorem 4.12. It remains to consider the case b).

To be more specific, assume that $E_1 \cap E_3 \neq 0$. We are going to show that in this case S is an operator quadruple of type 3^0.

We proceed by induction relative to the dimension $n(S) = \dim P$. If $n(S) = 1$, $\rho(S) = 0$ and $E_1 \cap E_3 \neq 0$, then $E_1 = E_3 = P$. Moreover, the equality $\rho(S) = 0$ is possible only if $E_2 = E_4 = 0$. An elementary verification shows that in a quadruple S of this kind the equalities $E_i \cap E_j = 0$, $E_i + E_j = P$ are valid for the pairs $(i,j) \in M(1,3) = \{(1,2),(3,4),(1,4),(2,3)\}$. By assumption we have $E_1 \cap E_3 \neq 0$. Therefore S is a non-decomposable quadruple of type 3^0.

Assuming that the theorem is proved for $\dim P = n$ let us show that it is valid also for $\dim P = n+1$. By assumption $E_1 \cap E_3 \neq 0$, i.e. $\dim(E_1 \cap E_3) \geq 1$. The quotient space $P/(E_1 \cap E_3)$ will be denoted by P' and the natural mapping $P \longrightarrow P'$ by τ. Furthermore, we set $S' = \{P'; \tau E_1, \ldots, \tau E_4\}$. Since $\dim(E_1 \cap E_3) \geq 1$, we have $\dim P' \leq (n+1) - 1 = n$. In Section 2 of this § we have proved that in a regular system S the subspace $E_1 \cap E_3$ is invariant, and that the quotient system S/F is regular provided F is an invariant subspace. Hence the system $S' = S/(E_1 \cap E_3)$ just constructed is regular.

Assume that the system S' can be decomposed into a direct sum $S' = \bigoplus_t S'_t$ of non-decomposable subsystems S'_t. Since $n(S') \leq n$, it follows that $n(S'_t) \leq n$ for every t. Thus, in view of the induction hypothesis, each subsystem S'_t is a non-decomposable operator system of one of the four types $1^0, 2^0, 3^0, 4^0$.

Denote the direct sum of non-decomposable subsystems of type 3^0 by S'' and the direct sum of non-decomposable subsystems of type 1^0, 2^0 or 4^0 by $\tilde{S}$. Set $S'' = \{P''; E''_1, \ldots, E''_4\}$, $\tilde{S} = \{\tilde{P}; \tilde{E}_1, \ldots, \tilde{E}_4\}$. So

$$S' \cong S'' \oplus \tilde{S},$$

where

$$P' \cong P'' \oplus \tilde{P},$$

$$\forall i\,(E'_i = E''_i \oplus \tilde{E}_i).$$

1. Properties of the subsystems S'', $\tilde{S}$. Since S'' is the direct sum of non-decomposable systems of type 3^0, it is natural that the subspaces E''_i satisfy the conditions

$$E''_1 \cap E''_3 \neq 0 \quad \text{or} \quad E''_2 \cap E''_4 \neq 0$$

and

(4.9) $$E''_i \cap E''_j = 0, \qquad E''_i + E''_j = P$$

for every pair $(i,j) \in M(1,3)$, where $M(1,3) = \{(1,2),(3,4),(1,4),(2,3)\}$.

Moreover,

$$m''_1 = m''_3, \qquad m''_2 = m''_4$$

and

$$m''_1 + m''_2 = n'',$$

where $m''_i = \dim E''_i$, $n'' = \dim P''$.

As to the system $\tilde{S}$, it is the direct sum of non-decomposable systems of type 1^0, 2^0 or 4^0. By definition, in every system of this kind the equalities $E_i \cap E_j = 0$, $E_i + E_j = P$ are valid for the pairs $(i,j) \in \{(1,3),(2,4)\}$. Consequently, the analogues of these equalities hold for the subspaces $\tilde{E}_1$, $\tilde{E}_3$ and $\tilde{E}_2$, $\tilde{E}_4$, i.e.

(4.10) $$\begin{aligned} \tilde{E}_1 \cap \tilde{E}_3 = 0, &\qquad \tilde{E}_1 + \tilde{E}_3 = \tilde{P} \\ \tilde{E}_2 \cap \tilde{E}_4 = 0, &\qquad \tilde{E}_2 + \tilde{E}_4 = \tilde{P}. \end{aligned}$$

Now, making use of the subsystems S'' and $\tilde{S}$ we construct two subsystems S°, $\bar{S}$ of the system S. These subsystems will be defined so that $\tau S^\circ = S''$, $\tau\bar{S} = \tilde{S}$, $S \cong S^\circ \oplus S''$, and that the subsystem S° contains the subspace $E_1 \cap E_3$. Having constructed these systems and investigated their properties the proof of the theorem will follow almost at once.

2. The subsystem S°. By $S^\circ = \{P^\circ; E^\circ_1, \dots, E^\circ_4\}$ we shall denote the

subsystem equal to the inverse image $\tau^{-1}S''$ of the subsystem S''. Thus $S^o = \{\tau^{-1}P''; (\tau^{-1}P'' \cap E_i)\}$. Set $\tau^{-1}P'' = P^o$, $\tau^{-1}P'' \cap E_i = E_i^o$. Clearly $\tau^{-1}P'' \supseteq \tau^{-1}\{0\} = \operatorname{Ker}\tau = E_1 \cap E_3$, i.e.

$$E_1 \cap E_3 \subseteq P^o . \tag{4.11}$$

As the subspace $E_1 \cap E_3 = \operatorname{Ker}\tau$ is different from zero, the subsystem S^o is also different from zero.

It is easy to see that

$$\tau S^o = S'' , \tag{4.12}$$

i.e. $\tau P^o = P''$, $\tau E_i^o = E_i'' = P'' \cap E_i'$.

Note that the subspace $P^o = \tau^{-1}P''$ is, according to Proposition 4.5, invariant.

3. The subsystem $\bar{S}$. Let $\tau^{-1}E_2$, $\tau^{-1}E_4$ be the inverse images of the subspaces $\tilde{E}_2$, $\tilde{E}_4$ appearing in the system $\tilde{S} = \{\tilde{P}; \tilde{E}_1, \dots, \tilde{E}_4\}$. Set $\bar{E}_2 = (\tau^{-1}\tilde{E}_2) \cap E_2$, $\bar{E}_4 = (\tau^{-1}\tilde{E}_4) \cap E_4$. Furthermore, introduce the notations $\bar{P} = \bar{E}_2 + \bar{E}_4$, $\bar{E}_1 = \bar{P} \cap E_1$, $\bar{E}_3 = \bar{P} \cap E_3$, $\bar{S} = \{\bar{P}; \bar{E}_1, \dots, \bar{E}_4\}$.

Let us prove that $\bar{P}$ has the following properties:

a) $\tau\bar{P} = \tilde{P}$,

b) $\bar{P} \cap \operatorname{Ker}\tau = 0$,

c) $\bar{P} \cap P^o = 0$,

d) $\bar{P} + P^o = P$,

e) $\bar{P}$ is invariant.

a) Before proving equality a) we show that $\tau\bar{E}_i = \tilde{E}_i$ for $i = 2,4$. First observe that if X_1, X_2 are two arbitrary subspaces of P, then $\tau(X_1 \cap X_2) \subseteq \tau X_1 \cap \tau X_2$. and $\tau(X_1 + X_2) = \tau X_1 + \tau X_2$. In particular, $\tau\bar{E}_2 = \tau((\tau^{-1}\tilde{E}_2) \cap E_2) \leq (\tau(\tau^{-1}\tilde{E}_2)) \cap \tau E_2 = \tilde{E}_2 \cap E_2' = \tilde{E}_2$. Hence

$$\tau\bar{E}_2 \subseteq \tilde{E}_2 . \tag{4.13}$$

Let, on the other hand, $\tilde{\xi}_2$ be any vector of $\tilde{E}_2$. Since $\tilde{E}_2 \subset E_2' = \tau E_2$, there exists a vector $\xi_2 \in E_2$ such that $\tau\xi_2 = \tilde{\xi}_2$. Evidently $\xi_2 \in (E_2 \cap \tau^{-1}\tilde{E}_2) = \bar{E}_2$.

Thus for every $\tilde{\xi}_2 \in \tilde{E}_2$ there is a $\xi_2 \in \bar{E}_2$ satisfying $\tau\xi_2 = \tilde{\xi}_2$. In other words, $\tau\bar{E}_2 \supseteq \tilde{E}_2$. Comparing this inclusion with (4.13) we obtain:

$$\tau\bar{E}_2 = \tilde{E}_2 .$$

The equality

$$\tau\bar{E}_4 = \tilde{E}_4$$

can be proved in a similar way.

Making use of these equalities we obtain: $\tau\bar{P} = \tau(\bar{E}_2 + \bar{E}_4) = \tau\bar{E}_2 + \tau\bar{E}_4 = \tilde{E}_2 + \tilde{E}_4$. For the system $\tilde{S}$ we have, by construction, $\tilde{E}_2 + \tilde{E}_4 = \tilde{P}$ (see (4.10)). Consequently, we have proved the equality $\tau\bar{P} = \tilde{P}$.

b) We are going to prove that $\bar{P} \cap \operatorname{Ker}\tau = 0$. Let $\xi \in (\bar{P} \cap \operatorname{Ker}\tau)$. By construction $\bar{P} = \bar{E}_2 + \bar{E}_4$. Hence ξ can be represented as a sum $\xi = \xi_2 + \xi_4$, where $\xi_i \in \bar{E}_i$. Let us find the image $\tau\xi = \tau\xi_2 + \tau\xi_4$ of the vector ξ. Since $\xi_i \in E_i$ $(i = 2,4)$ and $\tau\bar{E}_i = \tilde{E}_i$, therefore $\tau\xi_i \in E_i$ $(i = 2,4)$. Moreover, $\xi \in \operatorname{Ker}\tau$ implies $\tau\xi = 0$. Thus $\tau\xi_2 = -\tau\xi_4 = \tilde{\xi}_{2,4}$, where $\tilde{\xi}_{2,4} \in \tilde{E}_2 \cap \tilde{E}_4$. By construction the spaces $\tilde{E}_2, \tilde{E}_4$ do not intersect each other (see (4.10)). Hence $\tau\xi_2 = 0$ and $\tau\xi_4 = 0$, i.e. $\xi_2 \in (E_2 \cap \operatorname{Ker}\tau)$, $\xi_4 \in (E_4 \cap \operatorname{Ker}\tau)$. The system S being regular, the intersections $E_2 \cap (E_1 \cap E_3) = E_2 \cap \operatorname{Ker}\tau$ and $E_4 \cap \operatorname{Ker}\tau$ are equal to zero (see Corollary 4.8). Therefore $\xi_2 = 0$, $\xi_4 = 0$ and, consequently, $\xi = \xi_2 + \xi_4 = 0$. We proved that $\bar{P} \cap \operatorname{Ker}\tau = 0$.

The equalities $\bar{P} \cap \operatorname{Ker}\tau = 0$, $\tau\bar{P} = \tilde{P}$ imply that the restriction $\tau|_{\bar{P}} : \bar{P} \longrightarrow \tilde{P}$ is an isomorphism.

c) Let us prove that $\bar{P} \cap P^o = 0$. Let ξ be a non-zero vector of $\bar{P} \cap P^o$. Then $\tau\xi \in (\tau\bar{P} \cap \tau P^o) = \tilde{P} \cap P''$. As we have just proved, $\bar{P} \cap \operatorname{Ker}\tau = 0$. Therefore if $\xi \neq 0$ then also $\tau\xi \neq 0$. In other words: if $\bar{P} \cap P^o \neq 0$, then $\tilde{P} \cap P'' \neq 0$. But, according to the construction, $\tilde{P} \cap P'' = 0$. Hence $\bar{P} \cap P^o = 0$.

d) Let us prove that $\bar{P} + P^o = P$. Recall the notations $n = \dim P$, $n'' = \dim P''$, $n^o = \dim P^o$, $k_{1,3} = \dim(\operatorname{Ker}\tau)$ etc. By construction the mapping $P \longrightarrow P'$ is an epimorphism, and $P'' \oplus \tilde{P} \cong P'$. Therefore $n = n' + k_{1,3}$, $n' = n'' + \tilde{n}$.

The definition $P^o = \tau^{-1} P''$ yields: $n'' = n^o - k_{1,3}$. Moreover, we have proved that the restriction $\tau|_{\bar{P}} : \bar{P} \longrightarrow \tilde{P}$ is an isomorphism. Hence $\tilde{n} = \bar{n}$. Thus the following equalities are valid:

$$n = n' + k_{1,3},$$

$$n' = n'' + \tilde{n},$$

$$n'' = n' - k_{1,3},$$

$$\tilde{n} = \bar{n}.$$

From these equalities $n = n^\circ + \bar{n}$, i.e. $\dim P = \dim P^\circ + \dim \bar{P}$. Comparison with the equality $P^\circ \cap \bar{P} = 0$ yields: $\bar{P} + P^\circ = P$.

e) It remains to prove that $\bar{P}$ is invariant. To this end let us calculate the quantity $\rho(\bar{S}) = \sum_{i=1}^{4} \bar{m}_i - 2\bar{n}$. We have proved that the restriction $\tau|_{\bar{P}} : \bar{P} \longrightarrow \tilde{P}$ is an isomorphism and that $\tau\bar{E}_2 = \tilde{E}_2, \tau\bar{E}_4 = \tilde{E}_4$. Therefore $\bar{n} = \tilde{n}$, $\bar{m}_2 = \tilde{m}_2$, $\bar{m}_4 = \tilde{m}_4$. Now let us prove that $\tau\bar{E}_1 = \tilde{E}_1$, $\tau\bar{E}_3 = \tilde{E}_3$.

Let $\tilde{\xi}_1 \in \tilde{E}_1$. By construction $\tilde{E}_1 \subseteq \tilde{P} = \tau\bar{P}$, hence there exists a vector $\xi \in \bar{P}$ such that $\tau\xi = \tilde{\xi}_1$. We are going to show that $\xi \in E_1$. Since $\tilde{\xi}_1 \in \tilde{E}_1 \subset E_1'$, therefore $\xi \in \tau^{-1}E_1'$. From the relations $\tau E_1 = E_1'$, $\operatorname{Ker}\tau \subset E_1$ it follows that $\tau^{-1}E_1' = E_1$. Thus $\xi \in E_1$ and, consequently, $\xi \in \bar{P} \cap E_1 = \bar{E}_1$. We proved that for any vector $\tilde{\xi}_1 \in E_1$ there is a vector $\xi \in \bar{E}_1$ such that $\tau\xi = \tilde{\xi}_1$. In other words, $\tau\bar{E}_1 \supseteq \tilde{E}_1$.

On the other hand, from $\bar{E}_1 = E_1 \cap \bar{P}$ it follows that $\tau\bar{E}_1 \subseteq (\tau E_1 \cap \tau\bar{P}) = E_1' \cap \tilde{P} = \tilde{E}_1$. Finally we obtain: $\tau\bar{E}_1 = \tilde{E}_1$. The equality $\tau\bar{E}_3 = \tilde{E}_3$ can be proved in a similar fashion. From these equalities and the fact that $\tau|_{\bar{P}} : \bar{P} \longrightarrow \tilde{P}$ is an isomorphism we find: $\bar{m}_i = \tilde{m}_i$ $(i = 1, 3)$.

Thus $\bar{n} = \tilde{n}$, $\bar{m}_i = \tilde{m}_i$ $(i = 1, \dots, 4)$. Therefore $\rho(\bar{S}) = \sum_i \bar{m}_i - 2\bar{n} = \sum_i \tilde{m}_i - 2\tilde{n} = \rho(\tilde{S})$. The defect $\rho(\tilde{S})$ of the system $\tilde{S}$ is, by construction, equal to zero. Hence $\rho(\bar{S}) = 0$, i.e. the subspace $\bar{P}$ is invariant.

Concluding part of the **proof of Theorem 4.13.** We have proved that the subspaces $\bar{P}$, P° occuring in the subsystems $\bar{S} = \{\bar{P}; \bar{E}_1, \dots, \bar{E}_4\}$, $S^\circ = \{P^\circ; E_1^\circ, \dots, E_4^\circ\}$ are invariant, do not intersect each other, and their sum is P. Consequently, the system S is isomorphic to a direct sum:

$$S \cong \bar{S} \oplus S^\circ.$$

Owing to the construction we have $P^0 \supseteq \operatorname{Ker} \tau$, $\operatorname{Ker} \tau \neq 0$. Therefore the subsystem S^0 is different from zero. On account of the hypothesis of the theorem S is non-decomposable, hence $\bar{S} = 0$, $S = S^0$. The subsystems $\bar{S}$, S^0 have been constructed so that $\tau\bar{S} = \tilde{S}$, $\tau S^0 = S''$, where $\tilde{S} \oplus S'' \cong S'$. Therefore the equality $\bar{S} = 0$ yields $\tau\bar{S} = \tilde{S} = 0$, i.e. $S' = S''$, $P' = P''$, $E_i' = E_i''$. Recall that S'' was defined as the direct sum of the subsystems of type 3^0. Thus S' is the direct sum of the subsystems of type 3^0 and in S' the equalities

$$(4.14) \qquad E_i' \cap E_j' = 0, \qquad E_i' + E_j' = P'$$

are valid for the pairs $(i,j) \in M(1,3) = \{(1,2),(3,4),(1,4)(2,3)\}$.

Moreover, $S' = \tau S = \{\tau P; \tau E_1, \dots, \tau E_4\} = \{P'; E_1', \dots, E_4'\}$. We are going to show how the equalities (4.14) imply the analogous equalities for E_i. From $\tau E_i = E_i'$ it follows that $\tau(E_i \cap E_j) \subseteq E_i' \cap E_j'$. The system S is regular, hence the subspaces $E_i \cap E_j$ $(i<j,\ (i,j) \neq (1,3))$ do not intersect the subspace $E_1 \cap E_3 = \operatorname{Ker}\tau$, because in a regular system every intersection of the form $E_i \cap E_j \cap E_k$ $(i \neq j,\ j \neq k,\ i \neq k)$ is zero (see Corollary 4.8). This means that if $\xi_{i,j} \in E_i \cap E_j$, where $i < j$ and $(i,j) \neq (1,3)$, then the equality $\tau\xi_{i,j} = 0$ implies $\xi_{i,j} = 0$. As we have shown, $\tau\xi_{i,j} \in (E_i' \cap E_j')$ and in our case $E_i' \cap E_j' = 0$ for every $(i,j) \in M(1,3)$. Consequently, $\tau\xi_{i,j} = 0$ for every $\xi_{i,j} \in (E_i \cap E_j)$, where $(i,j) \in M(1,3)$. Hence $\xi_{i,j} = 0$, i.e. $E_i \cap E_j = 0$ for every $(i,j) \in M(1,3)$.

Let us show that these equalities yield

$$(4.15) \qquad E_i + E_j = P \quad \text{for} \quad (i,j) \in M(1,3).$$

In fact, the quadruple S is regular and has defect 0, i.e. $\sum_i m_i = 2n$, where $m_i = \dim E_i$. Moreover, we have shown that

$$(4.16) \qquad E_1 + E_2 = P, \qquad E_3 + E_4 = P.$$

Therefore $\dim E_1 + \dim E_2 \leq \dim P$, i.e. $m_1 + m_2 \leq n$ and similarly $m_3 + m_4 \leq n$. It is clear that the equality $\sum_i m_i = 2n$ can hold only if $m_1 + m_2 = n$, $m_3 + m_4 = n$. These equalities along with (4.16) yield:

$$E_1 + E_2 = P, \qquad E_3 + E_4 = P.$$

In an analogous way the equalities $E_1 \cap E_4 = 0$, $E_2 \cap E_3 = 0$ imply that $E_1 + E_4 = P$, $E_2 + E_3 = P$.

Thus we have shown that

$$E_i \cap E_j = 0, \qquad E_i + E_j = P$$

for every pair $(i,j) \in M(1,3) = \{(1,2),(3,4),(1,4),(2,3)\}$. At the same time, in view of the assumption, $E_1 \cap E_3 \neq 0$. So S is a non-decomposable operator system of type 3^0.

We carried out the proof of the theorem for the case $E_1 \cap E_3 \neq 0$. For the case $E_2 \cap E_4 \neq 0$ the proof can be obtained from the above one by interchanging the indices 1 and 2 as well as 3 and 4 among themselves. For the cases where some other subspace of the form $E_i \cap E_j$ is different from zero the proof can also be obtained from the proof performed above by a suitable change of indices.

5. In order to push further the study of operator quadruples, with each quadruple of this kind we associate a linear transformation $A: P \to P$. Then we show that if the quadruple S is non-decomposable, so is the transformation A, and conversely: if A is non-decomposable, so is S. After this, making use of the well-known canonical form of a non-decomposable transformation A we find the canonical form of non-decomposable operator quadruples.

Let S be an operator quadruple of type 1^0, 3^0, so that the equalities $E_i \cap E_j = 0$, $E_i + E_j = P$ are valid in S for the pairs $(i,j) \in M(1,3) = \{(1,2),(3,4),(1,4),(2,3)\}$. We are going to define a linear mapping $A(1,2,3,4): P \longrightarrow P$ corresponding to this quadruple. By the assumption we have $E_1 \cap E_2 = 0$, $E_1 + E_2 = P$. Therefore every vector $x \in P$ admits a unique representation of the form

$$x = x_1 + x_2, \quad \text{where} \quad x_i \in E_i. \tag{4.17}$$

Similarly, from the conditions $E_2 \cap E_3 = 0$, $E_2 + E_3 = P$ and $E_1 \cap E_4 = 0$, $E_1 + E_4 = P$ it follows that the vectors x_1, x_2 of (4.17) can be uniquely written in the form

$$\begin{aligned} x_1 &= z_3 - y_2, \\ x_2 &= z_4 - y_1, \end{aligned} \quad \text{where} \quad y_i, z_i \in E_2. \tag{4.18}$$

We set $y = y_1 + y_2$. The mapping $x \longrightarrow y$ is obviously linear. Denote it by $A(1,2,3,4)$. For the sake of brevity we shall write $A = A(1,2,3,4)$.

Let us show that the transformation A defined in this manner has the following properties:

1) $AE_1 \subseteq E_2$, $\qquad AE_2 \subseteq E_1$;

2) $(I+A)E_1 = E_3$, $\qquad (I+A)E_2 = E_4$,

where $I: P \longrightarrow P$ is the identity transformation;

3) the number $\lambda = 1$ is not an eigenvalue of A, i.e. $Ax = x$ only if $x = 0$.

1) Let $x \in E_1$. Then in the formula (4.17) $x_2 = 0$ and in (4.18) $y_1 = 0$. Therefore $Ax = Ax_1 = y_2 \in E_2$. The proof of the inclusion $AE_2 \subseteq E_1$ is similar.

2) We prove that $(I+A)E_1 = E_3$. Let $x_1 \in E_1$. Then, as we have just proved, $Ax_1 = y_2 \in E_2$ and, on account of (4.18), $z_3 = x_1 + y_2 = x_1 + Ax_1$, where $z_3 \in E_3$. Hence $(I+A)E_1 \subseteq E_3$. Now let z_3 be any vector of E_3. Then z_3 can uniquely be represented as $z_3 = x_1 + y_2$, where $x_1 \in E_1$, $y_2 \in E_2$. Thus $x_1 = z_3 + y_2$, i.e. $Ax_1 = y_2$, $z_3 = x_1 + Ax_1$. We have shown that $(I+A)E_1 \supseteq E_3$. Comparing this with the inclusion $(I+A)E_1 \subseteq E_3$ we obtain that $(I+A)E_1 = E_3$. The equality $(I+A)E_2 = E_4$ can be proved in a similar way.

3) We prove that if $Ax = x$, then $x = 0$. Let $Ax = x$. Then (cf. the equalities (4.17) – (4.18)) we have $x_1 = y_1$, $x_2 = y_2$. Therefore $x_1 + y_2 = x_2 + y_1$, $z_3 = z_4$ (see (4.18)). Owing to the assumption $E_3 \cap E_4 = 0$. Consequently $z_3 = z_4 = 0$, i.e. $x_1 + y_2 = 0$. Hence and from the condition $E_1 \cap E_2 = 0$ it follows that $x_1 = y_2 = 0$, $x_2 = y_1 = 0$, i.e. $x = x_1 + x_2 = 0$, $y = y_1 + y_2 = 0$. We proved that the equality $Ax = x$ holds only for the vector $x = 0$.

Proposition 4.14. *Let P be a finite-dimensional space and E_1, E_2 two subspaces of P such that $E_1 \cap E_2 = 0$, $E_1 + E_2 = P$. Let $A: P \longrightarrow P$ be a linear transformation with the properties that $\lambda = 1$ is not an eigenvalue of A and*

$$AE_1 \subseteq E_2, \qquad AE_2 \subseteq E_1.$$

Put $E_3 = (I+A)E_1$, $E_4 = (I+A)E_2$. Then the system $S = \{P; E_1, E_2, E_3, E_4\}$ is an operator system of type $1^0, 3^0$.

Proof. We must prove that

$$E_i \cap E_j = 0, \qquad E_i + E_j = P \tag{4.19}$$

for every pair $(i,j) \in M(1,3) = \{(1,2),(3,4),(1,4),(2,3)\}$.

For the pair $(i,j) = (1,2)$ the equalities (4.19) are satisfied by assumption.

Let us prove the equalities $E_2 \cap E_3 = 0$, $E_2 + E_3 = P$. By definition, the subspace E_3 is spanned by the vectors $x + Ax$, where $x \in E_1$. But $x \in E_1$ implies $Ax \in E_2$. From $E_1 \cap E_2 = 0$ it follows that $(x + Ax) \in E_2$ only if $x = 0$ i.e. $x + Ax = 0$. We proved that $E_2 \cap E_3 = 0$. Now choose a basis $e_1, \dots, e_k$ in E_1. Then E_3 is spanned by the vectors $e_i + Ae_i$; $i = 1, 2, \dots, k$. Since $Ae_i \in E_2$, from $E_1 \cap E_2 = 0$ it follows that the vectors $e_i + Ae_i$ $(i = 1, 2, \dots, k)$ are linearly independent, hence they form a basis in E_3. Thus $\dim E_3 = \dim E_1$. Comparing this with the equalities $E_1 \cap E_2 = 0$, $E_1 + E_2 = P$ satisfied by assumption, and with the equality $E_3 \cap E_2 = 0$ just proved we obtain that $E_2 + E_3 = 0$.

The equalities $E_1 \cap E_4 = 0$, $E_1 + E_4 = P$ can be proved in a similar way. The only difference is that one must choose a basis $f_1, \dots, f_\ell$ in E_2 and show that $\dim E_2 = \dim E_4$.

Let us now prove the equalities $E_3 \cap E_4 = 0$, $E_3 + E_4 = P$. Let $y \in (E_3 \cap E_4)$. Since $y \in E_3$, there exists a vector $x_1 \in E_1$ such that $y = x_1 + Ax_1$. Since $y \in E_4$, there exists a vector $z_2 \in E_2$ such that $y = z_2 + Az_2$. On account of the properties $AE_1 \subseteq E_2$, $AE_2 \subseteq E_1$ we have $Ax_1 \in E_2$, $Az_2 \in E_1$. Hence we may set $Ax_1 = x_2$, $Az_2 = z_1$. Then $y = x_1 + x_2 = z_1 + z_2$. In view of the assumption $E_1 \cap E_2 = 0$. Thus the decomposition $y = x_1 + x_2$, where $x_i \in E_i$, is unique and therefore $x_1 = z_1$, $x_2 = z_2$. Making use of these equalities we find: $Ay = A(x_1 + x_2) = A(x_1 + z_2) = Ax_1 + Az_2 = x_2 + z_1 = x_2 + x_1 = y$ i.e. $Ay = y$. But, according to the assumption $\lambda = 1$ is not an eigenvalue of A. Hence $y = 0$. We proved that $E_3 \cap E_4 = 0$.

Previously we have proved that $\dim E_3 = \dim E_1$ and $\dim E_4 = \dim E_2$. Moreover, $\dim E_1 + \dim E_2 = \dim P$ by assumption. Comparing these facts with the equality $E_3 \cap E_4 = 0$ we obtain that $E_3 + E_4 = P$. Proposition 4.14 is proved.

Now let S be an operator quadruple of type 1^0, K^0 $(K = 2, 3, 4)$. With a quadruple S of this kind we associate a linear transformation $A(1,i,k,\ell)$, where $i < \ell$, $\{i,k,\ell\} = \{2,3,4\}$. This transformation has the following properties:

$$A(1,i,k,\ell) E_1 \subseteq E_i ;$$

$$A(1,i,k,\ell) E_i \subseteq E_1 ;$$

$$(I + A(1,i,k,\ell))E_1 = E_k ;$$

$$(I + A(1,i,k,\ell))E_i = E_\ell ;$$

if $A(1,i,k,\ell)x = x$, then $x = 0$.

Note that if the number K occurring in the type $1^0, K^0$ of the operator quadruple is given then the conditions $\{i,k,\ell\} = \{2,3,4\}$, $i < \ell$ determine the indices i, ℓ uniquely. So to a quadruple of type $1^0, 2^0$ there corresponds the transformation $A(1,3,2,4)$, to one of type $1^0, 3^0$ the transformation $A(1,2,3,4)$, and to one of type $1^0, 4^0$ the transformation $A(1,2,4,3)$. Also the converse is true. Let us be given a space P along with a pair of subspaces E_1, E_i such that $E_1 \oplus E_i = P$. Let A be a linear transformation such that $AE_1 \subseteq E_i$, $AE_i \subseteq E_1$ and that $\lambda = 1$ is not an eigenvalue of A. Put $E_k = (I+A)E_1$, $E_\ell = (I+A)E_i$, where $\{i,k,\ell\} = \{2,3,4\}$. The quadruple $S = \{P; E_1, \dots, E_4\}$ so obtained is an operator quadruple.

Next we present a simple assertion concerning the properties of the transformation A that corresponds to an operator quadruple S.

Proposition 4.15. *Let S be an operator quadruple of type $1^0, K^0$, i.e. $S = \{P; E_1, \dots, E_4\}$, and the equalities $E_i \cap E_j = 0$, $E_i + E_j = P$ hold for every pair (i,j) with the possible exception of the pairs $(1,k)$ and (i,ℓ), where $\{i,k,\ell\} = \{2,3,4\}$. If the quadruple S is decomposable, so is the corresponding operator A.*

The proof, as usually, will be carried out for a quadruple of type $1^0, 3^0$. Let $S = \bigoplus_{i=1}^{r} S_t$ $(r > 1)$, where S_t are non-decomposable, non-zero subsystems. In Proposition 4.9 we have proved that in this case each quadruple S_t is either a non-decomposable quadruple of type 1^0, or a non-decomposable quadruple of type K^0. Naturally, to each subsystem $S_t = \{P_t; E_{t,1}, \dots, E_{t,4}\}$ there corresponds a linear transformation $A_t : P_t \quad P_t$. We omit the elementary verification of the equality A $A = \bigoplus_t A_t$.

Corollary 4.16. *Let S be an operator quadruple and A the corresponding linear transformation. If A is non-decomposable, so is the corresponding quadruple S.*

6. Let us study now the properties of the linear transformation A corresponding to an operator quadruple in more details.

A pair E_1, E_2 of subspaces in the space P will be called a graduation if $P \cong E_1 \oplus E_2$. A linear transformation $A: P \longrightarrow P$ will be called compatible with the graduation if $AE_1 \subseteq E_2$ and $AE_2 \subseteq E_1$. A subspace $F \subseteq P$ is said to be homogeneous relative to the graduation $P \cong E_1 \oplus E_2$ if

$$F = (F \cap E_1) + (F \cap E_2).$$

In other words, F is homogeneous if every vector $y \in F$ can uniquely be represented as a sum

$$y = y_1 + y_2,$$

where $y_i \in (F \cap E_i)$.

Lemma 4.17. *Let* P *be a space with a graduation* $P \cong E_1 \oplus E_2$ *(where* $E_1, E_2 \subseteq P$*), and* A *a linear transformation compatible with the graduation, so that* $AE_1 \subseteq E_2$, $AE_2 \subseteq E_1$. *If a subspace* F *is homogeneous relative to the graduation* $P \cong E_1 \oplus E_2$, *so are the subspaces* AF, $A^{-1}F$.

Proof. Let $y \in A^{-1}F$. Then the vector $x = Ay$ belongs to F. The subspace F being homogeneous, x can uniquely be represented as a sum $x = x_1 + x_2$, where $x_i \in F \cap E_i$. By assumption $P \cong E_1 \oplus E_2$. Consequently y can be written in the form $y = y_1 + y_2$, where $y_i \in E_i$. Then $Ay = Ay_1 + Ay_2$, and from the compatibility of the transformation A it follows that $Ay_1 \in E_2$, $Ay_2 \in E_1$. Therefore we set $Ay_2 = x'_1$, $Ay_1 = x'_2$. Then $x = Ay = x'_1 + x'_2 = x_1 + x_2$. The decomposition of the vector x into the sum $x = x_1 + x_2$ is unique; hence $x'_1 = x_1$, $x_2 = x'_2$, i.e. $x_1 = Ay_2$, $x_2 = Ay_1$. So $y_1, y_2 \in A^{-1}F$. We proved that any vector $y \in A^{-1}F$ can be represented as a sum $y = y_1 + y_2$, where $y_i \in (A^{-1}F \cap E_i)$. The homogeneity of the subspace $A^{-1}F$ is proved.

We omit the still simpler proof of the homogeneity of AF.

We now recall some facts from the theory of linear operators. Let A be a linear transformation of the space P. Then the kernels $\operatorname{Ker} A^i$ $(i = 1, 2, \ldots)$ form a chain:

$$0 \subseteq \operatorname{Ker} A \subseteq \operatorname{Ker} A^2 \subseteq \ldots \subseteq \operatorname{Ker} A^i \subseteq \ldots \subseteq P.$$

Moreover, for the kernels $\operatorname{Ker} A^i$ with $0 \leq i < \ell$ the strict inclusion $\operatorname{Ker} A^i \subset \operatorname{Ker} A^{i+1}$ is valid. We may formally take $\operatorname{Ker} A^0 = 0$. On the other hand, beginning with $i = \ell$ we have the equalities $\operatorname{Ker} A^\ell = \operatorname{Ker} A^{\ell+1} = \ldots = \operatorname{Ker} A^{\ell+j}$

for every $j \geq 0$. The subspace $\operatorname{Ker} A^{\ell}$ will be called the stable kernel.

The ranges $\operatorname{Im} A^{i}$ of the powers of the mapping A, as well, form a chain:

$$P \supseteq \operatorname{Im} A \supseteq \operatorname{Im} A^{2} \supseteq \ldots \supseteq \operatorname{Im} A^{i} \supseteq \ldots \supseteq 0 .$$

Moreover, for $0 \leq i < \ell$ the ranges $\operatorname{Im} A^{i}$ are strictly different: $\operatorname{Im} A^{i} \supset \operatorname{Im} A^{i+1}$ (we formally take $\operatorname{Im} A^{0} = P$). Beginning with $i = \ell$ (the same index as for the kernels) we have: $\operatorname{Im} A^{\ell} = \operatorname{Im} A^{\ell+1} = \ldots = \operatorname{Im} A^{\ell+j}$. The subspace $\operatorname{Im} A^{\ell}$ will be called the stable range.

Next we state, without proof, a well-known result concerning linear operators.

Proposition 4.18. *Let* A *be a linear transformation of the finite-dimensional space* P *with stable kernel* P_0 *and stable range* P_1. *Then* $P \cong P_0 \oplus P_1$, *and the operator* A *is decomposable on the pair* P_0, P_1, *so that the restriction* $A_0 = A|_{P_0}$ *is a nilpotent transformation, whereas* $A_1 = A|_{P_1}$ *is a non-degenerate linear transformation.*

Let us now return to the study of the connection between a transformation A and the corresponding quadruple S.

Proposition 4.19. *Let* $S = \{P; E_1, \ldots, E_4\}$ *be a quadruple of type* 1^0, 3^0, *i.e. in this quadruple the subspaces* E_i *satisfy the equalities*

$$E_i \cap E_j = 0, \qquad E_i + E_j = P$$

for every pair $(i,j) \in M(1,3) = \{(1,2),(3,4),(1,4),(2,3)\}$. *Let* $A = A(1,2,3,4)$ *be the corresponding linear transformation. If* A *is decomposable on the subspaces* F_j *and each* F_j *is homogeneous relative to the graduation* $P \cong E_1 \oplus E_2$, *then the quadruple* S *is decomposable on the subspaces* F_j.

Proof. In view of the assumption we have $P \cong \bigoplus_j F_j$ and

$$F_j = (F_j \cap E_1) + (F_j \cap E_2)$$

for every j.

Set $F_{j,1} = F_j \cap E_1$, $F_{j,2} = F_j \cap E_2$. It is easy to see that $P = \sum_j (F_{j,1} + F_{j,2})$ and the sum is a direct sum. Let us show that $\sum_j F_{j,i} = E_i$ for $i = 1,2$.

By definition, $F_{j,i} = F_j \cap E_i$. Therefore $\sum_j F_{j,i} \subseteq E_i$; $i = 1, 2$. Set $\sum_j F_{j,i} = \widetilde{E}_i$. Then the subspaces $\widetilde{E}_i$ have the following properties: a) $\widetilde{E}_i \subseteq E_i$; b) $\widetilde{E}_1 + \widetilde{E}_2 = P$. Comparing these relations with the equalities $E_1 \cap E_2 = 0$, $E_1 + E_2 = P$ one sees that $\widetilde{E}_i = E_i$ $(i = 1, 2)$ or, more explicitly:

$$E_i = \sum_j (F_j \cap E_i).$$

Next we prove that the subspaces E_3, E_4 are homogeneous.

We know that in an operator quadruple S of type 1^0, 3^0 the subspaces E_1 and E_3 have the same dimension. The transformation $A = A(1,2,3,4)$ satisfies the equality $(I + A)E_1 = E_3$ by construction. Consequently, the restriction $(I+A)|_{E_1} : E_1 \longrightarrow E_3$ is an isomorphism. Set $F_{j,3} = (I+A)F_{j,1}$. We have proved that $E_1 = \sum F_{j,1}$, and that this sum is a direct sum. Therefore $\sum_j F_{j,3} = E_3$, and also this sum is a direct one. According to the assumption F_j is invariant under A. Hence $AF_{j,1} \subset F_j$ and $F_{j,3} = (I+A)F_{j,1} = IF_{j,1} + AF_{j,1} \subseteq F_{j,1} + F_j = F_j$, i.e. $F_{j,3} \subseteq F_j$. Hence $F_{j,3} \subseteq (F_j \cap E_3)$. Thus from the equality $\sum_j F_{j,3} = E_3$ it follows that $\sum_j (F_j \cap E_3) \supseteq E_3$. Comparing this with the obvious inclusion $\sum_j (F_j \cap E_3) \subseteq$ $\subseteq E_3$ we obtain:

$$\sum_j (F_j \cap E_3) = E_3 .$$

The homogeneity of E_3 is proved. The homogeneity of E_4 can be proved similarly.

As a result, in the system $S = \{P; E_1, \dots, E_4\}$ each of the subspaces E_i is homogeneous relative to the family of the subspaces F_j. This means that the system S is decomposable. The proposition is proved.

Proposition 4.20. *Let* $S = \{P; E_1, \dots, E_4\}$ *be an operator quadruple such that the equalities*

$$E_i \cap E_j = 0, \qquad E_i + E_j = P$$

are satisfied for every pair $(i,j) \in M(1,3) = \{(1,2),(3,4),(1,4),(2,3)\}$. *In other words, let* S *be an operator quadruple of type* 1^0, 3^0. *Further let* $A = A(1,2,3,4)$ *be the linear transformation corresponding to* S. *Then there exist subspaces* P_0, P_1 *homogeneous relative to the graduation* $P \cong E_1 \oplus E_2$ *and such that the quadruple is decomposable on the pair* P_0, P_1, *the restriction* $A|_{P_0}$ *being nilpotent, while* $A|_{P_1}$ *is non-degenerate.*

We shall prove that P_0 and P_1 are uniquely determined and equal to the stable kernel $\operatorname{Ker} A^\ell$ and the stable range $\operatorname{Im} A^\ell$, respectively, of the transformation A. According to Proposition 4.18 A is decomposable on the subspaces P_0 and P_1, the restriction $A_0 = A|_{P_0}$ being nilpotent, while the restriction $A_1 = A|_{P_1}$ is non-degenerate.

Let us show that P_0 and P_1 are homogeneous relative to the graduation $P \cong E_1 \oplus E_2$. The subspaces $\{0\}$, P are homogeneous relative to any gradutation, hence, on account of Lemma 4.17, the subspaces $A^{-1}\{0\} = \operatorname{Ker} A$, $AP = \operatorname{Im} A$ are homogeneous relative to the pair E_1, E_2.

Furthermore, $\operatorname{Ker} A$ and $\operatorname{Im} A$ being homogeneous, it follows from the same lemma that the subspaces $A^{-1}(\operatorname{Ker} A) = \operatorname{Ker} A^2$ and $A(\operatorname{Im} A) = \operatorname{Im} A^2$ are also homogeneous. Continuing this argument we find that the stable kernel P_0 and the stable range P_1 are homogeneous, too.

Thus the conditions of Proposition 4.19 are fulfilled. Consequently, the quadruple S is decomposable on the subspaces P_0, P_1: $S \cong S_0 \oplus S_1$. The proposition is proved.

We shall say that the operator quadruple S is non-degenerate if the corresponding transformation A is non-degenerate. We shall say that the operator quadruple S is nilpotent if the corresponding transformation A is nilpotent.

Previously, in Section 4 of the present §, the attribute "non-degenerate" has been attached to operator quadruples of type 1^0, i.e. quadruples in which $E_i \cap E_j = 0$, $E_i + E_j = P$ for every pair (i,j); $i,j \in \{1,2,3,4\}$. We are going to show that these definitions are equivalent.

Proposition 4.21. *An operator quadruple* S *of type* 1^0, 3^0 *is a quadruple of type* 1^0 *if and only if the corresponding linear transformation* $A = A(1,2,3,4)$ *is non-degenerate.*

Proof. The condition is necessary. Let S be a quadruple of type 1^0. Further let x be a vector of P with $Ax = 0$. Since $P \cong E_1 \oplus E_2$, the vector x can uniquely be represented as a sum $x = x_1 + x_2$, where $x_i \in E_i$. For the transformation A we have $AE_1 \subseteq E_2$, $AE_2 \subseteq E_1$ by construction, i.e. $Ax = Ax_1 + Ax_2 = 0$, where $Ax_1 \in E_2$, $Ax_2 \in E_1$. Comparing all this with the condition $E_1 \cap E_2 = 0$ we obtain that $Ax_1 = 0$, $Ax_2 = 0$.

The transformation A is so behaved that $x_1 \in E_1$ implies $(x_1 + Ax_1) \in E_3$, whereas $x_2 \in E_2$ implies $(x_2 + Ax_2) \in E_4$. Let us now return to the vectors x_1, x_2 appearing in the decomposition $x = x_1 + x_2$. We have proved that $Ax_1 = Ax_2 = 0$. Therefore $(x_1 + Ax_1) = x_1 \in E_3$ and $(x_2 + Ax_2) = x_2 \in E_4$. Hence $x_1 \in (E_1 \cap E_3)$, $x_2 \in (E_2 \cap E_4)$. The quadruple is, by assumption, of type 1^0, i.e. every subspace $E_i \cap E_j$ is equal to 0. Consequently, $x_1 = x_2 = 0$, $x = x_1 + x_2 = 0$. Thus $Ax = 0$ implies $x = 0$, so that A is non-degenerate.

The condition is sufficient. Let the quadruple S be such that the corresponding linear transformation $A = A(1,2,3,4)$ is non-degenerate. We show that S is a quadruple of type 1^0. In proposition 4.14 we have established that if on a space P with graduation $P \cong E_1 \oplus E_2$ there is given a linear transformation A satisfying

$$AE_1 \subseteq E_2, \qquad AE_2 \subseteq E_1,$$

and $\lambda = 1$ is not an eigenvalue of A, then the quadruple $S = \{P; E_1, \dots, E_4\}$ corresponding to A is a quadruple of type 1^0, 3^0, i.e. the equalities $E_i \cap E_j = 0$, $E_i + E_j = P$ are valid in S for every pair $(i,j) \in M(1,3) = \{(1,2), (3,4), (1,4), (2,3)\}$. Let us prove that if, in addition, A is non-degenerate then $E_1 \cap E_3 = 0$, $E_2 \cap E_4 = 0$. Let $x_1 \in (E_1 \cap E_3)$. Then $y = (x_1 + Ax_1) \in E_3$. Since $x_1 \in E_3$, therefore $Ax_1 = (y - x_1) \in E_3$. Moreover, we know that $x_1 \in E_1$ implies $Ax_1 \in E_2$. We have proved that in our case $Ax_1 \in (E_2 \cap E_3)$. We noted above that to the transformation A there corresponds a quadruple S of type 1^0, 3^0, Consequently, $E_2 \cap E_3 = 0$. Thus we obtain: $Ax_1 = 0$. By assumption, the transformation A is non-degenerate, hence $x_1 = 0$. Owing to the construction, x_1 is an arbitrary element of $E_1 \cap E_3$. So $E_1 \cap E_3 = 0$. The equality $E_2 \cap E_4 = 0$ can be proved similarly.

Thus in S every intersection of the form $E_i \cap E_j$ is equal to 0. The quadruple S is an operator quadruple, hence it is regular. In Theorem 4.12 we have proved that if the quadruple S is regular and all subspaces of the form $E_i \cap E_j$ are zero in it then S is a quadruple of type 1^0. Then the proposition is proved.

Proposition 4.22. *An operator quadruple of type* 1^0, 3^0 *is of type* 3^0 *if and only if the corresponding linear transformation* A *is nilpotent.*

Proof. Let S be a quadruple of type 3^0, i.e. S equals the direct sum

$\bigoplus_t S^{(t)}$ of non-decomposable quadruples of type 3^0. In Proposition 4.20 we have proved that there exist subspaces P_0 and P_1 such that S is decomposable, $S \cong S_0 \oplus S_1$, on the pair P_0, P_1, the restriction $A_0 = A|_{P_0}$ being nilpotent, while $A_1 = A|_{P_1}$ is non-degenerate. From Proposition 4.21 it follows that to the non-degenerate transformation A_1 there corresponds an operator subsystem S_1 of type 1^0. But, according to the assumption, non-degenerate quadruples of type 1^0 do not occur in the decomposition $S = \bigoplus_t S^{(t)}$. Consequently, in the decomposition $P \cong P_0 + P_1$ we have $P_1 = 0$, i.e. the transformation A is nilpotent.

Let, conversely, the transformation A corresponding to the quadruple S be nilpotent. We shall prove that in the decomposition $S \cong \bigoplus_t S^{(t)}$ of S into the direct sum of non-decomposable systems each subsystem $S^{(t)}$ is an operator quadruple of type 3^0. Assume the converse: among the $S^{(t)}$ there are subsystems of type 1^0. In Proposition 4.15 we have proved that if the system S is decomposable then the corresponding transformation A is also decomposable: $A = \bigoplus_t A^{(t)}$. From Proposition 4.21 it follows that the transformations $A^{(t)}$ corresponding to subsystems $S^{(t)}$ of type 1^0 are non-degenerate. But, in view of the assumption, A is nilpotent. Consequently, all non-decomposable subsystems $S^{(t)}$ appearing in the decomposition $S = \bigoplus_t S^{(t)}$ are quadruples of type 3^0. Therefore S is also of type 3^0. The proposition is proved.

We do not present here the obvious generalizations of Propositions 4.21 – 4.22 to quadruples of type 1^0, K^0 $(K = 2, 3, 4)$.

§5. CLASSIFICATION OF NON-DEGENERATE OPERATOR QUADRUPLES

In the preceding § we have shown that in a non-degenerate operator quadruple the equalities $E_i \cap E_j = 0$, $E_i + E_j = P$ are valid for every pair (i, j); $i, j \in \{1, 2, 3, 4\}$. In a quadruple of this kind the number $n = \dim P$ is even, $n = 2k$, and the dimension of each subspace E_i is equal to k.

To an operator quadruple of type 1^0, 3^0 there corresponds a linear transformation $A = A(1,2,3,4) : P \longrightarrow P$ with the following properties:

(5.1) $$AE_1 \subseteq E_2; \qquad AE_2 \subseteq E_1;$$

(5.2) $$E_3 = (I+A)E_1 ; \qquad E_4 = (I+A)E_2 ;$$

$\lambda = 1$ is not an eigenvalue of A.

The quadruple S is non-degenerate if and only if the transformation A is so.

Since $\dim E_i = k$ and the mapping A is non-degenerate, for a non-degenerate operator quadruple S the inclusions (5.1) take the form

(5.3) $$AE_1 = E_2, \qquad AE_2 = E_1.$$

In this case, naturally, the restrictions $A_1 = A|_{E_1}: E_1 \to E_2$ and $A_2 = A|_{E_2}: E_2 \to E_1$ are isomorphisms. Note that $A^2E_1 = E_1$, $A^2E_2 = E_2$.

Proposition 5.1. *The non-degenerate operator quadruple S is non-decomposable if and only if the restriction $A_1^2 = A^2|_{E_1}: E_1 \longrightarrow E_1$ is a non-decomposable linear transformation.*

Proof. The condition is necessary. Let the quadruple S be decomposable: $S = \bigoplus_t S^{(t)}$, where $S^{(t)} = \{P^{(t)}; E_1^{(t)}, \dots, E_4^{(t)}\}$. As we have proved in Proposition 4.15, the transformation $A = A(1,2,3,4)$ is also decomposable: $A = \bigoplus_t A^{(t)}$. Evidently, the transformation $A^{(t)}$ has the following properties:

$$A^{(t)}E_1^{(t)} = E_2^{(t)}; \qquad A^{(t)}E_2^{(t)} = E_1^{(t)}.$$

Thus $(A^{(t)})^2E_1^{(t)} = E_1^{(t)}$ and $(A^{(t)})^2E_2^{(t)} = E_2^{(t)}$. Hence the transformation is decomposable on the subspaces $E_i^{(t)}$; $i = 1,2$; $t = 1,2,\dots$.

The condition is sufficient. Let the transformation $A_1^2 = A^2|_{E_1}$ be decomposable on the subspace E_1, i.e. there exists a family of subspaces $E_1^{(t)} \subset E_1$ such that $\bigoplus_t E_1^{(t)} \cong E_1$ and $A_1^2E_1^{(t)} = E_1^{(t)}$. Denote the subspace $AE_1^{(t)}$ by $E_2^{(t)}$. The quadruple S being non-degenerate, the restriction $A|_{E_1}: E_1 \to E_2$ is an isomorphism. Therefore the equality $\sum_t E_1^{(t)} = E_1$ and the fact that the latter sum is a direct sum imply $\sum_t E_2^{(t)} = E_2$, and this is also a direct sum. Set $P^{(t)} = E_1^{(t)} + E_2^{(t)}$.

Let us prove that S is decomposable on the subspaces $P^{(t)}$. From the definition of the subspaces $E_1^{(t)}$, $E_2^{(t)}$ it follows that $P \cong \bigoplus_t P^{(t)}$, where $P^{(t)}$ is homogeneous relative to the graduation $P \cong E_1 \oplus E_2$. By definition, $E_2^{(t)} = AE_1^{(t)}$, and the subspaces $E_1^{(t)}$ are, owing to the construction, invariant under A^2, i.e. $A^2 E_1^{(t)} = E_1^{(t)}$. Therefore $AE_2^{(t)} = A(AE_1^{(t)}) = A^2 E_1^{(t)} = E_1^{(t)}$. Hence we find:

$$AP^{(t)} = A(E_1^{(t)} + E_2^{(t)}) = AE_1^{(t)} + AE_2^{(t)} = E_2^{(t)} + E_1^{(t)} = P^{(t)}.$$

This means that the subspaces $P^{(t)}$ are invariant for the transformation A.

Thus A is decomposable on the subspaces $P^{(t)}$, and each of these subspaces is homogeneous relative to the graduation $P \cong E_1 \oplus E_2$. Consequently, on account of Proposition 4.19, S is decomposable on the subspaces $P^{(t)}$. The proposition is proved.

Making use of this proposition we are going to prove a classification theorem for non-degenerate operator quadruples.

Theorem 5.2. *Let S be a non-degenerate, non-decomposable operator quadruple. Then there is a number $\lambda \in K, \lambda \neq 0, \lambda \neq 1$ such that $S \cong S(2k, 0; \lambda)$, where $2k = \dim P$. In other words, in the space P one can choose a basis $\{e_1, \ldots, e_k, f_1, \ldots, f_k\}$ so, that the subspaces E_i are spanned by the following sets of vectors:*

$$E_1 = \{e_1, \ldots, e_k\},$$

$$E_2 = \{f_1, \ldots, f_k\},$$

$$E_3 = \{(e_1 + \lambda f_1), (e_2 + f_1 + \lambda f_2), \ldots, (e_k + f_{k-1} + \lambda f_k)\},$$

$$E_4 = \{(e_1 + f_1), \ldots, (e_k + f_k)\}.$$

Proof. From Proposition 5.1 it follows that to a non-decomposable non-degenerate quadruple S there corresponds a non-decomposable transformation $A_1^2 = A^2|_{E_1}$ on the subspace E_1. According to the assumption, P is a space over the algebraically closed field K. Hence in E_1 one can find a basis $e_1, \ldots, e_k$ with respect to which the matrix of A^2 is a Jordan cell, i.e.

(5.4) $A^2 e_1 = \lambda e_1,\ A^2 e_2 = e_1 + \lambda e_2,\ \ldots,\ A^2 e_k = e_{k-1} + \lambda e_k.$

As we know, for a non-degenerate quadruple S the restriction

$A|_{E_2}: E_2 \longrightarrow E_1$ is an isomorphism. Set $f_i = A^{-1}e_i$. Obviously, the vectors $f_i \in E_2$ form a basis in E_2 .

We are looking for representations of the vectors Ae_i . The properties of the transformation A imply that $Ae_i \in E_2$. Let $Ae_i = \sum_j \alpha_{i,j} f_j$. Then $A^2e_i = A\sum_j \alpha_{i,j} f_j = \sum_j \alpha_{i,j} A f_j = \sum_j \alpha_{i,j} e_j$. Comparing this expression with the equalities (5.4) we see that the matrix $\|\alpha_{i,j}\|$ coincides with the matrix of A^2 relative to the basis e_i . Therefore

$$Ae_1 = \lambda f_1, \; Ae_2 = f_1 + \lambda f_2, \ldots, Ae_k = f_{k-1} + \lambda f_k .$$

From the properties (5.2) of the transformation A it follows that the subspace E_3 is spanned by the vectors $x_1 + Ax_1$, where $x_1 \in E_1$, and E_4 is spanned by the vectors $x_2 + Ax_2$, where $x_2 \in E_2$.

Consequently, the vectors $e_i + Ae_i$ form a basis in E_3, while the vectors $f_i + Af_i$ form a basis in E_4 . Thus we may write

$$E_3 = \{(e_1 + \lambda f_1), (e_2 + f_1 + \lambda f_2), \ldots, (e_k + f_{k-1} + \lambda f_k)\},$$

$$E_4 = \{(f_1 + e_1), (f_2 + e_2), \ldots, (f_k + e_k)\}.$$

The theorem is proved.

The question arises as to what does $S(2k, 0; \lambda)$ represent when $\lambda = 0$ or $\lambda = 1$.

Making use of the expressions of the subspaces E_i in terms of bases it is not difficult to show that in the quadruple $S(2k, 0; \lambda)$ for $\lambda = 0$ the subspaces E_1, E_3 intersect each other, $\dim(E_1 \cap E_3) = 0$, whereas for every other pair (i, j) we have $E_i \cap E_j = 0$. Evidently, to a quadruple of this kind there corresponds a nilpotent transformation A. From the results of §6 it will follow that $S(2k, 0; 0)$ is non-decomposable and $S(2k, 0; 0) \cong S_{1,3}(2k, 0)$.

Similarly, it can be shown that for the quadruple $S(2k, 0; 1)$ we have $E_3 \cap E_4 \neq 0$ and $E_i \cap E_j = 0$ for every other pair (i, j). From the results of §6 it will follow that $S(2k, 0; 1)$ is non-decomposable and $S(2k, 0; 1) \cong S_{3,4}(2k, 0)$.

§6. CLASSIFICATION OF NILPOTENT OPERATOR QUADRUPLES

1. First we describe the canonical form of a non-decomposable nilpotent transformation A compatible with a graduation $P \cong E_1 \oplus E_2$ (i.e. having the properties $AE_1 \subseteq E_2, AE_2 \subseteq E_1$).

If A is a nilpotent non-decomposable linear transformation on the space P (without graduation), then, as it is well known, P has a basis $x_1, \dots, x_n$ such that $Ax_1 = 0$, $Ax_2 = x_1, \dots, Ax_n = x_{n-1}$. The matrix of A in this basis is a Jordan cell.

Moreover, the chain of kernels

$$0 \subset \mathrm{Ker}\, A \subset \mathrm{Ker}\, A^2 \subset \dots \subset \mathrm{Ker}\, A^n = \mathrm{Ker}\, A^{n+i} = P$$

satisfies the equalities $\dim \mathrm{Ker} A^i = i$; $i = 1, 2, \dots, n$. The basis $x_1, \dots, x_n$ can be constructed in the following way. The vector x_n must be chosen so that $x_n \in \mathrm{Ker} A^n$, $x_n \notin \mathrm{Ker} A^{n-1}$. Then is is not difficult to show that the vectors $x_i = A^{n-i} x_n$; $i = 1, 2, \dots, n-1$, together with x_n form a basis with the required properties in P. The action of the mapping A on the basis vectors x_i will be represented by the diagram Γ: $0 \longleftarrow x_1 \longleftarrow x_2 \longleftarrow \cdots \longleftarrow x_n$, which we shall call a "chain".

Let us examine, how the canonical form of the mapping A is altered when there is a graduation $P \cong E_1 \oplus E_2$ given on the space P and the mapping A is compatible with this graduation.

We shall say that the vector x_i is homogeneous relative to the graduation $P \cong E_1 \oplus E_2$, if either $x_i \in E_1$ or $x_i \in E_2$.

The chain Γ will be called homogeneous, if each of the vectors x_i is homogeneous.

Proposition 6.1. *Let A be a nilpotent linear transformation on the space P with graduation $P \cong E_1 \oplus E_2$. Assume that A is compatible with this graduation so that $AE_1 \subseteq E_2$ and $AE_2 \subseteq E_1$. If, in addition, the chain of kernels*

$$0 \subset \mathrm{Ker}\, A \subset \mathrm{Ker}\, A^2 \subset \dots \subset \mathrm{Ker}\, A^n = P$$

satisfies the equalities $\dim(\mathrm{Ker} A^i) = i$; $i = 1, 2, \dots, n$; $n = \dim P$, then in the space P one can find a homogeneous basis $x_1, \dots, x_n$ such that the action of A in this basis can be represented by the diagram

(6.1) $$0 \leftarrow x_1 \leftarrow x_2 \leftarrow \cdots \leftarrow x_n \,.$$

In the proof we shall make use of the following elementary lemma, which we state without proof.

Lemma 6.2. *Let* P *be a space with graduation* $P \cong E_1 \oplus E_2$, *where* $E_1, E_2 \subset P$, *and let* $F_1 \subset F_2$ *be two subspaces of* P *homogeneous relative to the graduation. Then there exists a homogeneous subspace* F *complementary to* F_1 *in* F_2, *i.e.* $F \cap F_1 = 0$, $F + F_1 = F_2$.

Proof of Proposition 6.1. In Proposition 4.20 we have proved that the kernels $\mathrm{Ker}\, A^i$ are homogeneous. Therefore we may apply Lemma 6.2 to the subspaces $\mathrm{Ker}\, A^{n-1}$, $\mathrm{Ker}\, A^n$ and find a homogeneous subspace F_n such that

$$\mathrm{Ker}\, A^{n-1} \cap F_n = n \,, \qquad \mathrm{Ker}\, A^{n-1} + F_n = \mathrm{Ker}\, A^n \,.$$

The equalities $\dim(\mathrm{Ker}\, A^i) = i$ $(i = 1, 2, \ldots, n)$ yield $\dim F_n = 1$. For subspaces of this kind homogeneity means that $F_n \subset E_i$ $(i \in \{1,2\})$. Denote a non-zero vector of F_n by x_n. Then either $x_n \in E_1$ or $x_n \in E_2$.

Set $x_{n-1} = Ax_n$, $x_{n-2} = Ax_{n-1}, \ldots, x_{i-1} = Ax_i, \ldots$. Note that in this case $x_1 = A^{n-1}x_n$. From the conditions $x_n \in \mathrm{Ker}\, A^n$, $x_n \notin \mathrm{Ker}\, A^{n-1}$ it follows that $x_1 \neq 0$ and $Ax_1 = A(A^{n-1}x_n) = A^n x_n = 0$, i.e. $x_1 \in \mathrm{Ker}\, A$.

It can be shown in a similar way that $x_i \neq 0$, $x_i \in \mathrm{Ker}\, A^i$, $x_i \notin \mathrm{Ker}\, A^{i-1}$. Consequently, the vectors $x_1, \ldots, x_n$ form a basis in P and the action of the transformation A on the basis $\{x_i\}$ is represented by the diagram (6.1).

Let us show that the basis $x_1, \ldots, x_n$ is homogeneous. By definition, the mapping A is compatible with the graduation, i.e. $AE_1 \subseteq E_2$, $AE_2 \subseteq E_1$ and, owing to the construction, $x_n \in E_i$; $i \in \{1,2\}$. Thus $x_{n-1} = Ax_n \in E_j$, where $\{i,j\} = \{1,2\}$. Furthermore, $x_{n-2} = Ax_{n-1} \in E_i$ etc. We obtain that $x_n, x_{n-2}, x_{n-4}, \ldots \ldots \in E_i$, while $x_{n-1}, x_{n-3}, x_{n-5}, \ldots \in E_j$; $\{i,j\} = \{1,2\}$. The proposition is proved.

Corollary 6.3. *There are four kinds of homogeneous chains*

$$0 \leftarrow x_1 \leftarrow \cdots \leftarrow x_n \; :$$

1. *The number* n *is odd*, $n = 2k+1$, *and* $x_n \in E_2$ *(in this case* $x_1 \in E_2$ *).*

2. *The number* n *is odd,* $n = 2k+1$, *and* $x_n \in E_2$ *(in this case* $x_1 \in E_2$ *).*

3. *The number* n *is even,* $n = 2k$, *and* $x_n \in E_2$ *(in this case* $x_1 \in E_1$ *).*

4. *The number* n *is even,* $n = 2k$, *and* $x_n \in E_1$ *(in this case* $x_1 \in E_2$ *).*

Theorem 6.4. *Let* A *be a nilpotent linear transformation on the space* P *with graduation* $P \cong E_1 \oplus E_2$, *compatible with this graduation, so that* $AE_1 \subseteq E_2$, $AE_2 \subseteq E_1$. *Then there exists a family of subspaces* P_t *whose members are homogeneous relative to the graduation and have the property that* A *is decomposable on the subspaces* P_t *into the direct sum of non-decomposable trnsformations. Moreover, in a suitable basis of* P_t *the transformation* $A_t = A|P_t$ *is represented by a "chain" belonging to one of the four types mentioned in Corollary 6.3.*

The proof will be accomplished is two steps. We first construct a family of subspaces F_i, $i = 1, 2, \ldots, \ell$, having the following properties: a) the subspaces are homogeneous relative to the graduation $P \cong E_1 \oplus E_2$; b) $\sum_{i=1}^{\ell} F_i = P$ and this is a direct sum; c) The subspaces F_i are "compatible" with the action of A, i.e. $AF_i = 0$, $AF_2 \subseteq F_1, \ldots, AF_\ell \subset F_{\ell-1}$, and every restriction $A|F_i : F_i \longrightarrow F_{i-1}$ is a monomorphism.

The family of subspaces F_i and the action of the transformation A on it will be represented by the diagram

$$0 \longleftarrow F_1 \longleftarrow F_2 \longleftarrow \cdots \longleftarrow F_\ell .$$

In the second step we shall show, how A can be decomposed into the direct sum of non-decomposable transformations, making use of the subspaces F_i.

Step 1. Construction of the family $F_1, \ldots, F_\ell$. Let $0 \subset \operatorname{Ker} A \subset \operatorname{Ker} A^2 \subset \ldots \ldots \subset \operatorname{Ker} A^\ell = P$ be the chain of the kernels $\operatorname{Ker} A^i$, the stable kernel being $\operatorname{Ker} A^\ell = P$. We choose the subspaces F_i so that

$$F_i \cap \operatorname{Ker} A^{i-1} = 0, \qquad F_i + \operatorname{Ker} A^{i-1} = \operatorname{Ker} A^i . \tag{6.2}$$

In this case $\sum_{i=1}^{\ell} F_i = P$ and the sum is a direct sum.

We are going to show that, in addition, the subspaces F_i can be chosen to be homogeneous and "compatible" with the transformation A (i.e. satisfying the inclusions $AF_i \subseteq F_{i-1}$).

In the case $\ell = 1$ we have $\mathrm{Ker}\, A = P$. Setting $F_1 = \mathrm{Ker}\, A$ we see that $AF_1 = 0$. In this case the family consists of the single subspace F_1.

If $\ell = 2$, then $\mathrm{Ker}\, A^2 = P$, and the kernel $\mathrm{Ker}\, A$ is strictly smaller then $\mathrm{Ker}\, A^2$. On account of Lemma 6.2 there exists a homogeneous subspace F_2 such that

$$F_2 \cap \mathrm{Ker}\, A = 0, \qquad F_2 + \mathrm{Ker}\, A = \mathrm{Ker}\, A^2 .$$

Put $\mathrm{Ker}\, A = F_1$. Then it is easy to see that $F_1 \oplus F_2 \cong P$ and $AF_2 \subseteq F_1$. The equality $F_2 \cap \mathrm{Ker}\, A = 0$ yields $\dim AF_2 = \dim F_2$, hence the restriction $A|F_2 : F_2 \longrightarrow F_1$ is a monomorphism. Thus for $\ell = 2$ the family F_1, F_2 is constructed.

Let us now constract the family $F_1, \dots, F_\ell$ for $\ell > 2$. Similarly to the procedure above, we must begin the selection of the subspaces F_i with F_ℓ, and F_ℓ can be chosen arbitrary except that it is homogeneous and satisfies the equalities $F_\ell \cap \mathrm{Ker}\, A^{\ell-1} = 0,\ F_\ell + \mathrm{Ker}\, A^{\ell-1} = \mathrm{Ker}\, A^\ell = P$.

Consider the image AF_ℓ of the subspace F_ℓ. a) As F_ℓ is homogeneous, so is AF_ℓ (cf. Lemma 4.17). b) By definition, F_ℓ consists of vectors x such that $A^\ell x = 0,\ A^{\ell-1}x \neq 0$. Consequently, AF_ℓ consists of vectors y such that $A^{\ell-1}y = 0,\ A^{\ell-2}y \neq 0$, i.e.

$$AF^\ell \subset \mathrm{Ker}\, A^{\ell-1}, \qquad AF^\ell \cap \mathrm{Ker}\, A^{\ell-2} = 0 .$$

Note also that the subspace $AF_\ell + \mathrm{Ker}\, A^{\ell-2}$, being the sum of homogeneous subspaces, is homogeneous, and we have

$$AF_\ell + \mathrm{Ker}\, A^{\ell-2} \subseteq \mathrm{Ker}\, A^{\ell-1} .$$

According to Lemma 6.2 we can choose a homogeneous subspace $\bar{F}_{\ell-1}$ such that

$$\bar{F}_{\ell-1} \cap (\mathrm{Ker}\, A^{\ell-2} + AF_\ell) = 0 ,$$

$$\bar{F}_{\ell-1} + (\mathrm{Ker}\, A^{\ell-2} + AF_\ell) = \mathrm{Ker}\, A^{\ell-1} .$$

Denote the subspace $AF_\ell + \bar{F}_{\ell-1}$ by $F_{\ell-1}$. It is easy to see that $F_{\ell-1}$ has the following properties: a) $F_{\ell-1}$ is homogeneous; b) $F_{\ell-1} \cap \mathrm{Ker}\, A^{\ell-2} = 0$; c) $F_{\ell-1} + \mathrm{Ker}\, A^{\ell-2} = \mathrm{Ker}\, A^{\ell-1}$; d) $AF_\ell \subseteq F_{\ell-1}$.

The subspaces $F_{\ell-2}, F_{\ell-3}, \dots, F_2$ can be constructed in a similar way. The subspace F_1 is chosen equal to $\mathrm{Ker}\, A$. As a result we obtain a family of homogeneous subspaces $F_1, \dots, F_\ell$, whose sum is a direct sum, and this family is compatible with the transformation A, i.e. $AF_i \subseteq F_{i-1}$. Hence this family can be represented by the diagram

$$0 \leftarrow F_1 \leftarrow F_2 \leftarrow \cdots \leftarrow F_{\ell-1} \leftarrow F_\ell .$$

It remains to show that in the latter diagram the mappings $F_i \to F_{i-1}$ $(i>1)$ are monomorphisms. As a matter of fact, $F_i \cap \mathrm{Ker}\, A^{i-1} = 0$ by construction. For $i > 1$ we have $\mathrm{Ker}\, A^{i-1} \supseteq \mathrm{Ker}\, A$ and, consequently, $F_i \cap \mathrm{Ker}\, A = 0$. Therefore AF_i has the same dimension as F_i, i.e. the mapping $A|F_i : F_i \longrightarrow AF_i$ is an isomorphism. In view of the construction $AF_i \subseteq F_{i-1}$. Hence the mapping $AF_i : F_i \longrightarrow$ $\to F_{i-1}$ is a monomorphism.

Step 2. We are going to utilize the family of subspaces $F_1, \dots, F_\ell$ for decomposing the mapping A into the direct sum of non-decomposable ones. The composition of A into the direct sum of non-decomposable transformations will be based on choosing compatible bases in every subspace F_i. (We say that the basis $\{x_{i,j}\}_{j \in \mathfrak{I}_i}$ of F_i and the basis $\{x_{i-1,j}\}_{j \in \mathfrak{I}_{i-1}}$ of F_{i-1} are compatible, if

$$Ax_{i,j} = x_{i-1,j}$$

for every $j \in \mathfrak{I}_i$.)

To begin with, let us find the basis of F_ℓ. Owing to the construction, F_ℓ is homogeneous, i.e. $F_\ell = (F_\ell \cap E_1) + (F_\ell \cap E_2)$ and this is a direct sum. Hence we can choose the basis $\{x_{\ell,j}\}_{j \in \mathfrak{I}_\ell}$ of F_ℓ to be homogeneous i.e. such that any vector $x_{\ell,j}$ belongs either to E_1 or to E_2.

The restriction $A|F_\ell : F_\ell \to F_{\ell-1}$ being a monomorphism, the vectors $Ax_{\ell,1}, Ax_{\ell,2}, \dots, Ax_{\ell,k_\ell}$ are linearly independent and form a basis in AF_ℓ. Moreover, from the condition of homogeneity of the vectors $x_{\ell,j}$, $j \in \mathfrak{I}_\ell$, and the compatibility of the mapping $A\,(AE_1 \subseteq E_2, AE_2 \subseteq E_1)$ it follows that the vectors $Ax_{\ell,j}$, $j \in \mathfrak{I}_\ell$, are also homogeneous. Set $Ax_{\ell,j} = x_{\ell-1,j}$. If AF_ℓ coincides with $F_{\ell-1}$, then the vectors $\{x_{\ell-1,j}\}$, $j = 1, 2, \dots, k_\ell$, form a basis in $F_{\ell-1}$. If, however, $AF_\ell \neq F_{\ell-1}$, then $F_{\ell-1} = AF_\ell + \bar{F}_{\ell-1}$ by construction, and the sum is a direct sum. In this case we choose any homogeneous basis $\{x_{\ell-1,j}\}$, $j = k_\ell + 1, \dots, k_{\ell-1}$ in

$\bar{F}_{\ell-1}$. Then the vectors $x_{\ell-1,1}, \dots, x_{\ell-1,k_{\ell-1}}$ form a basis in $F_{\ell-1}$.

In a similar way, starting with the basis $\{x_{\ell-1,j}\}_{j\in \mathcal{J}_{\ell-1}}$, we construct a basis in $F_{\ell-2}$, etc. As a result we obtain a basis $\{x_{i,j}\}$, $i=1,2,\dots,\ell$; $j\in\mathcal{J}_i$, of the whole space P. Here every vector $x_{i,j}$ is homogeneous, i.e. either $x_{i,j}\in E_1$ or $x_{i,j}\in E_2$. Moreover, the basis $\{x_{i,j}\}$ is compatible with the mapping A, i.e. $Ax_{i,j}=x_{i-1,j}$ for every $i>1$ and $j\in\mathcal{J}_i$.

The subspace spanned by the vectors $x_{1,t}, x_{2,t}, \dots, x_{m,t}$ having identical second subscript t will be denoted by P_t. These vectors satisfy the equalities $Ax_{i,t}=x_{i-1,t}$, $Ax_{1,t}=0$. Therefore they can be represented by the diagram

$$0 \leftarrow x_{1,t} \leftarrow x_{2,t} \leftarrow \cdots \leftarrow x_{m,t}. \tag{6.3}$$

Note that $x_{m,t}$ admits no inverse image, i.e. the basis $\{x_{m+1,j}\}_{j\in\mathcal{J}_{m+1}}$ does not contain a vector with subscript $j=t$. From the diagram (6.3) it is obvious that the subspace P_t is invariant. Moreover, P_t is homogeneous relative to the graduation $P\cong E_1\oplus E_2$, because all vectors $x_{i,j}$ are homogeneous.

It is clear that the homogeneous subspaces P_t, $t\in\mathcal{J}_1$ so defined yield a decomposition of P into the direct sum $P=\bigoplus_{t\in\mathcal{J}_1} P_t$. Since each of these subspaces is invariant, if follows that the operator A is decomposable on the subspaces P_t, $t\in\mathcal{J}_1$. The theorem is proved.

2. Now we give a canonical description of nilpotent operator quadruples.

Theorem 6.5. *Let $S=\{P; E_1,\dots,E_4\}$ be a non-decomposable operator quadruple of type 3^0, i.e. the equalities $E_i\cap E_j=0$, $E_i+E_j=P$ are valid for the pairs $(i,j)\in M(1,3)=\{(1,2),(3,4),(1,4),(2,3)\}$, but either $E_1\cap E_3\neq 0$ or $E_2\cap E_4\neq 0$. Then S is isomorphic to one of the canonical quadruples $S_{1,3}(n,0)$, $S_{2,4}(n,0)$, where $n=\dim P$.*

If the number $n=\dim P$ is odd, $n=2k+1$, $k\geq 0$, then the quadruples $S_{1,3}(2k+1,0)$, $S_{2,4}(2k+1,0)$ have the following canonical form:

$$S_{1,3}(2k+1,0)$$

$$\begin{aligned} P &= \{e_1,\dots,e_k,e_{k+1},f_1,\dots,f_k\}, \\ E_1 &= \{e_1,\dots,e_k,e_{k+1}\}, \\ E_2 &= \{f_1,\dots,f_k\}, \\ E_3 &= \{e_1,(e_2+f_1),\dots,(e_{k+1}+f_k)\}, \\ E_4 &= \{(e_1+f_1),\dots,(e_k+f_k)\}; \end{aligned}$$

$$S_{2,4}(2k+1,0)$$

$$\begin{aligned} P &= \{e_1,\dots,e_k,f_1,\dots,f_k,f_{k+1}\}, \\ E_1 &= \{e_1,\dots,e_k\}, \\ E_2 &= \{f_1,\dots,f_k,f_{k+1}\}, \\ E_3 &= \{(e_1+f_1),\dots,(e_k+f_k)\}, \\ E_4 &= \{f_1,(e_1+f_2),\dots,(e_k+f_{k+1})\}; \end{aligned}$$

On the other hand, if the number $n = \dim P$ *is even,* $n = 2k$, $k > 0$, *then* $S_{1,3}(2k,0)$ *and* $S_{2,4}(2k,0)$ *have the following canonical form:*

$$S_{1,3}(2k,0)$$

$$\begin{aligned} P &= \{e_1,\dots,e_k,f_1,\dots,f_k\}, \\ E_1 &= \{e_1,\dots,e_k\}, \\ E_2 &= \{f_1,\dots,f_k\}, \\ E_3 &= \{e_1,(e_2+f_1),\dots,(e_k+f_{k-1})\}, \\ E_4 &= \{(e_1+f_1),\dots,(e_k+f_k)\}; \end{aligned}$$

$$S_{2,4}(2k,0)$$

$$\begin{aligned} P &= \{e_1,\dots,e_k,f_1,\dots,f_k\}, \\ E_1 &= \{e_1,\dots,e_k\}, \\ E_2 &= \{f_1,\dots,f_k\}, \\ E_3 &= \{(e_1+f_1),\dots,(e_k+f_k)\}, \\ E_4 &= \{f_1,(e_1+f_2),\dots,(e_{k-1}+f_2)\}. \end{aligned}$$

In particular, for every quadruple $S_{1,3}(n,0)$ *we have* $E_1 \cap E_3 \neq 0, E_2 \cap E_4 = 0$, *while for the quadruples* $S_{2,4}(n,0)$ *the relations* $E_1 \cap E_3 = 0, E_2 \cap E_4 \neq 0$ *are valid.*

Proof. We know that to an operator quadruple S of type 3^0 there corresponds a nilpotent transformation $A = A(1,2,3,4)$ such that $AE_1 \subseteq E_2, AE_2 \subseteq E_1$, and $\lambda = 1$ is not an eigenvalue of A. In Theorem 6.4 we have proved that there exists a family of homogeneous subspaces P_t on which the transformation A is decomposable. Thus the conditions of Proposition 4.19 are satisfied and, consequently, the quadruple S is decomposable on the subspaces P_t. Therefore if the transformation A is decomposable, so is the quadruple S. Hence we deduce that if the quadruple is non-decomposable, so is the corresponding transformation A. In Corollary 4.16 we have proved the converse statement. Thus there is a one-to-one correspondence between non-decomposable quadruples and the corresponding operators.

In Corollary 6.3 we established that there are 4 kinds of non-decomposable transformations A compatible with the graduation: two kinds for the case where the dimension of the space P is odd, and two kinds when the dimension is even.

Let us now find the canonical form of quadruples corresponding to each of these kinds.

Let $\dim P = 2k+1$. Assume that in P one can find a homogeneous basis in which the transformation A is represented by the "chain"

$$0 \leftarrow x_1 \leftarrow x_2 \leftarrow \cdots \leftarrow x_{2k+1} .$$

Assume also that in this "chain" $x_{2k+1} \in E_1$. Then it follows that all vectors x_{2i+1} with odd subscript belong to E_1, while the vectors x_{2i} belong to E_2.

We set

$$x_1 = e_1, \; x_3 = e_2, \ldots, x_{2k+1} = e_{k+1},$$

$$x_2 = f_1, \; x_4 = f_2, \ldots, x_{2k} = f_k .$$

The subspace spanned by the vectors $e_1, \ldots, e_{k+1}$ will be denoted by $\tilde{E}_1$, and that spanned by $f_1, \ldots, f_k$ will be designated by $\tilde{E}_2$. According to the construction, all vectors x_i are linearly independent; therefore $\tilde{E}_1 \cap \tilde{E}_2 = 0$, $\tilde{E}_1 + \tilde{E}_2 = P$. Obviously $\tilde{E}_1 \subseteq E_1$, $\tilde{E}_2 \subseteq E_2$. Comparing these properties with the equalities $E_1 \cap E_2 = 0$, $E_1 + E_2 = P$ we obtain that $\tilde{E}_1 = E_1$, $\tilde{E}_2 = E_2$. Thus

$$P = \{e_1, \dots, e_k, e_{k+1}, f_1, \dots, f_k\},$$

$$E_1 = \{e_1, \dots, e_k, e_{k+1}\},$$

$$E_2 = \{f_1, \dots, f_k\}.$$

From the definition of the mapping A it follows that $x \in E_1$ implies $(x+Ax) \in E_3$, while $y \in E_2$ implies $(y+Ay) \in E_4$.

Hence we obtain that the vectors $e_{i+1} + Ae_{i+1} = e_{i+1} + f_i$ $(i=1,2,\dots,k)$ together with the vector $e_1 + Ae_1 = e_1$, belong to the subspace E_3, whereas the vectors $f_i + Af_i = f_i + e_i$ $(i=1,2,\dots,k)$ belong to E_4. The span of the vectors $e_1, (e_2+f_1), \dots, (e_{k+1}+f_k)$ will be denoted by $\tilde{E}_3$, and the span of the vectors $(e_1+f_1), (e_2+f_2), \dots, (e_k+f_k)$ by $\tilde{E}_4$. Clearly $\tilde{E}_3 \subseteq E_3$, $\tilde{E}_4 \subseteq E_4$. From the linear independence of the vectors $e_1, \dots, e_k, e_{k+1}, f_1, \dots, f_k$ it follows easily that $\dim \tilde{E}_3 = k+1$, $\dim \tilde{E}_4 = k$. It is also not hard to verify that $\tilde{E}_3 \cap \tilde{E}_4 = 0$, $\tilde{E}_3 + \tilde{E}_4 = P$. These equalities are compatible with the condition $E_3 \oplus E_4 \subseteq P$ if and only if $\tilde{E}_3 = E_3$, $\tilde{E}_4 = E_4$. Thus

$$E_3 = \{e_1, (e_2+f_1), \dots, (e_{k+1}+f_k)\},$$

$$E_4 = \{(e_1+f_1), \dots, (e_k+f_k)\}.$$

We obtained that to a chain of type 1) $n = 2k+1$; $x_{2k+1} \in E_1$, (cf. Corollary 6.3) there corresponds the quadruple $S_{1,3}(2k+1,0)$, in which $E_1 \cap E_3 \neq 0$, $E_2 \cap E_4 = 0$. It can be shown in a similar way that to a chain of type 2 $(n = 2k+1;$ $x_{2k+1} \in E_2)$ there corresponds the quadruple $S_{2,4}(2k+1,0)$, in which $E_1 \cap E_3 = 0$, $E_2 \cap E_4 \neq 0$, further to a chain of type 3 $(n = 2k,\ x_{2k} \in E_2)$ there corresponds the quadruple $S_{1,3}(2k,0)$, in which $E_1 \cap E_3 \neq 0$, $E_2 \cap E_4 = 0$ and, finally, to a chain of type 4 $(n = 2k,\ x_{2k} \in E_1)$ there corresponds the quadruple $S_{2,4}(2k,0)$, in which $E_1 \cap E_3 = 0$, $E_2 \cap E_4 \neq 0$. The theorem is proved.

Theorem 6.5 yields

Theorem 6.6. *Let $S = \{P; E_1, \dots, E_4\}$ be a non-decomposable operator quadruple of type 2^0, i.e. the equalities $E_i \cap E_j = 0$, $E_i + E_j = P$ are satisfied in this quadruple for the pairs $(i,j) \in M(1,2) = \{(1,3),(2,4),(1,4),(2,3)\}$, but either $E_1 \cap E_2 \neq 0$ or $E_3 \cap E_4 \neq 0$. Then S is isomorphic to one of the canonical quadruples*

$S_{1,2}(n,0), S_{3,4}(n,0)$.

The canonical quadruples $S_{1,2}(n,0)$, $S_{3,4}(n,0)$ *can be obtained from the quadruples* $S_{1,3}(n,0)$, $S_{2,4}(n,0)$ *by a change of indices of the subspaces* E_i : $E_1 \to E_1$, $E_3 \to E_2$, $E_2 \to E_3$, $E_4 \to E_4$.

Theorem 6.7. *Let* $S=\{P; E_1,\dots,E_4\}$ *be a non-decomposable operator quadruple of type* 4^0 *the equalities* $E_i \cap E_j = 0$, $E_i + E_j = P$ *are satisfied in* S *for the pairs* $(i,j) \in M(1,4) = \{(1,2),(3,4),(1,3),(2,4)\}$, *but either* $E_1 \cap E_4 \neq 0$ *or* $E_2 \cap E_3 \neq 0$. *Then* S *is isomorphic to one of the canonical quadruples* $S_{1,4}(n,0), S_{2,3}(n,0)$.

The canonical quadruples $S_{1,4}(n,0), S_{2,3}(n,0)$ *can be obtained from the quadruples* $S_{1,3}(n,0), S_{2,4}(n,0)$, *by a change of indices of the subspaces:* $E_1 \to E_1$, $E_2 \to E_2$, $E_3 \to E_4$, $E_4 \to E_3$.

The proof of these theorems follows from that of Theorem 6.5 if we perform the appropriate change of the indices of subspaces.

Corollary 6.8. *The invariants of the non-decomposable quadruple* $S_{i,j}(n,0)$, *where* $i<j$, *are the numbers* $n = \dim P$, $0 = \rho(S)$ *and the pair* (i,j), $i<j$; $i,j \in \{1,2,3,4\}$ *: In the quadruple* $S_{i,j}(n,0)$ *we have* $E_i \cap E_j \neq 0$ *and* $E_k \cap E_\ell = 0$ *for all other pairs* (k,ℓ).

Consequently, if a quadruple $S=\{P; E_1,\dots,E_4\}$ is non-decomposable and $\dim P = n$, $\rho(S)=0$, $E_i \cap E_j \neq 0$, then $S \cong S_{i,j}(n,0)$. Hence we obtain

Corollary 6.9. *Let* $S = S_{i,j}(n,0) = \{P; E_1,\dots,E_4\}$ *be a canonical non-decomposable quadruple. Let* σ *denote a permutation of the indices* $1,2,3,4$ *and let* $\sigma S = S' = \{P; E'_1,\dots,E'_4\}$ *be the system obtained from* S *by the change* $E'_i = E_{\sigma^{-1}(i)}$ *of the indices of subspaces* E_i . *Then*

$$\sigma S_{i,j}(n,0) \cong S_{\sigma(i),\sigma(j)}(n,0).$$

Let us summarize the results of §§ 3, 5 and 6.

Theorem 6.10. *Let* $S=\{P; E_1,\dots,E_4\}$ *be a system that consists of a finite-dimensional space* P *over the algebraically closed field* K *and a quadruple of subspaces* E_i . *Then the system can be decomposed into the direct sum of non-decomposable subsystems belonging to the following types:* $S(2m,0;\lambda)$, $\lambda \in K$, $\lambda \neq 0$, $\lambda \neq 1$; $S_{i,j}(m,0)$, $i<j$, $i,j \in \{1,2,3,4\}$; $S_i(m,-1)$, $S_i(m,+1)$, $i \in \{1,2,3,4\}$; $S(2m-1,-2)$, $S(2m-1,2)$, $m = 1,2,\dots$.

LITERATURE

[1] L. Kronecker, Algebraische Reduction der Schaaren bilinearer Formen, *Sitzungsber. Akademie Berlin*, 1890, 763-776.

[2] G. Birkhoff, *Lattice theory*, 3rd ed., (Providence, 1967).

[3] I.M. Gel'fand – V.A. Ponomarev, Non-decomposable representations of the Lorentz group, *Uspehi Mat. Nauk*, 23 (1968), 3-60 (in Russian).

[4] N.M. Dobrovol'skaja – V.A. Ponomarev, Pairs of confronted operators, *Uspehi Mat. Nauk*, 20 (1965), 81-86 (in Russian).

[5] S. MacLane, An algebra of additive relations, *Proc. Nat. Acad. Sci. USA*, 47 (1961), 1043-1051.

[6] V.A. Ponomarev, Elements of the theory of additive binary relations in a finite-dimensional vector space (in Russian).

[7] D. Puppe, Korrespondenzen in abelschen Kategorien, *Math. Ann.*, 148 (1962), 1-30.

COLLOQUIA MATHEMATICA SOCIETATIS JÁNOS BOLYAI
5. HILBERT SPACE OPERATORS, TIHANY (HUNGARY), 1970

Banach algebras generated by singular integral operators

I. C. GOHBERG and N. JA. KRUPNIK

Let Γ be an oriented contour in the complex plane consisting of a finite number of closed simple Ljapunov curves, and $\mathcal{L}(L_p(\Gamma))$ the Banach algebra of all linear bounded operators acting in the space $L_p(\Gamma)$.

We shall denote by $\Sigma(L_p, \Gamma, C)$ $(1 < p < \infty)$ the smallest closed subalgebra of $\mathcal{L}(L_p(\Gamma))$ containing all one-dimensional singular integral operators A of the form

$$(0.1) \qquad (A\varphi)(t) = c(t)\varphi(t) + \frac{d(t)}{\pi i}\int_\Gamma \frac{\varphi(\tau)}{\tau - t}\,d\tau \qquad (\varphi(t) \in L_p(\Gamma))$$

with arbitrary continuous coefficients $c(t)$, $d(t)$.

The following propositions are valid (see [1], [2], [5]):

1^0. *The algebra* $\Sigma(L_p, \Gamma, C)$ *is the linear hull of all operators of the form* (0.1) *and of the set* $\mathcal{T} = \mathcal{T}(L_p(\Gamma))$ *of all completely continuous operators belonging to* $\mathcal{L}(L_p(\Gamma))$.

2^0. *The factor algebra* $\Sigma(L_p,\Gamma,C)/\mathcal{T}$ *is isomorphic (and in the case* $p=2$ *isometric, too) to the algebra of all continuous functions defined on the compactum* $\Gamma\times J$ *where* J *is the two-point set* ± 1 . *The function corresponding to the operator* A *of the form* (0.1) *under this isomorphism is* $\mathfrak{A}(t,j)=c(t)+jd(t)$ $(t\in\Gamma,\ j\in J)$.

3^0. *The operator* A *of the form* (0.1) *is a* Φ *-or* $\Phi_\pm$ *-operator* in the space* $L_p(\Gamma)$ *if and only if the function* $\mathfrak{A}(t,j)$ *does not vanish at any point of* $\Gamma\times J$.

The function $\mathfrak{A}(t,j) = c(t)+jd(t)$ is called (see [1]) the symbol of the singular integral operator A .

The proof of these propositions is based on the following properties of the singular integration operator

$$(S\varphi)(t) = \frac{1}{\pi i}\int_\Gamma \frac{\varphi(\tau)}{\tau - t}\,d\tau$$

and the operator of multiplication by an arbitrary continuous function $\alpha(t)$:

(0.2) $$S^2 = I$$

(0.3) $$S - S^* \in \mathcal{T}$$

(0.4) $$\alpha(t)S - S\alpha(t)I \in \mathcal{T} .$$

Let us make clear that the operator S is considered in L_p, while S^* is understood to be the adjoint of S as an operator acting in L_q $(p^{-1}+q^{-1}=1)$.

The propositions 1^0 – 3^0 are no more valid in the case where the contour Γ is non-closed or non-simple, nor for algebras containing singular integral operators of the form (0.1) with discontinuous coefficients. In these cases not all of the properties (0.2) – (0.4) hold true. The latter circumstance leads to substantial complications. The

* For the definition of Φ (Φ_+,Φ_-)-operators see [3].

algebras generated by such operators consist not only of the linear hull of the operators (0.1) and those belonging to $\mathcal{T}$. The symbol and the set where its argument varies can be also constructed according to more complicated rules. Sometimes the symbol is a matrix function depending substantially on p .

Examination of the cases listed above leads to a number of commutative and non-commutative algebras generated by classes of one-dimensional singular integral operators under different restrictions concerning the contour and the coefficients. The investigation of these algebras and the construction of the symbol can be performed by a general scheme. The purpose of the present paper is to give an account of this scheme as well as various illustrations to it. It should be noted that the general scheme to be expounded makes possible also the study of algebras generated by other classes of operators. For instance, by singular integral operators with matrix coefficients, various operators of Wiener-Hopf type, and others.

Let us remark also that this time we will not dwell on establishing formulas for the index of the operators in question.

§ 1. ALGEBRAS WITH A SYMBOL. EXAMPLES

Let $\mathcal{B}$ be a Banach space and let $\mathcal{L} = \mathcal{L}(\mathcal{B})$ be the Banach algebra of all linear bounded operators acting in $\mathcal{B}$. In what follows, $\mathfrak{A}$ will denote a closed subalgebra of $\mathcal{L}$ containing the set $\mathcal{T} = \mathcal{T}(\mathcal{B})$ of all linear completely continuous operators.

Let $\mathfrak{V}$ be a subset of $\mathfrak{A}$ containing some generator system of the algebra $\mathfrak{A}$ as well as the operators cI , where c is an arbitrary complex number; further let X be a set in the n-dimensional euclidean space.

We shall say that on the set $\mathfrak{V}$ a symbol is defined, if to each operator $A \in \mathfrak{V}$ there corresponds a matrix function $\mathcal{A}(x)$ defined on the set X and having the following properties:

1.1. The order $n = n(x)$ of the matrix function $\mathcal{A}(x)$ depends, in general, on the point $x \in X$, but does not depend on the operator A .

1.2. The matrix corresponding to the operator cI is $c\mathfrak{I}_x$, where $\mathfrak{I}(x)$ stands for the unit matrix of order $n(x)$ and c is a complex number.

1.3. Let A_{jk} $(j,k = 1,2,\dots,s)$ be an arbitrary family of operators belonging to $\mathfrak{G}$, and $\mathcal{L}^s$ the direct sum of s copies of the space $\mathcal{L}$. In order that the operator $A = \|A_{jk}\|_{j,k=1}^{s}$ be a Φ-operator in the space $\mathcal{L}^s$ it is necessary and sufficient that for the determinant of the matrix $\mathcal{A}(x) = \|A_{jk}(x)\|_{j,k=1}^{s}$ (having the order $sn(x)$) the condition $\inf|\det \mathcal{A}(x)|>0$ $(x\in X)$ be satisfied.

In the sequel our basic problem is to extend the symbol to the whole algebra A so that the following conditions be fulfilled: 1) the correspondence $A \longrightarrow \mathcal{A}(x)$ is a homomorphism; 2) the operator $A (\in \mathfrak{A})$ is a Φ-operator if and only if $\inf|\det \mathcal{A}(x)| > 0$ $(x\in X)$.

The solution of this problem will be obtained in the next sections under additional restrictions. However, already here it is possible to extend the definition of the symbol to a set that is dense in $\mathfrak{A}$, namely to the linear hull $\mathfrak{L} = \mathfrak{L}(\mathfrak{A};\mathfrak{G})$ of all products of operators in $\mathfrak{G}$.

With an operator $A\in\mathfrak{L}$ represented as

$$A = \sum_{j=1}^{k} A_{j1} A_{j2} \cdots A_{jr} \qquad (A_{jl}\in\mathfrak{G}) \tag{1.1}$$

we associate a symbol $\mathcal{A}(x)$ according to the formula

$$\mathcal{A}(x) = \sum_{j=1}^{k} A_{j1}(x) A_{j2}(x) \cdots A_{jr}(x).$$

This definition of the symbol has the insufficiency that to one operator it associates several symbols corresponding to the different representations of the operator A in the form (1.1). In what follows, when speaking of the symbol of an operator $A\in\mathfrak{L}$, we shall keep in mind one of its symbols.

Theorem 1.1. *Let the algebra* $\mathfrak{A}$ *have the properties* 1.1 – 1.3. *An operator* $A\in\mathfrak{L}(\mathfrak{A};\mathfrak{G})$ *is a* Φ*-operator if and only if its symbol* $\mathcal{A}(x)$ *satisfies the condition* $\det \mathcal{A}(x) \neq 0$ $(x\in X)$.

The proof of this theorem is based on the following auxiliary assertion:

Lemma 1.1. *Let* A_{jl} $(j=1,2,\dots,k;\ l=1,2,\dots,r)$ *be arbitrary operators from* $\mathcal{L}$ *and let*

$$A = \sum^{k} A_{j1} A_{j2} \cdots A_{jr} .$$

Then we have the equality

$$(1.2) \qquad \begin{pmatrix} Z & X \\ Y & 0 \end{pmatrix} = \begin{pmatrix} I_{k(r+1)} & 0 \\ U & I \end{pmatrix} \begin{pmatrix} I_{k(r+1)} & 0 \\ 0 & A \end{pmatrix} \begin{pmatrix} Z & X \\ 0 & I \end{pmatrix},$$

where Z *is a quadratic matrix of order* $k(r+1)$,

$$Z = \left\| \begin{matrix} I_k & B_1 & 0 & \cdots & 0 \\ 0 & I_k & B_2 & \cdots & 0 \\ \cdot & \cdot & \cdot & \cdots & \cdot \\ 0 & 0 & 0 & \cdots & B_r \\ 0 & 0 & 0 & \cdots & I_k \end{matrix} \right\|$$

I_k *denotes the unit operator in the space* $\mathcal{L}^k$,

$$B_j = \left\| \begin{matrix} A_{1j} & 0 & \cdots & 0 \\ 0 & A_{2j} & \cdots & 0 \\ \cdot & \cdot & \cdots & \cdot \\ 0 & 0 & \cdots & A_{kj} \end{matrix} \right\|$$

$I = I_1$, *while* X, Y, U *are a cone-column and two one-row matrices, respectively, defined by the equalities*

$$X = \left\| \begin{matrix} 0 \\ \vdots \\ 0 \\ -I \\ \vdots \\ -I \end{matrix} \right\| \begin{matrix} \left.\vphantom{\begin{matrix}0\\ \vdots\\ 0\end{matrix}}\right\} kr \\ \left.\vphantom{\begin{matrix}-I\\ \vdots\\ -I\end{matrix}}\right\} k \end{matrix} \qquad Y = \| \overbrace{I, \ldots, I}^{k}, \overbrace{0, \ldots, 0}^{kr} \|,$$

$$U = \| M_0, \ldots, M_r \| ,$$

where $M_0 = \| \overbrace{I\ I \ldots I}^{k} \|$ *and*

$$M_j = \| A_{11}A_{12}\cdots A_{1j}, A_{21}A_{22}\cdots A_{2j}, \ldots, A_{k1}A_{k2}\cdots A_{kj} \| \qquad (j = 1, 2, \ldots, r).$$

The correctness of the lemma can be established by direct verification.

Proof of Theorem 1.1. Let us denote the matrix in the left-hand member of the equality (1.2) by B, and let us write it in the from $B = \| B_{j\ell} \|_{j,\ell=1}^{kr+k+1}$, where $B_{j\ell} \in \mathfrak{Y}$.

From the equality (1.2) it follows that B is a Φ-operator in $\mathcal{L}^{kr+k+1}$ if and only if A is a Φ-operator in $\mathcal{L}$. From the same equality it follows that $\det \| \mathcal{B}_{j\ell}(x) \| = \det \mathcal{A}(x)$, where $\mathcal{B}_{j\ell}(x)$ is the symbol of the operator $\mathcal{B}_{j\ell}$. It remains to make use of the properties of the symbol. The theorem is proved.

Below we shall consider the following examples.

1. The algebra $\Sigma(L_2, \Gamma_d, C)$. Let, as before, Γ be an oriented contour in the complex plane, consisting of a finite number of closed Ljapunov curves. By Γ_d we denote a contour obtained from Γ by adding a finite number of open Ljapunov curves.

We choose $\mathfrak{Y}$ to be the set of all singular integral operators in $L_2(\Gamma_d)$ of the form $A = c(t)I + d(t)S$ where $c(t)$ and $d(t)$ $(t \in \Gamma_d)$ are arbitrary continuous functions, and S stands for the operator of singular integration along Γ_d.

Let us consider the algebra $\Sigma(L_2, \Gamma_d, C)$, which is the least closed subalgebra of the algebra $\mathcal{L}(L_2(\Gamma_d))$ containing all operators of $\mathfrak{Y}$.

Under these assumptions, generally speaking, the property (0.2) does not hold.

As shown in [4], in this case the set X is a curve in the 3-dimensional real space, consisting of two copies of Γ_d and linear segments connecting the respective end-points of the open arcs. This curve is defined in the following way:

$$X = \{ x = (x_1, x_2, x_3) : x_1 + ix_2 \in \Gamma_d,\ -1 \le x_3 \le 1,\ (1 - x_3^2) \prod_{k=1}^{r} (x_1 + ix_2 - \alpha_k) = 0 \};$$

here $\alpha_1, \dots, \alpha_r$ are the end-points of the open arcs of the contour Γ_d.

The symbol of the operator $A = c(t)I + d(t)S$ is defined by the equality

$$\mathcal{A}(x) = c(t) + x_3 d(t),$$

where $t = x_1 + ix_2$.

2. The algebra $\Sigma(L_p, \Gamma_d, C)$. Replacing in the previous example the space L_2 by L_p $(1 < p < \infty)$ we obtain the algebra $\Sigma(L_p, \Gamma_d, C)$. In this algebra the sets $\mathfrak{G}$ and X are defined as in the case $p = 2$. Only the definition of the symbol is changed. The symbol $\mathcal{A}(x)$ of the operator $\mathcal{A} = c(t)I + d(t)S$ has the form

$$\mathcal{A}(x) = c(t) + \Omega_p(x)\, d(t),$$

where $t = x_1 + ix_2$ $(\in \Gamma_d)$,

$$\Omega_p(x) = \begin{cases} \dfrac{x_3(1+a^2) - i(1-x_3^2)a}{1+x_3a^2} & t = t_1, \dots, t_q, \\ x_3 & t \neq t_1, \dots, t_{2q}, \\ \dfrac{x_3(1+a^2) + i(1-x_3^2)a}{1+x_3a^2} & t = t_{q+1}, \dots, t_{2q} \end{cases}$$

$a = \operatorname{ctg}\frac{\pi}{p}$ and $t_1, \dots, t_q$ $(t_{q+1}, \dots, t_{2q})$ are the starting points (end-points) of the open arcs of the contour Γ_d.

3. The algebra $\Sigma(L_2, \Gamma, C_d)$. Let us denote by $C_d = C_d(\Gamma)$ the algebra of functions on Γ such that each point is either a point of continuity or a point of discontinuity of the first kind. In this example the role of $\mathfrak{G}$ will be played by the set of all singular integral operators from $\mathcal{L}(L_2(\Gamma))$ having the form $A = c(t)I + d(t)S$ where $c(t)$ and $d(t) \in C_d$.

The role of the algebra $\mathfrak{A}$ is played by the algebra $\Sigma(L_2, \Gamma, C_d)$ defined as the smallest subalgebra of $\mathcal{L}(L_2(\Gamma))$ containing $\mathfrak{G}$. In this case we may take X to be a cylinder in the 3-dimensional real space:

$$X = \{x = (x_1, x_2, x_3) : x_1 + ix_2 \in \Gamma,\ 0 \leqq x_3 \leqq 1\},$$

and the symbol $\mathfrak{A}(x)$ of the operator $A = c(t)I + d(t)S$ is given by the equality

$$\mathfrak{A}(x) = \begin{pmatrix} a(t+0)x_3 + a(t-0)(1-x_3) & \sqrt{x_3(1-x_3)}\,(b(t-0)-b(t-0)) \\ \sqrt{x_3(1-x_3)}\,(a(t+0)-a(t-0)) & b(t+0)(1-x_3)+b(t-0)x_3 \end{pmatrix},$$

where $t = x_1 + ix_2$, $a(t) = c(t)d(t)$ and $b(t) = c(t) - d(t)$ (see [5]).

4. The algebra $\Sigma(L_p, \Gamma, C_d)$ $(1 < p < \infty)$. This algebra can be obtained from the previous one if we replace $L_2(\Gamma)$ by $L_p(\Gamma)$. Here the sets $\mathfrak{Y}$ and X are the same as in the preceding example, but the definition of the symbol $\mathfrak{A}(x)$ of the operator $A = c(t)I + d(t)S$ is replaced by the following:

$$\mathfrak{A}(x) = \begin{pmatrix} \xi a(t+0) + (1-\xi)a(t-0) & \sqrt{\xi(1-\xi)}\,(b(t+0)-b(t-0)) \\ \sqrt{\xi(1-\xi)}\,(a(t+0)-a(t-0)) & \xi b(t-0)+(1-\xi)\,b(t+0) \end{pmatrix},$$

where

$$\xi = \frac{\sin(\Theta x_3)\exp(i\Theta x_3)}{\sin\Theta \exp(i\Theta)} \quad \text{and} \quad \Theta = \pi - \frac{2\pi}{p}.$$

It should be noted that in Examples 3 and 4 the property (0.4) fails.

A little more involved is the definition of the symbol in weighted spaces. We will not go into the details at this place.

5. The algebra $\Sigma(L_p, \Gamma_d, C_d)$. Let us complicate the preceding example replacing in it the contour Γ by Γ_d. For the algebra $\Sigma(L_p, \Gamma_d, C_d)$ so obtained the sets X and $\mathfrak{Y}$ are the same as in the preceding example. The symbol $\mathfrak{A}(x)$ of the operator $A = c(t)I + d(t)S$ is defined in the same way as in the pre-

ceding example for all points $x = (x_1, x_2, x_3) \in X$ such that $x_1 + ix_2$ does not coincide with the end-points of the open arcs of Γ_d. If for $x = (x_1, x_2, x_3)$ the point $\alpha = x_1 + ix_2$ is the starting point of an open arc, then

$$\mathfrak{A}(x) = \begin{pmatrix} \xi a(\alpha) + (1-\xi) b(\alpha) & 0 \\ 0 & b(\alpha) \end{pmatrix}$$

while if it is an end-point, then

$$\mathfrak{A}(x) = \begin{pmatrix} (1-\xi) a(\alpha) + \xi b(\alpha) & 0 \\ 0 & b(\alpha) \end{pmatrix}.$$

6. The algebra $WH(\ell_2, C_d)$. Let $a(\zeta)$ $(|\zeta|=1)$ be a measurable bounded function, and let a_j $(j = 0, \pm 1, \ldots)$ be its Fourier coefficients. We shall denote by T_a the linear bounded operator defined in the space ℓ_2 by the Toeplitz matrix $\| a_{j-k} \|_{j,k=0}^{\infty}$. In this example we retain the notation $C_d = C_d(\Gamma)$ for the case where Γ is the unit circle $(\Gamma = \{|\zeta| = 1\})$.

Let us define the algebra $WH(\ell_2, C_d)$ as the smallest subalgebra of $\mathcal{L}(\ell_2)$ containing the set of operators

$$\mathfrak{G} = \{T_a : a(\zeta) \in C_d\}.$$

Introduce a set X in the 3-dimensional real space, putting

$$X = \{x = (x_1, x_2, x_3) : |x_1 + ix_2| = 1, \ 0 \leqq x_3 \leqq 1\}.$$

The symbol of the operator T_a in the algebra $WH(\ell_2, C_d)$ is given by the equality

$$\mathcal{T}_a(x) = a(\zeta+0)x_3 + a(\zeta-0)(1-x_3).$$

where $\zeta = x_1 + ix_2$ (see [7]).

Algebras consisting of continual analogues of the operators T_a and other operators of Wiener-Hopf type can be introduced in a similar way.

7. The algebra $WH(h_p, C_d)$. Let H_p $(1<p<\infty)$ be the Hardy spaces, and h_p the Banach space, isometric to H_p, of numerical sequences $\xi = \{\xi_j\}_{j=0}^{\infty}$

composed of the Fourier coefficients of functions in H_p.

The algebra $WH(h_p, C_d)$ can be defined similarly to the previous example. The generating set $\mathfrak{U}$ and the set X are defined as in the ℓ_2 case. The symbol of the operator T_a in the space h_p is given by the new equality

$$\mathcal{T}_a(x) = \frac{\sin(1-x_3)\Theta \exp(ix_3\Theta)}{\sin\Theta} a(\zeta-0) +$$

$$+ \frac{\sin x_3\Theta \exp(i(x_3-1)\Theta)}{\sin\Theta} a(\zeta+0),$$

where $\Theta = \pi \dfrac{p-2}{p}$ (see [8]).

8. The algebra $\Sigma(L_2, \Lambda, C_\Lambda)$. Let Λ be a contour that consists of m simple closed curves $\Lambda_1, \dots, \Lambda_m$ having a point t_0 in common; we assume also that any two of these lines are not tangent to each other at t_0. We orient the contour Λ so that every curve Λ_j would be directed anti-clockwise. We denote by C_Λ the set of all functions $a(t)$ which: 1) are continuous everywhere on Λ with the possible exception of the point t_0 and 2) have finite (generally speaking, different) limits when t tends to t_0 along any line of the contour Λ.

Let us denote by a_{2k-1} $(k=1,\dots,m)$ respectively, the limits of the function $a(t)$ when t tends to t_0 along the curve Λ_k on the arc arriving at t_0, and by a_{2k} $(k=1,\dots,m)$, respectively, the limit of $a(t)$ along Λ_k on the arc leaving the point t_0.

By $\mathfrak{U}$ we denote the set of operators of the form $c(t)I + d(t)S$, where $c(t), d(t) \in C$. Let $\mathcal{L} = L_2(\Lambda)$. For symbol of the operator $A = c(t)I + d(t)S$ we can take the matrix function $\mathcal{A}(x)$ defined on the cylinder

$$X = \{x = (x_1, x_2, x_3) : x_1 + ix_2 \in \Lambda,\ 0 \leqq x_3 \leqq 1\}$$

as follows.

At points x such that $t = x_1 + ix_2$ does not coincide with t_0,

$$\mathcal{A}(x) = \left\| \begin{matrix} a(t) & 0 \\ 0 & b(t) \end{matrix} \right\|,$$

where $a(t) = c(t) + d(t)$, $b(t) = c(t) - d(t)$.

As for the points x with $x_1 + ix_2 = t_0$, let

$$\mathcal{A}(x) = \left\| u_{jk}(x_3 \right\|_{j,k=1}^{2n} .$$

The functions $u_{jk}(x_3)$ for $j < k$ are defined by the equalities

$$u_{jk}(x_3) = \begin{cases} (-1)^{j+1}(b_k - b_{k-1})x_3^{\frac{k-j}{2m}}(1-x_3)^{1-\frac{k-j}{2m}} & \text{if } k \text{ is even,} \\ (-1)^{j+1}(a_{k+1} - a_k)x_3^{\frac{k-j}{2m}}(1-x_3)^{1-\frac{k-j}{2m}} & \text{if } k \text{ is odd;} \end{cases}$$

for $j > k$ by the equalities

$$u_{jk}(x_3) = \begin{cases} (-1)^{j}(b_k - b_{k-1})x_3^{1-\frac{j-k}{2m}}(1-x_3)^{\frac{j-k}{2m}} & \text{if } k \text{ is even,} \\ (-1)^{j}(a_{k+1} - a_k)x_3^{1-\frac{j-k}{2m}}(1-x_3)^{\frac{j-k}{2m}} & \text{if } k \text{ is odd;} \end{cases}$$

and for $j = k$

$$u_{kk}(x_3) = \begin{cases} a_k x_3 + a_{k+1}(1-x_3) & \text{if } k \text{ is odd,} \\ b_k x_3 + b_{k+1}(1-x_3) & \text{if } k \text{ is even.} \end{cases}$$

The algebra $\Sigma(L_2, \Lambda, C_\Lambda)$ is defined as the smallest subalgebra of the algebra $\mathcal{L}(L_2(\Lambda))$ containing $\mathfrak{Y}$.

§ 2. SYMMETRIC ALGEBRAS

In this section we consider the case where the space $\mathcal{L} = \mathfrak{h}$ is a Hilbert space.

An algebra $\mathfrak{A} = \mathfrak{A}(\mathfrak{Y}, \mathfrak{h})$ satisfying the conditions 1.1 – 1.3 will be called symmetric, if it has the following properties:

2.1. Together with every operator A the set $\mathfrak{A}$ contains an operator A' such that the difference $A' - A^*$ is completely continuous.

2.2. The matrix $\mathcal{A}(x)$ is the adjoint of the matrix $\mathcal{A}(x)$: $\mathcal{A}'(x) = (\mathcal{A}(x))^*$.

It can be easily verified that the algebras $\Sigma(L_2, \Gamma_d, C)$, $\Sigma(L_2, \Gamma, C_d)$ $\Sigma(L_2, \Gamma_d, C_d)$ and $WH(\ell_2, C_d)$ satisfy the conditions 2.1 and 2.2.

Let us define the norm of the symbol $\mathcal{A}(x)$ setting

$$\| \mathcal{A}(x) \| = \sup_{x \in X} s_1(\mathcal{A}(x)),$$

where $[s_1(\mathcal{A}(x))]^2$ is the greatest eigenvalue of the matrix $\mathcal{A}(x)(\mathcal{A}(x))^*$.

We agree to denote the factor algebra $\mathfrak{A}/\mathfrak{T}$ by $\hat{\mathfrak{A}}$. The coset in $\hat{\mathfrak{A}}$ containing the operator $A (\in \mathfrak{A})$ will be denoted by $\hat{A}$. Let, finally,

$$| A | = \inf_{T \in \mathfrak{T}} \| A + T \| .$$

We shall repeatedly make use of the following well-known assertion.

The operator $A (\in \mathfrak{A})$ is a Φ-operator if and only if the element $\hat{A} (\in \hat{\mathfrak{A}})$ admits an inverse in the algebra $\hat{\mathcal{L}} = \mathcal{L}/\mathfrak{T}$ (see [3]). The spectrum $\hat{\sigma}(A)$ of the element $\hat{A}$ in $\hat{\mathcal{L}}$ consists of all points λ such that $A - \lambda I$ is not a Φ-operator.

Theorem 2.1. *Let* $\mathfrak{A}$ *be a symmetric algebra. Then for any operator* $A (\in \mathfrak{A})$ *we have the equality*

$$\| \mathcal{A}(x) \| = | A | . \tag{2.1}$$

Proof. According to an assertion in [9], (see [9], Corollary 7.1, p. 85), for any operator $Z \in \mathcal{L}(\mathfrak{H})$ the equality $|Z|^2 = |ZZ^*| = \sup \lambda \ (\lambda \in \hat{\sigma}(ZZ^*))$ holds. Let $A \in \mathfrak{A}$ and let the operator $A'(\in \mathfrak{A})$ satisfy the relation $A^* - A' \in \mathfrak{I}$. Then $\hat{\sigma}(AA^*) = \hat{\sigma}(AA')$. As $\hat{\sigma}(AA')$ coincides with the set of complex numbers λ such that the operator $AA' - \lambda I$ is not a Φ -operator, it consists (by Theorem 1.1) of the spectrum of the matrix $\mathfrak{A}(x)(\mathfrak{A}(x))^*$. Hence it follows that

$$\sup_{\lambda \in \hat{\sigma}(AA^*)} = \| \mathfrak{A}(x) \| .$$

Finally, taking into account that $\| \mathfrak{A}(x)(\mathfrak{A}(x))^* \| = \| \mathfrak{A}(x) \|^2$, we obtain the equality (2.1).

The theorem is proved.

Corollary 2.1. *The operator* $A(\in \mathfrak{A})$ *is completely continuous if and only if* $\mathfrak{A}(x) \equiv 0$.

Relation (2.1) enables us to extend the definition of the symbol by continuity to all operators in $\mathfrak{A}$. Obviously, equality (2.1) is preserved in the process.

Let us denote the Banach algebra of the symbols of all operators in $\mathfrak{A}$ by $\mathfrak{A}_S$. In the sequel we shall always assume that $\mathfrak{I} \subset \mathfrak{A}$.

The following theorem has been established:

Theorem 2.2. *The factor algebra* $\hat{\mathfrak{A}}$ *is isomorphic and isometric to the algebra* $\mathfrak{A}_S$ *. This isomorphism carries the class* $\hat{A}$ *, where* $A \in \mathfrak{A}$ *, into the symbol* $\mathfrak{A}(x) \ (x \in X)$.

The basic assertion of the present section is the following:

Theorem 2.3. *The operator* $A(\in \mathfrak{A})$ *is a* Φ_+*-or* Φ_-*-operator in* $\mathfrak{H}$ *if and only if its symbol* $\mathfrak{A}(x)$ *satisfies the condition*

(2.2) $$\inf |\det \mathfrak{A}(x)| > 0 \qquad (x \in X) .$$

If condition (2.2) *is fulfilled then* A *is a* Φ*-operator.*

Proof. If A is a Φ_+-(or Φ_- -) operator, then the operator $B = A^*A$ (or AA^*) is a Φ-operator, so that the coset $\hat{B}$ is invertible in the algebra $\hat{\mathcal{L}}(\mathfrak{H})$.

As the spectrum of the class $\hat{\mathcal{B}}$ lies on the real axis, this class is invertible in the algebra $\hat{\mathfrak{A}}$, too. From the isomorphism of the algebras $\hat{\mathfrak{A}}$ and $\hat{\mathfrak{A}}_S$ one obtains (2.2). The necessity of the condition (2.2) is proved. Let us prove the sufficiency of this condition. Suppose that for the operator $A \in \mathfrak{A}$ the condition (2.2) is satisfied. Then the matrix function $\mathcal{A}(x)(\mathcal{A}(x))^*$ is invertible in the algebra of all matrix functions bounded on X. Since the spectrum of the matrix $\mathcal{A}(x)(\mathcal{A}(x))^*$ lies on the real axis, $\mathcal{A}(x)(\mathcal{A}(x))^*$ is invertible in $\mathfrak{A}_S$; consequently, there exists an operator $B \in \mathfrak{A}$ such that $\mathcal{B}(x) = [\mathcal{A}(x)(\mathcal{A}(x))^*]^{-1}$. The operator $AA^*B - I$ has symbol zero, therefore it is completely continuous. The existence of an operator $C \in \mathfrak{A}$ with $CA^*A - I \in \mathcal{T}$ can be established in a similar way. Hence it follows that A is a Φ-operator. The theorem is proved.

We are going to exhibit sufficient conditions in order that the requirement $\inf |\det \mathcal{A}(x)| > 0$ in Theorem 2.3 might be replaced by the requirement $\det \mathcal{A}(x) \neq 0$. This is the case, for instance, when the set X can be equipped with a Hausdorff topology so that it becomes a compactum and the matrix functions $\mathcal{A}(x)$ $(\in \mathfrak{A}_S)$ are continuous on X.

A situation of this kind occurs in the algebras $\Sigma(L_2, \Gamma_d, C)$ and $WH(\ell_2, C_d)$ (see [7]). In the first example X is a space curve, and it is equipped with the usual topology. In the second one X is the cylinder

$$X = \{x = (x_1, x_2, x_3):\ x_1 + ix_2 \in \Gamma,\quad 0 \leqq x_3 \leqq 1\},$$

where Γ is the unit circle, oriented anti-clockwise. The topology on X is defined by neighbourhoods of the following three types. If $x = (x_1, x_2, x_3) \in X$ and $x_1 + ix_2 = t_0 \in \Gamma$, then

$$U(t_0, 0) = \{(t, x_3):\ |t - t_0| < \delta,\ t \prec t_0,\ 0 \leqq x_3 \leqq 1\} \cup \{(t_0, x_3):\ 0 \leqq x_3 < \varepsilon\},$$

$$U(t_0, 1) = \{(t, x_3):\ |t - t_0| < \delta,\ t \succ t_0,\ 0 \leqq x_3 \leqq 1\} \cup \{(t_0, x_3):\ \varepsilon < x_3 \leqq 1\}, \tag{2.4}$$

$$U(t_0, x_3^{(0)}) = \{(t_0, x_3):\ x_3^{(0)} - \delta_1 < x_3 < x_3^{(0)} + \delta_2\} \qquad (x_3^{(0)} \neq 0; 1),$$

where $0 < \delta_1 < x_3^{(0)}$, $0 < \delta_2 < 1 - x_3^{(0)}$, $0 < \varepsilon < 1$, and $t \prec t_0$ means that on the oriented contour Γ the point t precedes the point t_0.

For the two algebras considered above, the algebra $\mathfrak{A}_S$ coincides with the algebra $C(X)$ of all continuous functions on X .*

Let us consider a more complicated example. Let $\mathfrak{A}$ be a symmetric algebra, and suppose that $n = n(x)$ does not depend on x . We introduce n Hausdorff topologies on X that turn X into compact topological spaces $X_1, \dots, X_n$ respectively. Let X_0 be a non-void subset of X , closed in each X_j $(j = 1, 2, \dots, n)$ and having the following property: for any collection $U_1, \dots, U_n$ of neighbourhoods of the set X_0 , taken in the spaces $X_1, \dots, X_n$, respectively, there exists a set which is open in some topology X_r $(1 \leq r \leq n)$ and satisfies the condition $X_0 \subset \subset U \subset \bigcap_{j=1}^{n} U_j$.

We say that the algebra $\mathfrak{A}_S$ is compatible with the topologies $X_1, \dots, X_n$ and the set X_0 if A_S consists of matrix functions with the following properties:

a) For any matrix function $\mathcal{A}(x) = \| a_{jk}(x) \|_1^n \in \mathcal{A}_S$ every diagonal function $a_{jj}(x)$ is continuous in the respective space X_j , while the non-diagonal functions $a_{jk}(x)$ $(j \neq k)$ are continuous in all spaces $X_1, \dots, X_n$.

b) The functions $a_{jj}(x)$ $(j = 1, \dots, n)$ are continuous on the set $X \setminus X_0$ in each of the spaces $X_1, \dots, X_n$, and $a_{jk}(x) = 0$ $(x \in X_0, j \neq k)$.

Examples of algebras compatible with two topologies are furnished by the algebras $\Sigma(L_2, \Gamma, C_d)$ and $\Sigma(L_2, \Gamma_d, C_d)$. For these algebras, as already mentioned, the role of the set X is played by the cylinder

$$X = \{x = (x_1, x_2, x_3) : x_1 + ix_2 \in \Gamma, \; 0 \leq x_3 \leq 1\} .$$

On X one introduces two topologies. The first, X_1 is given by the neighbourhoods (2.4); the second, X_2 is defined by neighbourhoods of the same kind after changing the orientation of the contour Γ to the opposite.

In the first example the set X_0 is the union of the sets $\{x \in X; x_3 = 1\}$ and $\{x \in X : x_3 = 0\}$, in the second one it is the union of the same sets and also the sets

$$\{x \in X : \; x_1 + ix_2 = \alpha_k\} ,$$

where $\alpha_1, \dots, \alpha_s$ stand for the end-points of the open arcs contained in the contour

* One may examine a more general case replacing the algebras A_n just considered by the algebras $A_{n \times n}$, respectively, that consist of the operators $\| A_{mk} \|_1^n$ $(A_{mk} \in \mathfrak{A})$ acting in $\mathfrak{H}^n$. In this case the algebra of the symbols coincides with the algebra $C_{n \times n}(X)$.

Γ_d (see [5], [6]).

Theorem 2.4. *Let* $\mathfrak{A}$ *be a symmetric algebra such that the algebra* $\mathfrak{A}_S$ *of symbols is compatible with the topologies* $X_1, \ldots, X_n$ *and the set* X_0. *If* $\det \mathcal{A}(x) \neq 0 \quad (x \in X)$, *then* $\inf |\det \mathcal{A}(x)| > 0$.

Proof. We write the determinant of the matrix function $\mathcal{A}(x) = \|a_{jk}(x)\|_1^n$ in the form $\det \mathcal{A}(x) = \prod_{j=1}^{n} a_{jj}(x) + M(x)$. Since $M(x) \equiv 0$ on the set X_0, and $\det \mathcal{A}(x) \neq 0$, we have $\prod_{j=1}^{n} a_{jj}(x) \neq 0$ on X_0. Each of the functions $a_{jj}(x)$ is continuous on X_j and different from zero on the closed subset X_0. Consequently, for each $j = 1, \ldots, n$. there can be found an open set $U_j \quad (X_0 \subset U_j \subset X_j)$ and a positive number d_j such that $|a_{jj}(x)| > d_j$ for every $x \in U_j$. Let U be an open open set in X_r satisfying $X_0 \subset U \subset \bigcap_{j=1}^{n} U_j$, and let $d = d_1 \ldots d_n$. Then $\prod_{j=1}^{n} |a_{jj}(x)| > d \quad (x \in U)$.

Since each term of $M(x)$ contains at least one factor $a_{jk}(x)$ with $j \neq k$, the function $M(x)$ is continuous on every X_j and, in particular, on X_r. But $M(x) = 0$ on X_0. Therefore we can find a set $G \quad (X_0 \subset G \subset U)$, which is open in X_r and on which the inequality $|M(x)| < \frac{d}{2}$ holds. Hence it follows that $\inf |\det \mathcal{A}(x)| > 0 \quad (x \in G)$. On the set $X \setminus G$, which is closed in X_r the function $\det \mathcal{A}(x)$ is continuous and does not vanish. Thus $\inf |\det \mathcal{A}(x)| > 0$ $(x \in X \setminus G)$. The theorem is proved.

§ 3. ESTIMATION OF THE SYMBOL. EXTENSION OF THE DEFINITION OF THE SYMBOL

1. In this section we define a symbol for every operator belonging to the algebra $\mathfrak{A}$. Moreover, some new properties of the symbol will be established. We assume that the (non-symmetric) algebra $\mathfrak{A}$ fulfils the following additional requirements.

3.1. The number $n = n(x)$ does not depend on the point x.

3.2. There is an operator $P_j \; (j=1,\ldots,n)$ in $\mathfrak{A}$ such that every element of the symbol $\mathcal{P}_j(x)$ is equal to zero, excepting j-th element of the main diagonal, wich equals unity.

3.3. For the symbol $\mathcal{A}(x) = \|a_{jk}(x)\|_{j,k=1}^{n}$ of any operator $A \in \mathfrak{A}$ we have the relation

$$\sup |a_{jk}(x)| \leq \gamma |A| \qquad (j,k = 1,\dots,n;\ x \in X), \tag{3.1}$$

where γ is a constant depending on the space $\mathfrak{L}$ only. Recall that $|A| = \inf \|A+T\|$ $(T \in \mathfrak{T})$.

From the estimate (3.1) it follows immediately, that the symbol is uniquely determined by the operator in $\mathfrak{A}$, and also that the symbol of a completely continuous operator is zero.

Let us remark that the mapping $A \longrightarrow \mathcal{A}(x)$ is a homomorphism of the algebra $\mathfrak{A}$ onto an algebra of matrix functions of order n.

2. Checking of condition 3.3 in particular situations often involves considerable difficulties. We are going to present sufficient conditions ensuring that the inequality (3.1) holds.

Theorem 3.1. *Let the conditions* 1.1 – 1.3 *be satisfied and* $n = 1$. *Then for any operator* $A \in \mathfrak{A}$ *we have the relation*

$$\sup |\mathcal{A}(x)| \leq |A| \qquad (x \in X).$$

Proof. Theorem 1.1 implies that $\sup |\mathcal{A}(x)| = \max |\lambda|$ $(\lambda \in \hat{\sigma}(A))$. Recall that $\hat{\sigma}(A)$ denotes the spectrum of the coset $\hat{A}$ in $\hat{\mathfrak{L}} = \mathfrak{L} / \mathfrak{T}$. It remains to observe that $\max |\lambda| \leq |A|$ $(\lambda \in \hat{\sigma}(A))$. The theorem is proved.

From Theorem 3.1 it follows that property 3.3 is valid for the algebras $\Sigma(L_p, \Gamma_d, C)$ and $WH(h_p, C_d)$.

A linear operator M acting in $\mathfrak{A}$ will be called a permutation operator for the indices j_0, k_0 at the point x_0 provided for every operator $A \in \mathfrak{A}$ there can be found an operator $T \in \mathfrak{T}$ such that

a) the operator $B = MA - T \in \mathfrak{A}$;

b) for the symbols $\mathcal{A}(x) = \|a_{jk}(x)\|$ and $\mathcal{B}(x) = \|b_{jk}(x)\|$ we have the relation

$$a_{j_0k_0}(x_0) = b_{k_0j_0}(x_0); \tag{3.2}$$

c) $$|M| = \sup \frac{|MA|}{|A|} < \infty .$$

Theorem 3.2. *Let conditions* 1.1 – 1.3 *as well as conditions* 3.1 *and* 3.2 *be satisfied. If for every pair of numbers* j, k $(1 \leq k, j \leq n)$ *and every point* $x \in X$ *there exists a permutation operator* $M = M_{(j,k,x)}$ *for the indices* j, k *at the point* x *and the condition*

(3.3) $$\sup_{1 \leq j,k \leq n;\ x \in X} |M_{(j,k,x)}| < \infty$$

is fulfilled, then condition (3.1) *is also satisfied.*

Proof. Repeating, in essence, the proof of Theorem 3.1, it may be shown that the inequalities

$$|a_{jj}(x)| \leq |P_j A P_j| \ (\leq |P_j|^2 |A|)$$

hold for the diagonal elements of the symbol $\mathcal{A}(x) = \|a_{jk}(x)\|$ of the operator $A \in \mathfrak{S}$. Let M be a permutitation operator for the indices j_0, k_0 at the point $x_0 \in X$, and let $B = MA - T$ be the operator corresponding to A by the relation (3.2).

Let us consider the operator $C = P_{j_0} A P_{k_0}^2 B P_{j_0}$. For its symbol

$$\|c_{jk}(x)\| = \mathcal{P}_{j_0}(x) \mathcal{A}(x) \mathcal{P}_{k_0}^2(x) \mathcal{B}(x) \mathcal{P}_{j_0}(x)$$

we have the relation $|c_{jj}(x)| \leq |C|$ $(j = 1, \dots, n)$. It is easy to see that $c_{jj}(x) = a_{jk}(x) b_{kj}(x)$. Taking into account relation (3.2) we obtain

$$|a_{j_0 k_0}(x_0)|^2 \leq |C| .$$

Since $|C| < |P_{j_0}|^2 |P_{k_0}| |M| |A|^2$, the last two relations imply property 3.3.

The theorem is proved.

The conditions of this theorem are satisfied for the algebra $\Sigma(L_p, \Gamma, C_d)$. In this particular case $n(x) = 2$; $P_1 = \frac{1}{2}(I+S)$ and $P_2 = \frac{1}{2}(I+S)$. If the contour Γ is a circle, then the permutation operator $M_{(1,2,\zeta_0)}$ for the indices

1 and 2 at the point ζ_0 $(|\zeta_0| = 1)$ is defined on K by the equality

$$MA = -V_{\zeta_0} P_1 A P_2 V_{\zeta_0}$$

where $(V_{\zeta_0}\varphi)(\zeta) = \varphi(\frac{\zeta_0^2}{\zeta})$. For an arbitrary contour the operator M can be determined in a somewhat more complicated way.

3. Property 3.3 enables one to extend the definition of the symbol to arbitrary operators in $\mathfrak{A}$. Let A_n $(n = 1, 2, \ldots)$ be a sequence of operators belonging to $\mathfrak{S}$, which converges in norm to an operator $A \in \mathfrak{A}$. Then the matrix functions $\mathcal{A}_n(x)$ converge uniformly to a matrix function $\mathcal{A}(x)$. The latter is independent of the choice of the sequence $\mathcal{A}_n(x)$. The matrix function $\mathcal{A}(x)$ will be called the symbol of the operator A.

The correspondence $A \longrightarrow \mathcal{A}(x)$ is a homomorphism of the algebra $\mathfrak{A}$ into the algebra of matrix functions. Obviously, we have the estimate

$$|a_{jk}(x)| \leqq \gamma |A|.$$

It has to be noted that the symbols of all operators A from the same coset $\hat{A} \in \hat{\mathfrak{A}}$ coincide. We shall call this common symbol also the symbol of the coset. It will invariantly be denoted by $\mathcal{A}(x)$.

The mapping $\hat{A} \longrightarrow \mathcal{A}(x)$ is a homomorphism of the algebra $\hat{A}$ onto the algebra $\mathfrak{A}_S$ of the symbols of all operators belonging to $\mathfrak{A}$.

§ 4. THE FUNDAMENTAL THEOREM

In this section we consider the general case of a Banach space $\mathfrak{L}$. Concerning the algebra $\mathfrak{A}$ we assume that it satisfies the conditions 1.1 – 1.3 of § 1 and conditions 3.1 – 3.3 of § 3. As before, $\mathfrak{A}_S$ stands for the algebra of the symbols of all operators in $\mathfrak{A}$.

First let us suppose that $n = 1$ and the algebra $\mathfrak{A}$ has the following property:

4.1. The factor algebra $\hat{\mathfrak{A}} = \mathfrak{A}/\mathfrak{J}$ is commutative and all its maximal ideals are of the form

$$M_{x_0} = \{\hat{A} \in \mathfrak{A} : \mathcal{A}(x_0) = 0\} \qquad (x_0 \in X).$$

Under these assumptions it is easy to prove the following

Theorem 4.1. *The operator* $A \in \mathfrak{A}$ *is a* Φ*-operator if and only if*

$$\mathcal{A}(x) \neq 0 \qquad (x \in X).$$

In case this condition is satisfied, the operator A admits a regularizator also belonging to the algebra $\mathfrak{A}$.

Examples of algebras $\mathfrak{A}$ for which condition 4.1 is satisfied and, consequently, Theorem 4.1 holds, are furnished by the algebras $\Sigma(L_p, \Gamma_d, C)$ and $WH(h_p, C_d)$.

Consider the more general case where the algebra $\hat{\mathfrak{A}}$ is non-commutative and $n > 1$. Let us introduce new restrictions.

4.2. The operators P_j defined in condition 3.2 are projectors such that $\Sigma P_j = I$ and $P_j P_k = 0 \quad (j \neq k)$.

Denote by $\mathfrak{A}^{(j)}$ $(j = 1, \dots, n)$ the set of all operators of the form $P_j A P_j$ $(A \in \mathfrak{A})$ acting in the subspace $P_j \mathcal{L}$. Obviously, $\mathfrak{A}^{(j)}$ is a Banach algebra. With the operator $P_j A P_j$ $(A \in \mathfrak{A})$ we associate the function $a_{jj}(x)$ $(x \in X)$, where $\mathcal{A}(x) = \| a_{jk}(x) \|$ is the symbol of the operator A. We shall denote by $\hat{\mathfrak{A}}^{(j)}$ the factor algebra $\hat{\mathfrak{A}}^{(j)} = \mathfrak{A}^{(j)} / \mathcal{J}^{(j)}$, where $\mathcal{J}^{(j)} = \mathcal{J}(P_j \mathcal{L})$.

4.3. Let j be any of the numbers $1, 2, \dots, n$. If the condition $\inf |a_{jj}(x)| > 0$ is satisfied for the operator $B = P_j A P_j$ $(A \in \mathfrak{A})$, then the element $\hat{B} \, (\in \hat{\mathfrak{A}}^{(j)})$ is invertible in the algebra $\hat{\mathfrak{A}}^{(j)}$.

This condition is satisfied, for instance, for any $j = 1, \dots, n$ if the algebra $\mathfrak{A}^{(j)}$ $(j = 1, \dots, n)$ is commutative and all its maximal ideals are of the form

$$M_{x_0} = \{ \widehat{P_j A P_j} \in \hat{\mathfrak{A}}^{(j)} : \; a_{jj}(x_0) = 0 \} \qquad (x_0 \in X).$$

Finally, we shall need one more restriction.

4.4. For every $j = 1, \dots, n$ the set of invertible element of the algebra $\hat{\mathfrak{A}}^{(j)}$ is dense in $\hat{\mathfrak{A}}^{(j)}$.

From property 4.3 it is easy to deduce property 4.4 if we assume that in the set $\mathfrak{L}$ there exists a dense subset $\mathfrak{L}_0$ of operators whose symbols satisfy the following condition: none of the neighbourhoods of zero is fully contained in the closure of the range of the function $a_{jj}(x)$ $(j = 1, \dots, n)$.

Properties 4.2 – 4.4 are enjoyed, for example, by the algebra $\Sigma(L_p, \Gamma, C_d)$ $(1 < p < \infty)$. In the proof of property 4.4 for the algebra $\Sigma(L_p, \Gamma, C_d)$ the role of $\mathfrak{A}_0$ is played by the linear hull of all possible products of singular operators with piecewise rational coefficients.

In the sequel we shall make use of the following property of the symbol $\mathcal{A}(x)$. Let k be any of the numbers $1, 2, \dots, n-1$; $P = \sum_{j=1}^{k} P_j$, $Q = I - P$ and $A \in \mathfrak{A}$, where $\mathfrak{A}$ is an algebra with a symbol, and let the conditions 1.1 – 1.3, 3.1 – 3.3 and 4.2 – 4.3 be satisfied. Denote by $\mathfrak{A}_0$ the algebra of all operators of the form PAP $(A \in \mathfrak{A})$ acting in $P\mathcal{L}$. If $A_0 \in \mathfrak{A}_0$, we define its symbol to be the matrix $\mathcal{A}_0(x)$ of order k, standing in the left upper corner of the matrix function $\mathcal{A}(x)$. It is easy to see that the algebra $\mathfrak{A}_0$ is an algebra with a symbol, satisfying all of the conditions 1.1 – 1.3, 3.1 – 3.3 and 4.2 – 4.3.

Theorem 4.2. (Fundamental theorem) *Let the algebra* $\mathfrak{A}$ *have the properties* 1.1 – 1.3, 3.1 – 3.3 *and* 4.2 – 4.3. *The operator* $A\ (\in \mathfrak{A})$ *is a* Φ_+*-or* Φ_-*-operator if and only if*

$$\inf_{x \in X} |\det \mathcal{A}(x)| > 0 .$$

In case the latter condition is fulfilled, the operator A *is a* Φ*-operator, and its regularizators are contained in* $\mathfrak{A}$.

Before proving this theorem we present the following lemma:

Lemma 4.1. *Let* $A \in \mathfrak{A}$, *and denote by* $M_j(x)$ *the principal minor of order* j *of the matrix function* $\mathcal{A}(x)$. *If*

$$\inf |M_j(x)| > 0 \qquad (j = 1, \dots, n;\ x \in X),$$

then the element $\hat{A}$ *is invertible in* $\hat{\mathfrak{A}}$.

Proof. For $n = 1$ the conclusion of the lemma follows from condition 4.3. Let Q and P, respectively, denote the projectors $Q = P_n$, $P = I - Q$. The proof will be carried out by induction for n. From the induction hypothesis it follows easily that the coset $\hat{B} = \hat{P}\hat{A}\hat{P} + \hat{Q}$ is invertible in $\hat{\mathfrak{A}}$. The equality

$$\hat{A} = (\hat{I} + \hat{Q}\hat{A}\hat{P}\hat{B}^{-1}\hat{P})(\hat{P}\hat{A}\hat{P} + \hat{Q}(\hat{A} - \hat{A}\hat{P}\hat{B}^{-1}\hat{P}\hat{A})\hat{Q})(\hat{I} + \hat{P}\hat{B}^{-1}\hat{P}\hat{A}\hat{Q}) \tag{4.1}$$

can be verified directly. The extreme factors in the right-hand member of equality (4.1) are invertible in $\mathfrak{A}$. Their inverses are the elements $\hat{I} = \hat{Q}\hat{A}\hat{P}\hat{B}^{-1}\hat{P}$ and $\hat{I} - \hat{P}\hat{B}^{-1}\hat{P}\hat{A}\hat{Q}$, respectively. Let us show that also the middle factor is invertible in $\mathfrak{A}$. We notice that for the symbol $\mathcal{F}(x)$ of the coset

$$\hat{F} = \hat{Q}(\hat{A} - \hat{A}\hat{P}\hat{B}^{-1}\hat{P}\hat{A})\hat{Q}$$

the equality

$$\mathcal{F}(x) = \frac{\det \mathcal{A}(x)}{M_{n-1}(x)}$$

holds; consequently, $\inf |\mathcal{F}(x)| > 0 \quad (x \in X)$. From condition 4.3 it follows that $\hat{F}$ is invertible in $\hat{\mathfrak{A}}^{(n)}$. Since, in addition, the element $\hat{Q} + \hat{P}\hat{A}\hat{P}$ is invertible in $\hat{\mathfrak{A}}$, the middle factor in the right-hand member of equality (4.1) is invertible in $\hat{\mathfrak{A}}$. The lemma is proved.

Proof of Theorem 4.2. That the condition appearing in the theorem is sufficient will also be proved by induction. For $n = 1$ the conclusion follows from condition 4.3. Let n be arbitrary. Making use of the condition 4.4, we choose an operator $B \in \mathfrak{C}$ such that: 1) the number $|A - B|$ is sufficiently small and 2) the element $\hat{P}_n\hat{B}\hat{P}_n$ is invertible in $\hat{\mathfrak{A}}^{(n)}$. The equality

$$\hat{B} = (\hat{I} + \hat{P}\hat{B}\hat{Q}\hat{C}^{-1}\hat{Q})(\hat{Q}\hat{C}\hat{Q} + \hat{P}(\hat{B} - \hat{B}\hat{Q}\hat{C}^{-1}\hat{Q}\hat{B})(\hat{I} + \hat{Q}\hat{C}^{-1}\hat{Q}\hat{B}\hat{P}), \tag{4.2}$$

where $Q = P_n$, $P = I - Q$, $C = P + QBQ$, can be verified directly.

Let $\mathfrak{A}_0$ be the set of operators of the form PAP $(A \in \mathfrak{A})$ acting in $P\mathfrak{L}$, and $\mathcal{F}(x)$ the symbol of the coset

$$\hat{F} = \hat{P}(\hat{B} - \hat{B}\hat{Q}\hat{C}^{-1}\hat{Q}\hat{B})\hat{P} \in \hat{\mathfrak{A}}_0 .$$

As the number $|A - B|$ is sufficiently small and $\inf |\det \mathcal{A}(x)| > 0$, we have (in view of the inequality (3.1)) $\inf |\det \mathcal{B}(x)| > 0$. From Theorem 1.1 it follows that $\inf |\det \mathcal{C}(x)| > 0$. But

$$\det \mathcal{F}(x) = \frac{\det \mathcal{B}(x)}{\det \mathcal{C}(x)} .$$

Therefore

$$\inf |\det \mathfrak{F}(x)| > 0 \qquad (x \in X).$$

From the inducation hypothesis it follows that $\hat{F}$ is invertible in $\hat{\mathfrak{A}}_0$. From the invertibility of the element $\hat{Q}\hat{C}\hat{Q}$ in $\hat{\mathfrak{A}}^{(n)}$ we deduce that the coset $\hat{Q}\hat{C}\hat{Q}+\hat{F}$ is invertible in $\mathfrak{A}$. Thus we have shown that $\hat{B}$ is invertible in $\hat{\mathfrak{A}}$. Let $\hat{R} = \hat{B}^{-1}\hat{A}$; the symbol of the coset $\hat{R}$ can be written in the form $\mathfrak{R}(x) = \mathfrak{I}(x) + \mathfrak{B}^{-1}(x)(\mathfrak{A}(x) - \mathfrak{B}(x))$. Since $\inf |\det \mathfrak{B}(x)| > 0$, the estimate (3.1) implies that for sufficiently small values of $|A-B|$ the elements of the matrix $\mathfrak{B}^{-1}(x)(\mathfrak{A}(x)-\mathfrak{B}(x))$ have arbitrarily small absolute values. Hence $\inf |M_j(x)| > 0 \quad (j = 1, \ldots, n;\ x \in X)$, where $M_j(x)$ stand for the principal minors of the matrix $\mathfrak{R}(x)$. Owing to Lemma 4.1, the element $\hat{R}$ is invertible in $\hat{\mathfrak{A}}$. From the equality $\hat{A} = \hat{B}\hat{R}$ it follows that $\hat{A}$ is invertible in $\hat{\mathfrak{A}}$. The sufficiency of the condition appearing in the theorem is proved. We are going to prove that this condition is necessary, too. First we observe, that if the operator $A\ (\in \mathfrak{A})$ is a Φ_+-or Φ_--operator, then it is also a Φ-operator. This can be proved without difficulty, making essential use of condition 4.3.

Let A_n be a sequence of Φ-operators in $\mathfrak{L}$, uniformly converging to A. From Theorem 1.1. it follows that

$$\inf_X |\det \mathfrak{A}_n(x)| > 0,$$

and from what has been proved above we obtain, that the elements $\hat{A}_n$ are invertible in $\hat{\mathfrak{A}}$. Since $\hat{A}$ is invertible in $\hat{\mathfrak{L}}$ and $|A_n - A| \longrightarrow 0$, the norm sequence $|\hat{A}_n^{-1}|$ is bounded. The elements $\hat{A}_n$ belong to the algebra $\mathfrak{A}$, and their symbols are given by the matrix functions $\mathfrak{A}_n^{-1}(x)$. The inequality (3.1) implies that the sequence $\{\det \mathfrak{A}_n^{-1}(x)\}$ is uniformly bounded; thus

$$\inf |\det \mathfrak{A}_n(x)| > 0 \qquad (x \in X;\ n = 1, 2, \ldots).$$

Hence it follows that $\inf |\det \mathfrak{A}(x)| > 0 \ (x \in X)$. The theorem is proved.

§ 5. EXTENSION OF ALGEBRAS WITH A SYMBOL

In the preceding section we have studied algebras $\mathfrak{A}$ satisfying a number of conditions. However, not all of the examples presented in § 1 satisfy these conditions. An example of such an algebra $\Sigma(L_p, \Gamma_d, C_d)$. In this algebra, for instance, there is no operator having the symbol $\mathfrak{A}(x) = \begin{Vmatrix} 1 & 0 \\ 0 & 0 \end{Vmatrix}$, so that condition 3.2 (and, all the more, condition 4.2) is not fulfilled here. Nevertheless, the fundamental theorem (Theorem 4.2) is valid for this algebra. The proof of the latter fact is contained in the general scheme presented below.

Let us be given a space $\mathfrak{L}$, an algebra $\mathfrak{A}$, a set X and a symbol $\mathfrak{A}(x)$ which do not, generally speaking, satisfy the conditions of §§ 1 – 4. Assume that $\widetilde{\mathfrak{L}}$ is an extension of the space $\mathfrak{L}$, and there exists a projector P projecting $\widetilde{\mathfrak{L}}$ onto $\mathfrak{L}$. Furthermore, let $\widetilde{\mathfrak{A}}$ be a subalgebra of $\mathcal{L}(\widetilde{\mathfrak{L}})$ such that every operator $A \in \mathfrak{A}$ can be extended to an operator $\widetilde{A} \in \widetilde{\mathfrak{A}}$. We shall suppose that $\widetilde{\mathfrak{A}}$ is an algebra with a symbol, satisfying the conditions 1.1 – 1.3, 3.1 – 3.3 and 4.2 – 4.3, where the symbol associated to the operator $\widetilde{A} \in \widetilde{\mathfrak{A}}$ is $\widetilde{\mathfrak{A}}(x)$ $(x \in X)$. We also require that the following conditions would be fulfilled.

5.1. If the operator $A \in \mathfrak{A}$ is extended to the operator $\widetilde{A} \in \widetilde{\mathfrak{A}}$, then $Q\widetilde{A}Q$ is a Φ-operator in $Q\widetilde{\mathfrak{L}}$ $(Q = I - P)$.

5.2. The conditions $\inf|\det \mathfrak{A}(x)| = 0$ $(x \in X)$ and $\inf|\det \widetilde{\mathfrak{A}}(x) = 0$ $(x \in \widetilde{X})$ are equivalent.

Apparently, the operator A is a Φ- (Φ_+^-, Φ_-^-) operator in $\mathfrak{L}$ if and only if $\widetilde{A}$ is a Φ- (Φ_+^-, Φ_-^-) operator in $\widetilde{\mathfrak{L}}$. From Theorem 4.2 and condition 5.2 it follows that the operator A $(\in \mathfrak{A})$ is a Φ- (Φ_+^-, Φ_-^-) operator in $\mathfrak{L}$ if and only if $\inf|\det \mathfrak{A}(x)| > 0$ $(x \in X)$. If this condition is fulfilled then the operator A is a Φ-operator. As a result, under the above restrictions Theorem 4.2 is valid for the algebra $\mathfrak{A}$.

Let us return to the algebra $\Sigma(L_p, \Gamma_d, C_d)$, and show in this case how to extend the space $\mathfrak{L}$ and the operator A. Let us complete the contour Γ_d to a contour $\widetilde{\Gamma}$ that consists of closed curves only, satisfying the conditions of the first section. We choose $\widetilde{\mathfrak{L}}$ to be the space $L_p(\widetilde{\Gamma})$, and as $\widetilde{\mathfrak{A}}$ we choose the algebra $\Sigma(L_p, \widetilde{\Gamma}, C_d)$. The role of the projector P is played by the operator $\chi(t)I$ $(t \in \widetilde{\Gamma})$, where $\chi(t)$ stands for the characteristic function of the set Γ_d $(\subset \widetilde{\Gamma})$. Finally, the operator $\widetilde{A}$ will be defined in the following way. Let A_n be a sequence of operators belonging to $\mathfrak{Q}$ that converges in norm to the operator A. The operator A_n can be written in the form

$$A_n = \sum_{j=1}^{k_n} A_{j1}^{(n)} \cdots A_{jr_n}^{(n)},$$

where

$$A_{j\ell}^{(n)} = \alpha_{j\ell}^{(n)}(t)I + \beta_{j\ell}^{(n)}(t)S \qquad (\alpha_{j\ell}^{(n)}(t), \beta_{j\ell}^{(n)}(t) \in C_d(\Gamma_d)).$$

With the help of the functions $\alpha_{j\ell}^{(n)}(t)$ and $\beta_{j\ell}^{(n)}(t)$ $(t \in \Gamma_d)$ we construct functions $\gamma_{j\ell}^{(n)}(t)$ and $\delta_{j\ell}^{(n)}(t)$ so that 1) on Γ_d the functions $\gamma_{j\ell}^{(n)}(t)$ and $\delta_{j\ell}^{(n)}(t)$ coincide with the functions $\alpha_{j\ell}^{(n)}(t)$ and $\beta_{j\ell}^{(n)}(t)$ respectively; 2) the functions $\gamma_{j\ell}^{(n)}(t) - \delta_{j\ell}^{(n)}(t)$ are continuous on the closure of the contour $\tilde{\Gamma} \setminus \Gamma_d$, and 3) $\delta_{j\ell}^{(n)}(t) = 0$ on $\tilde{\Gamma} \setminus \Gamma_d$. Let us form the operators *

$$\tilde{A}_n = \sum_{j=1}^{k_n} \tilde{A}_{j1}^{(n)} \ldots \tilde{A}_{jr_n}^{(n)} P + Q .$$

It turns out that the subspaces $P\tilde{\mathcal{L}}$ and $Q\tilde{\mathcal{L}}$ are invariant relative to the operators $\tilde{A}_n$, and the restrictions of the operator $\tilde{A}_n$ to the subspaces $P\tilde{\mathcal{L}}$ and $Q\tilde{\mathcal{L}}$ coincide with the operators A and Q, respectively. This implies the existence of a uniform limit $\tilde{A}$ of the sequence $\tilde{A}_n$. It can be shown that the operator $\tilde{A}$ satisfies the conditions 5.1 and 5.2.

So the fundamental theorem (Theorem 4.2) holds for the algebra $\Sigma(L_p, \Gamma_d, C_d)$, too.

BIBLIOGRAPHY

[1] S.G. Mihlin, Singular integral equations, *Uspehi Mat. Nauk,* 3 (1948), 29-112.

[2] I.C. Gohberg, An application of the theory of normed rings to singular integral equations, *Uspehi Mat. Nauk,* 7 (1952), 149-156.

[3] I.C. Gohberg and M.G. Kreĭn, Basic propositions concerning the defect numbers, root numbers and indices of linear operators, *Uspehi Mat. Nauk,* 12 (1957), 44-118.

[4] I.C. Gohberg and N.Ja. Krupnik, Singular integral equations with continuous coefficients on a composite contour, *Mat. Issled. Kišinev,* 5 (1970), 89-103.

*By $\tilde{A}_{j\ell}^{(n)}$ we denote the operators $\gamma_{j\ell}^{(n)}(t)I + \delta_{j\ell}^{(n)}(t)S$.

[5] I.C. Gohberg and N.Ja. Krupnik, On the algebra generated by one-dimensional singular integral operators with piecewise continuous coefficients, *Funkc. Anal. i ego Prilož.*, 4 (1970), 27-38.

[6] I.C. Gohberg and N.Ja. Krupnik, Symbols of one-dimensional singular integral operators on an open contour, *Dokl. Akad. Nauk SSSR*, 191 (1970), 12-15.

[7] I.C. Gohberg and N.Ja. Krupnik, On the algebra generated by Toeplitz matrices, *Funkc. Anal. i ego Prilož.*, 3 (1969), 46-56.

[8] I.C. Gohberg and N.Ja. Krupnik, On the algebra generated by Toeplitz matrices in the spaces h_p, *Mat. Issled. Kišinev*, 4 (1969), 54-62.

[9] I.C. Gohberg and M.G. Kreĭn, *Introduction to the theory of linear non-selfadjoint operators* (Moscow, 1965).

COLLOQUIA MATHEMATICA SOCIETATIS JÁNOS BOLYAI
5. HILBERT SPACE OPERATORS. TIHANY (HUNGARY). 1970

Self-adjoint extensions of some Hermitian operators in a space with indefinite metric

V. I. GORBAČUK

A continuous function $f(x)$ $(x \in (-2\ell, 2\ell),\ \ell \leq \infty)$ is said to be Hermitian-indefinite with one negative square if the forms

$$\sum_{j,k=1}^{n} f(x_j - x_k)\xi_j \bar{\xi}_k \qquad (x_j \in (-\ell, \ell);\ n = 1, 2, \dots)$$

have no more than one negative square and at least one of them has exactly one negative square. With a function of this kind one can associate a space Π_1 with indefinite scalar product

$$(\varphi, r) = \int_{-\ell}^{\ell} \int_{-\ell}^{\ell} f(x-y)\varphi(x)\overline{r(y)}\,dx\,dy$$

of the set of all infinitely differentiable functions with bounded support. The operator $-i\frac{d}{dx}$ is Hermitian with respect to the above scalar product. In case it is not self-adjoint, it has, in the indefinite sense, an infinite number of self-adjoint extensions. In the sequel shall assume that $f(x)$ is real-valued and twice continuously differentiable.

Using L.S. Pontrjagin's theorem on the existence of a one-dimensional non-positive invariant subspace for a self-adjoint operator in a space Π_1 with indefinite metric, we may show that for $f(x)$ there exists a number $a = \sigma + i\tau$ such that the function $-f''(x) + (a)^2 f(x)$ is positive definite; here a depends on the choice of the self-adjoint extension of the operator $-i\frac{d}{dx}$. According to the Bochner – Hinčin theorem we have

$$-f''(x) + |a|^2 f(x) = \int_{-\infty}^{\infty} \cos\lambda x \, d\sigma(\lambda)$$

where $\sigma(\lambda)$ is a non-decreasing, bounded function on the real line with the norming conditions $\sigma(\lambda - 0) = \sigma(\lambda)$ and $\sigma(-\infty) = 0$. Hence for $f(x)$ we obtain the following representations:

a). in case $a = 0$ and $\int_{-\infty}^{\infty} \frac{d\sigma(\lambda)}{\lambda^2} = \infty$:

$$f(x) = f(0) + \int_{-\infty}^{\infty} \frac{\cos\lambda x - 1}{\lambda^2} \, d\sigma_p(\lambda) \qquad (\sigma_p(\lambda) = \sigma(\lambda))$$

(representation of parabolic type);

b) in case $a = 0$ and $\int_{-\infty}^{\infty} \frac{d\sigma(\lambda)}{\lambda^2} < \infty$:

$$f(x) = -c + \int_{-\infty}^{\infty} \cos\lambda x \, d\sigma_e(\lambda) \qquad \left(c > 0, \; d\sigma_e(\lambda) = \frac{d\sigma(\lambda)}{\lambda^2}\right)$$

(representation of elliptic type);

c) in case $a \neq 0$:

$$f(x) = \varrho \operatorname{ch}|a|x + \int_{-\infty}^{\infty} \cos\lambda x \, d\sigma_h(\lambda) \quad \left(\varrho \neq 0, \; d\sigma_h(\lambda) = \frac{d\sigma(\lambda)}{\lambda^2 + |a|^2}\right)$$

(representation of hyperbolic type).

When $f(x)$ is defined on the whole real axis, the unicity of one of these representations follows from the self-adjointness of the operator $-i\frac{d}{dx}$ in the space Π_1 and in the Hilbert space constructed with the help of the positive definite kernel $-f''(x-y)+(a)^2f(x-y)$; considering the behaviour of $f(x)$ at infinity, it is easy to decide which of the above representations applies. Thus, in case $f(x)$ is bounded, and only in this case, it has an elliptic representation; if at infinity $f(x)$ increases more quickly than any power of x then it has a hyperbolic representation; finally, if $f(x)$ tends to infinity, but its rate of increase does not exceed that of x^2 , then the parabolic representation applies.

In case of a finite interval too, $f(x)$ admits a unique representation under certain conditions (e.g. if $|f^{(n)}(0)| \leq m_n$ where $\{m_n\}_{n=0}^{\infty}$ $(m_n > m > 0)$ is a sequence of numbers with $\sum_0^{\infty} \frac{1}{\sqrt[2n]{m_{2n}}} = \infty)$.

But in general, for an $f(x)$ given on a finite interval the unicity of the representation does not follow. M.G. Kreǐn gave a condition in order that a Hermitian- indefinite function with one negative square have a non-hyperbolic representation. This condition stipulates that the kernel $f(x-y)+f(0)-f(x)-f(y)$ should be positive definite. Here we present a necessary and sufficient condition in order for $f(x)$ to have an elliptic representation. On this occasion, for the sake of a reasonable norming, it is practical to choose the constant c in the representation b) the least possible. We shall show how to find this least possible constant. To this end we make use of the following general fact:

Assume A is a self-adjoint operator with a pure point-spectrum in the separable Hilbert space H; denote by $-\lambda_1, \lambda_2, \ldots$ the points of the spectrum and by $\varphi_1, \varphi_2, \ldots$ a corresponding complete system of eigenvectors. Assume also that $\lambda_1 > 0, \lambda_i \geq 0 \;\; (i \geq 2)$ and that the eigen-space belonging to $-\lambda_1$ is one-dimensional. Consider simultaneously a one-dimensional positive bounded operator B (i.e. the range of B is one-dimensional).

Theorem 1. *In order that there exist a constant* $c > 0$ *such that the operator* $A + cB$ *is non-negative, it is necessary and sufficient that*

$$\sum_{\substack{i=2\\ \lambda_i \neq 0}}^{\infty} \frac{(B\varphi_i,\varphi_i)}{\lambda_i} < \frac{(B\varphi_1,\varphi_1)}{\lambda_1}$$

((.,.) denotes the scalar product in H *).*

The least such c *is given by the formula*

$$c_{min} = \frac{1}{\dfrac{(B\varphi_1,\varphi_1)}{\lambda_1} - \sum\limits_{\substack{i=2\\ \lambda_i \neq 0}}^{\infty} \dfrac{(B\varphi_i,\varphi_i)}{\lambda_i}}$$

From here, for a Hermitian-indefinite function $f(x)$ we deduce:

Theorem 2. *In order that the Hermitian-indefinite function* $f(x)$ *admit an elliptical representation it is necessary and sufficient that* $\int_{-\ell}^{\ell} \varphi_1(x)\,dx \neq 0$ *and*

$$\sum_{\substack{i=2\\ \lambda_i \neq 0}} \frac{\left(\int_{-\ell}^{\ell} \varphi_i(x)\,dx\right)^2}{\lambda_i} < \frac{\left(\int_{-\ell}^{\ell} \varphi_1(x)\,dx\right)^2}{\lambda_1}$$

The least constant c *in representation* b) *is given by*

$$c_{min} = \frac{1}{\dfrac{\left(\int_{-\ell}^{\ell} \varphi_1(x)\right)^2}{\lambda_1} - \sum\limits_{\substack{i=2\\ \lambda_i \neq 0}}^{\infty} \dfrac{\left(\int_{-\ell}^{\ell} \varphi_i(x)\,dx\right)^2}{\lambda_i}}.$$

Consider now the question what kinds of representation an $f(x)$ defined on a finite interval may possess in general. The next theorem shows that one and the same Hermitian-indefinite function given on a finite interval may have representations of all there types.

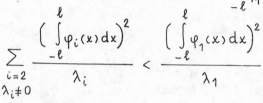

Theorem 3. *Let* $f(x)$ $(x \in (-2\ell, 2\ell),\ \ell < \infty)$ *be a Hermitian-indefinite function with one negative square. Assume that* $f(x)$ *is twice continuously*

differentiable and admits more than one representation, among these one of form

$$f(x) = f(0) + \int_{-\infty}^{\infty} \frac{\cos \lambda x - 1}{\lambda^2} \, d\sigma(\lambda) .$$

Then $f(x)$ *has a parabolic representation as well as an infinite number of elliptic and hyperbolic representations.*

As a corollary we obtain that in case $f(x)$ admits more than one representation it always admits a hyperbolic one.

It turns out that another case, in which $f(x)$ admits an infinite number of hyperbolic representations without having any elliptic or parabolic one is also possible. It is to be observed, too, that Theorem 3 and its corollary can be considered as an analogue of the results of I.S. Iohvidov concerning the discrete case.

Since each representation of $f(x)$ yields a natural extension of $f(x)$ to the whole line, all that has been said above holds for the extension of $f(x)$ as well. Now the operator $-i\frac{d}{dx}$, considered in the space $\tilde{\Pi}_1$ built up with the aid of the kernel $\hat{f}(x-y)$ ($\hat{f}(x)$ being an extension of $f(x)$ to the whole real line), is self--adjoint and it is an extension of the operator $-i\frac{d}{dx}$ considered in Π_1. Thus, with any representation of $f(x)$ on a finite interval we can associate a self-adjoint extension of a certain type of $-i\frac{d}{dx}$.

COLLOQUIA MATHEMATICA SOCIETATIS JÁNOS BOLYAI
5. HILBERT SPACE OPERATORS, TIHANY (HUNGARY), 1970

Invariant subspaces of the weighted shift

H. HELSON

1. This note is a restudy of theorems of Nikolskiĭ [5,6] about weighted shifts. Nikolskiĭ's work was based on an example given by Donoghue [3]. The problem is discussed also by Halmos [4].

In the space ℓ^p of one-sided sequences $\alpha = (\alpha_0, \alpha_1, \dots)$ we define an operator T by setting

(1) $$T\alpha = (0, \lambda_0\alpha_0, \lambda_1\alpha_1, \dots)$$

with a given bounded sequence of weights λ_j assumed to be strictly positive. Let B_k denote the subspace of all α such that $\alpha_j = 0$ for $j < k$. Obviously each B_k is invariant under T. The problem is to find hypotheses under which every proper closed invariant subspace has this form. In that case T is called a *unicellular* operator. Nikolskiĭ establishes sufficient conditions for T to have this property. My object is to show that Nikolskiĭ's theorems can be obtained from Banach algebra, and that the problem belongs to a type studied by Beurling and by Domar.

I am indebted to Domar for helpful suggestions, not all of which could be incorporated here.

The exponent $p = \infty$ is exceptional, and we assume that p satisfies $1 \leq p < \infty$. Define $\varrho(0) = 1$, $\varrho(n) = \prod_0^{n-1} \lambda_j^p$ for $n \geq 1$, and construct the weight spaces ℓ_ϱ^p with norms

(2) $$\|\alpha\|_{p,\varrho} = \Big[\sum_0^\infty |\alpha_n|^p \varrho_n\Big]^{1/p}.$$

Each λ_ϱ^p is isomorphic to ℓ^p in a natural way: to α in ℓ^p we associate β in ℓ_ϱ^p defined by

(3) $$\beta_n = \alpha_n \varrho(n)^{-1/p}.$$

This correspondence is an isometry. Furthermore $T\alpha$ corresponds to $S\beta = (0, \beta_0, \beta_1, \ldots)$. Thus the weighted shift T in ℓ^p is similar to the ordinary shift S in ℓ_ϱ^p. The subspaces B_k are defined in ℓ_ϱ^p as in ℓ^p, and we ask whether S is unicellular. We assume that $\varrho(n+1) \leq \varkappa \varrho(n)$ for some positive $\varkappa$, so that S is a bounded operator in each space ℓ_ϱ^p.

A relevant fact suggests itself when we pose the problem in terms of ℓ_ϱ^p and S. If ϱ satisfies

(4) $$\varrho(m+n) \leq \varkappa \varrho(m) \varrho(n)$$

for all $m, n \geq 0$ and some positive $\varkappa$, then ℓ_ϱ^1 is a Banach algebra with identity under convolution. An element is invertible, and therefore cyclic under S, if it belongs to no maximal ideal. If we assume

(5) $$\lim \varrho(n)^{1/n} = 0$$

there is only one maximal ideal, namely B_1. Thus we have the first element in an inductive proof that S is unicellular in ℓ_ϱ^1. We shall deduce results for other exponents from this special case.

2. Under hypotheses (4,5), every proper closed invariant subspace of ℓ_ϱ^1 is contained in B_1. We should like to repeat the argument to show that a closed invariant subspace properly contained in B_1 must be contained in B_2. For this and succeeding steps we construct in turn the Banach algebras based on weight

sequences $\{\varrho(k), \varrho(k+1), \ldots\}$, $k = 1, 2, \ldots$. But (4) is not automatically true for these sequences. In order to conclude that S is unicellular we have to know there are numbers $\varkappa(k)$ such that

(6) $$\varrho(m+n+k) \leq \varkappa(k)\varrho(m+k)\varrho(n+k) \qquad (m, n, k \geq 0).$$

For other values of p also, the conclusion that S is unicellular in ℓ^p_ϱ is proved by induction, only the first step being of interest. For simplicity we shall call ℓ^p_ϱ an N*-space* if each proper closed invariant subspace is contained in B_1 . This is the property we shall study.

3. Algebras a little more general than ℓ^1_ϱ have to be considered in order to obtain Nikolskiĭ's theorems. Suppose that ϱ is only required to satisfy (4) for all m, n at least equal to a given positive integer r . Then ℓ^1_ϱ is not in general an algebra. Denote by A_r the subspace of ℓ^1_ϱ consisting of all β such that $\beta_1 = \beta_2 = \ldots = \beta_{r-1} = 0$. A_r is a Banach algebra with identity. Under the hypothesis (5) it has only one maximal ideal: the set of β such that $\beta_0 = 0$. The operator S is not bounded in A_r , but its powers S^n are defined and continuous provided $n \geq r$.

Formally A_r is ℓ^1_ζ , where $\zeta_n = \varrho_n$ for $n = 0, r, r+1, \ldots$, and $\zeta_n = \infty$ for the other indices. We can form spaces ℓ^p_ζ in the same way.

4. The proof of Theorem 1 will operate in the dual of ℓ^p_ζ , where ζ has infinite values for indices $1, \ldots, r-1$. We have $1 < p < \infty$; let q satisfy $p^{-1} + q^{-1} = 1$. Define $\tau(n) = \zeta(n)^{1-q}$ (equal to 0 where ζ is infinite). Then ℓ^p_ζ and ℓ^q_τ are dual spaces in the pairing

(7) $$(\beta, \gamma) = \sum_0^\infty \beta_n \gamma_n \qquad (\beta \in \ell^p_\zeta,\ \gamma \in \ell^q_\tau).$$

With the same pairing ℓ^1_ζ is the dual of $\mathcal{L}_\zeta$, the space of sequence γ such that $\gamma_n \zeta(n)^{-1}$ tends to 0 , with norm $\max|\gamma_n|\zeta(n)^{-1}$. Note that two sequences are identified in ℓ^q_τ and in $\mathcal{L}_\zeta$ if their values differ only at indices where ζ is infinite.

The adjoint shift R is defined in ℓ^q_τ and in $\mathcal{L}_\zeta$ by

(8) $$R\gamma = (\gamma_1, \gamma_2, \ldots).$$

A similar formula defines powers of R . If $n \geq r$, then R^n is a bounded operator in each space, with adjoint S^n acting in the dual space. The annihilator of a closed subspace invariant under R^n is a closed subspace of the dual space invariant under S^n. Since ℓ^q_τ and ℓ^p_ζ are duals of each other, this correspondence relates all the closed invariant subspaces of one space with all those of the other. But not all closed subspaces of ℓ^1_ζ are annihilators, a fact that must be dealt with in the following proof.

5. Theorem 1. *For* $1 < p < \infty$, ℓ^p_ϱ *is an* N*-space if* ϱ *has these properties:* $\varrho(n+1) \leq \varkappa\varrho(n)$ *for some* $\varkappa$, (5) *holds, and*

(9) *to each* α *in* ℓ^q *there is a* μ *in* ℓ^q *such that* $|\alpha_n| < \mu_n$ *for all* n, *and* $\{\mu_n \varrho_n^{1/p}\}$ *is a weight sequence such that* (4) *holds for all* m, n *at least equal to some integer* r.

Let ζ be the weight sequence ϱ modified to have infinite values at indices $1, \ldots, r-1$. Let γ belong to ℓ^q_τ, $\gamma \neq 0$. Then $\alpha_n = \gamma_n \zeta(n)^{-1/p}$ is a sequence in ℓ^q. Choose μ by (9); we can also arrange that $\alpha_n \mu_n^{-1}$ tends to 0. If we set $\nu_n = \mu_n \xi(n)^{1/p}$, then the construction places γ in $\mathcal{L}_\nu$.

The hypotheses imply that ℓ^1_ν is a Banach algebra with a single maximal ideal. It follows that every closed non-trivial subspace of $\mathcal{L}_\nu$ invariant under R^n $(n \geq r)$ contains the sequence $\delta = (1, 0, 0, \ldots)$. Hence given $\delta \geq 0$ there is a linear combination γ of $\gamma, R^r\gamma, R^{r+1}\gamma, \ldots$ such that

$$(10) \qquad |\gamma'_n - \delta_n| \, \nu(n)^{-1} < \varepsilon \qquad (n = 0, r, r+1, \ldots).$$

Now $\nu(n)^{-1} = \mu_n^{-1} \zeta(n)^{-1/p} = \mu(n)^{-1} \tau(n)^{1/q}$. Set the last expression in (10), multiply both sides by μ_n, take the qth power and sum to obtain

$$(11) \qquad \| \gamma' - \delta \|_{q,\tau} \leq \varepsilon \| \mu \|_q .$$

This means the smallest closed subspace of ℓ^q_τ containing γ and invariant under R^n $(n \geq r)$ contains δ.

By duality, every proper closed subspace of ℓ^p_ζ invariant under S^n $(n \geq r)$ consists of sequences α such that $\alpha_0 = 0$. Let β be an element of ℓ^p_ϱ with $\beta_0 \neq 0$. A linear combination of $\beta, S\beta, \ldots, S^{r-1}\beta$ will give a sequence α such that $\alpha_0 \neq 0$, $\alpha_1 = \ldots = \alpha_{r-1} = 0$. Linear combinations of $\alpha, S^r\alpha, S^{r+1}\alpha, \ldots$ approximate δ (this is the statement proved about ℓ^p_ζ). Hence β is cyclic under S, and therefore ℓ^p_ϱ is an N-space.

The condition of Theorem 1 is not easy to verify directly. The next theorem gives a somewhat better condition.

Theorem 2. *If there is a positive sequence* η *in* ℓ^q *and a positive integer* r *such that*

$$\varrho(m+n)^{1/p} \leq \eta_m \eta_n \varrho(m)^{1/p} \varrho(n)^{1/p} \qquad (m,n \geq r) \tag{12}$$

then (9) *holds.*

Let α be any sequence in ℓ^q; we seek μ in ℓ^q to satisfy (9). Set $\mu_n = \max(|\alpha_n|, \eta_n)$ for $n \geq r$. We have to show for some $\varkappa$ that

$$\mu_{m+n} \varrho(m+n)^{1/p} \leq \varkappa \mu_m \mu_n \varrho(m)^{1/p} \varrho(n)^{1/p} \qquad (m,n \geq r). \tag{13}$$

If suffices to prove (13) with μ_m, μ_n replaced on the right side by η_m, η_n. By (12) and the fact that μ_n tends to 0, this is true with $\varkappa = 1$ for all but finitely many pairs (m,n). Taking $\varkappa$ large enough we obtain inequality for all $m, n \geq r$. (Note that $\varkappa$ depends on α.)

Finally we can give a condition without any quantifier.

Theorem 3. *The hypothesis of Theorem 2 holds if for some positive integer* r *we have*

$$\sum_{m,n \geq r} \left(\frac{\varrho(m+n)}{\varrho(m)\varrho(n)} \right)^{\frac{q-1}{2}} < \infty . \tag{14}$$

Define

$$\eta_n^q = \sum_{m \geq r} \left(\frac{\varrho(m+n)}{\varrho(m)\varrho(n)} \right)^{\frac{q-1}{2}} \qquad (n \geq r). \tag{15}$$

Multiplying the series for η_m and η_n gives trivially

$$(\eta_m \eta_n)^q \geq \left(\frac{\varrho(m+n)}{\varrho(m)\varrho(n)} \right)^{q-1} \qquad (m,n \geq r). \tag{16}$$

Thus (12) holds and the proof is finished.

The conditions (12) and (14) have been shown to imply (9); they easily imply (5) also. Thus they imply that ℓ^p_ϱ is an N-space, provided that S

operates in ℓ^p_ϱ (that is, $\varrho(n+1)\,\varrho(n)^{-1}$ is bounded).

6. Condition (14), expressed in terms of the weights λ_j, becomes

$$\sum_{m,n\geq r}\left(\frac{\lambda_m\cdots\lambda_{m+n-1}}{\lambda_0\cdots\lambda_{n-1}}\right)^{q/2}<\infty. \tag{17}$$

We shall prove (17) assuming that λ is a decreasing sequence and belongs to ℓ^s for some finite s. This is the main result of [5], and a similar argument will give Theorem 1 of [6].

Replace λ_j by $\lambda_j^{2/q}$ (giving a new value for s) and take $r\geq s$. We want to show

$$\sum_{m,n\geq r}\frac{\lambda_m\cdots\lambda_{m+n-1}}{\lambda_0\cdots\lambda_{n-1}}<\infty. \tag{18}$$

For $n=r$ the sum over m is finite, say equal to Q, because λ is in ℓ^r. The sum over m with $n=r+1$ does not exceed Q multiplied by λ_{2r}/λ_r since λ is monotonic. There are similar extimates for succeeding values of n, and the series (18) is dominated by

$$Q\left(\frac{\lambda_{2r}}{\lambda_r}+\frac{\lambda_{2r}\lambda_{2r+1}}{\lambda_r\lambda_{r+1}}+\frac{\lambda_{2r}\lambda_{2r+1}\lambda_{2r+2}}{\lambda_r\lambda_{r+1}\lambda_{r+2}}+\cdots\right). \tag{19}$$

After the rth term of this series, each new factor in the denominatoı cancels the first element of the numerator, so the denominator remains the same and the numerator has exactly r factors from that point on. The series converges again because λ is in ℓ^r, and the proof is finished.

The truncated sequences $\{\lambda_k,\lambda_{k+1},\cdots\}$ satisfy the same hypotheses, so the weighted shift is unicellular.

7. Let S be a linear operator in a topological vector space A. Say *the closure theorem holds for* S *in* A if every proper closed invariant subspace is contained in one of codimension one. This property has been studied in weight spaces with the shift operator by Beurling [1] and by Domar [2]; the terminology is that of Beurling but the general definition is due to Domar. Nikolskiǐ's theorems assert the closure theorem in certain weight spaces where there is just one maximal invariant subspace. Our hypothesis (5) was used only to make the maximal invariant subspace

unique. This leads to the desired unicellular property of the shift operator, but loses the interest of the general closure problem.

The essential point in our proof of Theorem 1 is to place each element of the dual of ℓ^p_ζ in an auxiliary space (depending on the element) possessing an adjoint closure property. Special circumstances enable us to draw a conclusion in the given norm (11). A similar argument should work more generally in spaces where the topology is given by a family of seminorms.

REFERENCES

[1] A. Beurling, A critical topology in harmonic analysis on semigroups, *Acta Math.,* 112 (1964), 215-228.

[2] Y. Domar, Spectral analysis in spaces of sequences summable with weights, *J. Functional Anal.,* 5 (1970), 1-13.

[3] W.F. Donoghue, The lattice of invariant subspaces of a completely continuous quasi-nilpotent transformation, *Pac. J. Math.,* 7 (1957), 1031-1035.

[4] P. Halmos, *A Hilbert Space Problem Book* (Princeton, 1967).

[5] N.K. Nikolskiǐ, Invariant subspaces of certain completely continuous operators, *Vestnik Leningrad Univ.,* 20 (1965), 68-77 (Russian).

[6] N.K. Nikolskiǐ, Invariant subspaces of weighted shift operators, *Math. of the USSR – Sbornik,* 3 (1967), 159-176 (translated from *Mat. Sbornik,* 74 (116), (1967)).

COLLOQUIA MATHEMATICA SOCIETATIS JÁNOS BOLYAI
5. HILBERT SPACE OPERATORS. TIHANY (HUNGARY), 1970

Operators with a representation as multiplication by x on a Sobolev space

H. HELTON

I shall speak on a structure theorem for a certain class of non selfadjoint operators based on the representation of such operators as multiplication by x on a type of Sobolev space.

Let T be a bounded operator on a Hilbert space and set $V(s) = e^{isT}$. If we define the operator-valued entire function $R(s) = V(\bar{s})^* V(s) = e^{-isT^*} e^{isT} = \sum_{n=0}^{\infty} A_n s^n$, then T is self adjoint if and only if $R(s) = 1$ and hence it seems reasonable to classify operators by the behavior of $R(s)$. In particular, a class of operators which seems nearly as natural and amenable to study as the self-adjoint operators is the operators T for which $R(s)$ is a polynomial in s, that is

$$\text{(POL)} \qquad R(s) = V(\bar{s})^* V(s) = \sum_{n=0}^{N} A_n s^n .$$

Under certain additional hypotheses, it will be proved that such operators have a representation as multiplication by x on a type of Sobolev space.

The type of Sobolev space we need is this. Let $\mu = \langle \mu_{ij} \rangle$ be a matrix of $(M+1)^2$ measures on the interval $[a,b]$ and define a bilinear form $(\,,\,)_\mu$ on $C^\infty[a,b]$ by

(REP) $$(f,g)_\mu = \sum_{i,j=0}^{M} \int_a^b f^{(j)} \bar{g}^{(j)} d\mu_{ij}$$

where $h^j(x) = \frac{d^j}{dx^j} h(x)$. If $(\,,\,)_\mu$ is positive definite we let $H(\mu)$ stand for the completion of $C^\infty[a,b]$ in $\|f\|_\mu = \sqrt{(f,f)_\mu}$. We will say that $H(\mu)$ has order M.

The object of this talk will be to sketch a proof of

Theorem. *If* T *satisfies* POL, *if* T *has a cyclic vector, and if* T *has spectrum* $[a,b]$, *then there is a space* $H(\mu)$ *and a unitary map* $U: H \to H(\mu)$ *such that* UTU^{-1} *is multiplication by* x *on* $H(\mu)$.

Before we begin the proof of this theorem, we make a few remarks. First of all, the theorem is much weaker than what can be proved, but for the sake of clear exposition we postpone a description of its generalizations to the end of this talk. Secondly, it is easy to prove that the degree N of the polynomial $R(s)$ is even. Undoubtedly the space $H(\mu)$ can be chosen to have order $N/2$. I have a heuristic proof for this. Lastly the POL condition can be expressed in more concerte terms. Namely, POL is equivalent to the condition $[T^*-T]^{[N+1]} = 0$, where $[A-B]^{[M]}$ is defined to be the operator $\sum_{K=0}^{M} \binom{M}{K} (-1)^{M-K} A^K B^{M-K}$. To check the equivalence of the two conditions note that for any bounded operator

$$\frac{d}{ds} e^{-isT^*} A e^{isT} = -ie^{-isT^*}[T^*A - AT] e^{isT}$$

and hence

$$\frac{d^{N+1}}{ds^{N+1}} e^{-isT^*} e^{isT} = (-i)^{N+1} e^{isT^*} C_T^{N+1}(I) e^{isT}$$

where C_T is the map of $\mathcal{L}(H)$, the set of all bounded operators on H, into $\mathcal{L}(H)$ defined by $C_T(A) = T^*A - AT$. The POL condition is equivalent to $\frac{d^{N+1}}{ds^{N+1}} R(s) = 0$ and consequently to $C_T^{N+1}(I) = 0$. A straightforward computation (cf. proof of

Theorem 2.2 [1]) shows that $[T^*-T]^{[M]} = (-1)^M C_T^M(I)$.

Now we sketch a proof of the representation theorem based on distribution theory. Recall (cf. Chapter 1 and 2 of Hörmander [2]) that $\mathcal{S}(R^n)$ is the space of infinitely differentiable functions on R^n which along with their derivatives decrease faster than any polynomial at infinity. Frequently we shall use $\mathcal{S}$ to denote $\mathcal{S}(R^1)$. The POL condition guarantees that $\|e^{isT}\| = O(|s|^N)$ and thus T has a C^n functional calculus (cf. Colojoară and Foiaş [1] for a discussion of such operators).

Define U_f by

$$U_f = \frac{1}{\sqrt{2\pi}} \int_{-\infty}^{\infty} \hat{f}(\zeta) e^{i\zeta T} d\zeta \tag{1}$$

for any $f \in \mathcal{S}$, where $\hat{f}$ denotes the Fourier transform of f. This is a slight modification of the usual construction of a C^n functional calculus for T. The integral is norm convergent since the function $\hat{f}$ is in $\mathcal{S}$, and U_f has the usual properties

$$U_{f+g} = U_f + U_g \quad \text{and} \quad U_{fg} = U_f U_g . \tag{2}$$

One can show that

$$e^{isT} U_{f(x)} = U_{e^{isx} f(x)} \quad \text{and} \quad T U_{f(x)} = U_{x f(x)} . \tag{3}$$

Let ψ_0 be a cyclic vector for T. Define a bilinear form $[\,,\,]$ on $\mathcal{S}$ by

$$[f, g] = (U_f \psi_0, U_g \psi_0).$$

The next few paragraphs will be devoted to proving that $[\,,\,]$ has the form $(\,,\,)_\mu$ given in REP. After this is accomplished we will show that the map $V\colon \mathcal{S} \to \mathcal{H}$ defined by

$$Vf = U_f \psi_0$$

induces a map V_r from $H(\mu) \cap \mathcal{S}$ into a dense subspace of H. The extension of V_f to $H(\mu)$ will be the map U^{-1} required by our theorem.

The Schwartz Nuclear Theorem (cf. Theorem 2.1 of Streater and Wightman [4]) implies that there is a continuous linear functional (distribution) on $\mathcal{S}(R^2)$ such that

$$\ell(f(x)\bar{g}(y)) = [f, g]. \tag{4}$$

The POL condition says that

$$\frac{d^{N+1}}{ds^{N+1}}[e^{isx} f(x), e^{isx} g(x)] = 0$$

for any $f, g \in \mathcal{S}$ or equivalently

$$\frac{d^{N+1}}{ds^{N+1}} \ell(e^{isx} f(x)\bar{g}(y)) = 0.$$

Thus

$$\ell([x-g]^{N+1} f(x)\bar{g}(y)) = 0$$

and since linear combinations of functions of the form $f(x)\bar{g}(y)$ are dense in $\mathcal{S}(R^2)$,

$$\ell([x-y]^{N+1} h(x,y)) = 0$$

for any function $h \in \mathcal{S}(R^2)$. This implies that if k is in $\mathcal{S}(R^2)$ and has a zero of order $N+1$ on $E = \{(x,y) \in R^2 :\ x = y\}$, then $\ell(k) = 0$.

Now we need a change of variables

$$x - y \longrightarrow \tau \qquad x + y \longrightarrow \beta.$$

Any function h in $\mathcal{S}(R^2)$ has a finite Taylor expansion

$$h(\tau, \beta) = \sum_{j=0}^{N} \frac{\partial^j}{\partial \tau^j} h(0, \beta)\tau^j + \psi(\tau, \beta),$$

where the remainder term $\psi(\tau, \beta)$ has a zero of order $N+1$ in τ at $\tau = 0$. Hence $\ell(\psi) = 0$* and thus we can obtain

$$\ell(h(\tau, \beta)) = \sum_{j=0}^{N} \ell_j\left(\frac{\partial^j}{\partial \tau^j} h(0, \beta)\right), \tag{5}$$

where each ℓ_j is a distribution on $\mathcal{S}(R^1)$. Now any distribution s on $\mathcal{S}(R^1)$ has a representation (Schwartz Kernel Theorem, sec. 2.3 Shilov [3] or equation 2-11 [4]) of the form

*Technically ψ is not in $\mathcal{S}(R^2)$, however a standard argument involving cutoff functions makes this proof of equation (5) rigorous.

$$s(f) = \sum_{\ell=0}^{K} \int_{-\infty}^{\infty} b_\ell(x) f^{(\ell)}(x)\,dx$$

for some continuous functions b_ℓ wich grow at worst like a polynomial at infinity. If we put together the two representations above and note that

$$\frac{\partial}{\partial\tau} m(x)n(y)\Big|_{x=y} = \frac{1}{2}\Big[\frac{\partial}{\partial x} - \frac{\partial}{\partial y}\Big] m(x)n(y)\Big|_{x=y} = \frac{1}{2} m'(x)n(x) - \frac{1}{2} m(x)n'(x)$$

we get

$$\ell(f(x)\bar{g}(y)) = \sum_{i,j=0}^{M} \int_{-\infty}^{\infty} C_{ij}(x) f^{(i)}(x) \bar{g}^{(j)}(x)\,dx$$

for $f, g \in \mathcal{S}$. We must restrict the interval of integration to $[a, b]$ and in order to do this we need the following lemmas:

Lemma I. *If* $f \in \mathcal{S}$ *and* $f \equiv 0$ *in a neighborhood of* $\sigma(T)$, *then* $U_f = 0$. *Moreover, the norm closure* $\bar{\mathcal{P}}$ *of* $\mathcal{P} = \{p(T) :\ p$ *is a polynomial*$\}$ *is equal to the norm closure* $\bar{S}$ *of* $S = \{U_f :\ f \in \mathcal{S}\}$.

Proof. First we show that $\bar{\mathcal{P}}$ contains S. Since for each s the operator e^{isT} is in $\bar{\mathcal{P}}$ and since the integral in (1) is norm convergent, $\bar{\mathcal{P}}$ contains each operator U_f and thus contains S.

It is easy to see from properties (2) and (3) of U_f that the Gelfand map of $\bar{\mathcal{P}}$ into the continuous functions on its maximal ideal space maps U_f into f. Thus U_f is an invertible operator if and only if f is an invertible function on the maximal ideal space of $\bar{\mathcal{P}}$ that is, on $\sigma(T)$. Now suppose that f is a function in $\mathcal{S}$ which is zero on a neighborhood $\sigma(T)$. There is certainly some function h in $\mathcal{S}$ which is invertible on $\sigma(T)$ and which is 0 on the support of f. Thus $U_f U_h = U_{fh} = 0$ and $U_f = 0$. The first half of Lemma I has been proved.

Now we show that $\bar{\mathcal{P}} = \bar{S}$. This is very easy for if p is a polynomial and the function h in $\mathcal{S}$ is never zero in a neighborhood of $\sigma(T)$, then $p(T) = [U_h]^{-1} U_{hp}$ and so $p(T)$ is the product of two operators in $\bar{S}$. This concludes the proof of Lemma I.

Lemma II. *If* $\sigma(T) = [a,b]$ *where* $a \neq b$, *we have that if* $f \in \mathcal{S}$ *and* $f \equiv 0$ *on* $[a,b]$ *then* $U_f = 0$.

Proof. Suppose that $f \equiv 0$ on $[a, b]$. Define $f^{\pm}$ by

$$f^{+} = \begin{cases} f(x) & x \geq b \\ 0 & x \leq b \end{cases}$$

$$f^{-} = \begin{cases} f(x) & x \leq a \\ 0 & x \geq a . \end{cases}$$

Since $a \neq b$ both f^{+} and f^{-} are in $\mathcal{S}$. The functions $f_n^{\pm}$ defined by $f_n^{\pm}(x) = f_n^{\pm}(x \mp 1/n)$ converge to $f^{\pm}$ in the $\mathcal{S}$ norm because

$$\lim_{n\to\infty} \left| x^{\ell} \frac{d^k}{dx^k} [f^{\pm}(x) - f^{\pm}(x \mp 1/n)] \right| =$$

$$= \lim_{n\to\infty} \frac{1}{n} \left| x^{\ell} \frac{d^k}{dx^k} [f^{\pm}(x) - f^{\pm}(x \mp 1/n)] / 1/n \right| = 0 .$$

Since $f_n^{\pm} \equiv 0$ on a neighborhood of $[a,b]$, the operators $\mathcal{U}_{f_n^{\pm}}$ are all 0. However, $\mathcal{U}_f$ is a continuous map of $\mathcal{S}$ into $\mathcal{L}(H)$ and hence $\mathcal{U}_{f^{\pm}} = 0$. Therefore $\mathcal{U}_f = \mathcal{U}_{f^+} + \mathcal{U}_{f^-} = 0$.

At this point let us remark that if $a = b$, then Lemma II fails to be true. For example, if T is an operator whose square is zero, then the spectrum of T is $\{0\}$ and $\mathcal{U}_f$ defined by (1) for T is equal to zero if and only if $f(0) = 0$ and $f'(0) = 0$.

We had shown, several paragraphs ago, that the distribution ℓ on $\mathcal{S}(R^2)$ with property (4) has support on E. Lemma I above implies further that ℓ has support on the set $E_{\sigma(T)} = \{(x,x) \in E : x \in \sigma(T)\}$. Now under the simplifying assumption that $\sigma(T)$ is an interval $[a,b]$ it is easy to show (with the help of the fact that any distribution with support at a point is a linear combination of derivatives of the dirac δ-function) that the representation REP holds for the form $[\ ,\ ]$.

Now we conclude the proof of the representation theorem. Define $V: \mathcal{S} \to H$ by

$$Vf = \mathcal{U}_f \psi_0 .$$

Since ψ_0 is cyclic for T, Lemma I implies that the orbit of ψ_0 under S is dense in H, and hence V maps onto a dense subspace of H. By Lemma II we may define a map V_r of $C^\infty[a,b]$ into H by

$$V_r g = V\tilde{g}$$

where g is a function in $C^\infty[a,b]$ and $\tilde{g}$ is any extension of g which belongs to $\mathcal{S}$. From the statement that $[\ ,\ ]$ has a representation $(\ ,\ )_\mu$ which was prove above we can conclude that

$$(f,g)_\mu = (V_r f, V_r g).$$

The bilinear form $(\ ,\ )_\mu$ must be positive since $(\ ,\ )$ is positive. Moreover, it is positive definite since $(f,f)_\mu = 0$ implies that $V_r f = 0$, which implies that $U_f\psi_0 = Vf = 0$ which implies that $U_f U_g \psi_0 = 0$ for any g. This in turn implies that $U_f \equiv 0$, which (by the proof of Lemma I) implies that $f \equiv 0$ on $\sigma(T)$. Thus $C^\infty[a,b]$ can be completed in $(\ ,\ )_\mu$ to produce a space $H(\ )$. Naturally V_r extends to an isometry $\bar{V}$ from $H(\mu)$ onto H. If $f \in C^\infty[a,b]$, then $\bar{V}T\bar{V}^{-1}f(x) = \bar{V}TV_r f(x) = \bar{V}TU_{\tilde{f}} = \bar{V}U_{x\tilde{f}} = xf(x)$. The equation $\bar{V}T\bar{V}^{-1}f(x) = xf(x)$ extends by continuity to any $f \in H(\mu)$. This completes the proof of the representation theorem.

The above theorem can be generalized considerably. In fact most of its hypotheses can be dropped an still a strong conclusion holds. Firstly, we can replace the assumption that T has a cyclic vector ψ_0 with the assumption that T has a finite cyclic set $\{\psi_1, \dots, \psi_n\}$. In this case we study the bilinear form $[f,g] = (U_{f_1}\psi_1 + \dots + U_{f_n}\psi_n, U_{g_1}\psi_1 + \dots + U_{g_n}\psi_n)$ where F and G are C^∞ n-vector valued functions with components $F = (f_1, \dots, f_n)$ and $G = (g_1, \dots, g_n)$. Under the POL assumption this bilinear form will have a representation which is the natural generalization of REP to vector valued functions. Note that the form $[\ ,\ ]$ in this vector valued case must be positive but not necessarily positive definite. Secondly, suppose that $T_1, \dots, T_k$ is a commuting family each operator of which satisfies POL. We then are forced to consider functions f not of one variable but of k variables and to build a functional calculus

$$U_f = \frac{1}{\sqrt{2\pi}^k} \int_{R^k} \hat{f}(\zeta_1, \dots, \zeta_k) e^{iT_1\zeta_1 + \dots + iT_k\zeta_k} d\zeta_1 \dots d\zeta_k .$$

The bilinear form $[\ ,\ ]$ on $\mathcal{S}(R^k)$ defined by

$$[f,g] = (U_f\psi_0, U_g\psi_0)$$

will have a representation which is the natural generalization of REP to R^k. Thirdly, the operator T need not be bounded. All that we really need to consider is a one parameter group of operators $V(s)$ which satisfies the POL condition. If we modify the cyclic vector assumption property, then REP holds with $b=\infty$, $a=-\infty$ and $V(s)$ maps into multiplication by e^{isx}. Fourthly, the three generalizations above can be put together to give the following theorem:

Theorem. *Suppose that*

a) $V_1(s),\dots,V_k(s)$ *are commuting one parameter groups of operators each of which satisfies the* POL *condition*

and

b) *there are* n*-vectors* $\psi_1,\dots,\psi_n$ *such that the orbit of the subspace spanned by them under the algebra generated by* $V_1(s),\dots,V_k(s)$ *is dense in* H.

Construct a map $\mathcal{S}(R^k)\to\mathcal{L}(H)$ *by*

$$U_f=\frac{1}{\sqrt{2\pi}^{\,k}}\int_{R^k}\hat{f}(\zeta_1,\dots,\zeta_k)\,e^{i[\zeta_1T_1+\dots+\zeta_kT_k]}\,d\zeta_1\dots d\zeta_k.$$

Define $[\ ,\ ]$ *by*

$$[F,G]=(U_{f_1}\psi_1+\dots+U_{f_n}\psi_n,\ U_{g_1}\psi_1+\dots+U_{g_n}\psi_n)\quad \textit{for}\quad (f_1,\dots,f_n)$$

and $G=(g_1,\dots,g_n)$ n*-vector valued functions with components* f_i, g_i *in* $\mathcal{S}(R^k)$. *Then* $[\ ,\ ]$ *has a representation*

$$[F,G]=\sum_{|\beta|,|\alpha|<M}\ \sum_{i,j=0}^{n}\int_{R^k}^{\infty}\frac{\partial^{|\alpha|}}{\partial x^{\alpha}}f_i(x)\,\frac{\partial^{|\beta|}}{\partial x^{\beta}}\bar{g}_j(x)\,d\mu_{\alpha\beta ij}$$

where $\mu_{\alpha\beta ij}$ *are measures on* R^k *and where* $\frac{\partial^{\alpha}}{\partial x^{\alpha}}$ *stands for* $\frac{\partial^{|\alpha|}}{\partial x_1^{\alpha_1}\dots\partial x_k^{\alpha_k}}$ *for each* k *tuple* $\alpha=(\alpha_1,\dots,\alpha_k)$ *of positive integers. Here* $|\alpha|=\alpha_1+\dots+\alpha_k$.

Furthermore, $V_j(s)$ *acting on* H *corresponds to multiplication by* e^{isx_j} *on the space of* n*-vector valued functions with components in* $\mathcal{S}(R^k)$.

BIBLIOGRAPHY

[1] I. Colojoară and C. Foiaş, *Theory of generalized spectral operators* (New York, 1968).

[2] L. Hörmander, *Linear partial differential operators* (Berlin, 1963).

[3] G.E. Shilov, *Generalized functions and partial differential equations* (New York, 1968).

[4] R.F. Streater and A.S. Wightman, *PCT Spin and Statistics and all that* (New York, 1964).

COLLOQUIA MATHEMATICA SOCIETATIS JÁNOS BOLYAI
5. HILBERT SPACE OPERATORS. TIHANY (HUNGARY), 1970

On the structure of the spectrum of *G*-selfadjoint and *G*-unitary operators in Hilbert space

I. S. IOHVIDOV

1. Let a G-metric $[x, y]$ in Hilbert space $\mathfrak{h}$ be introduced by the formula

$$[x,y] = (Gx,y) \qquad (x,y \in \mathfrak{h}),$$

where G is a bounded selfadjoint operator (*Gram operator* [1]). The G-metric $[x,y]$ will be supposed *non-degenerate*, i.e. $0 \notin \sigma_p(G)$ (here and below for an arbitrary linear operator T let $\rho(T)$ be the set of its regular points, $\sigma(T)$ the spectrum, $\sigma_p(T)$, $\sigma_c(T)$ and $\sigma_r(T)$ the point, continuous and residual spectra, respectively).

Several problems of mathematical physics lead to the study of linear operators in $\mathfrak{h}$ which are selfadjoint or unitary with respect to some G-metric (see, for instance, [2], [3], [4] etc.). But while such operators with respect to a *regular* G-metric (i.e. under the condition $0 \in \varrho(G)$) have been studied relatively well, for the *non-regular* case ($0 \in \sigma_c(G)$) little is known, although that case often appears in applications ([2], [4] etc.).

2. Only one topic will be discussed here: the structure of the spectrum of G-selfadjoint and G-unitary operators.

Let A be a linear operator in $\mathfrak{h}$ with domain $\mathfrak{D}_A$ satisfying the "G-density" condition (in $\mathfrak{h}$):

(1) $$[x_0, y] = 0 \quad (\forall y \in \mathfrak{D}_A) \quad \Longrightarrow \quad x_0 = 0 .$$

(The condition (1) is always fulfilled if $\overline{\mathfrak{D}_A} = \mathfrak{h}$, but not vice-versa). Then, as it is easy to see, the *G-adjoint* A^c of the operator A is uniquely defined:

$$[Ax, y] = [x, z] \quad (\forall x \in \mathfrak{D}_A) \Rightarrow y \in \mathfrak{D}_{A^c}, \ A^c y = z .$$

If $A^c = A$, the operator A will be called *G-selfadjoint.* If $U^c U = U U^c = I$, we call the operator U *G-unitary*. This definition is equivalent to the requirement:

$$\mathfrak{D}_U = \mathfrak{R}_U = \mathfrak{h}, \quad [Ux, Uy] = [x, y] \quad (x, y \in \mathfrak{h})$$

which automatically implies linearity and boundedness of the operator U ([5], [6]); $\mathfrak{R}_U$ denotes the range of the operator U). Evidently, $U^c = U^{-1}$ is G-unitary together with U .

For a long time only the following theorem of H. Langer [7] about the spectra of G-selfadjoint operators has been known:

Let $\overline{\mathfrak{D}_A} = \mathfrak{h}$ *and* $A^c = A$. *Then*

$$\lambda \in \sigma_p(A) \Longrightarrow \bar{\lambda} \in \sigma_p(A) \cup \sigma_r(A) .$$

If, additionally, the G*-metric is regular* ($0 \in \varrho(G)$), *then also:*

α) $\lambda \in \sigma(A) \Longrightarrow \bar{\lambda} \in \sigma(A)$ *(the symmetry of the spectrum with respect to the line* $\operatorname{Im} \lambda = 0$ *);*

$\beta)$ $\lambda \in \sigma_r(A) \Rightarrow \lambda \in \sigma_p(A)$.

This theorem has the following corollaries *in the case* $0 \in \varrho(G)$:

1) $\sigma_p(A) \cup \sigma_r(A)$ *is symmetric with respect to* $\Im m\,\lambda = 0$;
2) $\sigma_r(A)$ *does not contain real points;*
3) $\sigma_c(A)$ *is symmetric with respect to* $\Im m\,\lambda = 0$.

Langer's theorem and its corollaries 1)-3) left open several questions:

a) In the case of a *regular* G -metric $(0 \in \varrho(G))$, do there exist operators $A\,(=A^c)$ with non-empty $\sigma_r(A)$ or with non-real $\sigma_c(A)$ (i.e. do the corollaries 2) and 3) have a real sense)?

b) Is the main assertion of the theorem true when $0 \in \sigma_c(G)$, but the requirement $\overline{\mathcal{D}_A} = \mathfrak{h}$ is replaced by condition (1)?

c) Does the symmetry of the spectrum $\sigma(A)$ (see α)) still hold if $0 \in \sigma_c(G)$ (at least for the case $\overline{\mathcal{D}_A} = \mathfrak{h}$)?

d) The same for the implication β).

e) Which of the corollaries 1)-3) remain valid and which not if $0 \in \sigma_c(G)$?

Now we can answer all these questions.

3. Examples of a G -selfadjoint operator A $(0 \in \varrho(G))$ with $\pm i \in \sigma_c(A)$ and a G -selfadjoint operator A $(0 \in \varrho(G))$ with $\sigma_r(A)$ containing the whole upper half-plane have been constructed by T.J. Azizov (the answer to question a)).

The question b) can be answered positively in a trivial way, but the question c) is answered negatively – and not quite trivially. The matter is that in the general case (i.e. without the requirement $0 \in \varrho(G)$) only the following assertion can be made:

Let $\lambda \in \varrho(A)$ $(A = A^c)$ *; then for* $\lambda \in \varrho(A)$ *it is necessary and sufficient that*

(2) $$\mathcal{R}_{[(A-\lambda I)^{-1}]^* G} \subset \mathcal{R}_G .$$

At the same time, examples can be constructed where the symmetry of the spectrum $\sigma(A)$ falls, i.e. condition (2) fails for some $\lambda \in \varrho(A)$ (even if $\overline{\mathcal{D}_A} = \mathfrak{h}$).

This question is closely connected to the other problem. As it is known, for selfadjoint $(n \times n)$-matrices $A(=A^*)$ and $B(=B^*)$ the spectrum $\sigma(AB)$ is always symmetric with respect to the real line. It was E.A. Larionov who suggested in 1969 (3 rd. Voronež Winter Mathematical School) that the same is true for arbitrary selfadjoint operators in $\mathfrak{H}$. But this suggestion is false. This can be shown by an example (also due T.J. Azizov). By the way, in this example the operator AB is bounded and $\mathcal{D}_{AB} = \mathfrak{H}$ although A and B is unbounded selfadjoint operator.

Question d) is answered negatively, too. As to the question e), there are examples showing that all the corollaries 1)-3) of Langer's theorem fail if $0 \notin \rho(G)$.

4. There are analogues of Langer's theorem and its corollaries for G-unitary operators. The same is true for the analogues of the question a)-e) and their answers. But most of these results cannot be derived from their analogues for G-selfadjoint operators by means of the Cayley transformation, since (as it was mentioned above) examples of G-selfadjoint operators A are known with $\sigma(A)$ containing the whole upper halfplane (or even the whole complex plane). Also G-unitary operators U are known with $\sigma_p(U)$ covering the whole circle $|\lambda| = 1$ (an example is due to the late M.L. Brodskiĭ). In the light of this, it is curious that an example of a G-unitary operator U with $\sigma(U)$ non-symmetric (relative to the circle $|\lambda| = 1$) was constructed by D.I. Derguzov [4] already in 1964 and this operator U admits Cayley transformation.

A more detailed exposition of the above and related results together with other topics will be found in the survey by T.J. Azizov and I.S. Iohvidov [8].

REFERENCES

[1] Ju.P. Ginzburg and I.S. Iohvidov, Investigations in the geometry of infinite-dimensional spaces with a bilinear metric, *Uspehi Mat. Nauk,* 17:4 (106) (1962) 3-56 [Russian].
English translation: *Russian Math. Surveys* (London) 17:4 (1962).

[2] M.G. Kreĭn and G.J. Lubarskiĭ, On the theory of transmission bands of periodic waveguides, *Prikl. Mat. Mech.,* 25 (1961), 29-48 [Russian].
English translation: *J. Appl. Mat. Mech.,* 25 (1961), 29-48.

[3] V.I. Derguzov, Sufficient conditions for the stability of the Hamilton equations with unbounded periodic coefficients, *Mat.* Sb., 64 (106):3 (1964), 419-435 [Russian].

[4] V.I. Derguzov , On the stability of the Hamilton equations with unbounded periodic coefficients, *Mat.Sb.*, 63 (105) :4 (1964), 591-619 [Russian].

[5] I.S. Iohvidov and M.G. Kreĭn , Spectral theory of operators in spaces with an indefinite metric. I, *Trudy Mosk. Mat. Obšč.*, 5 (1956), 367-432 [Russian]. English translation: *Amer. Math. Soc. Transl.*, (2) 13 (1960), 105.

[6] I.S. Iohvidov, On the boundedness of $\mathfrak{J}$-isometric operators, *Uspehi Mat. Nauk*, 16:4 (100) (1961), 167-170 [Russian]. English translation: *Amer. Math. Soc. Transl.*, (2) 47 (1965), 67.

[7] H. Langer, Zur Spektraltheorie $\mathfrak{J}$-selbstadjungierter Operatoren, *Math. Ann.*, 146 (1962), 60-85.

[8] T.J. Azizov and I.S. Iohvidov , Linear operators in a Hilbert space with a G-metric, *Uspehi Mat. Nauk*, 26 (1971) [Russian].

COLLOQUIA MATHEMATICA SOCIETATIS JÁNOS BOLYAI
5. HILBERT SPACE OPERATORS. TIHANY (HUNGARY), 1970

Best approximation on the unitary group

H. JOHNEN

0. Let G be a compact group and P the representative ring of G, i.e., the set of functions f on G which arise by finite linear combinations of the coefficients of irreducible unitary representations of G. Then the famous theorem of Peter – Weyl states:

If $f \in C(G)$, G *a compact group, then to each* $\varepsilon > 0$ *there exists an* $f_\varepsilon \in P$ *such that* $\| f - f_\varepsilon \| = \sup_{p \in G} |f(p) - f_\varepsilon(p)| \leq \varepsilon$.

The aim of this paper is to study the convergence expressed here in more detail for the unitary group.

1. To this end one firstly orders the function set P into a sequence of sets $P_n \subset P_{n+1}$, $n \in N$, P_n being called the set of trigonometric polynomials of degree n, as follows:

If $u \in \mathfrak{U}(m)$, the unitary $m \times m$ group, then we define the unitary representation

$$P_n(u) = (t_{ij}(u)) = \otimes^n (e \oplus u \oplus \bar{u}),$$

where $\otimes^n$ denotes the nth Kronecker power, $\oplus$ the direct sum, and P_n the linear span of the functions $t_{ij}(u)$ (see Ragozin [10]).

It is not difficult to show that $P_n \subset P_{n+1}$, $n \in \mathbb{N}$. For $f \in C(U(m))$ and P_n one defines

$$E(f, P_n) = \inf\{\|f - p_n\|;\ p_n \in P_n\}.$$

By general results on best approximation there exists a $p_n^*(f) \in P_n$ such that $E(f, P_n) = \|f - p_n^*(f)\|$.

2. In the classical theory of best approximation one tries to connect the behaviour of $E(f, P_n)$ with smoothness conditions upon f. The smoothness conditions upon f will be expressed in terms of properties of the moduli of smoothness

(2.1) $$\omega_r(t, f) = \sup\{\|\Delta_v^r f(u)\|;\ \varrho(e,v) \le t\},$$

where Δ_v^r denotes the rth right difference of f with respect to v, given by

(2.2) $$\Delta_v^r f(u) = \sum_{k=0}^{r} (-1)^{r-k} \binom{r}{k} f(uv)^k,$$

$\varrho(e,v)$ is the arc length of the shortest path connecting e and v, corresponding to the Riemannian structure which arises by the bilinear form

(2.3) $$B(X,Y) = -\mathrm{tr}(XY)$$

on $u(m)$, the Lie algebra of $U(m)$. The moduli of smoothness have the properties collected in

Lemma 1. $\omega_r(t,f)$ *is monotonely increasing on* $0 \le t \le \infty$ *such that*

(i) $\lim_{t \to 0} \omega_r(t,f) = 0 \Longleftrightarrow f$ *is continuous of* $U(m)$;

(ii) $\omega_r(t,f) \le 2^k \omega_{r-k}(t,f) \qquad (0 \le k \le r)$;

(iii) $\omega_r(\lambda t, f) \le (1+\lambda)^r \omega_r(t,f) \qquad (\lambda > 0)$;

(iv) *if* $f \in C^k(U(m))$ *and* $X_1, \dots, X_{m^2}$ *is an orthonormal basis of* $u(m)$ *with respect to* B, *then*

$$\omega_r(t,f) \le t^k \sum_{i_1 \dots i_k} \omega_{r-k}(t, X_{i_1} \dots X_{i_k} f) \qquad (r \ge k).$$

Another way to characterize smoothness properties of a function f is by means of the K-functional. If $f \in C^k(U(m))$, a norm is given by

$$\|f\|_k = \|f\| + \sum_{i_1 \dots i_k} \|X_{i_1} \cdots X_{i_k} f\|, \tag{2.4}$$

where $X_1, \dots, X_{m^2}$ is an orthonormal system with respect to B. We define

$$K(t,f,k,k+r) = \inf\{\|f-g\|_k + t\|g\|_{k+r}\}, \tag{2.5}$$

the infimum being extended over all $g \in C^{k+r}(U(m))$. The connection with the moduli of smoothness is given by

Lemma 2. *There exists a constant* c *independent of* f, *such that*

a) $\sum_{i_1 \dots i_k} \omega_r(t, X_{i_1} \dots X_{i_k} f) \le c K(t^r, f, k, k+r)$,

b) $$K(t^r, f, k, k+r) \le c\Big[\max[t^2, t^r]\|f\|_k + \sum_{i_1 \dots i_k} \omega_r(t, X_{i_1} \dots X_{i_k} f) + \sum_{i_1 \dots i_k} t^2 \int_t^1 \frac{\omega_r(s, X_{i_1} \dots X_{i_k} f)}{s^2}\, ds\Big].$$

c) *If in addition*

$$\sum_{i_1 \dots i_k} \int_0^1 \omega_r(s, X_{i_1} \dots X_{i_k} f)\, s^{-(j+1)}\, ds < \infty \qquad (1 \le j \le r-1)$$

then

$$K(t^r, f, k, k+r) \le c\Big[\max[t^{2j+2}, t^r]\|f\|_k + \sum_{i_1 \dots i_k} \omega_r(t, X_{i_1} \dots X_{i_k} f) + \sum_{i_1 \dots i_k} \Big(t^{2j+2} \int_t^1 \frac{\omega_r(s, X_{i_1} \dots X_{i_k} f)}{s^{j+2}}\, ds + t^{2j} \int_0^{t^2} \frac{\omega_r(s, X_{i_1} \dots X_{i_k} f)}{s^{j+1}}\, ds \Big)\Big].$$

For a proof of these extimates see Johnen [9].

3. Now we are going to compare $E(f,P_n)$ with the smoothness conditions upon f. First we prove the Jackson-type

Theorem 1. *Let* $X_1,\dots,X_{m^2}$ *be an orthonormal basis of* $u(m)$ *with respect to* B. *There exists a sequence of operators* $I_{n,k}$ *mapping* $C^k(U(m))$ *into* P_n *such that*

$$E(f,P_n) \le \| I_{n,k}f - f\| \le cn^{-k} \sum_{i_1\dots i_k} \omega_2(n^{-1}, X_{i_1}\dots X_{i_k}f)\,. \tag{3.1}$$

Proof. We begin to prove (3.1) for $k = 0$. To this end one defines the Jackson-kernel (see Gong-Sheng [6])

$$J_n(u) = \lambda_{n'} \left| \frac{\det[e-u^{n'+1}]}{\det[e-u]} \right|^{4m},$$

where n' is the greatest integer less than or equal to $n/4m^2$ and

$$\lambda_{n'} = \int_{U(m)} \left| \frac{\det[e-u^{n'+1}]}{\det[e-u]} \right|^{4m} du$$

(du the normalized Haar measure on $U(m)$).

$J_n(u)$ satisfies the following conditions:

(i) $J_n(u) = J_n(u^{-1}) = J_n(vuv^{-1})$ $\quad (u,v \in U(m))$,

(ii) $\int J_n(u)\,du = 1$,

(iii) $\int J_n(u)\varrho^2(e,u)\,du = \mathcal{O}(n^{-2})$.

Thus if one sets $I_{n,0}f(v) = \int J_n(vu)^{-1} f(u)\,du$, then

$$\|I_{n,0}f - f\| = \sup_v \left| \int J_n(vu^{-1}) f(u)\,du - f(v)\right| =$$

$$= \sup_v \left| \frac{1}{2}\int [J_n(vu^{-1}v^{-1}) f(vu) + J_n(vuv^{-1}) f(vu^{-1}) - 2f(v)\,du \right| \le$$

$$\leq \frac{1}{2} \int J_n(u)\,\omega_2(\rho(e,u),f)\,du\,,$$

the latter inequality following by (i) – (ii). Since $\omega_2(\rho(e,u),f) \leq (1+n(e,u))^2 \cdot$ $\cdot\,\omega_r(1/n,f)$ by Lemma 1 and $I_{n,0}f \in P_n$, the inequality (3.1) follows in case $k=0$ by (iii).

If $k \geq 1$, then one may set $-I_{n,k} = [I - I_{n,0}]^{k+1} - I$ and it follows by Lemma 1 and the results for $k = 0$ that

$$\| [I - I_{n,0}]^{k+1} f \| \leq c n^{-1} \sum_{i_1} \| [I - I_{n,0}]^k X_{i_1} f \|\,.$$

Hence (3.1) follows by induction.

$I_{n,0}$ corresponds to the usual Jackson integral for higher order differences in $U(1)$. Preliminary results here using only orders of the first difference are due to Gong-Sheng [6]. D.L. Ragozin [10] treats the case of first differences of derivatives in his thesis using other methods.

Corollary. *If the modulus of smoothness satisfies*

$$\int_0^1 \frac{\omega_r(s,f)}{s^{j+1}}\,ds < \infty \qquad (1 \leq j \leq r-1)\,,$$

then there exists a constant c *such that*

(3.2) $$E(f,P_n) \leq c\big[\max[n^{-(2j+2)}, n^{-r}] + \omega_r(1/n,f) +$$

$$+\, n^{-(2j+2)} \int_{n^{-1}}^{1} \frac{\omega_r(s,f)}{s^{j+2}}\,ds + n^{-2j} \int_0^{n^{-2}} \frac{\omega_r(s,f)}{s^{j+1}}\,ds\big]\,.$$

Proof. By (3.1) and the linearity of $I_{n,r}$ one has for $g \in C^r(U(m))$

$$\| I_{n,r} f - f \| \leq \| (I_{n,r} - I(f-g) \| + \| I_{n,r} g - g \| \leq c(\|f-g\| + n^{-r} \|g\|_r)$$

and hence $E(f,P_n) \leq \|I_{n,r}f - f\| \leq cK(n^{-r}, f, 0, r)$. Thus (3.2) follows by Lemma 2.

4. The converse theorem to Theorem 1 is proved by the Bernstein-type inequality

(4.1) $$\| X p_n \| \leq n \| p_n \| \qquad (p_n \in P_n)\,,$$

where $X \in u(m)$ satisfies $B(X,X) = 1$. To prove this inequality note that X is a skew hermitian matrix, whence

$$B(X,X) = \sum_{i=1}^{m} \vartheta_i^2 ,$$

where $i\vartheta_1, \ldots, i\vartheta_m$ are the characteristic roots of X . Since

$$X = V \begin{pmatrix} i\vartheta_1 & & 0 \\ & \ddots & \\ 0 & & i\vartheta_m \end{pmatrix} V^{-1}$$

for some unitary matrix V, $p_n(u \exp tX)$ is equal to

$$p_n\left(uV \begin{pmatrix} e^{it\vartheta_1} & & 0 \\ & \ddots & \\ 0 & & e^{it\vartheta_m} \end{pmatrix} V^{-1}\right) .$$

Hence one finds easily that it is an entire function of exponential type in t not exceeding $n \cdot \max \cdot |\vartheta_i| \leq n[B(X,X)]^{1/2} = n$, and (4.1) follows by the Bernstein inequality for entire functions of this type (see Ragozin [10]).

Thus one can prove the Bernstein-type

Theorem 2. *If* f *belongs to* $C(U(m))$, *such that*

(4.2) $$E(f, P_n) = \mathcal{O}(\omega(1/n))$$

where $\omega(1/n)$ *is nondecreasing and satisfies the condition*

$$t^k \int_t^1 \frac{\omega(s)}{s^{k+1}} ds = \mathcal{O}(\omega(t)) \qquad (t \to 0),$$

then

(4.3) $$\omega_r(t,f) = \mathcal{O}(\omega(t)) \qquad (r \geq k) .$$

Proof. For $0 < t \leq 1$ choose $k \in \mathbf{N}$ such that $2^{k+1} < t < 2^k$. Then by the definition of the K-functional

$$K(t^r, f, 0, r) \leq \| f - p^*_{2^k}(f) \| + t^r \| p^*_{2^k}(f) \|_r$$

and (4.3) follows by the usual method (see [4]) of proof, estimating $\| p^*_{2k}(f)\|_r$ by the inequality (4.1), and Lemma 2.

Using the inequalities (3.1) and (4.1) one shows that for a nondecreasing positive function $\omega(t)$ satisfying the conditions

$$\int_0^t \frac{\omega(s)}{s}\,ds \le c\omega(t), \qquad t^2\int_t^1 \frac{\omega(s)}{s^3}\,ds \le c\omega(t) \tag{4.4}$$

the following theorem also holds:

Theorem 3. *The following assertions are equivalent:*

a) $E(f,P_n) = O(n^{-k}\omega(1/n))$,

b) $f\in C^k(U(m))$ *and* $\|f-p^*_n(f)\|_k = O(\omega(1/n))$,

c) $\|p^*_n(f)\|_k = O(n^{k'-k}\omega(1/n)) \qquad (k'>k+1)$,

d) $f\in C^k(U(m))$ *and* $\sum_{i_1\ldots i_k} \omega_2(t,X_{i_1}\ldots X_{i_2}f) = O(\omega(t))$

where $X_1,\ldots,X_{m^2}$ *is an orthonormal basis of* $u(m)$ *and the sum is extended over all* k *-tuples* $(i_1,\ldots,i_k)$, $1\le i_1,\ldots,i_k\le m^2$.

A proof of this theorem may be found in [3] or [9].

Remark. If one strengthens the second postulate of (4.4) to $t\int_t^1 \frac{\omega(s)}{s^2}\,ds \le c\omega(t)$, then the second modulus of smoothness in d) can be replaced by the first one and in c) the natural number k may be chosen so that $k'>k$.

5. Finally some remark concerning possible generalizations. If G is a closed subgroup of $U(m)$, then we may define T'_n and P'_n as the restriction of T'_n and P'_n to G. In the case when G is a rotation group $O^+(m)$, a special unitary group $SU(m)$ or a symplectic group $SpU(2m)$, one may use the same kernel function as in Theorem 1 but suitably normalized and restricted. In general, every function f defined on a closed subgroup G of $U(m)$ may be extended to a function $f^\sim$ on $U(m)$ such that (see [2])

$$K_{U(m)}(t^r,f^\sim,k,k+r) \le c_{r,k}K_G(t^r,f,k,k+r).$$

Since Lemma 2 holds on every closed subgroup G of $U(m)$, we also obtain in this case a version of Jackson's theorem that includes higher moduli of smoothness. Also Theorems 2 and 3 are valid. But every compact Lie group may be imbedded isomorphically in some $U(m)$. Hence Theorems 1-3 also give information on the degree of approximation by trigonometric polynomials in this case.

6. The author wants to express his thanks to P.L. Butzer and K. Scherer for many discussions and helpful suggestions, and to the Deutsche Forschungsgemeinschaft which kindly supported this research.

REFERENCES

[1] P.L. Butzer, and H. Berens, *Semi-Groups of Operators and Approximation* (Berlin 1967).

[2] P.L. Butzer, and H. Johnen, Lipschitz spaces on compact manifolds, *J. Functional Analysis* (in print).

[3] P.L. Butzer, and K. Scherer, Über die Fundamentalsätze der klassischen Approximationstheorie in abstracten Räumen. Abstract Spaces and Approximation, *Proceedings of the Oberwolfach Conference,* 1968. ISNM 10 (1969), 113-125.

[4] P.L. Butzer, and K. Scherer, Jackson and Bernstein type inequalities for families of commutative operators in Banach spaces, *J. Approx. Theory* (in print).

[5] Ch. Chevalley, *Theory of Lie Groups.* I (Princeton 1946).

[6] Gong-Sheng, Fourier Analysis on the unitary group. IV. On the Peter – Weyl theorem, *Chinese Math. Acta, 4* (1963), 351-359.

[7] S. Helgason, *Differential Geometry and Symmetric Spaces* (New York, 1962).

[8] H. Johnen, Klassen stetiger Funktionen und Approximation auf kompakten Mannigfaltigkeiten, *Forschungsber. des Landes Nordrhein-Westfalen* 2078, 1970.

[9] H. Johnen, Stetigkeitsmoduli und Approximationstheorie auf kompakten Lie-Gruppen (Dissertation, Aachen, 1970) (unpublisched).

[10] D.L. Ragozin, Approximation theory on compact manifolds and Lie groups with applications to harmonic analysis (Ph.D. Thesis, Harvard Univ. Cambridge, Mass., 1967).

[11] D.L. Ragozin, Approximation theory on $SU(2)$, *J. Approx. Theory,* 1 (1968), 464-475.

[12] A.F. Timan, *Theory of Approximation of Functions of a Real Variable* (London, 1963).

[13] H. Weyl, *The Classical Groups* (Princeton, 1946).

COLLOQUIA MATHEMATICA SOCIETATIS JÁNOS BOLYAI
5. HILBERT SPACE OPERATORS, TIHANY (HUNGARY), 1970

On the boundedness of certain singular integral operators

R. K. JUBERG

Consider functions from the interval $[0,1]$ into the reals. The operations J^β (Riemann – Liouville fractional integral) and $J^{*\beta}$ are defined by

$$J^\beta \varphi(t) = \Gamma(\beta)^{-1} \int_0^t (t-s)^{\beta-1} \varphi(s)\,ds$$

and

$$J^{*\beta} \varphi(t) = \Gamma(\beta)^{-1} \int_t^1 (s-t)^{\beta-1} \varphi(s)\,ds .$$

For $0 < \beta < 1$ their inverses are defined by

* Partially supported during the course of this work under National Science Foundation Grant GP–13288.

$$J^{-\beta}\varphi(t) = \frac{d}{dt} J^{1-\beta}\varphi(t)$$

and

$$J^{*-\beta}\varphi(t) = -\frac{d}{dt} J^{*1-\beta}\varphi(t) .$$

As is suggested by the notation these operations when considered as defining operators in $L^p(=L^p(0,1))$ spaces are adjoints.

The question we consider here is: *for* $0 < \alpha < 1$ *does the operation* $\varphi \to J^{\alpha}J^{*-\alpha}\varphi$ *define a bounded operator in* L^p, $1 \leqq p \leqq \infty$ *?*

This question was encountered by Kalisch [2] in the course of his study of similarity and/or equivalence relations in a class of operators on L^p. The author came upon it in a very different context; an attempt to use fractional order derivatives in the definition of function spaces in which to consider evolutionary type problems.

The result is presented in Section 1. A quite detailed proof of the lemma, of which the result is a consequence, is given in Section 2. In Section 3 we show how the lemma can be obtained using an identity in [4]. A mild generalization is presented in the Appendix together with a discussion of the analogue using the Weyl formulation of fractional order integration.

1. The Result. The answer to the question is given in the following

Theorem. *The operation* $\varphi \to J^{\alpha}J^{*\alpha}\varphi$ $(0<\alpha<1)$ *defines a bounded operator in* L^p *if and only if* $(1-\alpha)^{-1} < p < \infty$.

Denote by R the operation $R\varphi(t) = \varphi(1-t)$. One observes readily the properties: $RJ^{\pm\beta} = J^{*\pm\beta}R$, $DR = -RD$ $(D = d/dt)$, and $R^2 =$ identity. Then restricting to "smooth" functions, note that

$$J^{*\alpha}J^{-\alpha} = J^{*\alpha}R^2J^{-\alpha} = RJ^{\alpha}J^{*-\alpha}R .$$

This leads to the following

Corollary 1. *If* $J^{\alpha}J^{*-\alpha}$ *defines a bounded operator in* L^p *, then the operator is invertible.*

Let

$$<f,g> = \int_0^1 f(t)g(t)dt ,$$

the usual Banach space pairing of L^p and $L^{p*}(1/p + 1/p^* = 1)$. Then for smooth φ and ψ an elementary calculation leads to

$$< J^\alpha J^{*-\alpha}\varphi, \psi > = J^{*1-\alpha}\varphi(0)\cdot J^{*\alpha}\psi(0) - <\varphi, J^{1-\alpha}J^{*\alpha-1}\psi>$$

and, therefore,

$$(J^\alpha J^{*-\alpha})^*\psi(t) = t^{-\alpha}\Gamma(1-\alpha)^{-1}\Gamma(\alpha)^{-1}\int_0^1 s^{\alpha-1}\psi(s)ds - J^{1-\alpha}J^{*\alpha-1}\psi(t).$$

This combined with the fact that for smooth f

$$J^{-\alpha}f(t) = \Gamma(1-\alpha)^{-1}t^{-\alpha}f(0) + J^{1-\alpha}f'(t)$$

shows that

$$(J^\alpha J^{*-\alpha})^* = J^{-\alpha}J^{*\alpha} ;$$

which establishes the following

Corollary 2. *The operation* $\varphi \to J^{-\alpha}J^{*\alpha}$ $(0 < \alpha < 1)$ *defines a bounded operator in* L^p *if and only if* $1 < p < \alpha^{-1}$.

The theorem follows mainly by using the representation provided by the identity.

Lemma. *For sufficiently smooth* φ *, say* $\varphi \in C^1[0,1]$, $0 < \alpha < 1$

(1) $$J^\alpha J^{*-\alpha}\varphi(t) = (\cos\pi\alpha)\varphi(t) + \pi^{-1}(\sin\pi\alpha)t^\alpha(\text{p.v})\int_0^1 \varphi(s)s^{-\alpha}(t-s)^{-1}ds$$

the integral being the Cauchy principal value (p.v.).

Remark. (1) is valid for $\alpha = 0$, giving the identity mapping; hence bounded in L^P, $1 \leqq p \leqq \infty$. At the other extreme we take the limit on the right in (1) as $\alpha \uparrow 1$ giving $-\varphi(t) + \varphi(0)$; the associated operator being unbounded for all p . Thus the operation $\varphi \to J^\alpha J^{*-\alpha}\varphi$, $0 \leqq \alpha \leqq 1$, defines a bounded operator in L^p, $1 \leqq p \leqq \infty$ if and only if (i) $0 < \alpha < 1$, $(1-\alpha)^{-1} < p < \infty$, or (ii) $\alpha = 0$, $1 \leqq p \leqq \infty$.

Proof of theorem. We choose a smooth function φ and then commencing from (1) we rearrange it into the form

$$
\begin{aligned}
J^{\alpha}J^{*-\alpha}\varphi(t) &= (\cos\pi\alpha)\varphi(t) + \frac{\sin\pi\alpha}{\pi}t^{\alpha-1}\int_0^t \varphi(s)s^{-\alpha}ds + \\
&+ \frac{\sin\pi\alpha}{\pi}\left\{\int_0^t \varphi(s)\frac{(s/t)^{1-\alpha}-1}{t-s}ds + \int_t^1 \varphi(s)\frac{(t/s)^{\alpha}-1}{t-s}ds + (\text{p.v.})\int_0^1 \frac{\varphi(s)}{t-s}ds\right\}.
\end{aligned} \tag{2}
$$

That the first integral defines a bounded operator in L^p for $(1-\alpha)^{-1} < p \leqq \infty$ follows from an inequality in [1], p. 245.

The first two integrals inside the brackets are bounded, respectively, by

$$t^{-1}\int_0^t |\varphi(s)|\,ds$$

and

$$\int_0^t s^{-1}|\varphi(s)|\,ds .$$

That the p-norm of the former is dominated by the p-norm of φ for $1 < p \leqq \infty$ is just Hardy's inequality ([1], p. 240). Then the bounding of the p-norm of the latter in terms of the p-norm of φ for $1 \leqq p < \infty$ follows by duality.

Finally, the last term in brackets is just the finite Hilbert transform providing a bounded operator in L^p for $1 < p < \infty$.

In summary then, for a smooth function φ,

$$\|J^{\alpha}J^{*-\alpha}\varphi\|_p \leqq (\text{Const.})\|\varphi\|_p, \qquad (1-\alpha)^{-1} < p < \infty .$$

The necessity can be obtained in various ways. Indeed, Kalisch has observed the lower limitation. This follows also from the Riesz theorem on the representation of linear functionals on L^p. It is also a consequence of the sharpness of the inequalities used above. The necessity of the limitations follows as well from an examination of the examples: (I) $\varphi(t) = t^{\alpha-1}(\log t^{-1})^{\alpha-2}$, $0 < t \leqq e^{-1}$, $\varphi(t) = 0$, $t > e^{-1}$;

and (II) $\varphi(t) \equiv 1$.

2. Proof of the Lemma. Choose a smooth function φ, say $\varphi \in C_0^\infty(0,1)$. Then

(3) $$\Gamma(\alpha)\Gamma(1-\alpha)J^\alpha J^{*-\alpha}\varphi(t) = \int_0^t (t-x)^{\alpha-1}\Big(-\frac{d}{dx}\int_x^1 (y-x)^{-\alpha}\varphi(y)dy\Big)dx.$$

For t fixed, $0 < t < 1$, define the function $F : \mathbf{C}\to\mathbf{C}$, $\mathbf{C}$ the field of complex numbers, by

(4) $$F(z) = \int_0^t (t-x)^{z+\alpha-1}\Big(-\frac{d}{dx}\int_x^1 (y-x)^{-\alpha}\varphi(y)dy\Big)dx.$$

Clearly F is analytic in $\mathrm{Re}\, z > -\alpha$.

Now restrict z in (4) to have sufficiently large real part, at least $\mathrm{Re}\, z > 1-\alpha$. Then an integration by parts leads to

(5) $$F(z) = t^{z+\alpha-1}\int_0^1 y^{-\alpha}\varphi(y)\,dy - (z+\alpha-1)\int_0^1 (t-x)^{z+\alpha-2}\Big(\int_x^1 (y-x)^{-\alpha}\varphi(y)dy\Big)dx.$$

According to Fubini's theorem, the iterated integral in (5) can be replaced by

(6) $$\int_0^t \varphi(y)\Big(\int_0^y (y-x)^{-\alpha}(t-x)^{z+\alpha-2}dx\Big)dy + \int_t^1 \varphi(y)\Big(\int_0^t (y-x)^{-\alpha}(t-x)^{z-\alpha-2}dx\Big)dy.$$

Now a partial integrating of the inner integral in the second term in (6) (differentiating the factor $(y-x)^{-\alpha}$) and substituting into (5) yields

(7) $$F(z) = t^{z+\alpha-1}\int_0^1 y^{-\alpha}\varphi(y)dy + \int_0^t \varphi(y)\Big((1-\alpha-z)\int_0^y (y-x)^{-\alpha}(t-x)^{z+\alpha-2}dx\Big)dy - \int_t^1 \varphi(y)\Big(\alpha\int_0^1 (y-x)^{-\alpha-1}(t-x)^{z+\alpha-1}dx\Big)dy.$$

Next adding and subtracting the term $|t-y|^{z-1}$ to the inner integrals in (7) gives the expression

$$F(z) = t^{z+\alpha-1}\int_0^t y^{-\alpha}\varphi(y)\,dy + \int_0^t \varphi(y)H(y,t;z)\,dy - \int_t^1 \varphi(y)K(y,t;z)\,dy$$

(8)

$$+\int_t^1 \varphi(y)|t-y|^{z-1}\,\mathrm{sgn}\,(t-y)\,dy$$

where

(9) $$H(y,t;z) = (1-\alpha-z)\int_0^y (y-x)^{-\alpha}(t-x)^{z+\alpha-2}dx - (t-y)^{z-1},$$

(10) $$K(y,t;z) = \alpha\int_0^t (y-x)^{-\alpha-1}(t-x)^{z+\alpha-1}dx - (y-t)^{z-1},$$

and $\mathrm{sgn}(t-y) = 1,\ t > y;\ = 0,\ t = y;\ = -1,\ t < y$.

On examining (9) and (10) one observes readily that H and K are well-defined for $\mathrm{Re}\, z > 0$. Hence the right side of (8) defines a function analytic in $\mathrm{Re}\, z > 0$, and by the uniqueness property of analityc functions (8) is valid in $\mathrm{Re}\, z > 0$.

Now suppose $\mathrm{Re}\, z > 0$ and rearrange (8) into the form

$$F(z) = t^{z+\alpha-1}\int_0^t y^{-\alpha}\varphi(y)\,dy + \int_0^1 \varphi(y)|t-y|^{z-1}\,\mathrm{sgn}\,(t-y)\,dy +$$

(11) $$+\int_0^t [\varphi(y)-\varphi(t)]\,H(y,t;z)\,dy - \int_t^1 [\varphi(y)-\varphi(t)]\,K(y,t;z)\,dy +$$

$$+\varphi(t)\Big[\int_0^t H(y,t;z)\,dy - \int_t^1 K(y,t;z)\,dy\Big].$$

The change of variables $y-x = (t-y)s^{-1}$ in (9) gives

$$H(y,t;z) = (t-y)^{z-1}\Big[(1-\alpha-z)\int_{(t-y)/y}^{\infty} s^{-z}(1+s)^{z+\alpha-2}\,ds - 1\Big]. \tag{9'}$$

From (9') it is clear that

$$H(y,t;z) \longrightarrow \frac{(y/t)^{1-\alpha}-1}{t-y} \quad \text{as } z\to 0,\ \operatorname{Re} z>0. \tag{12}$$

Similarly, the substitution $t-x = (y-t)s^{-1}$ yields

$$K(y,t;z) = (y-t)^{z-1}\Big[\alpha\int_{(y-t)/t}^{\infty} s^{-z}(1+s)^{-\alpha-1}\,ds - 1\Big], \tag{10'}$$

and, as above, from (10')

$$K(y,t;z) \longrightarrow \frac{(t/y)^{\alpha}-1}{y-t} \quad \text{as } z\to 0,\ \operatorname{Re} z>0. \tag{13}$$

Then the integrand of the third term in (11) arrayed in the form

$$\left(\frac{\varphi(y)-\varphi(t)}{t-y}\right)(t-y)H(y,t;z)$$

and using (9') is found to be dominated by

$$(\text{Const.})\,|(t-y)^{z}|\left[1+\left(\frac{t}{t-y}\right)^{\operatorname{Re} z}\right]\max|\varphi'|$$

or

$$(\text{Const.})\max|\varphi'|. \tag{14}$$

Thus by the Dominated Convergence Theorem, the limit of the third term in (11) exist as $z\to 0$, $\operatorname{Re} z>0$, and it is computed with the aid of (12).

Following exactly the same procedure and producing the same form of domination (14), the limit of the fourth is shown to exist and it is computed using (13).

Through straightforward computations and use of the Fubini theorem, one finds that

$$\int_0^t H(y,t;z)\,dy - \int_t^1 K(y,t;z)\,dy = -\frac{t^z}{t-\alpha} - \frac{t^z-(1-t)^z}{z} + \tag{15}$$
$$+\int_0^t (t-x)^{z+\alpha-1}(1-x)^{-\alpha}\,dx\,.$$

That the second term in (11) produces in the limit the finite Hilbert transform (up to a constant factor) becomes apparent on writing

$$\int_0^1 \varphi(y)|t-y|^{z-1}\operatorname{sgn}(t-y)\,dy = \int_0^1 [\varphi(y)-\varphi(t)]\,|t-y|^{z-1}\operatorname{sgn}(t-y)\,dy +$$
$$+\varphi(t)\int_0^1 |t-y|^{z-1}\operatorname{sgn}(t-y)\,dy\,;$$

the last term being computed easily to $z^{-1}[t^z-(1-t)^z]$. Hence

$$\int_0^1 \varphi(y)|t-y|^{z-1}\operatorname{sgn}(t-y)\,dy \to -\int_0^1 [\varphi(y)-\varphi(t)](y-t)^{-1}dy + \varphi(t)\log(t/1-t) =$$
$$= -(\text{p.v.})\int_0^1 [\varphi(y)-\varphi(t)](y-t)^{-1}dy + \varphi(t)\log(t/1-t) = \tag{16}$$
$$= (\text{p.v.})\int_0^1 \varphi(y)(t-y)^{-1}dy + \varphi(t)(\text{p.v.})\int_0^1 (y-t)^{-1}dy + \varphi(t)\log(t/1-t) =$$
$$= (\text{p.v.})\int_0^1 \varphi(y)(t-y)^{-1}dy\,.$$

Thus on passing to the limit in (11) as $z \to 0$, $\mathrm{Re}\, z > 0$ we find using (12), (13), (15), and (16) that

$$F(0) = t^{\alpha-1} \int_0^t y^{-\alpha}\varphi(y)dy + (\mathrm{p.v.})\int_0^1 \varphi(y)(t-y)^{-1}dy + \int_0^t \varphi(y)[(y/t)^{1-\alpha}-1](t-y)^{-1}dy -$$

$$(17) \qquad - \int_t^1 \varphi(y)[(t/y)^{\alpha}-1](y-t)^{-1}dy + \varphi(t)\Big[(\alpha-1)^{-1} - \log(t/1-t) + \int_0^t (t-x)^{\alpha-1}(t-x)^{-\alpha}dx -$$

$$- \int_0^1 [(y/t)^{1-\alpha}-1](t-y)^{-1}dy + \int_t^1 [(t/y)^{\alpha}-1](y-t)^{-1}dy\Big].$$

The first four terms in (17) combine to produce, with the exception of the constant factor, the second term in (1). Denote by $g(t)$ the lengthy expression in square brackets in (17). It can be seen easily that $t \to g(t)$ is a differentiable function and on computing that $g'(t) \equiv 0$. Hence $g(t) \equiv$ const. Taking $t = 1/2$ and after considerable computation, though direct and simple, one finds that

$$g(t) \equiv \text{const.} = \sum_{k \geq 1} \left(\frac{1}{k+\alpha-1} - \frac{1}{k-\alpha+1} \right) - \frac{1}{1-\alpha} .$$

Finally, from the formula (see [3], p.310)

$$\sum_{k \geq 1} \left(\frac{1}{k-x} - \frac{1}{k+x} \right) = \frac{1}{x} - \pi \cot \pi x, \quad 0 < x < 1,$$

$$(18) \qquad g(t) = \pi \cot \pi\alpha .$$

Substituting from expression (18) for the square bracket in (17) and using the identity

$$\Gamma(\alpha)\Gamma(1-\alpha) = \pi \csc \pi\alpha$$

(1) follows on composing (2), (3), (4), and (17).

3. Alternate proof of the Lemma. After having discovered the indentity (1) (which unfolded as displayed in (17)) we found an identity, presented without proof, in a note by Samko [4]. It is possible to derive (1) commencing with this identity, but it requires some care to avoid a subtle error.

The identity in [4] in the form that we need and presented in our notation is

(19) $$J^{*1-\alpha}\varphi(t) = -(\cos\pi\alpha)J^{1-\alpha}\varphi(t) - (\sin\pi\alpha)J^{1-\alpha}M^{\alpha-1}SM^{1-\alpha}\varphi(t).$$

where φ is a smooth function on $[0,1]$, M is the operator defined by $(Mg)(t) = tg(t)$, and S is the finite Hilbert transform.

It is true that for φ smooth each term in (19) is smooth; in particular for $\varphi \in C^\infty$ each term is in C^∞. Hence applying $-d/dt$ to (19) we are lead to

(A) $$J^{*-\alpha}\varphi(t) = (\cos\pi\alpha)J^{-\alpha}\varphi(t) + (\sin\pi\alpha)J^{-\alpha}M^{\alpha-1}SM^{1-\alpha}\varphi(t) .$$

On applying J^α to (A) we find that

(B) $$J^\alpha J^{*-\alpha}\varphi(t) = (\cos\pi\alpha)\varphi(t) + (\sin\pi\alpha)M^{\alpha-1}SM^{1-\alpha}\varphi(t),$$

or

(B') $$J^\alpha J^{*-\alpha}\varphi(t) = (\cos\pi\alpha)\varphi(t) + \pi^{-1}(\sin\pi\alpha)t^{\alpha-1}(\text{p.v.})\int_0^1 \varphi(s)s^{1-\alpha}(t-s)^{-1}ds .$$

One deduces from (B') that it generates a bounded operator in L^p if and only if $1 < p < (1-\alpha)^{-1}$. This stands in stark contradiction to the theorem. The error appears in (A): The second term on the right is not the result of applying $-d/dt$ to the second term on the right in (19).

A correct derivation of (1) starting with (19) procedes as follows. Rearrange (19), in particular the second term, to read

(19') $$J^{*1-\alpha}\varphi(t) = -(\cos\pi\alpha)J^{1-\alpha}\varphi(t) - (\sin\pi\alpha)J^{1-\alpha}\left[M^\alpha SM^{-\alpha}\varphi - \pi^{-1}M^{\alpha-1}\int_0^1 s^{-\alpha}\varphi(s)ds\right](t).$$

Thus

$$J^{*1-\alpha}\varphi(t) = -(\cos\pi\alpha)\,J^{1-\alpha}\varphi(t) -$$

(19'')
$$-(\sin\pi\alpha)J^{1-\alpha}M^{\alpha}SM^{-\alpha}\varphi(t) + \Gamma(1-\alpha)^{-1}\int_0^1 s^{-\alpha}\varphi(s)ds.$$

Now an application of $-d/dt$ to (19'') produces

(20)
$$J^{*-\alpha}\varphi(t) = (\cos\pi\alpha)J^{-\alpha}\varphi(t) + (\sin\pi\alpha)J^{-\alpha}M^{\alpha}SM^{-\alpha}\varphi(t).$$

Then (1) follows from (20) on applying J^{α}.

APPENDIX

1. A mild generalization. Suppose $g \in C^1[0,1]$ and is monotonically increasing; $g'(t) \not\equiv 0$. Consider operations K^{β} and $K^{*\beta}$ defined by

$$K^{\beta}f(t) = \Gamma(\beta)^{-1}\int_{g^{-1}(0)}^{t} [g(t)-g(s)]^{\beta-1} f(s)g'(s)ds$$

and

$$K^{*\beta}f(t) = \Gamma(\beta)^{-1}\int_{t}^{g^{-1}(1)} [g(s)-g(t)]^{\beta-1} f(s)g'(s)ds .$$

The inverses of these operations are given, respectively, by

$$K^{-\beta}f(t) = \frac{1}{g'(t)}\frac{d}{dt}K^{1-\beta}f(t)$$

and

$$K^{*-\beta}f(t) = \frac{1}{g'(t)}\left(-\frac{d}{dt}\right)K^{*1-\beta}f(t).$$

As before these operations when considered as defining operators in the spaces $L^p[(g^{-1}(0), g^{-1}(1)); g'(t)dt]$ are adjoints.

For the operation $K^{\alpha}K^{*-\alpha}$ $(0<\alpha<1)$ one finds that

(i) $$K^{\alpha}K^{*-\alpha}(\ ) = g \circ J^{\alpha}J^{*-\alpha}g^{-1} \circ (\ )$$

where

$$F \circ G(t) = G(F(t)).$$

Now it follows from (i) for a smooth f that

(ii) $$\int_{g^{-1}(0)}^{g^{-1}(1)} |K^{\alpha}K^{*-\alpha}f(t)|^p g'(t)\,dt = \int_0^1 |(J^{\alpha}J^{*-\alpha}g^{-1} \circ f(s)|^p ds.$$

Hence, from (i) and (ii), the operation $f \to K^{\alpha}K^{*-\alpha}f$ $(0<\alpha<1)$ defines a bounded operator in $L^p[(g^{-1}(0), g^{-1}(1)); g'(t)dt]$ if and only if $(1-\alpha)^{-1} < p < \infty$.

2. The Weyl Analogue. We now consider periodic functions of period one and with integral over a period equal to zero. Denote by ϑ^{γ} the Weyl formulation of the operation on these functions that corresponds to J, see [5], Chapter XII, Section 8. The operations ϑ^{α} and $\vartheta^{*\alpha}$, . $(0<\alpha<1)$, have the representations

$$\vartheta^{\alpha}\varphi(t) = \int_0^1 \varphi(s)\Psi_{\alpha}(t-s)\,ds$$

and

$$\vartheta^{*\alpha}\varphi(t) = \int_0^1 \varphi(s)\Psi_{\alpha}(s-t)\,ds,$$

where Ψ_{α} is periodic of period one and

$$\Gamma(\alpha)\Psi_{\alpha}(x) = \lim_{n\to\infty}\left\{\sum_{k=0}^{n}(x+k)^{\alpha-1} - \frac{n^{\alpha}}{\alpha}\right\}$$

for $0<x<1$. The inverses of these operations are defined by

$$\vartheta^{-\alpha}\varphi(t) = \frac{d}{dt}\vartheta^{1-\alpha}\varphi(t)$$

and

$$\vartheta^{*-\alpha}\varphi(t) = -\frac{d}{dt}\vartheta^{*1-\alpha}\varphi(t).$$

Let φ be smooth and denote its Fourier coefficients by

$$\hat{\varphi}(n) = c_n, \quad n = \pm 1, \pm 2, \ldots .$$

Now (see [5], Chapter XII, Section 8)

$$\hat{\Psi}_\alpha(n) = |2\pi n|^{-\alpha} \exp(-i2^{-1}\pi\alpha \operatorname{sgn} n), \quad n \neq 0; \; \hat{\Psi}(0) = 0 .$$

Thus

$$(\vartheta^\alpha \vartheta^{*-\alpha})^\wedge(n) = c_n \exp(-i\pi\alpha \operatorname{sgn} n), \quad n \neq 0 .$$

From this, utilizing the theory of conjugate series and functions (see [0], Section 12.8; or [5], Chapter VII), it follows that the operation $\varphi \to \vartheta^\alpha \vartheta^{*-\alpha}$ $(0 < \alpha < 1)$ decomposes into a multiple of the identity plus a multiple of the operation on periodic functions that is analogous to the Hilbert transform. According to a theorem of M. Riesz (see above references), the associated operator is bounded in L^p for $1 < p < \infty$.

Note. The identity (1) (Lemma, p. 4) presents an interesting relation involving the finite Hilbert transform. That is, letting H denote this transform

$$Hf(t) = \pi^{-1}(\text{p.v.}) \int_0^1 f(s)(t-s)^{-1} ds ,$$

the identity (1) provides the relation

$$(\cos \pi\alpha) I + (\sin \pi\alpha) H = M^{-\alpha} J^\alpha J^{*-\alpha} M^\alpha ;$$

M denotes the operation defined by $Mf(t) = tf(t)$.

This relation, when looked at with a focus of attention on the one parameter groups of operators $\{M^\alpha\}$ and $\{J^\alpha\}$, gives the boundedness in L^p of a sort of commutator. This is similar to a result for certain related groups on the whole line proved and used by R. Kunze and E. Stein (see *Amer. J. Math.*, 82 (1960), 1-62; *ibid.*, 83 (1961), 723-786; *ibid.*, 89 (1967), 385-442; *Bull. Amer. Math. Soc.*, 67 (1961), 593-596). In this case the groups are dual in the sense of the Fourier transform. Here there is a similarity: functional integration/differentiation in one case and multiplication by a power of the variable in the other.

REFERENCES

[0] R.E. Edwards, *Fourier Series* (New York, 1967).

[1] G.H. Hardy, J.E. Littlewood, and G. Pólya, *Inequalities* (Cambridge, 1959).

[2] G.K. Kalisch, On the similarity of certain operators (following this paper).

[3] S. Saks and A. Zygmund, *Analytic Functions* (Warsaw, 1965).

[4] S.G. Samko, Solution of generalized Abel equation by means of an equation with Cauchy kernel, *Dokl. Akad. Nauk SSSR,* 176 (1967), 1019-1022; English transl., *Soviet Math. Dokl.,* 8 (1967), 1259-1262.

[5] A. Zygmund, *Trigonometric Series* (Cambridge, 1968).

COLLOQUIA MATHEMATICA SOCIETATIS JÁNOS BOLYAI
5. HILBERT SPACE OPERATORS, TIHANY (HUNGARY), 1970

On the peripheral spectrum of an element in a strict closed semi-algebra

M. A. KAASHOEK

1. INTRODUCTION

In this paper B denotes a complex Banach algebra with identity e . A non-empty subset A of B is called a *semi-algebra* if a, b in A and $\alpha \geq 0$ imply that $a+b$, ab and αa are in A . A semi-algebra A is said to be *closed* if A is a closed subset of B ; A is said to be *strict* if

$$A \cap (-A) = \{0\},$$

i.e., if a and $-a$ in A implies $a = 0$.

Suppose that t is a non-zero element in B such that the peripheral spectrum of t is a set of poles of t . The main results of the present paper show that the smallest closed semi-algebra $A(t)$ generated by such an element t is strict if and only if

(i) the spectral radius $r(t)$ of t belongs to the spectrum of t,

(ii) the pole $r(t)$ is of maximal order in the peripheral spectrum.

The proof of the necessity of these conditions is given in section 2; it goes in two steps. First of all we prove (i); this part uses standard spectral theory only. For compact linear operators (i) has been proved by Bonsall in [2] under a slightly different assumption. Our proof differs considerably of that of [2]. Having proved (i) we may conclude from [4] that $A(t)$ is locally compact, i.e.,

$$\{x \in A(t) : \|x\| \leq 1\}$$

is a compact subset of B . To prove (ii) we observe that a strict locally compact semi-algebra is a normal cone. This implies that we can use Schaefer's vector valued version of Pringsheim's theorem to get the desired results.

The results of section 2 include a semi-algebra proof of the Krein-Rutman theorem on compact positive linear operators ([6], Theorem 6.1) and Maibaum's theorem on positive linear operators with a peripheral spectrum consisting of poles ([7], Satz 1 (1.3)).

The sufficiency of the conditions (i) and (ii) is proved in section 3. The result of this section is related to Satz 3.1 in L. Elsner [3].

The paper is an elaboration of the talk the author gave at the conference.

2. ELEMENT IN A STRICT CLOSED SEMI-ALGEBRA

Let t be an element of the Banach algebra B . The symbol $\sigma(t)$ denotes the spectrum of t and $r(t)$ its spectral radius. The set

$$\sigma(t) \cap \{\lambda : |\lambda| = r(t)\}$$

is called the *peripheral spectrum* of t . A point λ in $\sigma(t)$ is said to be a *pole* of t of *order* n if it is a pole of order n of the analytic function

$$z \to R(z;t).$$

Here $R(z;t)$ denotes the resolvent $(ze-t)^{-1}$. A pole of t is called *simple* if the order of the pole is one.

For any t in B the smallest closed semi-algebra generated by t is denoted by $A(t)$. Thus $A(t)$ is the closure in B of the set

$$\{\alpha_1 t + \dots + \alpha_n t^n : \alpha_i \geq 0 \ (1 \leq i \leq n),\ n = 1,2,\dots\}.$$

Obviously, $A(t)$ is a commutative set. Note that $A(t)$ is strict if and only if belongs to a strict closed semi-algebra.

Theorem 1. *Let* t *be an element of the Banach algebra* B *. Suppose that*

(i) *the peripheral spectrum of* t *is a set of poles of* t *.*

(ii) $A(t)$ *is strict.*

Then $r(t) \in \sigma(t)$.

Proof. If $r(t) = 0$, then the theorem is trivially true. Therefore suppose that $r(t) > 0$. Let s be αt for some $\alpha > 0$. Then $A(s) = A(t)$, $r(s) \equiv \alpha r(t)$ and s and t have the same spectral properties. Hence, without loss of generality, we may suppose that $r(t) = 1$. Now assume that

(1) $$1 \notin \sigma(t).$$

We shall show that (1) leads to a contradiction.

From our hypothesis it follows that the peripheral spectrum of t is a spectral (i.e., closed and open) subset of $\sigma(t)$. Let p be the corresponding spectral idempotent. We shall prove that (1) implies

(2) $$-p \in A(tp).$$

Write $t_1 = tp$. From (1) and the definition of t_1 it follows that

(3) $$\sigma(t_1) \cap \{\alpha \in \mathbf{R} : \alpha > 0\} = \emptyset .$$

Since $t_1 R(\lambda; t_1) \in A(t_1)$ for $\lambda > r(t_1)$, and since

$$R(\mu; t_1) = \sum_{n=0}^{\infty} (\lambda - \mu)^n R(\lambda; t_1)^{n+1}$$

for μ close to λ , (3) implies that $t_1 R(\mu; t_1) \in A(t_1)$ for $\mu > 0$. Note that either 0 is a simple pole of t_1 with spectral idempotent $e-p$ or 0 is in the resolvent set of t_1 and then $e - p = 0$. In both cases $R(\mu; t_1)$ admits a Laurent expansion of the form

$$R(\mu; t_1) = \sum_{n=0}^{\infty} \mu^n a_n + \frac{1}{\mu}(e-p)$$

with $t_1 a_0 = -p$. But then

$$t_1 R(\mu; t_1) = \sum_{n=0}^{\infty} \mu^n t_1 a_n$$

for $|\mu| > 0$ and sufficiently small. Thus

$$-p = t_1 a_0 = \lim_{\mu \downarrow 0} t_1 R(\mu; t_1) \in A(t_1).$$

This completes the proof of (2).

If t has equibounded iterates, then it follows from our hypotheses that

$$\{\lambda \in \sigma(t) : |\lambda| = 1\}$$

is a set of simple poles of t. So we can apply Theorems 2.1 and 2.3 in [5] to show that the spectral idempotent p is the uniform limit of a subsequence of $\{t^n : n = 1, 2, \ldots\}$. This implies $p \in A(t)$. But then $tp \in A(t)$, and thus, by (2), $-p \in A(t)$. However, this contradicts the strictness of $A(t)$.

Hence there exists a subsequence $\{n_i\}$ of the sequence of all positive integers such that $\| t^{n_i} \| \geq i$ for $i = 1, 2, \ldots$. Put

$$s_i = \| t^{n_i} \|^{-1} t^{n_i} \qquad (i = 1, 2, \ldots).$$

From the definition of p it follows that $r(t(e-p)) < 1$. Hence

$$t^n (e-p) = \{t(e-p)\}^n \to 0 \quad (n \to +\infty).$$

In particular, $s_i(e-p) \to 0$ if $i \to +\infty$. The sequence $\{s_i p\}$ is a bounded sequence in the closed subalgebra B_0 of B generated by tp. From what we know about the peripheral spectrum of t, it follows that B_0 is finite dimensional. So $\{s_i p\}$ has a converging subsequence. By passing to this subsequence, we may suppose that $\{s_i p\}$ has a limit, q say. Note that

$$s_i = s_i(e-p) + s_i p \to q \quad (i \to +\infty).$$

This implies $q \in A(t)$. Further it follows that $qp = pq = q$, because p is an idempotent. Since $\| s_i \| = 1$ for $i = 1, 2, \cdots$ we have $\| q \| = 1$, in particular $q \neq 0$.

From $q \in A(t)$ and $qp = pq = q$ we may conclude that

$$\{qx : x \in A(tp)\} \subset A(t).$$

In particular, because of (2), $-q = q(-p) \in A(t)$. Thus q and $-q$ are in $A(t)$. Since $A(t)$ is strict this implies $q = 0$. Contradiction. So (1) must be false, i.e., $r(t) = 1 \in \sigma(t)$.

A semi-algebra A in B is said to be *locally compact* if A contains non-zero elements and if

$$A \cap \{x \in B : \|x\| \leq 1\}$$

is a compact subset of B. It is easily seen that a locally compact semi-algebra in B is a closed subset of B. If t is a non-zero element is B such that the peripheral spectrum of t is a set of poles of t, then

$$r(t) \in \sigma(t)$$

implies that the smallest closed semi-algebra generated by such an element t is locally compact (see [4], Theorem 9). Hence if t is a non-zero element in B satisfying the conditions of Theorem 1, $A(t)$ is locally compact. We will use this fact to prove that $r(t)$ is a pole of maximal order in the peripheral spectrum of t, if t satisfies the conditions of Theorem 1. We begin with a simple observation.

Lemma 2. *A strict locally compact semi-algebra in* B *is a closed normal cone in* B.

Proof. Let A be a strict locally compact semi-algebra in B. Clearly A is a closed cone in B. To prove the normality of A, write $x \leq y$ or $y \geq x$ whenever $y - x \in A$. We have to show that

(4) $$0 \leq x_n \leq y_n \ (n = 1, 2, \dots), \quad \lim y_n = 0$$

implies $\lim x_n = 0$.

Firstly, assume the sequence $\{x_n\}$ to be bounded. Then, because of local compactness, $\{x_n\}$ has at least one converging subsequence. Suppose

$$\lim_i x_{n_i} = x_0$$

for some subsequence $\{x_{n_i}\}$ of $\{x_n\}$. It suffices to show that $x_0 = 0$. Since

A is closed, (4) implies that x_0 and $-x_0$ are in A. But A is strict, and so $x_0 = 0$.

Next assume that $\{x_n\}$ is unbounded. Let $\{x_{n_i}\}$ be a subsequence such that $\|x_{n_i}\| \geq i$ $(i = 1, 2, \ldots)$. Put

$$z_i = \|x_{n_i}\|^{-1} x_{n_i}, \quad w_i = \|x_{n_i}\|^{-1} y_{n_i}$$

for $i = 1, 2, \ldots$. From (4) we get

$$0 \leq z_i \leq w_i \quad (i = 1, 2, \ldots), \quad \lim w_i = 0 .$$

Since $\{z_i\}$ is bounded, we can apply the result of the previous paragraph to show that $\lim z_i = 0$. However, this contradicts the fact that $\|z_i\| = 1$ for $i = 1, 2, \ldots$. Thus the sequence $\{x_n\}$ cannot be unbounded.

Since every normal cone is weakly normal (see [8], V. 3.3 Corollary 3), and since B is complete and, therefore, semi-complete, we can apply [8], Theorem App. 2.1 to show that a generalized version of Pringsheim's theorem is available in this context.

Generalized version of Pringsheim's theorem: *Let* A *be a strict locally compact semi-algebra in* B. *If* $a_n \in A$ *for* $n = 0, 1, 2, \ldots$ *and if the power series*

$$\sum_{n=0}^{\infty} z^n a_n$$

has radius of convergence 1, *then the analytic function represented by the power series is singular at* $z = 1$. *In addition, if this singularity is a pole, it is of maximal order under the poles on* $|z| = 1$.

We want to apply this result to the Neumann series of an element t in a strict locally compact semi-algebra. However, if t belongs to a semi-algebra A, then all the coefficients of the Neumann series of t are in A except maybe the first one. Therefore we do need the following extension of A. Put

$$A_e = \{a + \alpha e : a \in A, \ \alpha \geq 0\} .$$

Clearly, A_e is the smallest semi-algebra in B containing A and the identity e. The next lemma shows that strictness and local compactness of A carry over to A_e. The proof of this lemma is straightforward and is, therefore, omitted.

Lemma 3. *Let* A *be a semi-algebra in* B. *Then*

(i) A_e *is strict if and only if* A *is strict.*

If $A \neq \{0\}$ *and* A *is closed in* B *, then*

(ii) A_e *is locally compact if and only if* A *is locally compact.*

Theorem 4. *Let* A *be a strict locally compact semi-algebra in* B *, and let* $t \in A$ *. Then*

(i) $r(t) \in \sigma(t)$,

(ii) *if* $r(t)$ *is a pole of* t *, it is of maximal order under the poles in the peripheral spectrum of* t *.*

Proof. As in the proof of Theorem 1, we may suppose that $r(t) = 1$. Then the power series

$$\sum_{n=0}^{\infty} z^n t^{n-1} \tag{5}$$

has radius of convergence 1. Note that all its coefficients are in A_e, and that is a strict locally compact semi-algebra because of Lemma 3. So we can apply the generalized version of Pringsheim's theorem to show that the analytic function

$$z \to R(\tfrac{1}{z}; t)$$

(that is the analytic function represented by the power series (5)) is singular at $z = 1$. Thus $r(t) = 1 \in \sigma(t)$. Further it follows that, if 1 is a pole, it is of maximal order under the poles on $|z| = 1$. Clearly, this implies (ii).

The next theorem shows that the spectral radius $r(t)$ in Theorem 4 is an "eigenvalue" with a corresponding eigenvector in A. The elementary techniques used in the proof are based on a method used earlier by F.F. Bonsall (see e.g. [1] and [2]).

Theorem 5. *Let* $t \in B$ *, and suppose that* $A(t)$ *is strict and locally compact. Then there exists* $z \neq 0$ *in* $A(t)$ *such that*

$$tz = r(t)z.$$

Proof. Let A_e be the smallest semi-algebra in B containing $A(t)$ and e, i.e.,

$$A_e = \{a + \alpha e : a \in A(t),\ \alpha \geq 0\}.$$

Note that $R(\lambda;t) \in A_e$ for $\lambda > r(t)$. Let $\{\lambda_n\}$ be a sequence of positive real numbers such that $\lambda_n \downarrow r(t)$. Put $\beta_n = \|R(\lambda_n;t)\|$ for $n = 1,2,\ldots$. Clearly, $\beta_n > 0$ for $n = 1,2,\ldots$. Since $r(t) \in \sigma(t)$ (Theorem 4 (i)), $\lim \beta_n = +\infty$. Write $z_n = \beta_n^{-1}R(\lambda_n;t)$ $(n = 1,2,\ldots)$. Then

$$\|z_n\| = 1, \quad tz_n = \lambda_n z_n - \beta_n^{-1}e.$$

Now $\{z_n\}$ is a bounded sequence in the semi-algebra A_e. From our hypotheses, it follows that A_e is locally compact (cf. Lemma 3). Hence $\{z_n\}$ has a converging subsequence with limit z, say. From (6) we conclude that

$$z \neq 0, \qquad tz = r(t)z.$$

Further, $z \in A_e$. We shall prove that $z \in A(t)$.

If $r(t) > 0$, then $z = r(t)^{-1}tz$, and hence $z \in A(t)$. Suppose $r(t) = 0$. Since $z \in A_e$, we have $z = w + \alpha e$ for some w in $A(t)$ and $\alpha \geq 0$. From $tz = 0$, it follows that $tw = \alpha t = 0$. Since $A(t)$ is strict and $t \neq 0$, this implies $\alpha = 0$. Thus $z = w$ and $z \in A(t)$. This completes the proof.

If we summarize the preceding results for element in B with a peripheral spectrum consisting of poles, we get the following theorem.

Theorem 6. *Let t be an element in the Banach algebra B such that the peripheral spectrum of t is a set of poles of t. If, in addition, $A(t)$ is strict, then*

(i) *$r(t)$ is a pole of maximal order in the peripheral spectrum of t,*

(ii) *there exists $z \neq 0$ in $A(t)$ such that $tz = r(t)z$.*

Theorem 6 includes the Kreĭn-Rutman theorem on compact positive linear operators ([6], Theorem 6.1) and Maibaum's theorem on positive linear operators with a peripheral spectrum consisting of poles ([7], Satz 1. (1.3)). To see this let E be the complexification of a real ordered Banach space F. Suppose that the cone K of positive element in F is a closed and total subset of F. Further, let T be a bounded linear operator on E which leaves invariant the cone K. Since K is closed and total in F and thus in E, it follows that $A(t)$ is a strict semi-algebra in the complex Banach algebra of all bounded linear operators on E. Hence, if T is a compact linear operator with $r(T) > 0$ (i.e., T satisfies the conditions of the Kreĭn-Rutman theorem) or if the peripheral spectrum of T is a set of poles of T (i.e., T satisfies the conditions of Maibaum's theorem), then T satisfies the conditions of Theorem 6.

Hence in these cases

(i) $r(T) \in \sigma(T)$

and there exists $Z \neq 0$ in $A(T)$ such that $TZ = r(T)Z$. From the last fact it is eacy to conclude (see [2], section 4) that

(ii) there exists $x_0 \neq 0$ in K such that $Tx_0 = r(T)x_0$,
(iii) there exists $f_0 \neq 0$ in K' such that $T'f_0 = r(T)f_0$.

Here T' is the adjoint of T acting on the dual F' of F, and

$$K' = \{f \in F' : f(u) \geq 0 \quad (u \in K)\},$$

The content of the Kreĭn-Rutman theorem and that of Maibaum's theorem is precisely given by (i), (ii) and (iii).

For compact linear operators with non-zero spectral radius the second part of Theorem 6 has been proved by Bonsall in [2] under a slightly different assumption. Bonsall's proof is based on the fact that a compact linear operator acts compactly on its centraliser. The application of Theorem 6 mentioned above follows the exposition given in section 4 of [2].

In the next section we shall prove the converse of Theorem 6.

3. A SUFFICIENT CONDITION FOR STRICTNESS

Theorem 7. *Let* t *be an element in the Banach algebra* B *such that the peripheral spectrum of* t *is a set of poles of* t. *If, in addition,* $r(t)$ *is a pole of maximal order in the peripheral spectrum of* t, *then* $A(t)$ *is strict.*

Proof. The proof consists of two parts. 1. First of all we consider the case $r(t) > 0$. As in the proof of Theorem 1, we may suppose that $r(t) = 1$. Then 1 is a pole of t of order n_0, say. Let e_0 be the spectral idempotent associated with the spectral set $\{1\}$. Put $x_0 = (e-t)^{n_0-1}e_0$. Then

$$x_0 \neq 0, \qquad tx_0 = x_0.$$

Suppose that $p(t) = \alpha_1 t + \dots + \alpha_n t^n$ with $\alpha_i \geq 0 \quad (i = 1, \dots, n)$. By the spectral mapping theorem

$$\sigma(p(t)) = \{p(\lambda) : \lambda \in \sigma(t)\}.$$

First of all this implies that $r(p(t)) \le p(r(t)) = p(1)$ and secondly that $p(1) \in \sigma(p(t))$. Both results together show that

$$r(p(t)) = p(1).$$

Further we have

$$p(t)x_0 = p(1)x_0 = r(p(t))\,x_0 .$$

Take s in $A(t) \cap (-A(t))$. We want to show that $s = 0$. Since $s \in A(t)$, there exists a sequence $\{b_n\}$ in $A(t)$ such that $s = \lim b_n$ and

$$b_n = \sum_{i=1}^{\infty} \alpha_i(n)\,t^i \qquad (n = 1,2,\dots)$$

with $\alpha_i(n) \ge 0$ and $\alpha_i(n) = 0$ for sufficiently large i . From what we proved in the previous paragraph it follows that

$$r(b_n) = \sum_{i=1}^{\infty} \alpha_i(n)\,, \quad b_n x_0 = r(b_n)\,x_0$$

for $n = 1,2,\dots$. By the continuity of the spectral radius on commutative sets, the last equality implies

(7) $$s x_0 = r(s)\,x_0 .$$

Since $-s$ also belongs to $A(t)$, we can repeat the argument to show that

(8) $$-s x_0 = r(-s)\,x_0 .$$

Now $r(-s) = r(s)$. Hence the formulas (7) and (8) imply that $r(s) = 0$ and so

(9) $$\lim_n \sum_{i=1}^{\infty} \alpha_i(n) = \lim_n r(b_n) = r(s) = 0 .$$

Let p be the spectral idempotent associated with the spectral subset

$$\sigma(t) \cap \{\lambda : |\lambda| = r(t) = 1\}.$$

Then $t(e-p)$ has spectral radius less than one, and so

$$t^n(e-p) = \{t(e-p)\}^n \longrightarrow 0 \quad (n \to +\infty).$$

In particular this implies that the sequence $\{t^n(e-p)\}$ is bounded in B, and so by (9)

$$\lim_n \sum_{i=1}^{\infty} \alpha_i(n) t^i (e-p) = 0 .$$

That is $s(e-p) = \lim b_n(e-p) = 0$. Hence in order to prove that $s = 0$ it is sufficient to show that $sp = 0$

Suppose that

$$\{\lambda_0 = 1, \lambda_1, \ldots, \lambda_r\}$$

is the peripheral spectrum of t. Then $|\lambda_i| = 1$, λ_i is a pole of t of order n_i say and $n_i \le n_0$ $(i = 0, 1, \ldots, r)$. Let e_i be the spectral idempotent associated with the spectral subset $\{\lambda_i\}$. By elementary spectral theory

$$p = e_0 + e_1 + \cdots + e_r .$$

Hence it suffices to show that $se_i = 0$ for $i = 0, 1, \ldots, r$.

Put $\alpha_0(n) = 0$ for $n = 1, 2, \ldots$ and let

$$\beta_j^{\ell}(n) = \sum_{i=j}^{\infty} \alpha_i(n) \binom{i}{j} \lambda_{\ell}^{i-j}$$

for $\ell = 0, 1, \ldots, r$; $j = 0, 1, \ldots$ and $n = 1, 2, \ldots$. Note that for fixed n and ℓ and for j sufficiently large $\beta_j^{\ell}(n) = 0$. Consider the complex polynomial

$$q_n^{\ell}(X) = \sum_{j=0}^{\infty} \beta_j^{\ell}(n) X^j$$

for $\ell = 0, 1, \ldots, r$ and $n = 1, 2, \ldots$. A simple computation shows that

$$q_n^{\ell}(X) = \sum_{i=0}^{\infty} \alpha_i(n)(\lambda_{\ell} + X)^i .$$

In particular

(10) $$b_n = q_n^{\ell}(t - \lambda_{\ell})$$

for $\ell = 0, 1, \ldots, r$ and $n = 1, 2, \ldots$. Note that

(11) $$|\beta_j^{\ell}(n)| \le \sum_{i=j}^{\infty} \alpha_i(n) \binom{i}{j} |\lambda_{\ell}|^{i-j} = \beta_j^0(n)$$

for $\ell = 0,1,\dots,r$; $j = 0,1,2,\dots$ and $n = 1,2,\dots$

Since $\lambda_0 = 1$ is a pole of order n_0, the elements

(12) $$e_0,\ (t-\lambda_0)e_0,\ \dots,\ (t-\lambda_0)^{n_0-1}e_0$$

are linearly independent. Further $(t-\lambda_0)^{n_0}e_0 = 0$ and so by (10)

$$b_n e_0 = q_n^0(t-\lambda_0)e_0 = \sum_{j=0}^{n_0-1} \beta_j^0(n)(t-\lambda_0)^j e_0 .$$

Now $se_0 = \lim b_n e_0$. Hence

(13) $$se_0 = \sum_{j=0}^{n_0-1} \beta_j^0(\infty)(t-\lambda_0)^j e_0$$

where

$$\beta_j^0(\infty) = \lim_n \beta_j^0(n) \quad (j = 0,1,\dots,n_0-1) .$$

Note that formula (11) implies that $\beta_j^0(\infty) \geq 0$ for $j = 0,1,\dots,n_0-1$.

Also, $-s \in A(t)$. So repeating the argument we may conclude that

(14) $$-se_0 = \sum_{j=0}^{n_0-1} \gamma_j^0(t-\lambda_0)^j e_0$$

for some $\gamma_j^0 \geq 0$ $(j = 0,1,\dots,n_0-1)$. Since the elements (12) are linearly independent, formulas (13) and (14) together imply that $\beta_j^0(\infty) = 0$ for $j = 0,1,\dots$ $\dots, n_0-1$. But then we can use formula (11) to show that

(15) $$\lim_n \beta_j^\ell(n) = 0$$

for $\ell = 0,1,\dots,r$ and $j = 0,1,\dots,n_0-1$.

Since λ_ℓ is a pole of order n_ℓ with $n_\ell \leq n_0$ for $\ell = 0,1,\dots,r$, we have

$$(t-\lambda_\ell)^{n_0}e_\ell = 0$$

for $\ell = 0,1,\dots,r$, and so, by formula (10),

$$b_n e_\ell = q_n^\ell(t-\lambda_\ell)e_\ell = \sum_{j=0}^{n_0-1} \beta_j^\ell(n)(t-\lambda_\ell)^j e_\ell .$$

Hence using formula (15) we see that

$$se_\ell = \lim_n b_n e_\ell = 0 \qquad (\ell = 0,1,\dots,r).$$

This completes the proof of part 1.

2. Next we consider the case $r(t) = 0$. The assumption $r(t) = 0$ together with the hypothesis of the theorem implies that t is nilpotent. Let n_0 be the order of nilpotence. If $n_0 = 1$, then $t = 0$ and the theorem is trivially true. Therefore suppose that $n_0 \geq 2$.

Take s in $A(t) \cap (-A(t))$. We want to show $s = 0$. Let the sequence $\{b_n\}$ in $A(t)$ be as in part 1. Then

$$b_n = \sum_{i=1}^{n_0-1} \alpha_i(n) t^i \qquad (n = 1,2,\dots).$$

Using the linear independency of the elements $t, t^2, \dots, t^{n_0-1}$ we see that

$$s = \sum_{i=1}^{n_0-1} \alpha_i(\infty) t^i \tag{16}$$

where $\alpha_i(\infty) = \lim \alpha_i(n) \geq 0$ for $i = 1,\dots,n_0-1$. Also, $-s$ in $A(t)$. So we can repeat the argument to show that

$$-s = \sum_{i=1}^{n_0-1} \delta_i t^i \tag{17}$$

for some $\delta_i \geq 0$ $(i = 1,\dots,n_0-1)$. Since the elements $t, t^2, \dots, t^{n_0-1}$ are linearly independent, the formulas (16) and (17) imply that $\alpha_i(\infty) = 0$ for $i = 1,\dots$ $\dots, n_0 - 1$. Thus $s = 0$. This completes the proof.

For compact linear operators, Theorem 7 may be deduced from Satz 3.1 in L. Elsner [3]. In fact, the following theorem holds.

Theorem 8. *Let* T *be a compact linear operator on a complex Banach space* E *with* $r(T) > 0$. *Equivalent are*

(i) $A(T)$ *is strict;*

(ii) $r(T)$ *is a pole of maximal order in the peripheral spectrum of* T;

(iii) *there exists a closed and total cone* K *in* E *such that* $TK \subset K$.

Proof. The implication (i) $\Longrightarrow$ (ii) follows from Theorem 6. The implication

(ii) $\Longrightarrow$ (iii) is a little variation on Satz 3.1 in L. Elsner [3]. In Elsner's theorem (ii) $\Longrightarrow$ (iii) is stated for the case that E is a real Banach space, but the proof deals with the complex case first. The implication (ii) $\Longrightarrow$ (i) is trivial.

Theorem 8 also holds for a bounded linear operator T with $r(T) > 0$ and with a peripheral spectrum consisting of poles of T of finite rank. Whether or not the theorem holds without the finite rank condition remains an unsolved problem.

REEERENCES

[1] F.F. Bonsall, Positive operators compact in an auxiliary topology, *Pacific J. Math.*, 10 (1960), 1131-1138.

[2] F.F. Bonsall, Compact linear operators from an algebraic standpoint, *Glasgow Math., J.*, 8 (1967), 41-49.

[3] L. Elsner, Monotonie und Randspektrum bei vollstetigen Operatoren, *Arch. Rational Mech. Anal.*, 36 (1970), 356-365.

[4] M.A. Kaashoek and T.T. West, Locally compact monothetic semi-algebras, *Proc. London Math. Soc.*, 18 (1968), 428-438.

[5] M.A. Kaashoek and T.T. West, Compact semigroups in commutative Banach algebras, *Proc. Camb. Phil. Soc.*, 66 (1969), 265-274.

[6] M.G. Kreĭn and M.A. Rutman, Linear operators leaving invariant a cone in a Banach space, *Uspehi Mat. Nauk*, 3 (1948), 3-95 (Russian); *Amer. Math. Soc. Transl.*, 26.

[7] G. Maibaum, Über Scharen positiver Operatoren, *Math. Ann.*, 184 (1970), 238-256.

[8] H.H. Schaefer, *Topological vector spaces* (New York, 1966).

COLLOQUIA MATHEMATICA SOCIETATIS JÁNOS BOLYAI
5. HILBERT SPACE OPERATORS. TIHANY (HUNGARY), 1970

On the similarity of certain operators

R. G. KALISCH

We consider $L_p(0,1) = L_p$ with $1 < p < \infty$. In L_p we consider the following bounded operators: I (the indentity, $zI = z$ for complex z, $M, M^* = M$, $J^\varkappa$ and $J^{*\varkappa}$ for complex $\varkappa$ with non-negative real part. If $f \in L_p$, these operators are defined as follows: $(Mf)(x) = xf(x)$, $(J^\varkappa f)(x) = 1/\Gamma(\varkappa)\int_0^x (x-y)^{\varkappa-1} f(y)\,dy$, $(J^{*\varkappa} f)(x) = 1/\Gamma(\varkappa)\int_x^1 (y-x)^{\varkappa-1} f(y)\,dy$. We write $J = J^1$ and $J^* = J^{*1}$. The adjoint of $J^\varkappa$ in L_p is $J^{*\varkappa}$ in $L_{p^*} = L^*$ $(1/p + 1/p^* = 1)$ – of course both $J^\varkappa$ and $J^{*\varkappa}$ are bounded in all L_p considered here.

In the present paper we discuss unitary invariants of the operators $M + \varkappa J$ and $M + \varkappa J^*$ in L_2 for complex $\varkappa$ and λ, and similarity invariants of these operators in L_p where we call two operators A and B similar if there exists a bounded and boundedly invertible operator P such that $AP = PB$.

We show first that $M + \varkappa J$ is similar (in all L_p) to $M + (\mathrm{Re}\,\varkappa)J$ with an analogous result for $M + \varkappa J^*$. As mentioned in [3] the proof of this result is ultimately based on the multiplier theorem for Fourier transforms, an easy result in L_2 but very deep in L_p for $p \neq 2$. We next consider unitary equivalence in L_2. It turns out that $\varkappa$ is the unitary invariant of $M+\varkappa J$ and also of $M+\varkappa J^*$ and that $M+\varkappa J$ and $M+\varkappa J^*$ are never unitarily equivalent with the surprising single exception that $M+i\beta J$ and $M+i\beta J^*$ for real β are unitarily equivalent. The proof of this unitary equivalence uses the idea of M.S. Livšic [1] but is self-contained and does not use the extensive machinery of [1].

The question of similarity in L_p of the operators $M+\varrho J$ and $M+\sigma J^*$ for real ϱ and σ lies much deeper and also contains a surprising exception. It turns out that ϱ is the similarity invariant of $M+\varrho J$ and also of $M+\varrho J^*$. The operators $M+\varrho J$ and $M-\sigma J^*$ are never similar with the exception that $M+\varrho J$ and $M-\varrho J^*$ are similar in L_p if $\varrho \in (-1/p^*, 1/p)$; the cases $\varrho = -1/p^*$ and $\varrho = 1/p$ remain unsettled. The proof of the similarity of $M+\varrho J$ and $M-\varrho J^*$ is based on the difficult paper of R.K. Juberg [2] which precedes this paper. He shows that if $0 \leq \gamma < 1/p^*$, then $J^{\gamma} J^{*-\gamma}$ is bounded and boundedly invertible in L_p and that if $0 \leq \gamma < 1/p$, then $J^{-\gamma} J^{*\gamma}$ is bounded and boundedly invertible in L_p.

Taking the view that adding a complex multiple of J to M perturbs it in a fairly harsh manner, the last theorem shows that adding a complex multiple of J^{μ} with $Re(\mu) > 1$ has a much milder perturbing effect: the operators $M+\varkappa J+\lambda J^{\mu}$ and $M+\varkappa J$ are similar.

Part of Theorem 4 was also proved by S. Kantorovitz [4]; his methods are entirely different from ours and are based on his functional calculus results; see [4] for references. I wish to thank my friend B.R. Gelbaum for stimulating discussions. Roger Stafford has simplified part of my original proof of Theorem 4. I wish to express my gratitude to the National Science Foundation (grants GP 13288, 21334) for their support.

The well-known formulas $J^{\varkappa} J^{\lambda} = J^{\varkappa+\lambda}$ and $J^{*\varkappa} J^{*\lambda} = J^{*\varkappa+\lambda}$, valid for appropriate complex $\varkappa$ and λ (viz., with non-negative real parts), express the semi-group property of $\{J^{\varkappa}\}$. They are used throughout the paper. A related semi-group, defined by

$$(K_g^{\varkappa} f)(x) = 1/\Gamma(\varkappa) \int_0^x (g(x)-g(y))^{\varkappa-1} f(y)\,dg(y),$$

will be explored elsewhere. – We have the formulas

(1) $$J^{\varkappa}M = (M - \varkappa J)J^{\varkappa}$$

(2) $$MJ^{\varkappa} = J^{\varkappa}(M + \varkappa J)$$

(3) $$MJ^{*\varkappa} = J^{*\varkappa}(M - \varkappa J^*)$$

(4) $$J^{*\varkappa}M = (M + \varkappa J^*)J^{*\varkappa}$$

valid in L_p for all complex $\varkappa$ with non-negative real part (if $\varkappa$ is pure imaginary, see [3]). – We use the Greek letter chi for characteristic functions of sets.

Theorem 1. *Let* $\varkappa$ *be a complex number. Then* $M + \varkappa J$ *is similar in* L_p *to* $M + (\mathrm{Re}\,\varkappa)J$ *and* $M + \varkappa J^*$ *is similar in* L_p *to* $M + (\mathrm{Re}\,\varkappa)J^*$.

Proof. This is a direct consequence of (2). The crucial fact is that $J^{i\beta}$ for real β is bounded with bounded inverse $J^{-i\beta}$ in L_p for all p with $1 < p < \infty$ [3]. This shows that $M + i\beta J$ for real β is similar to M. Thus if $\varkappa = \alpha + i\beta$ with real α and β we have $J^{i\beta}(M + \alpha J + i\beta J)J^{-i\beta} = M + \alpha J$ since $J^{i\beta}$ commutes with J. The proof for $M + \varkappa J^*$ is similar and the theorem is proved.

Theorem 2. *The operators* $S = M + (\alpha + i\beta)J$ *and* $S' = M + (\alpha' + i\beta')J$ *are unitarily equivalent if and only if they are equal. The operators* S *and* $T = M + {}$ ${}+ (\gamma + i\delta)J^*$ *are unitarily equivalent if and only if* $\alpha = \gamma = 0$ *and* $\beta = \delta$.

Proof. 1. The unitary equivalence of S and S'. If $\alpha = 0$, the only eigenvalue of $S - S^*$ is $i\beta$, otherwise it has infinitely many eigenvalues. Hence $\alpha = 0$ implies $\alpha' = 0$ and $\beta = \beta'$ so that in this case the unitary equivalence of S and S' implies their equality. – If $\alpha \neq 0$ and $\beta = 0$, then the only eigenvalue of $S + S^*$ is $2/(1 - \exp(-2/\alpha))$, otherwise it has infinitely many eigenvalues. Hence again, the unitary equivalence of S and S' implies their equality. – If α and β and hence α' and β' are different from λ_n, then the eigenvalues λ_n of $S - S^*$ satisfy the equations

(5) $$\cos(2\alpha i/\lambda_n) = -(\alpha^2 - \beta^2)/(\alpha^2 + \beta^2)$$

(6) $$\sin(2\alpha i/\lambda_n) = -2\alpha\beta/(\alpha^2 + \beta^2)$$

(7) $$2\alpha i/\lambda_n = 2\alpha i/\lambda_0 + 2n\pi .$$

Equation (7) implies that if S and S' are unitarily equivalent, then $|\alpha| = |\alpha'|$ and equation (6) implies then that $\beta = \beta'$. If now $S = M+(\alpha+i\beta)J$ were unitarily equivalent to $S' = M+(-\alpha+i\beta)J$, a comparison of the negative eigenvalues of $S+S^*$ and $S'+S'^*$ of largest absolute value shows that $\alpha = 0$, contradicting our present hypothesis that $\alpha \neq 0$. If $\beta > 0$, the eigenvalue in question for $S+S^*$ is $2/(1-\exp(2(\pi-\Theta)/\beta))$ and for $S'+S'^*$ is $2/(1-\exp(2\Theta/\beta))$ where $\alpha+i\beta = = \varrho\exp(i\Theta)$ with $0 \leq \Theta < 2\pi$. These two eigenvalues can coincide only if $\Theta = \pi/2$, that is, if and only if $\alpha = 0$. If $\beta < 0$, the two eigenvalues are $2/(1-\exp(2(\pi-\Theta)/\beta))$ and $2/(1-\exp(2(\Theta-2\pi)/\beta))$. Again they can coincide only if $\alpha = 0$. This completes the proof of the first part of the theorem.

2. The unitary equivalence of S and T. If $\alpha = 0$, the only eigenvalue of $S-S^*$ is $i\beta$, otherwise it has infinitely many eigenvalues; analogously, the only eigenvalue of $T-T^*$ is $i\delta$ if $\gamma = 0$. This shows that the unitary equivalence of S and T implies that $S = M+i\beta J$ and $T = M+i\beta J^*$. We shall show at the and of this proof that these two operators are unitarily equivalent. – If $\alpha \neq 0$ and $\beta = 0$ then the only eigenvalue of $S+S^*$ is $2/(1-\exp(-2/\alpha))$, otherwise it has infinitely many eigenvalues; analogously, the only eigenvalue of $T+T^*$ is $2/(1-\exp(-2/\gamma))$ if $\delta = 0$. This shows that the unitary equivalence of S and T implies that $S = M+\alpha J$ and $T = M+\alpha J^*$. We shall show at the end of the proof that these two operators are unitarily equivalent if and only if $\alpha = 0$. – If α and β and hence γ and δ are different from 0, then the eigenvalues λ_n of $S-S^*$ satisfy equations (5), (6), (7) and the eigenvalues μ_n of $T-T^*$ satisfy the same equations with α and β replaced by γ and δ. Thus as in the first part of the proof, $|\alpha| = |\gamma|$ and $\beta = \delta$ if S and T are unitarily equivalent. This gives rise to two cases:

(i) $$S = M+(\alpha+i\beta)J, \quad T = M+(\alpha+i\beta)J^*.$$

In this case a comparison of the eigenvalues of $S+S^*$ and $T+T^*$ does not settle the unitary equivalence question of S and T; we show toward the end of the proof that for $\alpha \neq 0$, S and T are never unitarily equivalent.

(ii) $$S = M+(\alpha+i\beta)J, \quad T = M+(-\alpha+i\beta)J^*$$

As in the first part of the proof, a comparison of the negative eigenvalues of largest absolute value of $S+S^*$ and $T+T^*$ shows that for $\alpha \neq 0$, the operators S and T are never unitarily equivalent.

We now show that if $\alpha \neq 0$, then $S = M+(\alpha+i\beta)J$ and $T = M+(\alpha+i\beta)J^*$

are not unitarily equivalent, indeed are not even similar. Theorem 1 shows that S and T are similar respectively to $M+\alpha J$ and $M+\alpha J^*$. Suppose now that there existed a bounded non-zero operator P such that $(M+\alpha J)P = P(M+\alpha J^*)$. If $\alpha > 0$ and using equations (2) and (4) we have $J^*(M+\alpha J)PJ^{*\alpha} = J^{\alpha}P(M+\alpha J^*)J^{*\alpha} =$ $= MJ^{\alpha}PJ^{*\alpha} = J^{\alpha}PJ^{*\alpha}M$, an impossibility since $J^{\alpha}PJ^{*\alpha}$ is compact and different from zero. The case $\alpha < 0$ is handled similarly using equations (1) and (3).

We show last that the operators $S = M+i\beta J$ and $T = M+i\beta J^*$ are unitarily equivalent for all real β . To that end we check that the function e identically equal to 1 is cyclic for both S and T and that the operators $S - S^*$ and $T - T^*$ (wich are equal) map $L_2(0,1)$ onto the one-dimensional subspace generated by e . The crucial next step is to show that

$$(S^m e, S^n e) = (T^m e, T^n e) \tag{8}$$

for all non-negative integers m and n . This will be done presently. Once (8) is established, we can define an operator U of L_2 into itself by writing $U\sum x_j S^j e =$ $= \sum x_j T^j e$ for all finite sums. Equation (8) implies that U is well-defined and isometric; the cyclic nature of e for S and T shows that U can be extended to an isometry of L_2 onto itself and that $US = TU$. The last equation results from the equations $US\sum x_j S^j e = \sum x_j T^{j+1} e = T\sum x_j T^j e = TU\sum x_j S^j e$.
– To show (8) we show first that

$$(S^m e, e) = (T^m e, e) \tag{9}$$

for all non-negative integers m . This is best done by showing that for all large enough complex z we have $((S-z)^{-1}e,e) = ((T-z)^{-1}e,e)$; the resolvent $R_S = (S-z)^{-1}$ of S is given by the formula $(R_S f)(x) = (x-z)^{-1}f(x) - i\beta(x-z)^{-1-i\beta} \cdot \int_0^x (y-z)^{-1+i\beta} f(y)\,dy$ for all $f \in L_2$; the resolvent R_T of T is given by the formula $(R_T f)(x) = (x-z)^{-1}f(x) - i\beta(x-z)^{-1+i\beta}\int_x^1 (y-z)^{-1-i\beta} f(y)\,dy$.
The proof of (8) is then based on the operator identity valid for all suitable complex z and w

$$(z-\bar{w})(S^*-\bar{w})^{-1}(S-z)^{-1} = (S-z)^{-1} - (S^*-\bar{w})^{-1} +$$

$$+ (S^*-\bar{w})^{-1}(S-S^*)(S-z)^{-1}$$

applied to e and the equation valid for all $f \in L_2$ asserting $(S-S^*)f = i\beta(f,e)e$ This implies

$$(z-w)((S^*-\bar{w})^{-1}(S-z)^{-1}e,e) = ((S-z)^{-1}e,e) - ((S^*-\bar{w})^{-1}e,e) +$$

$$+ i\beta((S-z)^{-1}e,e)((S^*-w)^{-1}e,e).$$

This shows that $((S-z)^{-1}e,(S-w)^{-1}e)$ is determined by $((S-z)^{-1}e,e)$ and $((S-w)^{-1}e,e)$ so that (9) implies (8). This shows that $M+i\beta J$ and $M+i\beta J^*$ are unitarily equivalent and the proof of the theorem is complete.

The next theorem contains a compilation of elementarily verifiable facts about spectral properties in L_p of the operators $M+\varkappa J$ and $M-\varkappa J^*$ with complex $\varkappa = \varrho + i\sigma$ (ϱ and σ real) that are used in the proof of the theorem following this one.

Theorem 3. *Consider the operators* $M+\varkappa J$ *and* $M-\varkappa J^*$ *in* L_p . 1) *If* $\varrho \in [-1/p^*, 1/p]$ *then their point spectra and residual spectra are empty and their continuous spectra are the unit interval* $[0,1]$. *If* $\varrho \in (-\infty, -1/p^*)$ *then their residual spectra are empty; the point spectrum of* $M+\varkappa J$ *is the half-open unit interval* $[0,1)$ *with simple eigenfunctions* $\chi_{[\lambda,1]}(x-\lambda)^{\varkappa-1}$ *for all* $\lambda \in [0,1]$ *and the continuous spectrum is* $\{1\}$ *; the point spectrum of* $M-\varkappa J^*$ *is* $(0,1]$ *with simple eigenfunctions* $\chi_{[0,\lambda]}(\lambda-x)^{\varkappa-1}$ *for all* $\lambda \in (0,1]$ *and the continuous spectrum is* $\{0\}$.

2) *If* $\varrho \in (1/p,\infty)$ *then their point spectra are empty; the residual spectrum of* $M+\varkappa J$ *is* $(0,1]$ *with defect* 1 *and the continuous spectrum is* $\{0\}$ *; the residual spectrum of* $M-\varkappa J^*$ *is* $[0,1)$ *with defect* 1 *and the continuous spectrum is* $\{1\}$.

Theorem 4. *Consider in* L_p *the operators* $S = M+(\alpha+i\beta)J$, $S' = M+$ $+(\alpha'+i\beta')J$ *and* $T = M+(\gamma+i\beta)J^*$. 1) *The operator* S *is similar to the operator* S' *if and only if* $\alpha = \alpha'$. 2i) *If* $\alpha \neq -\gamma$ *, or if* $\alpha = -\gamma$ *with* α *or* $\gamma \notin [-1/p^*, 1/p]$ *then the operators* S *and* T *are not similar.* 2ii) *If* $\alpha \in (-1/p^*,1/p)$ *and* $\alpha = -\gamma$ *then the operators* S *and* T *are similar.*

Proof. First of all, Theorem 1 shows that we can confine ourselves to the cases $\beta = \beta' = \delta = 0$.

1) (a) Let $\alpha' = 0$ and $\alpha > 0$. Then if there existed an operator $P \neq 0$ such that $PM = (M+\alpha J)P$ we would have, using (2), the equations $J^{\alpha}PM=$

$= J^{\alpha}(M+\alpha J)P = J^{\alpha}PM = MJ^{\alpha}P$; the compact non-zero operator $J^{*}P$ cannot commute with M. Equation (1) is similarly used to treat the case $\alpha < 0$ so that M and $M+\alpha J$ cannot be similar if $\alpha \neq 0$.

(b) Let α and α' be different from 0 but of opposite signs; say $\alpha > 0$ and write $\varepsilon = -\alpha' > 0$. Then if there existed an operator $P \neq 0$ such that $P(M-\varepsilon J) = (M+\alpha J)P$, we would have, using (1) and (2), $J^{\alpha}P(M-\varepsilon J)J^{\varepsilon} = J^{\alpha}(M+\alpha J)PJ^{\varepsilon}$ so that $J^{\alpha}PJ^{\varepsilon}M = MJ^{\alpha}PJ^{\varepsilon}$ and again the compact non-zero operator $J^{\alpha}PJ^{\varepsilon}$ cannot commute with M and so S and S' cannot be similar in the present case.

(c) We show next that S and S' are not similar if $\alpha' < \alpha < 0$. Write $\varrho = -\alpha$ and $\sigma = -\alpha'$ so that $0 < \varrho < \sigma$ and let $\sigma - \varrho = \tau > 0$. Suppose now that there existed a bounded and boundedly invertible operator P such that

(10) $$P(M-\sigma J) = (M-\varrho J)P.$$

Then (1) implies that $J^{\sigma}M = (M-\sigma J)J^{\sigma} = J^{\tau}J^{\varrho}M = J^{\tau}(M-\varrho J)J^{\varrho}$ so that $J^{\tau}(M-\varrho J) = (M-\sigma J)J^{\tau}$. This and (10) imply that $(M-\sigma J)J^{\sigma} = P^{-1}(M-\varrho J)PJ^{\tau} = J^{\tau}(M-\varrho J)$ or

(11) $$PJ^{\tau}(M-\varrho J) = (M-\varrho J)PJ^{\tau}.$$

We now distinguish two cases.

Case I: $\varrho > 1/p^{*}$. In this case Theorem 3 says that $M = \varrho J$ has $I = [0,1)$ as its simple point spectrum. If $\lambda \in I$, let $\varphi_{\lambda} \neq 0$ be a corresponding eigenfunction. Equation (11) gives

$$PJ^{\tau}(M-\varrho J)\varphi_{\lambda} = \lambda PJ^{\tau}\varphi_{\lambda} = (M-\varrho J)PJ^{\tau}\varphi_{\lambda}$$

so that $PJ^{\tau}\varphi_{\lambda}$ is a non-zero eigenfunction of $M-\varrho J$ corresponding to λ and since the point spectrum is simple there exists a complex number $\varkappa_{\lambda}$ such that $PJ^{\tau}\varphi_{\lambda} = \varkappa_{\lambda}\varphi_{\lambda}$. We must have $\varkappa_{\lambda} \neq 0$ since otherwise $PJ^{\tau}\varphi_{\lambda}$ and hence φ_{λ} would be zero, a contradiction. Since the set $\{\varphi_{\lambda}\}$ for all $\lambda \in I$ is linearly independent, the compact operator PJ^{τ} has too many, namely continuously many, linearly independent eigenfunctions corresponding to non-zero eigenvalues which is impossible. Therefore the operators $M-\varrho J$ and $M-\varrho J$ are not similar in this case.

Case II: $0 < \varrho \leq 1/p^{*}$. We show first that under the assumption of the existence of P the operator PJ^{τ} is generalized nilpotent, that is, since it is compact, that it has no non-zero eigenvalues. Suppose on the contrary that PJ^{τ} had the non-zero eigenvalue λ with corresponding finite-dimensional eigen-space $L(\lambda)$. If $f \in L(\lambda)$, then equation (11) implies that

$$PJ^{\tau}(M-\varrho J)f = (M-\varrho J)PJ^{\tau}f = \lambda(M-\varrho J)f$$

so that $(M-\varrho J)f \in L(\lambda)$ and therefore $(M-\varrho J)L(\lambda) \subset L(\lambda)$. The finite dimensionality of $L(\lambda)$ implies that $M-\varrho J$ has eigenvalues, contradicting Theorem 3 in the present case. Thus PJ^{τ} is generalized nilpotent and its generalized nilpotency will ultimately lead to the contradiction that will then establish the validity of our theorem in the present case. To that end we establish next

$$(12_n) \quad (J^{*\sigma}P^{*}f^{*})^{n}(J^{*\varrho}g^{*}) = (J^{*\varrho}f^{*})^{n}J^{*\varrho}(J^{*\tau}P^{*})^{n}g^{*}$$

for all f^* and g^* in $L_{p^*} = L^*$ and all integers $n \geq 1$. We prove (12_1); the general case (12_n) follows by induction. If n is a positive integer set $f_n = (\Gamma(n+1)/\Gamma(n-\varrho+1))(1-x)^{n-\varrho}$ so that $J^{*\varrho}f_n = (1-x)^n$ and, if we set $M_0 = 1-M$,

$$(13) \quad (1-x)^{n+m} = (1-x)^{n}J^{*\varrho}f_m = (1-x)^{m}J^{*\varrho}f_n = M_0^{n}J^{*\varrho}f_m = M_0^{m}J^{*\varrho}f_n .$$

Equations (3) and (10) imply that for all positive integers m we have

$$(M_0+\sigma J^{*})^{m}P^{*} = P^{*}(M_0+\varrho J^{*})^{m}$$

and

$$M_0^{m}J^{*\varrho} = J^{*\varrho}(M_0+\varrho J^{*})^{m}$$

so that (13) implies that

$$J^{*\varrho}(M_0+\varrho J^{*})^{m}f_n = J^{*\varrho}(M_0+\varrho J^{*})^{n}f_m .$$

Since the operator $J^{*\varrho}$ is 1–1, this equation in turn implies that $(M_0+\varrho J^{*})^{m}f_n = (M_0+\varrho J^{*})^{n}f_m$. Now

$$J^{*\sigma}P^{*}f_m J^{*\varrho}f_n = M_0^{n}J^{*\sigma}P^{*}f_m = J^{*\sigma}(M_0+\sigma J^{*})^{n}P^{*}f_m =$$

$$= J^{*\sigma}P^{*}(M_0+\varrho J^{*})^{n}f_m$$

and similarly

$$J^{*\sigma}P^{*}f_n J^{*\varrho}f_m = J^{*\sigma}P^{*}(M_0+\varrho J^{*})^{m}f_n$$

so that we have

$$J^{*\sigma}P^{*}f_m J^{*\varrho}f_n = J^{*\sigma}P^{*}f_n J^{*\varrho}f_m .$$

But the functions f_n and their linear combinations are dense in L^* and therefore (13) is valid for all f^* and g^* in L^* instead of f_m and f_n and so (12_1) is established. The induction proceeds as follows:

$$(J^{*\varrho}f^*)^{n+1}J^{*\varrho}(J^{*\tau}p^*)^{n+1}g^* =$$

$$= J^{*\varrho}f^*((J^{*\varrho}f^*)^n J^{*\varrho}(J^{*\tau}p^*)^n((J^{*\tau}p^*)g^*)) =$$

$$= (J^{*\varrho}f^*)(J^{*\sigma}p^*f^*)^n J^{*\varrho}((J^{*\tau}p^*)g^*) =$$

$$= (J^{*\sigma}p^*f^*)^n(J^{*\varrho}f^*)(J^{*\sigma}p^*g^*) =$$

$$= (J^{*\sigma}p^*f^*)^n(J^{*\sigma}p^*f^*)(J^{*\varrho}g^*) \quad .$$

Thus (12_n) is completely established. Let us now write L_n and R_n for the left and right sides of (12_n) and let us choose $f^* \neq 0$ so that $|J^{*\varrho}f^*|$ is bounded by $B > 0$ a.e. and g^* identically equal to 1 say. We shall then see that L_n approaches zero (if at all) at most geometrically while R_n approaches zero faster than geometrically since $J^{*\tau}p^* = (pJ^\tau)^*$ is generalized nilpotent as we saw earlier. Let Δ be a set of positive measure where $|J^{*\sigma}p^*f^*| \geq a > 0$ for some a . Then

$$\|L_n\|^p \geq a^{np}\int_\Delta |J^{*\varrho}g^*|^p = K_L^p a^{np}$$

for some positive K_L that is independent of n . On the other hand

$$\|R_n\| \leq B^n \|J^{*\varrho}\| \, \|(J^{*\tau}p^*)^n\| \, \|g^*\| .$$

Since $J^{*\tau}p^*$ is generalized nilpotent, $\|(J^{*\tau}p^*)^n\|^{1/n} = \delta_n$ is a null sequence and so $\|R_n\|^{1/n} \leq B\|J^{*\delta}\|^{1/n}\|g^*\|^{1/n}\varepsilon_n \longrightarrow 0$ while $\|L_n\|^{1/n} \geq K_L^{1/n}a \to$ $\to a > 0$, a contradiction. This completes the proof of c) Case II.

(d) To complete the proof of the first part of the theorem, it remains to show that S and S' are not similar if $0 < \alpha < \alpha'$. This case is easily reduced to case (c). Define the bounded and boundedly invertible operator R by $(Rf)(x) = f(1-x)$ We have $R^2 = 1$, $RMR = 1 - M$, $RJ^\alpha R = J^{*\alpha}$. Thus if there existed a bounded and boundedly invertible operator P such that $P(M+\alpha J) = (M+\alpha')P$, we would have $RPR(M-\alpha J^*) = (M-\alpha' J^*)RPR$. Thus $M-\alpha J^*$ and $M-\alpha' J^*$ would be similar in L_p and so their adjoints $M-\alpha J$ and $M-\alpha' J$ would be similar in L_{p^*} contradicting case (c) of the proof. This then completes the proof of the first part of the theorem.

2) We are now considering the similarity relationships of $S = M + \alpha J$ and $T = M + \gamma J^*$. The cases α or γ equal to zero are covered in part 1) (a).

(a) Consider the case $\alpha > 0$, $\gamma > 0$. It is handled like 1) (b): If there existed a non-zero operator P such that $(M + \alpha J)P = P(M + \gamma J^*)$, we would have, using equations (2) and (4), upon multiplying our alleged equation on the right by $J^{*\gamma}$ and on the left by J^{α}, the impossible equation $MJ^{\alpha}PJ^{*\gamma} = J^{\alpha}PJ^{*\gamma}M$, since a non-zero compact operator cannot commute with M. The case $\alpha < 0, \gamma < 0$ is handled similarly.

(b) We shall now consider the case where α and γ have opposite signs and $\alpha \neq -\gamma$. Upon suitable renaming of α and γ, we have the following four cases:

(A) $M + \rho J$ and $M - \sigma J^*$ with $0 < \rho < \sigma$,

(B) $M + \rho J$ and $M - \sigma J^*$ with $0 < \sigma < \rho$,

(C) $M - \rho J$ and $M + \sigma J^*$ with $0 < \rho < \sigma$,

(D) $M - \rho J$ and $M + \sigma J^*$ with $0 < \sigma < \rho$.

We shall prove (A) in detail; the other cases are simple consequences of it: (D) results from it by starring and (C) results from (B) by starring; (B) reduces to (A) as 1) (d) was reduced to 1) (c). Thus we turn to the proof of (A).

Suppose that there existed a bounded and boundedly invertible operator P such that

$$P(M + \rho J) = (M - \rho J^*)P.$$

Let $\tau = \sigma - \rho > 0$. We show first that

$$(J^{\tau}RP)^2(M + \rho J) = (M + \rho J)(J^{\tau}RP)^2. \tag{14}$$

Writing $N = 1 - M$, we have

$$P(M+\rho J) = (M - \sigma J^*)P = (M - \sigma RJR)P = R(RMR - \sigma J)RP = R(N - \sigma J)RP,$$

$$J^{\sigma}RP(M+\varrho J) = J^{\sigma}(N-\sigma J)RP = NJ^{\varrho}J^{\tau}RP = J^{\varrho}(N-\varrho J)J^{\tau}RP$$

where we have used $NJ^{\varrho} = J^{\varrho}(N-\varrho J)$. This implies

$$J^{\tau}RP(M+\varrho J) = (N-\varrho J)J^{\tau}RP,$$

$$(J^{\tau}RP)^2(M+\varrho J) = J^{\tau}RP(N-\varrho J)J^{\tau}RP = (J^{\tau}RP)^2 - J^{\tau}RP(M+\varrho J)J^{\tau}RP =$$

$$= (J^{\tau}RP)^2 - (N-\varrho J)(J^{\tau}RP)^2 = (M+\varrho J)(J^{\tau}RP)^2.$$

We show next that $(J^{\tau}RP)^2$ is generalized nilpotent, that is, since it is compact, that it has no non-zero eigenvalues. Suppose on the contrary that it had the non-zero eigenvalue λ with corresponding finite-dimensional eigenspace $L(\lambda)$. If $f \in L(\lambda)$ then (14) implies that

$$(J^{\tau}RP)^2(M+\varrho J)f = \lambda(M+\varrho J)f$$

so that $(M+\varrho J)f \in L(\lambda)$ and therefore $(M+\varrho J)L(\lambda) \subset L(\lambda)$. The finite dimensionality of $L(\lambda)$ implies that $M+\varrho J$ has eigenvalues, contradicting Theorem 3 in the present case. Thus $(J^{\tau}RP)^2$ is generalized nilpotent. In analogy with (12_n) we establish

(15_n) $$J^{\varrho}((J^{\tau}RP)^2)^n f(RJ^{\varrho}gJ^{\varrho}h)^n = (RJ^{*\sigma}PgJ^{*\sigma}Ph)^n J^{\varrho}f$$

for all f, g, and h in L. Equation (15_n) follows (15_1) by induction thus:

$$J^{\varrho}((J^{\tau}RP)^2)^n(J^{\tau}RP)^2 f(RJ^{\varrho}gJ^{\varrho}h)^n RJ^{\varrho}gJ^{\varrho}h =$$

$$= (RJ^{*\sigma}PgJ^{*\sigma}Ph)^n J^{\varrho}(J^{\tau}RP)^2 fRJ^{\varrho}gJ^{\varrho}h =$$

$$= (RJ^{*\sigma}PgJ^{*\sigma}Ph)^n(RJ^{*\sigma}PgJ^{*\sigma}Ph)J^{\varrho}f.$$

In order to prove (15_1) we first show that

(16) $$J^{*\sigma}PfJ^{\varrho}g = J^{*\sigma}PgJ^{\varrho}f$$

for all f and g in L. Let f_m in L be such that $J^{\varrho}f_m = x^m$; they exist for all sufficiently large n.
Then

$$J^{*\sigma}Pf_nJ^{\varrho}f_m = x^m J^{*\sigma}Pf_n = M^m J^{*\sigma}Pf_n = J^{*\sigma}(M-\sigma J^*)^m Pf_n = J^{*\sigma}P(M+\varrho J)^m f_n$$

and similarly $J^{*\sigma}Pf_m J^{\varrho} f_n = J^{*\sigma}P(M-\varrho J)^n f_m$. But $J^{\varrho}(M+\varrho J)^m f_n = M^m J^{\varrho} f_n = x^{m+n} = J^{\varrho}(M+\varrho J)^n f_m$, , so that, since J^{ϱ} is 1 – 1, we have $(M+\varrho J)^m f_n = (M+\varrho J)^n f_m$ which implies the truth of (16) for $f = f_m$ and $g = f_n$; but this implies the validity of (16) for all f and g in L.

To prove (15_1), we observe that (16) implies

$$J^{*\sigma}PfJ^{\varrho}g = RJ^{\sigma}RPfJ^{\varrho}g = J^{*\sigma}PgJ^{\varrho}f \text{ or}$$

$$R(RJ^{\sigma}RPfJ^{\varrho}g) = J^{\varrho}J^{\tau}RPfRJ^{\varrho}g = R(J^{*}PgJ^{\varrho}f) = RJ^{*\sigma}PgRJ^{\varrho}f$$

and so

$$J^{\varrho}(J^{\tau}RP)(J^{\tau}RP)fRJ^{\varrho}gJ^{\varrho}h = RJ^{*\sigma}PgRJ^{\varrho}(J^{\tau}RP)fJ^{\varrho}h =$$

$$= R(J^{*\sigma}PgJ^{\varrho}(J^{\tau}RP)fRJ^{\varrho}h) = R(J^{*\sigma}PgRJ^{*\sigma}PhRJ^{\varrho}f) =$$

$$= RJ^{*\sigma}PgJ^{*\sigma}PhJ^{\varrho}f .$$

that is, (15_1).

Just as in the case (c) of the first part of the proof, we shall use the presence of a generalized nilpotent operator in the left side L_n of (15_n) occurring with exponent n ; the right side R_n contains no such factors but only powers of functions. The growth estimates in question require us to find a function g in L such that $J^{\varrho}g$ is bounded and $RJ^{*\sigma}PgJ^{*\sigma}Pg \neq 0$ on a set of positive measure. Let Δ_m be the set where $J^{*\sigma}Pf_m = 0$. Equation (16) says that $J^{*\sigma}Pf_m x^m = J^{*\sigma}Pf_n x^m$ so that $\Delta_m = \Delta_n = \Delta$ say. Thus $J^{*\sigma}Pf_m = 0$ on Δ for all relevant m so that $J^{*\sigma}Pf = 0$ for all $f \in L$ on Δ. Since P is invertible, there exist functions f such that $J^{*\sigma}Pf \neq 0$ so that the measure of Δ must be zero. Hence to find a function g as specified above is now easily accomplished by letting it be equal to some f_m .

In (15_n) let $g = h$ be the function just found; f will be specified presently. Let the positive real number B be a bound for $|RJ^{\varrho}gJ^{\varrho}g|$. Then

$$\| L_n \| \leq B^n \| J^{\varrho} \| \, \| (J^{\tau}RP)^{2n} \| \, \| f \| .$$

Let Γ be a set of positive measure where $|RJ^{*\sigma}gJ^{*\sigma}g| \geq a \geq 0$ for some a. Then

$$\| R_n \|^p \geq a^{np} \int_{\Gamma} |J^{\varrho}f|^p .$$

Now choose f such that $J^{\rho}f \neq 0$ on Γ. Then there exists $b > 0$ so that $\int_{\Gamma} |J^{\rho}f|^p = b^p$ and

(17) $$\|R_n\|^{1/n} \geq ab^{1/n} \to a > 0 .$$

Since $(J^{\tau}RP)^2$ is generalized nilpotent, $\|(J^{\tau}RP)^{2n}\|^{1/n} = \varepsilon_n$ is a null sequence and we have

$$\|L_n\|^{1/n} \leq B\|J^{\rho}\|^{1/n}\|f\|^{1/n}\varepsilon_n \to 0 ,$$

contradicting (17). This completes the proof that if $\alpha \neq -\gamma$, the operators S and T are not similar.

(c) The only remaining part of 2i), namely, if α or γ is not in the interval $[-1/p^*, 1/p]$, is settled by considering the various kinds of spectra of S and T as described in Theorem 3.

(d) We lastly prove 2ii), namely, if $\alpha = -\gamma \in (-1/p^*, 1/p)$, then S and T are similar. R.K. Juberg [2] has proved that if $0 \leq \gamma < 1/p^*$, then the equation

$$J^{\gamma} = QJ^{*\gamma}$$

has a bounded and boundedly invertible solution Q in L_p. We have the equations $QJ^{*\gamma}M = Q(M+\gamma J^*)J^{*\gamma} = J^{\gamma}M = (M-\gamma J)J^{\gamma} = (M-\gamma J)QJ^{*\gamma}$ which imply that $Q(M+\gamma J^*) = (M-\gamma J)Q$, i.e., S and T are similar if $\alpha \in (-1/p^*, 0)$. The case $\alpha \in [0, 1/p)$ is handled similarly: still referring to Juberg's paper, we use the fact that the equation

$$J^{*\alpha} = J^{\alpha}Q$$

has a bounded and boundadly invertible solution Q in L_p if $0 \leq \alpha < 1/p$. We now have the equations $MJ^{\alpha}Q = J^{\alpha}(M+\alpha J)Q = J^{*\alpha}(M-\alpha J^*) = J^{\alpha}Q(M-\alpha J^*)$ which imply that $(M+\alpha J)Q = Q(M-\alpha J^*)$ in L_p for $\alpha \in [0, 1/p)$. This completes the proof of the theorem.

If we view $M+\varkappa J$ and $M+\varkappa J^*$ for complex $\varkappa$ as perturbations of M, our preceding work shows that they are rather harsh perturbations. Our last theorem says that perturbing $M+\varkappa J$ or $M+\varkappa J^*$ by multiples of J^{μ} or $J^{*\mu}$ is much less harsh, for $\mathrm{Re}(\mu) > 1$.

Theorem 5. *Let $\varkappa, \lambda$ and μ be complex numbers where $\mathrm{Re}(\mu) > 1$. Then the bounded operator $M+\varkappa J+\lambda J^{\mu}$ in $L_p(0,1)$ is similar in L_p to $M+\varkappa J$.*

The same conclusion holds if J *or* J^μ *or both are replaced by* J^* *and* $J^{*\mu}$ *respectively.*

Proof. This is merely a matter of calculation. Write $(I+T_k)(M+\varkappa J+\lambda J^\mu)=$ $=(M+\varkappa J)(I+T_k)$ where $(T_k f)(t)=(k*f)(t)=\int_0^t k(t-s)f(s)ds$.

This leads to the equation $tk-\lambda/\Gamma(\mu)(t^{\mu-1}+t^{\mu-1}*k)=0$ which has the absolutely convergent infinite series solution

$$k(t)=\sum_{j=1}^{\infty}\lambda^j/(j!(\mu-1)^j\Gamma(j(\mu-1)))t^{j(\mu-1)-1}$$

in $L_1(0,1)$. The proof for the remaining cases is similar.

REFERENCES

[1] M.S. Brodskiĭ and M.S. Livšic, Spektral'nyĭ analiz nesamosopražonnyh operatorov..., *Usp. Mat. Nauk,* XIII, 1,3 – 85, 1958 = *Am. Math. Soc. Transl.,* (2) 13, (1960) 265-346. (Spectral analysis of non-selfadjoint operators...).

[2] R.K. Juberg, On the boundedness of certain singular integral operators; preceding this paper.

[3] G.K. Kalisch, On fractional integral of pure imaginary order in L_p, *Proc. Amer. Math. Soc.,* 18, (1967), 136-139.

[4] S. Kantorovitz, The C^k-classification of certain operators in L_p. II, *Trans. Amer. Math. Soc.,* 146, (1969), 61-67.

COLLOQUIA MATHEMATICA SOCIETATIS JÁNOS BOLYAI
5. HILBERT SPACE OPERATORS, TIHANY (HUNGARY), 1970

On the location of invariant subspaces of dissipative operators

G. E. KISILEVSKIĬ

M.S. Brodskiĭ in his paper [1] studies the mutual location of those invariant subspaces for operators of the class $\Lambda^{(\exp)}$ * in which the induced operators belong to Λ_1, the class of Volterra-type dissipative operators with a one-dimensional imaginary component. The following problem naturally arises: how should the subspaces $\{\mathfrak{h}_k\}_{k=1}^{m}$ of a Hilbert space be located in order that there exist a dissipative operator $\mathfrak{A}$ in the space $\tilde{\bigcup}_{k=1}^{m} \mathfrak{h}_k$ satisfying the requirements:

$$\mathfrak{A}\mathfrak{h}_k \subset \mathfrak{h}_k, \quad \mathfrak{A}_k \in \Lambda_1 \qquad (\mathfrak{A}_k = \mathfrak{A} \mid \mathfrak{h}_k; \quad k = 1, 2, \ldots, m).$$

In principle, this problem is solved in the present paper. For the sake of simplicity, we confine ourselves to the case $m = 2$.

Theorem. *Assume that* $\mathfrak{h}_1$ *and* $\mathfrak{h}_2$ *are subspaces of a separable Hilbert space and that* τ_1 *and* τ_2 *are positive real numbers* $(\tau_1 \leq \tau_2)$. *In order that in the space* $\mathfrak{h}_1 \tilde{\cup} \mathfrak{h}_2$ *there exist an operator* $\mathfrak{A}$ *of the class* $\Lambda^{(\exp)}$ *satisfying the requirements:*

* We use the notations and terminology of the above-mentioned paper [1] .

$$\mathfrak{A}\mathfrak{h}_k \subset \mathfrak{h}_k, \quad \mathfrak{A}_k \in \Lambda_1, \quad \mathrm{sp}\,\mathfrak{Im}\,\mathfrak{A}_k = \tau_k \qquad (k = 1, 2),$$

it is necessary and sufficient that $\mathfrak{h}_2$ *can be decomposed as* $\mathfrak{h}_2^{(1)} \oplus \mathfrak{h}_2'$, *where* $\mathfrak{h}_2'$ *is different from zero if and only if* $\tau_1 \neq \tau_2$, *and, moreover, that there exist orthonormal bases* $\{e_i^{(1)}\}$ *and* $\{e_i^{(2)}\}$ *in* $\mathfrak{h}_1$ *and* $\mathfrak{h}_2^{(1)}$, *respectively, such that*

$$\mathfrak{h}_2' \perp \mathfrak{h}_1, \qquad (e_i^{(1)}, e_j^{(2)}) = 0 \qquad (i \neq j)$$

(1) $$(e_1^{(1)}, e_1^{(2)}) = (e_2^{(1)}, e_2^{(2)}) = \ldots = \Theta .$$

Proof. Consider the Hilbert space $L_{\mathfrak{G}}^{(2)}(0, \ell)$ with $\ell \geq \max \tau_k$. Obviously, the inner product of any two vectors of the form

$$f_k = \varphi_k(x) h_k \quad (h_k \in \mathfrak{G}, \ \varphi_k(x) \in L^2(0, \ell); \ \ k = 1, 2)$$

of $L_{\mathfrak{G}}^{(2)}(0, \ell)$ can be expressed as

(2) $$(f_1, f_2) = (h_1, h_2)_{\mathfrak{G}} \cdot \int_0^{\ell} \varphi_1(x)\, \bar{\varphi}_2(x)\, dx .$$

Put

$$H_k = L(h_k, \tau_k) = \{\varphi^{(k)}(x) h_k\} \ (h_k \in \mathfrak{G}, \ \|h_k\|_{\mathfrak{G}} = 1, \ \varphi^{(k)}(x) \in L^{(2)}(0, \tau_k)^{*}; \ \ k = 1, 2).$$

Take an arbitrary orthonormal basis $\{\varphi_i(x)\}$ of the space $L^{(2)}(0, \tau_1)$ and, in case $\tau_1 < \tau_2$, extend it to an orthonormal basis of the space $L^{(2)}(0, \tau_2)$ by adding the vectors $\{\varphi_\gamma'(x)\}$ $(\gamma \in \Gamma)$. Consider the vectors

$$g_i^{(k)} = \varphi_i(x) h_k \quad (i = 1, 2, \ldots; \ k = 1, 2), \quad g_\gamma' = \varphi_\gamma'(x) h_2 \quad (\gamma \in \Gamma).$$

On account of (2) the following conditions are satisfied:

1) $\{g_i^{(1)}\}_{i=1}^{\infty}$ is an orthonormal basis of H_1;

2) $\{g_i^{(2)}\}_{i=1}^{\infty} \cup \{g_\gamma'\}_{\gamma \in \Gamma}$ is an orthonormal basis of H_2;

3) $(g_i^{(1)}, g_i^{(2)}) = (h_1, h_2)$ $(i = 1, 2, \ldots)$; $(g_i^{(1)}, g_j^{(2)}) = 0$ $(i \neq j)$;

* We consider the space $L^{(2)}(0, \tau)$ $(\tau \leq \ell)$ as a subspace of $L^{(2)}(0, \ell)$.

4) $(g_i^{(1)}, g'_\gamma) = 0 \quad (i = 1, 2, \dots; \ \gamma \in \Gamma)$.

Taking now the spaces spanned by the vectors $\{g_1^{(2)}, g_2^{(2)}, \dots\}$ and $\{g'_1, g'_2, \dots\}$ as $H_2^{(1)}$ and H'_2, respectively, we derive from what has been said above that

$$H_2 = H_2^{(1)} \oplus H'_2, \qquad H'_2 \perp H_1 .$$

Since according to [2] each operator $\mathfrak{A}$, $\sigma(\mathfrak{A}) \leq 2\ell$, of the class $\Lambda^{(exp)}$ is unitarily equivalent to the restriction of the integration operator $\mathfrak{J}$ to one of its invariant subspaces $L \subset L^{(2)}_{\mathfrak{G}}(0, \ell)$, and since in view of Lemma 2.1 of [1] every invariant subspace of $\mathfrak{J}$ in which the induced operator belongs to Λ_1 coincides with one of the subspaces $L(h, \tau)$, the necessity part of the theorem is established.

In order to prove the sufficiency part, choose a sequence of vectors $h_k \in \mathfrak{G}$ ($\|h_k\|_{\mathfrak{G}} = 1, \ k = 1, 2$) satisfying the equality

(3) $$(h_1, h_2) = \Theta$$

and consider the subspaces $H_k = L(h_k, \tau_k)$ $(k = 1, 2)$. As shown above, there exist a decomposition $H_2 = H_2^{(1)} \oplus H'_2$ and three orthonormal bases $\{g_i^{(1)}\}, \{g_i^{(2)}\}, \{g'_j\}$ corresponding to the spaces $H_1, H_2^{(1)}$, and H'_2 such that relations 1)–4) are satisfied.

Owing to (1) and (3) the formulae

$$U e_i^{(k)} = g_i^{(k)}, \quad U e'_\gamma = g'_\gamma \qquad (i = 1, 2, \dots; \ \gamma \in \Gamma, \ k = 1, 2),$$

where $\{e'_\gamma\}$ is any orthonormal basis in $\mathfrak{H}'_2$, obviously determine an isometric mapping U of the space $\mathfrak{H}_1 \tilde{\cup} \mathfrak{H}_2$ onto $H_1 \tilde{\cup} H_2$ such that the subspace $\mathfrak{H}_k$ is mapped onto H_k $(k = 1, 2)$. It remains to set

$$\mathfrak{A} = U^* (\mathfrak{J} \mid H_1 \tilde{\cup} H_2) U .$$

REFERENCES

[1] M.S. Brodskiĭ, *Some invariant subspaces of dissipative operators of exponential type* (Moscow, 1969).

[2] M.S. Brodskiĭ, *Triangular and Jordan-type representations of linear operators* (Moscow, 1969).

COLLOQUIA MATHEMATICA SOCIETATIS JÁNOS BOLYAI
5. HILBERT SPACE OPERATORS, TIHANY (HUNGARY), 1970

Power-bounded operators and finite type von Neumann algebras

I. KOVÁCS

There are certain facts about linear operators of a finite-dimensional Hilbert space which can be formulated and proved in the infinite dimensional case, too, provided that the operators are elements of a finite type von Neumann algebra. One of these facts was observed by C. Foiaş and me several years ago. In [1] we proved that, if a completely non-unitary contraction T (the terminology will be that of [2]) belongs to a finite type von Neumann algebra A on a Hilbert space H, then T^n tends strongly to zero as $n \to \infty$. (This property turned out even to be characteristic for von Neumann algebras of this kind.) This is evidently true in the finite-dimensional case since then the spectrum of a completely non-unitary contraction is situated strictly inside the unit circle of the complex plane. Let me recall the idea of the proof for the general case. Consider an arbitrary contraction T in A and take the bounded monotonic sequences $T^{*n} T^n$ and $T^n T^{*n}$ of selfadjoint elements of A. Making heavily use of the existence of sufficiently many normal traces on A, we show that $\lim T^{*n} T^n = \lim T^n T^{*n}$, and that this common limit, say E is a projection of A. Moreover,

$EH = \{x \in H : T^n x \to 0\} = \{x \in H : T^{*n} x \to 0\}$, and EH is that part of H, in the canonical decomposition of T, on wich T is completely non-unitary. Form this the assertion is immediate.

This theorem has two consequences which are worth being mentioned (and can be directly proved in the finite-dimensional case).

1) The minimal unitary dilation of a completely non-unitary contraction belonging to a finite type von Neumann algebra is a bilateral shift (cf. [2], chap. II, Thm. 1.2).

2) Any contraction of class $C_{1\cdot}$ belonging to a finite type von Neumann algebra is necessarily unitary. (Dr. J. Szücs has called my attention to the fact that if in a von Neumann algebra A every contraction of class C_{11} is unitary, then A is necessarily of finite type.)

All these suggest to ask the following question:

Let A be a finite type von Neumann algebra and T a power-bounded element of A which is of class C_{11}. Is T innerly similar to a unitary element of A, i.e. do there exist a boundedly invertible selfadjoint element S and a unitary element U of A such that $T = S^{-1} U S$? This is certainly the case if the underlying Hilbert space is of finite dimension. One way to check this fact may consist in performing the argument used to prove [2], chap. II. Prop. 5.3. Howerer, this argument relying upon Banach limits does not seem helpful with the above problem.

REFERENCES

[1] C. Foiaş – I. Kovács, Une caractérisation nouvelle des algèbres de von Neumann finies, *Acta Sci. Math.*, 23 (1962), 274-278.

[2] B. Sz.-Nagy – C. Foiaş, *Analise harmonique des opérateurs de l'espace de Hilbert* (Budapest, 1967).

COLLOQUIA MATHEMATICA SOCIETATIS JÁNOS BOLYAI
5. HILBERT SPACE OPERATORS, TIHANY (HUNGARY), 1970

Über die verallgemeinerten Resolventen und die charakteristische Funktion eines isometrischen Operators im Raume $\Pi_\varkappa$

M. G. KREĬN und H. LANGER

In unserem vorangegangenen Artikel [1] gaben wir die allgemeine Form der verallgemeinerten Resolvente eines π-hermiteschen Operators A im Pontrjaginschen Raum $\Pi_\varkappa$ $(0 < \varkappa < \infty)$ an unter der Voraussetzung, dass A gleiche Defektzahlen hat.

In der vorliegenden Arbeit wird die analoge Aufgabe für einen π-isometrischen Operator V in $\Pi_\varkappa$ mit gleichen Defektzahlen gelöst; dabei setzen wir zusätzlich voraus, dass der Definitionsbereich $\mathfrak{D}(V)$ und der Wertebereich $\mathfrak{R}(V)$ des Operators V abgeschlossene nichtentartete Teilraume von $\Pi_\varkappa$ sind.*

* In [1] wurden die verallgemeinerten Resolventen des Operators A ohne zusätzliche Voraussetzungen dieser Art beschrieben. Wir könnten uns auch hier auf allgemeinere Voraussetzungen beschränken, dann würden jedoch die Aussagen komplizierter (siehe z.B. die Arbeit von V.M. Adamjan, D.Z. Arov und M.G. Kreĭn [4], wo für eine spezielle Klasse π-isometrischer Operatoren mit Defektzahlen eins im wesentlichen die Beschreibung aller verallgemeinerten Resolventen auch für den Fall erhalten wurde, dass die Teilräume $\mathfrak{D}(V)$ und $\mathfrak{R}(V)$ entartet sind).

Gleichzeitig entwickeln wir unter denselben Voraussetzungen eine Theorie der charakteristischen Funktion $X_V(z)$ des π-isometrischen Operators V. Diese Operatorfunktion $X_V(z)$ ist das Analogon der von M.S. Livšic ([2], [3]) eingeführten charakteristischen Funktion eines isometrischen Operators im Hilbertraum. Es zeigt sich dabei, dass ein enger Zusammenhang zwischen der Theorie der verallgemeinerten Resolventen und der Theorie der charakteristischen Funktionen solcher Operatoren V besteht. Möglicherweise sind einige Resultate (z.B. Satz 4.2) auch neu für den Fall eines Hilbertraumes $\mathfrak{H} = \Pi_0$ (alle unsere Betrachtungen gelten auch für $\varkappa = 0$, wobei sie sich jedoch wesentlich vereinfachen).

Beim Beweis der charakteristischen inneren Eigenschaften der Funktion $X_V(z)$ benutzte M.S. Livšic die Spektralzerlegung eines unitären Operators im Hilbertraum. In unserem Falle würde ein solcher Weg zu den bekannten Schwierigkeiten führen, die jedoch nicht dem Wesen der Sache entsprächen. An Stelle der Spektralzerlegung benutzen wir deshalb ein bei Fragestellungen von der Art des Momentenproblemes bekanntes Vorgehen (mit gewissen, durch die Indefinitheit bedingten Modifizierungen); es besteht darin, dass man einem positiv definiten Kern von zwei Veränderlichen einen hermiteschen oder isometrischen Operator in einem von dem Kern erzeugten Hilbertraum zuordnet. Unsere Konstruktion ist dabei eine unmittelbare Verallgemeinerung der in den Arbeiten von B.Sz.-Nagy und A. Korányi (siehe [5]) angewandten Methode.

An anderer Stelle werden wir auf den Zusammenhang zwischen der Funktion $X_V(z)$ und der Funktion $Q_A(z)$ aus [1] für einen π-isometrischen Operator V und einen π-hermiteschen Operator A, die beide durch die Cayleytransformation verbunden sind, eingehen.

Wie im definiten Fall hat man die Funktion $X_V(z)$ als Spezialfall der charakteristischen Funktion $X_T(z)$ einer allgemeineren Klasse von Operatoren T (z.B. der J-kontrahierenden Operatoren) anzusehen. Erste Untersuchungen solcher Funktionen finden sich in den Arbeiten von A.V. Kužel' (siehe [6]). Ihre weitere Entwicklung (insbesondere der Beweis eines hinreichend allgemeinen Multiplikationssatzes für charakteristische Funktionen) wird es möglicherweise gestatten, einige Betrachtungen in § 3 zu vereinfachen. Wir umgehen hier die Benutzung dieses Apparates, indem wir eine Reihe wichtiger Begriffe und Methoden aus den Arbeiten von V.P. Potapov [7] und Ju.P. Ginzburg [8] benutzen.

Schliesslich bemerken wir, dass man die hier und in [1] erhaltenen

Formeln für die verallgemeinerten Resolventen als Verallgemeinerung der Formeln von Kreĭn-Saakjan (siehe z.B. [9]) für Operatoren im Hilbertraum anzusehen hat. Wie im definiten Falle folgt aus ihnen unmittelbar die Gültigkeit der Regeln von Štraus-Čumakin (siehe z.B. [10], [11]) zur Konstruktion der verallgemeinerten Resolventen für die von uns betrachteten Operatoren.

§ 1. BEZEICHNUNGEN UND TERMINOLOGIE

Wir benutzen in der vorliegenden Arbeit die in [1] eingeführte Terminologie. Dabei seien die wichtigsten Definitionen hier noch einmal wiederholt.

Unter einem $\Pi_\varkappa$*-Raum* oder *Pontrjaginraum mit* $\varkappa$ *negativen Quadraten,* $0 \leq \varkappa < \infty$, verstehen wir einen komplexen linearen Raum $\Pi_\varkappa$, der versehen ist mit einem Skalarprodukt [. , .], dem sogenannten π*-Skalarprodukt,* und in der Form

(1.1) $$\Pi_\varkappa = \Pi_- [+] \Pi_+$$

dargestellt werden kann; dabei bezeichnet Π_- einen $\varkappa$-dimensionalen negativen Teilraum* von $\Pi_\varkappa$, Π_+ einen endlich-oder unendlichdimensionalen Teilraum, der bezüglich des π-Skalarproduktes einen Hilbertraum bildet, und $[+]$ die direkte und π-orthogonale Summe, d.h., jedes Element $f \in \Pi_\varkappa$ lässt sich eindeutig in der Form $f = f_- + f_+$ mit $f_\pm \in \Pi_\pm$, $[f_+, f_-] = 0$ darstellen. Ein Π_0-Raum ist also insbesondere ein Hilbertraum bezüglich des π-Skalarproduktes [. , .].

Im Falle $\varkappa > 0$ gibt es unendlich viele Zerlegungen (1.1) des Raumes $\Pi_\varkappa$. Für jede solche Zerlegung definieren wir Projektionen P_+ und P_- auf Π_+ bzw. Π_- durch die Gleichungen $P_\pm f = f_\pm$ $(f = f_+ + f_-,\ f_\pm \in \Pi_\pm)$, und setzen noch $J = P_+ - P_-$. Versehen mit dem Skalarprodukt

(1.2) $$(f, g) = [Jf, g], \quad f, g \in \Pi_\varkappa,$$

* Ein Element f eines Raumes mit indefiniten Skalarprodukt $[\cdot,\cdot]$ heisst positiv, negativ, neutral usw., wenn $[f, f] > 0$, < 0, $= 0$ usw. gilt; ein Teilraum eines solchen Raumes heisst positiv, negativ, neutral usw., wenn alle seine vom Nullelement verschiedenen Elemente diese Eigenschaft haben.

ist Π_κ ein Hilbertraum und P_+, P_- sind darin orthogonale Projektionen mit $P_+ + P_- = I$. Die Normtopologie dieses Hilbertraumes ist unabhängig von der speziellen Wahl der Zerlegung (1.1). Wenn nichts anderes vermerkt ist, beziehen sich alle topologischen Begriffe in Π_κ auf diese Normtopologie.

Eine Folge $(f_n) \subset \Pi_\kappa$ ist genau dann konvergent in Π_κ gegen ein Element $f \in \Pi_\kappa$, wenn für $n \to \infty$ folgendes gilt:

a) $[f_n, f_n] \to [f, f]$

b) $[f_n, g] \to [f, g]$ für alle g aus einer in Π_κ totalen Menge. Dabei heisst eine Teilmenge $\mathfrak{M} \subset \Pi_\kappa$ *total* in Π_κ, falls kein Element aus Π_κ π-orthogonal ist auf $\mathfrak{M}$, d.h., aus $[f_0, g] = 0$ für ein $f_0 \in \Pi_\kappa$ und alle $g \in \mathfrak{M}$ folgt $f_0 = 0$.

Es sei jetzt $\mathcal{L}$ ein unendlichdimensionaler komplexer linearer Raum, der versehen ist mit einem indefiniten und i.a. entartenden* Skalarprodukt [. , .], das genau κ negative Quadrate hat. Letzteres bedeutet, dass jeder negative Teilraum $\mathcal{L}_- \subset \mathcal{L}$ eine Dimension $\leqq \kappa$ hat und unter allen negativen Teilräumen mindestens einer der Dimension κ existiert. Dann entartet das π-Skalarprodukt auf dem Faktorraum $\mathcal{L}/\mathcal{L}_0$, $\mathcal{L}_0 = \mathcal{L} \cap \mathcal{L}^{[\perp]}$, nicht und dieser lässt sich zu einem Π_κ-Raum Π_κ vervollständigen. Dadurch wird jedem $f \in \mathcal{L}$ in natürlicher Weise ein Element aus Π_κ zugeordnet, und die Menge dieser Bilder liegt dicht in Π_κ. Der Raum Π_κ ist bis auf Isomorphie eindeutig bestimmt. Für ein Element $f \in \mathcal{L}$ bezeichnen wir das ihm zugeordnete Element aus Π_κ ebenfalls mit f und sagen, der Raum $\mathcal{L}$ sei in den Raum Π_κ *kanonisch eingebettet.*

Ein linearer Operator V, der eine Teilmenge $\mathfrak{D}(V) \subset \Pi_\kappa$ in Π_κ abbildet, heisst π*-isometrisch,* wenn

$$[Vf, Vg] = [f, g], \quad f, g \in \mathfrak{D}(V),$$

gilt; ist dabei $\mathfrak{D}(V) = \Pi_\kappa$ und auch der Wertebereich $\mathfrak{R}(V) = \Pi_\kappa$, so heisst

* d.h., es kann der isotrope Teil $\mathcal{L}_0 = \mathcal{L} \cap \mathcal{L}^{[\perp]} \neq \{0\}$ sein, wobei für eine beliebige Teilmenge $\mathfrak{N} \subset \mathcal{L}$ mit $\mathfrak{N}^{[\perp]}$ das π-orthogonale Komplement von $\mathfrak{N}$ bezeichnet wird: $\mathfrak{N}^{[\perp]} = \{f \in \mathcal{L} \mid [f, g] = 0$ für alle $g \in \mathfrak{N}\}$.

V π*-unitär.* Ein π-isometrischer Operator, für den das π-Skalarprodukt auf den Abschliessungen $\overline{\mathfrak{D}(V)}$ und $\overline{\mathfrak{R}(V)}$ nicht entartet, ist beschränkt ([12], Satz 4.3); insbesondere ist also jeder π-unitäre Operator beschränkt.

Für einen beschränkten linearen Operator $A \in [\Pi_\varkappa, \Pi_\varkappa]$ definieren wir den π*-adjungierten Operator* A^+ durch die Gleichung

$$[Af, g] = [f, A^+ g], \quad f, g \in \Pi_\varkappa .$$

Ein Operator $T \in [\Pi_\varkappa, \Pi_\varkappa]$ heisse π*-kontrahierend,* wenn

$$[Tf, Tf] \leqq [f, f], \qquad f \in \mathfrak{D}(T) = \Pi_\varkappa$$

gilt. Bekanntlich ([13], S. 150) ist mit T auch T^+ π-kontrahierend. Ein π-kontrahierender Operator T in $\Pi_\varkappa$ hat einen $\varkappa$-dimensionalen nicht-positiven invarianten Teilraum $\mathfrak{I}$, so dass alle Punkte z von $\sigma(T|\mathfrak{I})$ – dem Spektrum der Einschränkung von T auf $\mathfrak{I}$ – die Eigenschaft $|z| \geqq 1$ haben und mit Ausnahme solcher Punkte $z \in \sigma(T|\mathfrak{I})$ das Spektrum $\sigma(T)$ ganz im abgeschlossenen Einheitskreis liegt. Die Funktion $(I - zT)^{-1}$ ist folglich im Inneren $\mathcal{E}$ des Einheitskreises meromorph * mit höchstens $\varkappa$ Polen.

Ist $\mathfrak{G}$ ein Π_k-Raum, $0 \leqq k < \infty$, mit dem π-Skalarprodukt $[\,.\,,.\,]$ und $\mathfrak{U}$ eine nichtleere Menge, so nennen wir eine Abbildung $K(\,.\,,.\,)$ von $\mathfrak{U} \times \mathfrak{U}$ in $[\mathfrak{G}, \mathfrak{G}]$ einen $\mathfrak{G}$*-Kern* (auf $\mathfrak{U}$). Wir sagen, ein solcher $\mathfrak{G}$-Kern habe auf $\mathfrak{U}$ genau $\varkappa$ *negative Quadrate,* wenn folgendes gilt:

1) $K(z, \zeta) = K^+(\zeta, z), \qquad z, \zeta \in \mathfrak{U};$

2) für beliebige natürliche Zahl n, beliebige Elemente $f_1, f_2, \ldots, f_n \in \mathfrak{G}$ und beliebige $z_1, z_2, \ldots, z_n \in \mathfrak{U}$ hat die (hermitesche) Matrix

$$([K(z_i, z_j) f_i, f_j])$$

höchstens $\varkappa$ negative Eigenwerte und für mindestens eine solche Wahl von n, $f_1, f_2, \ldots, f_n$ und $z_1, z_2, \ldots, z_n$ genau $\varkappa$ negative Eigenwerte. Ist dabei $\varkappa = 0$, so nennen wir den $\mathfrak{G}$-Kern auch *positiv definit.*

* Dabei nennen wir eine Funktion in einem (offenen) Gebiet der komplexen Ebene meromorph, wenn sie dort mit evtl. Ausnahme isolierter Punkte definiert und holomorph ist und diese isolierten Punkte Pole der Funktion sind.

§ 2. FUNKTIONEN MIT DER EIGENSCHAFTEN $\mathcal{R}_\varkappa(\mathfrak{G})$

1. Es sei $\mathfrak{G}$ ein Π_k-Raum mit dem π-Skalarprodukt [. , .]. Wir sagen, eine Funktion F habe die *Eigenschaft* $\mathcal{R}^i_\varkappa(\mathfrak{G})$, wenn sie definiert ist auf einer im Inneren $\mathfrak{E}$ des Einheitskreises gelegenen Umgebung $\mathfrak{U}_F$ des Punktes $z = 0$, ihre Werte in $[\mathfrak{G}, \mathfrak{G}]$ liegen, sie im Punkte $z = 0$ schwach stetig ist und der $\mathfrak{G}$-Kern

$$\Phi(z,\zeta) = \frac{F(z) + F^+(\zeta)}{1 - z\bar{\zeta}} \tag{2.1}$$

genau $\varkappa$ negative Quadrate auf $\mathfrak{U}_F$ hat.

Der Raum $\mathfrak{G}$ wurde dabei mit einem indefiniten Skalarprodukt versehen, um in der Theorie der charakteristischen Funktionen und der verallgemeinerten Resolventen in §§ 3,4 einige Aussagen einfacher formulieren zu können. Für das Studium der Funktionen F mit der Eigenschaft $\mathcal{R}^i_\varkappa(\mathfrak{G})$ (oder der auf Seite 15 eingeführten Eigenschaft $\mathcal{R}_\varkappa(\mathfrak{G})$) ist die Indefinitheit des Skalarproduktes unwesentlich: Hat nämlich eine Funktion F z.B. die Eigenschaft $\mathcal{R}^i_\varkappa(\mathfrak{G})$, so hat für den Operator J aus der Beziehung (1.2) die Funktion JF die Eigenschaft $\mathcal{R}^i_\varkappa(\mathfrak{G}_0)$, falls $\mathfrak{G}_0$ den Hilbertraum bezeichnet, der sich ergibt, wenn wir $\mathfrak{G}$ mit dem positiv definiten Skalarprodukt (1.2) versehen.*

Jede Funktion F mit der Eigenschaft $\mathcal{R}^i_\varkappa(\mathfrak{G})$ erzeugt auf natürliche Weise einen $\Pi_\varkappa$-Raum, den wir mit $\Pi^i_\varkappa(F)$ oder kurz mit $\Pi^i_\varkappa$ bezeichnen. In der Tat, wir ordnen jedem Punkt $z \in \mathfrak{U}_F$ ein Symbol ε_z zu und bilden die lineare Menge $\mathcal{L}^i = \mathcal{L}^i(F)$ aller formalen Summen

$$f = \sum_{z \in \mathfrak{U}_F} \varepsilon_z x_z, \quad x_z \in \mathfrak{G}, \quad z \in \mathfrak{U}_F, \tag{2.2}$$

* Diese Bemerkung gilt offensichtlich noch, wenn für den Operator J aus (1.2) der Wertebereich der Projektion P_- unendlichdimensional ist.

wobei nur endlich viele der x_z vom Nullelement verschieden sein sollen. Für zwei Elemente f und $g = \sum_{z \in \mathfrak{U}_F} \varepsilon_z y_z \in \mathcal{L}^i$ definieren wir ein (möglicherweise entartendes) Skalarprodukt $[f, g]$ durch die Gleichung

$$[f, g] = \sum_{z, \zeta \in \mathfrak{U}_F} [\Phi(z, \zeta)\, x_z, y_\zeta] ; \tag{2.3}$$

dieses hat auf Grund der Bedingung $\mathcal{R}^i_\varkappa(\mathfrak{G})$ genau $\varkappa$ negative Quadrate. Wir betten den Raum $\mathcal{L}^i$ kanonisch in einen $\Pi_\varkappa$-Raum ein; das ist der gesuchte Raum $\Pi^i_\varkappa = \Pi^i_\varkappa(F)$. Identifizieren wir die Elemente aus $\mathcal{L}^i$ mit den ihnen zugeordneten Elementen von $\Pi^i_\varkappa$, so liegt $\mathcal{L}^i$ dicht in $\Pi^i_\varkappa$.

Auf der Menge

$$\hat{\vartheta} = \{ f \mid f = \sum_{z \in \mathfrak{U}_F} \varepsilon_z x_z \in \mathcal{L}^i \quad \text{mit} \quad \sum_{z \in \mathfrak{U}_F} x_z = 0 \}$$

definieren wir einen Operator $\hat{V} = \hat{V}_F$ durch die Gleichung

$$\hat{V} f = \sum_{z \in \mathfrak{U}_F} z \varepsilon_z x_z .$$

Man sieht leicht, dass $\hat{V}$ π-isometrisch ist: $[\hat{V} f, \hat{V} g] = [f, g], f, g \in \hat{\vartheta}$. Sein Wertebereich $\mathcal{R}(\hat{V})$ besteht aus allen $g \in \mathcal{L}^i$, $g = \sum_{z \in \mathfrak{U}_F} \varepsilon_z y_z$ mit $y_0 = 0, \sum_{z \in \mathfrak{U}_F} \frac{y_z}{z} = 0$. Er liegt dicht* in $\mathcal{L}^i$, denn für $\zeta \longrightarrow 0$ und alle $x \in \mathfrak{G}$ gilt $\varepsilon_\zeta x \to \varepsilon_0 x, \zeta \varepsilon_\zeta x \to 0$, und es gehört mit $\sum_{z \in \mathfrak{U}_F} \varepsilon_z x_z \in \mathcal{L}^i$ $x_0 = 0$, das Element $\sum_{z \in \mathfrak{U}_F} \varepsilon_z x_z - \varepsilon_\zeta \sum_{z \in \mathfrak{U}_F} \frac{x_z}{z} \zeta$ zu $\mathcal{R}(\hat{V})$. Damit ergibt sich ohne Schwierigkeit, dass der Operator $\hat{V}$ sogar stetig ist: Aus

* Die Begriffe "dicht" und "stetig" in $\mathcal{L}^i$ beziehen sich zunächst auf die in [1] für ein beliebiges $\Pi_\varkappa$-Lineal erklärte Konvergenz einer Elementfolge.

$(f_n) \subset \hat{\vartheta}$, $f_n \to 0$ folgt nämlich $[\hat{V} f_n, \hat{V} f_n] \to 0$ und $[\hat{V} f_n, \hat{V} g] = [f_n, g] \to 0$ für alle $g \in \hat{\vartheta}$, $n \to \infty$.

Ist $h \in \hat{\vartheta}$ ein Element des isotropen Teiles von $\mathcal{L}^i$, so gilt $[\hat{V} h, \hat{V} f] = 0$ für alle $f \in \hat{\vartheta}$, also gehört auch $\hat{V} h$ zum isotropen Teil von $\mathcal{L}^i$. Deshalb geht $\hat{V}$ bei der kanonischen Einbettung von $\mathcal{L}^i$ in $\Pi_\varkappa^i$ in einen stetigen π-isometrischen Operator in $\Pi_\varkappa^i$ über, dessen Abschliessung wir mit $V^i = V_F^i$ bezeichnen.

Das π-Skalarprodukt entartet auf $\vartheta(V^i)$ nicht: Wäre dies nämlich der Fall, so könnten wir $\vartheta(V^i)$ mit seinem isotropen Teil ϑ_0 in der Form $\vartheta(V^i) = \vartheta_0 [+] \vartheta_1$ darstellen. Der Operator V^i würde dann ϑ_1 π-isometrisch und eineindeutig auf einen dichten Teil von $\Pi_\varkappa^i$ abbilden, also hätte das π-Skalarprodukt auf ϑ_1 genau $\varkappa$ negative Quadrate. Das ist im Falle $\vartheta_0 \neq \{0\}$ unmöglich. Aus [12], Satz 4.3 folgert man damit leicht, dass der Wertebereich von V^i der ganze Raum $\Pi_\varkappa$ ist.

Wir definieren jetzt einen Operator $\Gamma_F | \mathfrak{G} \to \mathcal{L}^i$ durch die Gleichung

$$\Gamma_F x = \varepsilon_0 x, \quad x \in \mathfrak{G}.$$

Da für $x, y \in \mathfrak{G}$, $z \in \mathfrak{N}_F$

(2.4) $$[\Gamma_F x, \Gamma_F y] = [\operatorname{Re} F(0) x, y], \quad [\Gamma_F x, \varepsilon_z y] = \frac{1}{2}[(F(0) + F^+(z)) x, y]$$

gilt, ist Γ_F stetig. Wir fassen ihn als Abbildung von $\mathfrak{G}$ in $\Pi_\varkappa^i(F)$ auf. Dann existiert ein stetiger Operator $\Gamma_F^+ \in [\Pi_\varkappa^i, \mathfrak{G}]$ mit der Eigenschaft

$$[f, \Gamma_F x] = [\Gamma_F^+ f, x], \quad f \in \Pi_\varkappa^i, \quad x \in \mathfrak{G}.$$

Aus den Beziehungen (2.4) folgt für $x \in \mathfrak{G}$, $z \in \mathfrak{N}_F$

(2.5) $$\Gamma_F^+ \Gamma_F = \operatorname{Re} F(0), \quad \Gamma_F^+ \varepsilon_z x = \frac{1}{2}(F^+(0) + F(z)) x.$$

Das Spektrum des Operators $(V^i)^{-1} \in [\Pi_\varkappa^i, \Pi_\varkappa^i]$ besteht ausserhalb des Einheitskreises aus höchstens $\varkappa$ Eigenwerten, jeder entsprechend seiner Vielfachheit oft gezählt. Folglich existiert für $|z| < 1$ der Operator

$$(V^{i} - zI)^{-1} = (V^{i})^{-1}(I - z(V^{i})^{-1})^{-1}$$

mit Ausnahme von höchstens $\varkappa$ Punkten, die Pole dieser Resolvente sind. Aus der Gleichung $(V^{i} - zI)(\varepsilon_z x - \varepsilon_0 x) = z\varepsilon_0 x$ folgt für $z \in \mathfrak{A}_F$ mit $\frac{1}{z} \bar{\in} \sigma((V^{i})^{-1})$:

$$\Gamma_F^{+}\varepsilon_z x - \Gamma_F^{+}\varepsilon_0 x = z\Gamma_F^{+}(V^{i} - zI)^{-1}\varepsilon_0 x,$$

und daraus auf Grund von (2.5)

$$F(z) = i\,\mathrm{Im}\,F(0) + \Gamma_F^{+}(V^{i} + zI)(V^{i} - zI)^{-1}\Gamma_F. \tag{2.6}$$

Mit $\mathbf{R}_{\varkappa}^{i}(\mathfrak{G})$ bezeichnen wir die Gesamtheit aller im Inneren des Einheitskreises definierten und meromorphen Funktionen F mit Werten in $[\mathfrak{G},\mathfrak{G}]$, die bei $z = 0$ holomorph sind und für die der $\mathfrak{G}$-Kern (2.1) im Holomorphiegebiet $\mathfrak{C}_F$ von F genau $\varkappa$ negative Quadrate hat.

Satz 2.1. *Zu jeder Funktion F mit der Eigenschaft $\mathcal{R}_{\varkappa}^{i}(\mathfrak{G})$ gibt es eine Funktion $F_1 \in \mathbf{R}_{\varkappa}^{i}(\mathfrak{G})$, so dass die Werte $F(z)$ und $F_1(z)$ für $z \in \mathfrak{A}_F$ übereinstimmen mit Ausnahme von höchstens endlich vielen Punkten z.*

Zum Beweis des Satzes bezeichnen wir die rechte Seite von (2.6) mit $F_1(z)$ und brauchen dann nur noch zu zeigen, dass der mit $F_1(z)$ gemäss (2.1) gebildete Kern Φ_1 genau $\varkappa$ negative Quadrate hat. Das folgt aus dem

Satz 2.2. a) *Es sei V ein maximaler $(\mathcal{R}(V) = \Pi_{\varkappa})$ π-isometrischer Operator im $\Pi_{\varkappa}$-Raum $\Pi_{\varkappa}$, $\mathfrak{G}$ ein Π_k-Raum, $S = S^{+} \in [\mathfrak{G},\mathfrak{G}]$ und $\Gamma \in [\mathfrak{G},\Pi_{\varkappa}]$. Dann gehört die Funktion*

$$F(z) = iS + \Gamma^{+}(V + zI)(V - zI)^{-1}\Gamma, \quad |z| < 1, \quad z \bar{\in} \sigma(V), \tag{2.7}$$

zur Klasse $\mathbf{R}^i_{\varkappa'}(\mathfrak{G})$ *für ein* $\varkappa'$ *mit* $0 \leqq \varkappa' \leqq \varkappa$ *. Sind die Operatoren* V *und* Γ *eng* i *-verbunden, so ist* $\varkappa' = \varkappa$ *.*

b) *Jede Funktion* $F \in \mathbf{R}^i_{\varkappa}(\mathfrak{G})$ *gestattet die Darstellung* (2.7) *mit* $\Pi_\varkappa = \Pi^i_\varkappa(F)$, $S = \operatorname{Im} F(0)$, $\Gamma = \Gamma_F$ *und dem maximalen* $(\mathcal{R}(V) = \Pi_\varkappa)$ π *-isometrischen Operator* $V = V^i_F$ *, wobei* V *mit* Γ *eng* i *-verbunden ist.*

Dabei nennen wir einen maximalen $(\mathcal{R}(V) = \Pi_\varkappa)$ π-isometrischen Operator V in $\Pi_\varkappa$ *eng* i *-verbunden* mit $\Gamma \in [\mathfrak{G}, \Pi_\varkappa]$, wenn $\Pi_\varkappa$ die abgeschlossene lineare Hülle der Elemente $V(V - zI)^{-1}\Gamma x = (I - zV^{-1})^{-1}\Gamma x$, $x \in \mathfrak{G}$, $\frac{1}{z} \bar{\in} (V^+)$, $|z| < 1$, ist. Das ist bekanntlich gleichbedeutend damit, dass $\Pi_\varkappa$ die abgeschlossene lineare Hülle aller Elemente

(2.8) $$V(V - zI)^{-1}\Gamma x, \quad x \in \mathfrak{G}, \quad z \in \mathfrak{U},$$

ist, wobei $\mathfrak{U}$ eine beliebige nichtleere offene Menge aus dem Inneren des Einheitskreises bezeichnet, für deren Punkte z die Inverse in (2.8) existiert.

Zum Beweis der Aussage b) von Satz 3.2 haben wir nur noch zu zeigen, dass der Operator V^i_F mit Γ eng i -verbunden ist. Das folgt aus der für $z \in \mathfrak{L}_F$, $\frac{1}{z} \bar{\in} \sigma((V^i)^{-1})$ gültigen Beziehung

(2.9) $$V^i(V^i - zI)^{-1}\Gamma x = \varepsilon_z x .$$

Um die Aussage a) zu beweisen, überlegen wir uns, dass die Funktion F aus (2.7) auch die Darstellung

(2.10) $$F(z) = iS + \Gamma^+(\tilde{U} + z\tilde{I})(\tilde{U} - z\tilde{I})^{-1}\Gamma, \quad |z| < 1, \quad z \bar{\in} \sigma(\tilde{U})$$

mit einem π -unitären Operator $\tilde{U}$ in einem $\Pi_\varkappa$ -Raum $\tilde{\Pi}_\varkappa \supset \Pi_\varkappa$ gestattet. Das ist trivial, wenn V selbst π -unitär ist. Anderenfalls ist auf $\mathfrak{D}(V)^{[\perp]}$ das π -Skalarprodukt positiv definit, und wir erhalten die gewünschte π -unitäre Erweiterung $\tilde{U}$ von V , indem wir in einem Π_0 -Raum $\tilde{\mathfrak{H}}$ einen π -isometrischen Operator $\tilde{V}_0$ mit $\mathfrak{D}(\tilde{V}_0) = \tilde{\mathfrak{H}}$ und $\dim \mathcal{R}(\tilde{V}_0)^{[\perp]} = \dim \mathfrak{D}(V)^{[\perp]}$

wählen und anschliessend den π-isometrischen Operator $\tilde{V}_0[+]V$ in $\tilde{\Pi}_\varkappa = \tilde{\mathfrak{H}}[+]\Pi_\varkappa$ zu einem π-unitären Operator $\tilde{\mathfrak{U}}$ in diesem Raum erweitern. Die Operatoren Γ und Γ^+ werden dabei in natürlicher Weise zu Abbildungen von $\mathfrak{G}$ in $\tilde{\Pi}_\varkappa$ bzw. $\tilde{\Pi}_\varkappa$ in $\mathfrak{G}$ fortgesetzt.

Aus der Beziehung (2.10) folgt leicht für $z, \zeta \in \sigma(\tilde{\mathfrak{U}})$, $|z|<1$, $|\zeta|<1$:

$$(2.11) \qquad \Phi(z,\zeta) = \frac{1}{2}\,\frac{F(z)+F^+(\zeta)}{1-z\bar{\zeta}} = \Gamma^+(\tilde{\mathfrak{U}}^+-\bar{\zeta}\tilde{I})^{-1}(\tilde{\mathfrak{U}}-z\tilde{I})^{-1}\Gamma,$$

also ist

$$(2.12) \qquad [\Phi(z,\zeta)x,y] = [(\tilde{\mathfrak{U}}-z\tilde{I})^{-1}\Gamma x, (\tilde{\mathfrak{U}}-\zeta\tilde{I})^{-1}\Gamma y] =$$

$$= [(V-zI)^{-1}\Gamma x, (V-zI)^{-1}\Gamma y].$$

Somit hat der $\mathfrak{G}$-Kern Φ höchstens $\varkappa$ negative Quadrate. Sind V und Γ eng i-verbunden, so fällt die abgeschlossene lineare Hülle der Elemente

$$\tilde{\mathfrak{U}}(\tilde{\mathfrak{U}}-z\tilde{I})^{-1}\Gamma x = V(V-zI)^{-1}\Gamma x, \quad x \in \mathfrak{G}, \quad z \in \mathfrak{U}_0$$

mit $\Pi_\varkappa$ zusammen, wenn $\mathfrak{U}_0$ eine beliebige hinreichend kleine Umgebung von $z=0$ bezeichnet. Dann hat aber der $\mathfrak{G}$-Kern Φ auf Grund von (2.12) genau $\varkappa$ negative Quadrate.

Zwei Operatoren A und A' im $\Pi_\varkappa$-Raum $\Pi_\varkappa$ bzw. $\Pi'_\varkappa$ heissen *π-unitäräquivalent,* wenn ein Operator T existiert, der den Raum $\Pi_\varkappa$ π-isometrisch (d.h. unter Erhaltung des π-Skalarproduktes) auf $\Pi'_\varkappa$ abbildet *, so dass $T\mathfrak{D}(A) = \mathfrak{D}(A')$ und $A'Tf = TAf$, $f \in \mathfrak{D}(A)$, gilt.

Satz 2.3. *In der Darstellung* (2.7) *der Funktion* $F \in \mathbf{R}^i_\varkappa(\mathfrak{G})$ *mit einem maximalen* $(\mathfrak{R}(V) = \Pi_\varkappa)$ *π-isometrischen Operator* V *und eng i-ver-*

* Ein solcher Operator T bildet $\Pi_\varkappa$ linear, stetig und eineindeutig auf $\Pi'_\varkappa$ ab. Man sieht auch leicht, dass sich jeder Operator T_0, der eine dichte Menge von $\Pi_\varkappa$ π-isometrisch auf eine dichte Menge von $\Pi'_\varkappa$ abbildet, stets zu einem solchen Operator T abschliessen lässt.

bundenen Operatoren V *und* Γ *ist* V *bis auf* π*-unitäre Äquivalenz eindeutig bestimmt.*

Beweis. Die Funktion $F \in \mathbf{R}^{i}_{\varkappa}(\mathcal{O}\!\mathit{f})$ sei dargestellt in der Form (2.7) mit eng verbundenen Operatoren V und Γ. Daneben besteht die oben konstruierte Darstellung (2.6) mit V^{i}_{F}, Γ_F in $\Pi^{i}_{\varkappa}(F)$. Für $\frac{1}{z} \bar{\in} \sigma_p(V)^{-1} \cup \sigma_p((V^{i}_{F})^{-1})$, $|z| < 1$ und $x \in \mathcal{O}\!\mathit{f}$ setzen wir

$$T_0 V(V - zI)^{-1} \Gamma x = V^{i}_{F}(V^{i}_{F} - zI)^{-1} \Gamma_F x .$$

Aus der Beziehung (2.12) und einer entsprechenden Beziehung für V^{i}_{F} an Stelle von V ergibt sich, dass T_0 einen dichten Teil von $\Pi_{\varkappa}$ π-isometrisch auf einen dichten Teil von $\Pi^{i}_{\varkappa}(F)$ abbildet. Also lässt sich T_0 zu einem π-isometrischen Operator T von $\Pi_{\varkappa}$ auf $\Pi^{i}_{\varkappa}(F)$ fortsetzen, für den noch

$$TV(V - zI)^{-1} = V^{i}_{F}(V^{i}_{F} - zI)^{-1} T$$

gilt. Daraus folgt leicht $T\mathfrak{D}(V) = \mathfrak{D}(V^{i}_{F})$ und $TVF = V^{i}_{F} Tf$, $f \in \mathfrak{D}(V)$.

2. Für eine Funktion F, welche die Darstellung (2.7) gestattet, haben wir aus den Beziehungen (2.11) und (2.12) die Eigenschaft $\mathcal{R}^{i}_{\varkappa}(\mathcal{O}\!\mathit{f})$ abgeleitet. Wir extrapolieren jetzt die Funktion F auf das Spiegelbild $\mathfrak{h}^{*}_{F} = \{z \mid \frac{1}{z} \in \mathfrak{h}_F\}$ von $\mathfrak{h}_F$ bezüglich der Einheitskreislinie durch die Festsetzung

(2.13) $$F(z) = -F^{+}\left(\frac{1}{\bar{z}}\right), \quad z \in \mathfrak{h}^{*}_{F}$$

und bilden für diese extrapolierte Funktion F wieder den Kern Φ aus (2.1); dabei setzen wir für $z = \frac{1}{\bar{\zeta}}$

$$\Phi(z,\zeta) = -zF'(z), \quad z \in \mathfrak{h}_F ;$$

$$\Phi(z,\zeta) = -\bar{\zeta} F'^{+}(z), \quad z \in \mathfrak{h}^{*}_{F} .$$

Man sieht ohne Schwierigkeit, dass die Beziehungen (2.11) und (2.12) auch für beliebige $z, \zeta \in \mathfrak{h}_F \cup \mathfrak{h}^{*}_{F}$ richtig bleiben. Daraus ergibt sich die folgende Aussage:

Der extrapolierte Kern Φ *hat auf* $\mathfrak{k}_F \cup \mathfrak{k}_F^*$ *genau* $\varkappa$ *negative Quadrate.* Wir sagen für diesen letzten Sachverhalt auch, die Funktion F habe die *Eigenschaft* $\mathcal{R}_\varkappa(\mathcal{Y})$.

Die Beziehungen (2.11) und (2.12) wurden oben bewiesen unter der Voraussetzung, dass für F die Darstellung (2.10) mit einem π-unitären Operator $\tilde{\mathcal{U}}$ in einem $\Pi_\varkappa$-Raum $\tilde{\Pi}_\varkappa$ besteht. Wir zeigen, dass es die Eigenschaft $\mathcal{R}_\varkappa(\mathcal{Y})$ andererseits gestattet, eine natürliche solche π-unitäre Erweiterung $\tilde{\mathcal{U}} = \mathcal{U}_F$ zu erhalten auf folgende Weise. Jedem Punkt $z \in \mathfrak{k}_F \cup \mathfrak{k}_F^*$ wird ein Symbol ε_z zugeordnet, die lineare Menge $\mathcal{L} = \mathcal{L}(F)$ aller formalen endlichen Summen (2.2) mit $\mathfrak{U}_F$ ersetzt durch $\mathfrak{k}_F \cup \mathfrak{k}_F^*$ gebildet und zwischen zwei Elementen $f, g \in \mathcal{L}$ ein Skalarprodukt durch die Beziehung (2.3) definiert. Auf der Menge

$$\vartheta = \{ f \mid f = \sum_{z \in \mathfrak{k}_F \cup \mathfrak{k}_F^*} \varepsilon_z x_z \in \mathcal{L} \text{ mit } \sum_{z \in \mathfrak{k}_F \cup \mathfrak{k}_F^*} x_z = 0 \}$$

erklären wir einen Operator $\hat{\mathcal{U}}$ durch die Gleichung $\hat{\mathcal{U}} f = \sum_{z \in \mathfrak{k}_F \cup \mathfrak{k}_F^*} z \varepsilon_z x_z$. Offensichtlich ist $\mathcal{L}^i \subset \mathcal{L}$ und $\hat{V} \subset \hat{\mathcal{U}}$. Die kanonische Einbettung von $\mathcal{L}_i$ in $\Pi_\varkappa^i$ lässt sich erweitern zu einer kanonischen Einbettung von $\mathcal{L}_i$ in einen $\Pi_\varkappa$-Raum $\Pi_\varkappa \supset \Pi_\varkappa^i$. Dabei induziert der Operator $\hat{\mathcal{U}}$ einen π-isometrischen Operator in $\Pi_\varkappa$, dessen Abschliessung $\mathcal{U}_F$ der gesuchte π-unitäre Operator ist. Man zeigt nämlich ebenso wie für $\hat{V}$, dass der Wertebereich von $\hat{\mathcal{U}}$ in $\mathcal{L}$ dicht liegt. Auch der Definitionsbereich ϑ hat diese Eigenschaft, denn es gilt $\varepsilon_\zeta x \to 0$ für $\zeta \to \infty$, $x \in \mathcal{Y}$ und für $\sum \varepsilon_z x_z \in \mathcal{L}$ gehört das Element

$$\sum_{z \in \mathfrak{k}_F \cup \mathfrak{k}_F^*} \varepsilon_z x_z - \varepsilon_\zeta \sum_{z \in \mathfrak{k}_F \cup \mathfrak{k}_F^*} x_z$$

zu ϑ .

Einen π-unitären Operator $\mathcal{U}$ in $\Pi_\varkappa$ nennen wir *eng verbunden* mit dem Operator $\Gamma \in [\mathcal{Y}, \Pi_\varkappa]$, wenn $\Pi_\varkappa$ die abgeschlossene lineare Hülle der Elemente

$$\mathcal{U}(\mathcal{U}-zI)^{-1}\Gamma x, \quad x \in \mathfrak{G}, \quad z \,\overline{\in}\, \sigma(\mathcal{U})$$

ist. Das ist gleichbedeutend damit, dass $\Pi_\varkappa$ die abgeschlossene lineare Hülle aller Elemente $\mathcal{U}(\mathcal{U}-zI)^{-1}\Gamma_x$ ist, wenn x ganz $\mathfrak{G}$ und z nur eine beliebige Umgebung von $z = 0$ und $z = \infty$ durchläuft. Man sieht leicht, dass analog (2.6) für $\mathcal{U} = \mathcal{U}_F$ die Darstellung

(2.14) $$F(z) = i \operatorname{Im} F(0) + \Gamma_F^+(\mathcal{U}+zI)(\mathcal{U}-zI)^{-1}\Gamma_F,$$

$$z \in (\mathfrak{k}_F \cup \mathfrak{k}_F^*) \setminus \sigma_p(\mathcal{U})$$

besteht und $\mathcal{U}_F$ mit Γ_F eng verbunden ist (vgl. (2.9)).

Mit $\mathbf{R}_\varkappa(\mathfrak{G})$ bezeichnen wir die Klasse aller im Inneren und Äusseren des Einheitskreises meromorphen Funktionen F mit Werten in $[\mathfrak{G},\mathfrak{G}]$ und der Eigenschaft (2.13), die bei $z = 0$ holomorph sind und die Eigenschaft $\mathcal{R}_\varkappa(\mathfrak{G})$ haben. Damit lässt sich die folgende, Satz 2.2 entsprechende Aussage formulieren:

Satz 2.4. a) *Es sei* $\mathcal{U}$ *ein* π *-unitärer Operator im* $\Pi_\varkappa$*-Raum* $\Pi_\varkappa$, $\mathfrak{G}$ *ein* Π_k *-Raum,* $S = S^+ \in [\mathfrak{G},\mathfrak{G}]$ *und* $\Gamma \in [\mathfrak{G},\Pi_\varkappa]$ *. Dann gehört die Funktion*

(2.15) $$F(z) = iS + \Gamma^+(\mathcal{U}+zI)(\mathcal{U}-zI)^{-1}\Gamma, \quad z \,\overline{\in}\, \sigma(\mathcal{U}),$$

zur Klasse $\mathbf{R}_{\varkappa'}(\mathfrak{G})$ *für ein* $\varkappa'$ *mit* $0 \leqq \varkappa' \leqq \varkappa$ *. Sind die Operatoren* $\mathcal{U}$ *und* Γ *eng verbunden, so ist* $\varkappa' = \varkappa$ *.*

b) *Jede Funktion* $F \in \mathbf{R}_\varkappa(\mathfrak{G})$ *gestattet die Darstellung* (2.15) *mit* $\Pi_\varkappa = \Pi_\varkappa(F)$, $S = \operatorname{Im} F(0)$, $\Gamma = \Gamma_F$ *und dem* π *-unitären Operator* $\mathcal{U} = \mathcal{U}_F$, *wobei* $\mathcal{U}$ *mit* Γ *eng verbunden ist.*

Dabei ist die Aussage b) nur eine Zusammenfassung der vorangegangenen Überlegungen. Der erste Teil der Aussage a) folgt unmittelbar aus Satz 2.2, a), den Beweis des zweiten Teiles überlassen wir dem Leser.

Analog zu Satz 2.3 beweist man:

Satz 2.5. *Ist in der Darstellung* (2.15) *der Funktion* $F \in \mathbf{R}_\varkappa(\mathfrak{G})$ *der* π*-unitäre Operator* $\mathfrak{U}$ *mit* Γ *eng verbunden, so ist* $\mathfrak{U}$ *bis auf* π*-unitäre Äquivalenz eindeutig bestimmt, genauer,* $\mathfrak{U}$ *ist dem Operator* $\mathfrak{U}_F$ π*-unitär äquivalent.*

Wir vermerken schliesslich ohne Beweis noch den folgenden

Satz 2.6. *Das Spektrum der Funktion* $F \in \mathbf{R}_\varkappa(\mathfrak{G})$ *stimmt überein mit dem Spektrum des Operators* $\mathfrak{U}_F$.

Zum *Spektrum* der Funktion F sollen dabei definitionsgemäss alle Pole von F sowie diejenigen Punkte z mit $|z| = 1$ gehören, in die sich F nicht so holomorph fortsetzen lässt, dass $F^+(z) = -F(z)$ gilt.

Aus den Sätzen 2.4 und 2.6 ergibt sich, dass in (2.14) die Bedingung $z \in (\mathfrak{k}_F \cup \mathfrak{k}_F^*) \setminus \sigma_p(\mathfrak{U}_F)$ mit $z \in \mathfrak{k}_F \cup \mathfrak{k}_F^*$ gleichbedeutend ist.

Im nächsten Paragraphen hat die betrachtete Funktion F die zusätzliche Eigenschaft $F(0) = I$. Dann folgt aus (2.4)

$$[\Gamma_F x, \Gamma_F y] = [x, y], \quad x, y \in \mathfrak{G},$$

also vermittelt Γ eine π-Isometrie zwischen $\mathfrak{G}$ und dem Teilraum aller $\varepsilon_0 x$, $x \in \mathfrak{G}$, von $\mathfrak{L}$. Wir können deshalb $\mathfrak{G}$ mit diesem Teilraum identifizieren. Erweitern wir dann den Operator Γ_F auf ganz $\Pi_\varkappa$ durch die Festsetzung $\Gamma_F f = 0$ für $f \in \mathfrak{G}^{[\perp]}$, d.h., setzen wir

$$\Gamma_F \varepsilon_z x = \frac{1}{2} \varepsilon_0 (F(z) + I) x, \quad z \in \mathfrak{k}_F \cup \mathfrak{k}_F^*, \quad x \in \mathfrak{G},$$

so ist Γ_F die π-orthogonale Projektion von $\Pi_\varkappa$ auf $\mathfrak{G}$.

§ 3. DIE CHARAKTERISTISCHE FUNKTION EINES π-ISOMETRISCHEN OPERATORS

1. Es sei $\mathfrak{G}$ wieder ein Π_k-Raum mit dem π-Skalarprodukt [. , .]. Weiter bezeichne $\mathbf{K}(\mathfrak{G})$ im folgenden die Klasse aller Funktionen X mit Werten in $[\mathfrak{G}, \mathfrak{G}]$, die definiert und holomorph sind auf einer Menge $\mathfrak{k}_X = \mathfrak{k} \setminus \gamma_X$; dabei sei $\mathfrak{k}$ wieder das Innere des Einheitskreises und γ_X eine (evtl. leere) Menge isolierter Punkte in $\mathfrak{k}$. Mit $\mathbf{K}_\varkappa(\mathfrak{G})$ bezeichnen wir die Menge aller Funktionen aus $\mathbf{K}(\mathfrak{G})$, für die jeder der $\mathfrak{G}$-Kerne

$$K_X(z,\zeta) = \frac{I - X^+(\zeta)X(z)}{1 - z\bar{\zeta}}, \quad \bar{K}_{X^+}(z,\zeta) = \frac{I - X(\bar{\zeta})X^+(\bar{z})}{1 - z\bar{\zeta}}$$

in $\mathfrak{k}_X$ bzw. $\bar{\mathfrak{k}}_X = \{\bar{z}:\ z \in \mathfrak{k}_X\}$ genau $\varkappa$ negative Quadrate hat.

In allen uns interessierenden Fällen folgt dabei aus der Tatsache, dass K_X genau $\varkappa$ negative Quadrate hat, dasselbe für den $\mathfrak{G}$-Kern $\bar{K}_{X^+}$ und umgekehrt. Für den Fall $\varkappa = 0$ ist diese Aussage enthalten in dem

Lemma 3.1 *Für eine Funktion* $X \in \mathbf{K}(\mathfrak{G})$ *sind die folgenden Aussagen äquivalent:*

a) *Für alle* $z \in \mathfrak{k}_X$ *ist* $X(z)$ π*-kontrahierend;*
b) *für alle* $z \in \bar{\mathfrak{k}}_X$ *ist* $X^+(\bar{z})$ π*-kontrahierend;*
c) *der* $\mathfrak{G}$*-Kern* K_X *ist positiv definit auf* $\mathfrak{k}_X$ *;*
d) *der* $\mathfrak{G}$*-Kern* $\bar{K}_X$ *ist positiv definit auf* $\bar{\mathfrak{k}}_X$.

Beweis. Die Äquivalenz der Aussagen a) und b) wurde von Ju.P. Ginzburg (siehe [13], S. 150) bewiesen. Wir zeigen, dass aus a) auch c) folgt. Zu diesem Zweck wählen wir in $\mathfrak{G}$ ein positiv definites Skalarprodukt (. , .), so dass

$$[f,g] = (Jf,g), \quad f,g \in \mathfrak{G}, \quad J = P_+ - P_- ,$$

gilt mit zwei orthogonalen Projektionen P_+, P_-; $P_+ = P_+^2 = P_+^*$, $P_+ + P_- = I$. Wie im Beweis von Satz 1 aus [8] führen wir die Operatorfunktion $X_\varrho(z) = (\varrho P_+ + \frac{1}{\varrho} P_-)X(z)$ ein. Nach den Ergebnissen von Ju. P. Ginzburg hat eine Cayleytransformierte

$$F_\varrho(z) = (I + X_\varrho(z))(I - X_\varrho(z))^{-1}$$

die für $|z| < r < 1$ die Darstellung

$$(3.1)\quad JF_\varrho(z) = i\,\mathrm{Im}(JF_\varrho(0)) + \frac{1}{2\pi}\int_0^{2\pi} \frac{re^{i\vartheta} + z}{re^{i\vartheta} - z}\, Q(re^{i\vartheta})\, d\vartheta$$

mit $Q(re^{i\vartheta}) = \operatorname{Re}(JF_\varrho(re^{i\vartheta})) \geqq 0$ gestattet. Daraus folgt

$$\frac{JF_\varrho(z) + F_\varrho^*(z)J}{2(r^2 - z\bar{\zeta})} = \frac{1}{2\pi}\int_0^{2\pi} \frac{1}{(re^{i\vartheta} - z)(re^{-i\vartheta} - \bar{\zeta})} Q(re^{i\vartheta})\,d\vartheta.$$

Die linke Seite dieser Gleichung lässt sich in der Form

$$(I - X_\varrho^*(\zeta))^{-1} \frac{J - X_\varrho(\zeta)JX_\varrho^*(z)}{r^2 - z\bar{\zeta}} (I - X_\varrho(z))^{-1}$$

schreiben, also ist der $\mathfrak{G}$-Kern K_{X_ϱ} positiv definit. Strebt $\varrho \uparrow 1$, so folgt die Aussage c).

Umgekehrt ergibt sich aus c) offensichtlich a); ersetzt man in a) und c) den Operator $X(z)$ durch $X^+(\bar{z})$ so folgt die Äquivalenz von b) und d).

In § 4 benötigen wir eine einfache Verallgemeinerung einer Ungleichung von Ju. P. Ginzburg ([8], [14]): Ist $X(z) \in K_0(\mathfrak{G})$, so gilt für beliebiges natürliches n; $f_1, f_2, \ldots, f_n, g_1, g_2, \ldots, g_n \in \mathfrak{G}$, $z_1, z_2, \ldots, z_n \in \mathfrak{E}_X$, $\zeta_1, \zeta_2, \ldots, \zeta_n \in \bar{\mathfrak{E}}_X$:

$$(3.2) \qquad \left| \sum_{j,k=1}^{n} \left[\frac{X(z_j) - X(\bar{\zeta}_k)}{z_j - \bar{\zeta}_k} f_j, g_k \right] \right|^2 \leqq$$

$$\leqq \sum_{j,k=1}^{n} [K_X(z_j, z_k) f_j, f_k] \sum_{j,k=1}^{n} [\bar{K}_{X^+}(\zeta_j, \zeta_k) g_j, g_k];$$

für $z_j = \bar{\zeta}_k$ hat man dabei den Quotienten auf der linken Seite durch $X'(z_j)$ zu ersetzen.

Zum Beweis von (3.2) beachten wir zunächst die aus (3.1) für $r > \max(|z_j|, |\zeta_k|)$ folgende Beziehung

$$JF_\varrho(z) - JF_\varrho(\bar{\zeta}) = \frac{z-\bar{\zeta}}{\pi} \int_0^{2\pi} \frac{re^{i\vartheta}}{(re^{i\vartheta}-z)(re^{i\vartheta}-\bar{\zeta})} Q(re^{i\vartheta})\,d\vartheta .$$

Mit $f_j' = (I - X_\varrho(z_j)) f_j,\ g_k' = (I - X_\varrho^+(\bar{\zeta}_k))\, g_k$ ergibt sich dann

$$\left| \sum_{j,k=1}^{n} \left[\frac{X_\varrho(z_j) - X_\varrho(\bar{\zeta}_k)}{z_j - \bar{\zeta}_k} f_j, g_k \right] \right|^2 = \left| \sum_{j,k=1}^{n} \left[\frac{F_\varrho(z_j) - F_\varrho(\bar{\zeta}_k)}{2(z_j - \bar{\zeta}_k)} f_j', g_k' \right] \right|^2 =$$

$$= \left| \sum_{j,k=1}^{n} \frac{1}{2\pi} \int_0^{2\pi} \frac{re^{i\vartheta}}{(re^{i\vartheta}-z_j)(re^{i\vartheta}-\bar{\zeta}_k)} (Q(re^{i\vartheta}) f_j', g_k')\,d\vartheta \right|^2 \leqq$$

$$\leqq \frac{1}{2\pi} \int_0^{2\pi} \Big(Q(re^{i\vartheta}) \sum_{j=1}^{n} \frac{f_j'}{re^{i\vartheta} - z_j},\ \sum_{j=1}^{n} \frac{f_j'}{re^{i\vartheta} - z_j} \Big)\,d\vartheta \ .$$

$$\cdot \frac{1}{2\pi} \int_0^{2\pi} \Big(Q(re^{i\vartheta}) \sum_{k=1}^{n} \frac{g_k'}{re^{-i\vartheta} - \zeta_k},\ \sum_{k=1}^{n} \frac{g_k'}{re^{-i\vartheta} - \zeta_k} \Big)\,d\vartheta =$$

$$= \sum_{j,k=1}^{n} \frac{((JF_\varrho(z_j) + F_\varrho^*(z_k)J) f_j', f_k')}{2(r^2 - z_j \bar{z}_k)} \sum_{j,k=1}^{n} \frac{((JF_\varrho(\bar{\zeta}_k) + F_\varrho^*(\bar{\zeta}_j)J) g_j', g_k')}{2(r^2 - \zeta_j \bar{\zeta}_k)} =$$

$$= \sum_{j,k=1}^{n} \left[\frac{I - X_\varrho^+(z_k) X_\varrho(z_j)}{r^2 - z_j \bar{z}_k} f_j, f_k \right] \sum_{j,k=1}^{n} \left[g_j, \frac{I - X_\varrho(\bar{\zeta}_j) X_\varrho^+(\bar{\zeta}_k)}{r^2 - \bar{\zeta}_j \zeta_k} g_k \right],$$

woraus leicht die Behauptung folgt.

Lemma 3.2. *Ist die Funktion* $X \in \mathbf{K}(\mathfrak{G})$ *bei* $z = 0$ *holomorph und hat* $I - X(0)$ *eine Inverse in* $[\mathfrak{G}, \mathfrak{G}]$ *, so haben die* $\mathfrak{G}$ *-Kerne* K_X *und* $\bar{K}_{X^+}$ *die gleiche Anzahl negativer Quadrate.*

Beweis. Wir führen wieder die Cayleytransformierte

$$F(z) = (I + X(z))(I - X(z))^{-1}, \quad z \in \mathfrak{U}_0 , \tag{3.3}$$

ein, wenn $\mathfrak{U}_0$ eine hinreichend kleine Umgebung von $z = 0$ bezeichnet. Dann besteht die Beziehung

$$\Phi(z, \zeta) = \frac{F(z) + F^+(\zeta)}{1 - z\bar{\zeta}} = 2(I - X^+(\zeta))^{-1} K_X(z, \zeta)(I - X(z))^{-1}. \tag{3.4}$$

Hat der $\mathfrak{G}$ -Kern K_X in $\mathfrak{U}_0$ genau $\varkappa$ negative Quadrate, so gilt folglich dasselbe für den $\mathfrak{G}$ -Kern Φ in $\mathfrak{U}_0$. Dann hat aber auch der $\mathfrak{G}$ -Kern

$$\frac{F^+(\bar{z}) + F(\zeta)}{1 - z\bar{\zeta}}$$

auf Grund von § 2 diese Eigenshaft, was wiederum gleichbedeutend damit ist, dass dasselbe für den $\mathfrak{G}$ -Kern $\bar{K}_{X^+}$ zutrifft. Da sich diese Überlegungen umkehren lassen, ist das Lemma bewiesen.

Schliesslich haben die $\mathfrak{G}$ -Kerne K_X und $\bar{K}_{X^+}$ auch dieselbe Anzahl negativer Quadrate, wenn $\mathfrak{G}$ ein Π_0 -Raum ist; das ist der Inhalt der unten bewiesenen Folgerung 3.3.

2. Uns interessieren im folgenden π -isometrische Operatoren V in einem unendlichdimensionalen Raum $\Pi_\varkappa$ mit den Eigenschaften:

1) $\mathfrak{D}(V)$ und $\mathfrak{R}(V)$ sind abgeschlossene Teilräume, auf denen das π -Skalarprodukt nicht entartet;
2) die Defektzahlen von V , das sind die Dimensionen der π -Orthogonalkomplemente von $\mathfrak{D}(V)$ und $\mathfrak{R}(V)$, sind gleich; dabei kann der Wert dieser Dimensionen eine beliebige endliche oder unendliche Kardinalzahl sein.

Offensichtlich stimmen dann die Signaturen von $\mathfrak{D}(V)$ und $\mathfrak{R}(V)$ und damit auch die ihrer π-Orthogonalkomplemente überein. Eine ausgezeichnete Rolle spielt der Teilraum $\mathfrak{D}(V)^{[\perp]}$; wir bezeichnen diesen Π_k-Raum, $0 \leqq k \leqq \varkappa$, mit $\mathfrak{G}$, und die Menge aller π-isometrischen Operatoren V in $\Pi_\varkappa$, welche die Eigenschaften 1), 2) und $\mathfrak{D}(V)^{[\perp]} = \mathfrak{G}$ besitzen, mit $\mathfrak{V}(\Pi_\varkappa ; \mathfrak{G})$.

Man sieht leicht, dass unter den Voraussetzungen 1) und 2) π-unitäre Erweiterungen von V im Ausgangsraum $\Pi_\varkappa$ existieren; wir wählen eine solche π-unitäre Erweiterung $\mathfrak{U}$ und halten diese im folgenden fest.

Es sei weiter $\mathfrak{R}_z = (zV - I)\mathfrak{D}(V)$ und $\mathfrak{N}_z = \mathfrak{R}_{\bar{z}}^{[\perp]}$. Ist Γ_0 die π-orthogonale Projektion von $\Pi_\varkappa$ auf $\mathfrak{G} = \mathfrak{N}_0$, so setzen wir für $z \bar{\in} \sigma(\mathfrak{U})$

$$\Gamma_z = (I - z\mathfrak{U}^+)^{-1}\Gamma_0 = \mathfrak{U}(\mathfrak{U}z - I)^{-1}\Gamma_0, \quad \Gamma_\infty = 0.$$

Dann bildet Γ_z den Teilraum $\mathfrak{G}$ eineindeutig und in beiden Richtungen stetig auf $\mathfrak{N}_z$ ab, denn es gilt einerseits für $f \in \mathfrak{D}(V)$, $g \in \Pi_\varkappa$

$$[(\bar{z}V - I)f, \Gamma_z g] = [(I - \bar{z}\mathfrak{U})^{-1}(\bar{z}V - I)f, \Gamma_0 g] = 0,$$

andererseits gehört für $h \in \mathfrak{N}_z$ das Element $f = \mathfrak{U}^+(\mathfrak{U} - zI)h$ zu $\mathfrak{G}$, und es ist $h = \Gamma_z f$.

Die Operatoren $\Gamma_z \in [\mathfrak{G}, \Pi_\varkappa]$ hängen für $z \in \varrho(\mathfrak{U})$ holomorph von z ab. Bezeichnet $\mathfrak{D}_V$ die Menge derjenigen komplexen Zahlen, für die das π-Skalarprodukt auf $\mathfrak{N}_z$ entartet, so sieht man analog wie in [1], Satz 3.3, dass $\mathfrak{D}_V$ keinen Kreis enthält und das nicht auf der Einheitskreislinie gelegene Spektrum von $\mathfrak{U}$ zu $\mathfrak{D}_V$ gehört.

Der Operator $V \in \mathfrak{V}(\Pi_\varkappa ; \mathfrak{G})$ heisst *einfach,* wenn $\Pi_\varkappa$ die abgeschlossene lineare Hülle aller Teilräume $\mathfrak{N}_z$, $z \in (\mathfrak{U}_0 \cup \mathfrak{U}_0^*) \setminus \mathfrak{D}_V$ ist; dabei bezeichnet $\mathfrak{U}_0$ eine geeignete Umgebung des Punktes $z = 0$. Man sieht leicht, dass für einen einfachen Operator $V \in \mathfrak{V}(\Pi_\varkappa ; \mathfrak{G})$ der Raum $\Pi_\varkappa$ die abgeschlossene lineare Hülle aller $\mathfrak{N}_z$ ist, wenn z eine beliebige Umgebung von $z = 0$ und $z = \infty$ durchläuft.

Ein einfacher Operator $V \in \mathfrak{V}(\Pi_\varkappa ; \mathfrak{G})$ hat keine Eigenwerte. Gilt nämlich $Vf_0 - \alpha f_0 = 0$, $f_0 \in \mathfrak{D}(V)$, so ist auch $\mathfrak{U}f_0 - \alpha f_0 = 0$ und damit

$(I-\bar{z}U)^{-1}f_0 = \frac{1}{1-\bar{z}\alpha} f_0$. Daraus folgt für beliebige $g \in \mathfrak{G}$

$$[\Gamma_z g, f_0] = [g, (I-\bar{z}U)^{-1} f_0] = \frac{1}{1-z\bar{\alpha}} [g, f_0] = 0,$$

also ist f_0 π-orthogonal auf allen Teilräumen $\mathfrak{N}_z$.

Neben $V \in \mathfrak{W}(\Pi_\kappa; \mathfrak{G})$ betrachten wir die Erweiterung $T = T_V$ von V , definiert durch die Beziehung

(3.5)
$$Tf = \begin{cases} Vf & f \in \mathfrak{D}(V) \\ 0 & f \in \mathfrak{D}(V)^{[\perp]} \end{cases} .$$

Ist $\mathfrak{G}$ ein Π_0-Raum, d.h. ein Hilbertraum bezüglich des π-Skalarproduktes, so ist der Operator T π-kontrahierend: Für $f \in \Pi_\kappa$, $f = f' + f''$, $f' \in \mathfrak{D}(V)$, $f'' \in \mathfrak{G}$ gilt nämlich

(3.6)
$$[Tf, Tf] = [Vf', Vf'] = [f', f'] \leqq [f, f];$$

dabei steht das Gleichheitszeichen zwischen den letzten beiden Ausdrücken genau dann, wenn f zu $\mathfrak{D}(V)$ gehört. Ebenso ergibt sich, dass in der Beziehung $[T^+f, T^+f] \leqq [f, f]$, $f \in \Pi_\kappa$, das Gleichheitszeichen genau dann steht, wenn f zu $\mathfrak{R}(V) = \mathfrak{D}(V^{-1})$ gehört.

Lemma 3.3. *Der Operator* $V \in \mathfrak{W}(\Pi_\kappa; \mathfrak{G})$ *sei einfach und* $\mathfrak{G}$ *ein* Π_0 *-Raum. Dann gestattet der Raum* Π_κ *die eindeutige Zerlegung*

(3.7)
$$\Pi_\kappa = \Pi_+ + \Pi_- ,$$

wobei beide Teilräume auf der rechten Seite invariant sind für den Operator T *aus* (3.5); Π_+ (Π_-) *ist ein positiver (negativer) Teilraum und* $\sigma(T \mid \Pi_+)$ *liegt ganz in* $\{z \mid \ |z| \leqq 1\}$, $\sigma(T \mid \Pi_-)$ *ganz in* $\{z \mid \ |z| > 1\}$.

Beweis. Der gemäss (3.6) π-kontrahierende Operator T hat einen κ-dimensionalen invarianten nichtpositiven Teilraum Π_- , so dass für alle Punkte $z \in \sigma(T \mid \Pi_-)$ gilt: $|z| \geqq 1$. Hätte T einen Eigenwert auf der

Einheitskreislinie: $Tf_0 = \alpha f_0$, $f_0 \neq 0$, $|\alpha| = 1$, so wäre $[Tf_0, Tf_0] = [f_0, f_0]$ also $f_0 \in \mathfrak{D}(V)$ und damit auch $Vf_0 = \alpha f_0$. Das ist wegen der Einfachheit des Operators V unmöglich. Also liegt $\sigma(T\,|\,\Pi_-)$ im Äusseren des Einheitskreises. Einhilte Π_- ein neutrales Element, dann gehörte dieses auch zu dem bezüglich T^+ invarianten nichtnegativen Teilraum $\Pi_-^{[\perp]}$. Die Menge der neutralen Elemente g_0 von $\Pi_-^{[\perp]}$ wird von T^+ auf Grund der Beziehung

$$0 \leqq [T^+g_0, T^+g_0] \leqq [g_0, g_0] = 0$$

in sich abgebildet, also hat T^+ und damit auch V^{-1} ein neutrales Eigenelement. Das ist wiederum auf Grund der Einfachheit des Operators V unmöglich.

Bezeichnen wir schliesslich mit Π_+ denjenigen invarianten Teilraum von T, der zu dem in der abgeschlossenen Einheitskreisscheibe gelegenen Spektrum von T gehört, so besteht die Zerlegung (3.7). Den Nachweis der Positivität des Teilraumes Π_+ überlassen wir dem Leser.

Auch für einen nicht notwendig einfachen Operator $V \in \mathfrak{H}(\Pi_\varkappa; \mathfrak{G})$ mit einem π_k-Raum $\mathfrak{G}$, $0 \leqq k \leqq \varkappa$, besteht das Spektrum des zugehörigen Operators T ausserhalb des Einheitskreises aus lauter isolierten normal abspaltbaren Eigenwerten. Um das zu sehen, stellen wir $\mathfrak{G}$ als π-orthogonale Summe eines positiven Teilraumes $\mathfrak{G}_+$ und eines (k-dimensionalen) negativen Teilraumes $\mathfrak{G}_-$ dar und erweitern V zu einem auf $\mathfrak{D}(V) + \mathfrak{G}_-$ definierten π-isometrischen Operator $\tilde{V}$. Dann ist $\tilde{T} = T_{\tilde{V}}$ π-kontrahirend, also besteht $\sigma(\tilde{T})$ ausserhalb des Einheitskreises aus höchstens $\varkappa$ normal abspaltbaren Eigenwerten. Die Operatoren $T = T_V$ und $\tilde{T}$ unterscheiden sich aber nur durch einen k-dimensionalen Operator, deshalb ist auch $\sigma(T)$ ausserhalb des Einheitskreises diskret.

3. Es sei wieder $V \in \mathfrak{H}(\Pi_\varkappa; \mathfrak{G})$ und $\mathfrak{U}$ eine π-unitäre Erweiterung von V in $\Pi_\varkappa$. Die für alle Punkte z mit $\frac{1}{z} \in \varrho(T^+)$ sowie für $z = 0$ definierte Operatorfunktion $X = X_V$:

$$X_V(z) = z\Gamma_0\, \mathfrak{U}^+ (I - zT^+)^{-1} \Gamma_0$$

mit Werten in $[\mathfrak{G}, \mathfrak{G}]$ heisse die *charakteristische Funktion* des Operators V. Sie ist nach den Bemerkungen am Ende des vorangehenden Abschnittes in $\mathfrak{L}$

meromorph und bei $z = 0$ holomorph. Wir überlegen uns, dass sie der folgenden Beziehung genügt:

$$(3.8) \qquad (I+X(z))(I-X(z))^{-1} = \Gamma_0(\mathcal{U}+zI)(\mathcal{U}-zI)^{-1}\Gamma_0 ;$$

diese besteht für diejenigen Punkte $z \in \mathcal{C}$, für die alle auftretenden Operatoren sinnvoll sind.

Zum Beweis von (3.8) gehen wir aus von den Gleichungen $\mathcal{U}^+ - T^+ = \mathcal{U}^+P_|$ und

$$(3.9) \qquad (I-zT^+)^{-1} - (I-z\mathcal{U}^+)^{-1} = -z(I-z\mathcal{U}^+)^{-1}\mathcal{U}^+P(I-zT^+)^{-1} ;$$

dabei bezeichnet P die π-orthogonale Projektion auf $\mathcal{R}(V)^{[\perp]}$. Durch Multiplikation von (3.9) mit Γ_0 von rechts und mit $z\Gamma_0\mathcal{U}^+$ von links ergibt sich

$$X(z) - z\Gamma_0\mathcal{U}^+(I-z\mathcal{U}^+)^{-1}\Gamma_0 = -z\Gamma_0\mathcal{U}^+(I-z\mathcal{U}^+)^{-1}X(z),$$

also

$$X(z) = z\Gamma_0(\mathcal{U}-zI)^{-1}\Gamma_0(I-X(z)),$$

d.h.

$$X(z)(I-X(z))^{-1} = z\Gamma_0(\mathcal{U}-zI)^{-1}\Gamma_0 ,$$

woraus leicht (3.8) folgt.

Das Hauptergebnis dieses Paragraphen ist der folgende Satz, dessen Beweis nach den Vorbereitungen in § 2.2 keine Schwierigkeiten bereitet.

Satz 3.1 a) *Die charakteristische Funktion* X_V *des* π*-isometrischen Operators* $V \in \mathcal{M}(\Pi_\varkappa ; \mathfrak{g})$ *hat die folgenden Eigenschaften:*

1) $X_V(0) = 0$,

2) $X_V \in \mathbf{K}_{\varkappa'}(\mathfrak{g}),\quad 0 \leqq \varkappa' \leqq \varkappa$.

Ist der Operator V *einfach, so gilt* $\varkappa' = \varkappa$.

b) *Zu jeder in einer Umgebung* $\mathfrak{U}_0 \subset \mathfrak{C}$ *von* $z = 0$ *holomorphen Funktion* X *, mit Werten in* $[\mathfrak{G}, \mathfrak{G}]$ *und* $X(0) = 0$ *, für die der* $\mathfrak{G}$*-Kern*

$$K_X(z,\zeta) = \frac{I - X^+(\zeta)X(z)}{1 - z\bar{\zeta}}, \qquad z, \zeta \in \mathfrak{U}_0$$

genau $\varkappa$ *negative Quadrate hat, gibt es einen* $\pi_\varkappa$*-Raum* $\Pi_\varkappa$ *, der* $\mathfrak{G}$ *enthält, und einen einfachen* π*-isometrischen Operator* $V \in \mathfrak{V}(\Pi_\varkappa; \mathfrak{G})$ *, so dass* $X(z) = X_V(z)$ *,* $z \in \mathfrak{U}_0$ *, gilt.*

Beweis. a) Die Eigenschaft 1) ist offensichtlich. Aus (3.8) ergibt sich weiter

$$(1 - \bar{\zeta}z)\Gamma_0(U^+ - \bar{\zeta}I)^{-1}(U - zI)^{-1}\Gamma_0 =$$

$$= \Gamma_0\{(U^+ + \bar{\zeta}I)(U^+ - \bar{\zeta}I)^{-1} + (U + zI)(U - zI)^{-1}\}\Gamma_0 =$$

$$= (I + X_V^+(\zeta))(I - X_V^+(\zeta))^{-1} + (I + X_V(z))(I - X_V(z))^{-1} =$$

$$= (I - X_V^+(\zeta))^{-1}\{(I - X_V^+(\zeta)X_V(z)\}(I - X_V(z))^{-1},$$

also hat der $\mathfrak{G}$-Kern $K_{X_V}(z,\zeta) = \dfrac{I - X_V^+(\zeta)X_V(z)}{1 - z\bar{\zeta}}$ höchstens $\varkappa$ negative Quadrate; diese Anzahl sei $\varkappa'$, $0 \leqq \varkappa' \leqq \varkappa$. Dann hat auf Grund von Lemma 3.2 auch der $\mathfrak{G}$-Kern $\bar{K}_{X_V^+}$ genau $\varkappa'$ negative Quadrate, womit die Eigenschaft 2) bewiesen ist.

Für die Cayleytransformierte (3.3) von $X = X_V$ gilt gemäss (3.8)

$$F(z) = \Gamma_0(U + zI)(U - zI)^{-1}\Gamma_0 .$$

Ist V einfach, so ist U mit Γ_0 eng verbunden, also hat gemäss Satz 2.4, a) der $\mathfrak{G}$-Kern Φ aus (3.4) genau $\varkappa$ negative Quadrate. Dann trifft aber dasselbe für den $\mathfrak{G}$-Kern K_{X_V} zu.

b) Die Cayleytransformierte F von X gestattet nach den Ergebnissen von § 2.2 die Darstellung

(3.10) $$F(z) = \Gamma_F(\mathcal{U}_F + zI)(\mathcal{U}_F - zI)^{-1}\Gamma_F$$

mit einem π-unitären Operator $\mathcal{U}_F$ im $\Pi_\varkappa$-Raum $\Pi_\varkappa(F)$, der $\mathfrak{G}$ enthält, und der π-orthogonalen Projektion Γ_F von $\Pi_\varkappa(F)$ auf $\mathfrak{G}$; dabei ist $\mathcal{U}_F$ eng verbunden mit Γ_F. Letzteres besagt, dass die Einschränkung V von $\mathcal{U}_F$ auf $\mathfrak{G}^{[\perp]}$ ein einfacher Operator ist. Offensichtlich gilt $V \in \mathfrak{M}(\Pi_\varkappa(F); \mathfrak{G})$, und für seine charakteristische Funktion, gebildet mit $\mathcal{U} = \mathcal{U}_F$, besteht gemäss (3.8) die Beziehung

$$\Gamma_F(\mathcal{U}_F + zI)(\mathcal{U}_F - zI)^{-1}\Gamma_F = (I + X_V(z))(I - X_V(z))^{-1}.$$

Zusammen mit (3.3) und (3.10) folgt daraus $X(z) = X_V(z)$ für alle z aus einer hinreichend kleinen Umgebung von $z = 0$.

Folgerung 3.1. *Jede Funktion* X, *die den Voraussetzungen von Satz* 3.1 b) *genügt, lässt sich zu einer im* $\mathfrak{C}$ *meromorphen Funktion fortsetzen.*

Folgerung 3.2 *Die Funktionen der Klasse* $\mathbf{K}_\varkappa(\mathfrak{G})$ *sind meromorph in* $\mathfrak{C}$.

Ist nämlich $X \in \mathbf{K}_\varkappa(\mathfrak{G})$, so können wir o.B.d.A. voraussetzen, dass X im Punkte $z = 0$ holomorph ist, da bei einer gebrochen linearen Abbildung von $\mathfrak{C}$ auf sich die Anzahl der negativen Quadrate der Kerne K_X und $\bar{K}_{X^+}$ unverändert bleibt. Die Funktion $zX(z)$ ist dann gemäss Satz 3.1 b) die charakteristische Funktion eines π-isometrischen Operators, also meromorph in $\mathfrak{C}$.

In der Aussage b) von Satz 3.1 lässt sich die Voraussetzung der Holomorphie von X abschwächen. Setzt man z.B. für die Funktion X nur voraus, dass sie bei $z = 0$ in der gleichmässigen Operatorentopologie stetig ist und der $\mathfrak{G}$-Kern K_X in einer Umgebung $\mathfrak{U}_0$ von $z = 0$ genau $\varkappa$ negative Quadrate hat, dann ergibt sich die Gleichung $X(z) = X_V(z)$ mit einem einfachen π-isometrischen Operator $V \in \mathfrak{M}(\Pi_\varkappa; \mathfrak{G})$ wieder für $z \in \mathfrak{U}_0$ mit evtl. Ausnahme von höchstens endlich vielen Punkten z (vgl. Satz 2.1).

4. In diesem Abschnitt sei $\mathfrak{G}$ ein Π_0-Raum, d.h. ein Hilbertraum mit

dem Skalarprodukt $[\,.\,,\,.\,]$; den zu $A \in [\mathfrak{G},\mathfrak{G}]$ bezüglich dieses Skalarproduktes adjungierten Operator bezeichnen wir auch mit A^* (an Stelle von A^*). Als eine Anwendung von Satz 3.1 leiten wir in Satz 3.2 eine Darstellung für Operatorfunktionen der Klasse $\mathbf{K}_\varkappa(\mathfrak{G})$ her.

Lemma 3.4. *Die Funktion* X *mit Werten in* $[\mathfrak{G},\mathfrak{G}]$ *sei definiert und holomorph im Inneren des Einheitskreises,* $X(0) = 0$ *und der Kern* K_X *habe dort genau* k, $0 \leqq k < \infty$, *negative Quadrate. Dann ist* $k = 0$.

Beweis. Nach Satz 3.1 ist X die charakteristische Funktion eines einfachen π -isometrishen Operators $V \in \mathcal{W}(\Pi_k;\mathfrak{G})$ in einem Π_k-Raum Π_k , d.h., sie gestattet die Darstellung $X(z) = z\Gamma_0 U^+(I - zT^+)^{-1}\Gamma_0$. Im Falle $k > 0$ hat der Operator T gemäss Lemma 3.3 mindestens einen Eigenwert β mit $|\beta| > 1$. Wir bezeichnen mit $P_\beta = \frac{1}{2\pi i}\oint_{C_\beta}(\zeta I - T)^{-1}d\zeta$ die zugehörige Rieszsche Projektion. Dann ist sicher $P_\beta \mathcal{R}(V)^{[\perp]} \neq \{0\}$. Anderenfalls wäre nämlich $P_\beta^+ \Pi_\varkappa \subset \mathcal{R}(V)$, also gehörten insbesondere die Eigenelemente f von T^+ zum Eigenwert $\bar{\beta}$ zu $\mathcal{R}(V)$. Dann würde aber $V^+ f = \bar{\beta} f$ gelten, im Widerspruch zur Einfachheit des Operators V .

Es gibt somit eine natürliche Zahl r mit den Eigenschaften

$$(3.11)\quad (T-\beta I)^r P_\beta \mathcal{R}(V)^{[\perp]} = \{0\},\quad (T-\beta I)^{r-1} P_\beta \mathcal{R}(V)^{[\perp]} \neq \{0\}.$$

Auf Grund der Holomorphie von X gilt für einen hinreichend kleinen Kreis C_β um β und $f, g \in \mathfrak{G}$:

$$0 = \frac{1}{2\pi i}\oint_{C_\beta^*}(1-\bar{\beta}z)^{r-1} z^{-r-1}[X(z)f, g]\,dz =$$

$$= \frac{1}{2\pi i}\oint_{C_\beta^*}(1-\bar{\beta}z)^{r-1} z^{-r}[U^+(I - zT^+)^{-1} f, g]\,dz =$$

$$= [f, \frac{1}{2\pi i}\oint_{C_\beta}(1-\beta\bar{z})^{r-1}\bar{z}^{-r}(I - \bar{z}T)^{-1}d\bar{z}\, Ug] =$$

$$= [f, \frac{1}{2\pi i} \oint_{C_\beta} (\zeta - \beta)^{r-1} (\zeta I - T)^{-1} d\zeta\, \mathcal{U} g] =$$

$$= [f, (T - \beta I)^{r-1} P_\beta \mathcal{U} g],$$

also $\Gamma_0 (T - \beta I)^{r-1} P_\beta h = 0$ für alle $h \in \mathcal{R}(V)^{[\perp]}$. Das ist aber gleichbedeutend damit, dass $(T - \beta I)^{r-1} P_\beta \mathcal{R}(V)^{[\perp]} \subset \vartheta(V)$ gilt. Auf Grund von (3.11) gibt es folglich ein $h \in \mathcal{R}(V)^{[\perp]}$ mit $h' = (T - \beta I)^{r-1} P_\beta h \in \vartheta(V)$, $h' \neq 0$ und $(V - \beta I) h' = (T - \beta I) h' = 0$, d.h., β ist Eigenwert des Operators V. Das widerspricht seiner Einfachheit.

Unter einem *Blaschke-Potapov-Produkt* * verstehen wir eine Operatorfunktion B der Gestalt

(3.12) $$B(z) = \mathcal{U}_0 \overset{\curvearrowleft}{\prod_{i=1}^{m}} \left(\frac{z - \gamma_i}{1 - \bar{\gamma}_i z} P_i + Q_i \right);$$

dabei sind die P_i und Q_i orthogonale Projektionen im Hilbertraum $\mathfrak{G}$, $Q_i = I - P_i$, $0 \leqq |\gamma_i| < 1$, $i = 1, 2, \ldots, m$, und $\mathcal{U}_0$ ein unitärer Operator in $\mathfrak{G}$. Die Zahl $\varkappa = \sum_{i=1}^{m} \dim P_i$, $0 \leqq \varkappa \leqq \infty$, nennen wir die *Ordnung* des Blaschke-Potapov-Produktes (3.12); ist $\varkappa < \infty$, so heisse dieses von *endlicher Ordnung.*

Bei geeigneter Wahl der P_i, Q_i und $\mathcal{U}_0$ kann man B stets so schreiben, dass die γ_i auf der rechten Seite von (3.12) in beliebig vorgegebener Reihenfolge auftreten, insbesondere also so, dass alle Faktoren mit gleichen γ_i nebeneinander stehen. Dann hat (3.12) die Form

(3.13) $$B(z) = \mathcal{U}_0 \overset{\curvearrowleft}{\prod_{j=1}^{n}} B_j(z), \quad B_j(z) = \overset{\curvearrowleft}{\prod_{k=1}^{k_j}} \left(\frac{z - \alpha_j}{1 - \bar{\alpha}_j z} P_{jk} + Q_{jk} \right),$$

* Für Räume mit indefinitem Skalarprodukt wurden diese im endlichdimensionalen Fall von V.P. Potapov [7], im unendlichdimensionalen Fall von Ju.P. Ginzburg [14] eingeführt.

wobei $0 \leqq |\alpha_j| < 1$ und $\alpha_j \neq \alpha_{j'}$ für $j \neq j'$ gilt und die P_{jk} und Q_{jk} wieder orthogonale Projektionen in $\mathfrak{G}$ mit $Q_{jk} = I - P_{jk}$ $(k = 1, 2, \ldots, k_j;$ $j, j' = 1, 2, \ldots, n)$ sind.

Wir betrachten im folgenden Funktionen Y der Gestalt

$$Y(z) = B^{-1}(z)\, Y_0(z) \tag{3.14}$$

mit einem Blaschke-Potapov-Produkt B endlicher Ordnung und einer in $\mathcal{E}$ holomorphen und kontrahierenden * Operatorfunktion Y_0, d.h. $Y_0 \in \mathbf{K}_0(\mathfrak{G})$. Hat das Blaschke-Potapov-Produkt dabei insbesondere die Form (3.13) und gilt für jedes $j = 1, 2, \ldots, n$

$$P_{j1} \geqq P_{j2} \geqq \ldots \geqq P_{jk_j} \tag{3.15}$$

sowie

$$\mathcal{R}(P_{j1} Y_{j+1}(\alpha_j)) = \mathcal{R}(P_{j1}),$$

dann nennen wir die Darstellung (3.14) *regulär;* dabei haben wir

$$Y_k(z) = \Big(\overset{\curvearrowright}{\prod_{j=k}^{n}} B_j^{-1}(z) \Big)\, \mathcal{U}_0^{-1}\, Y_0(z), \qquad k = 1, 2, \ldots, n$$

und $Y_{n+1} = \mathcal{U}_0^{-1} Y_0$ gesetzet.

Lemma 3.5 *Gestattet die Funktion* Y *eine Darstellung* (3.14) *mit einem Blaschke-Potapov-Produkt* B *der Ordnung* $\varkappa$, $0 \leqq \varkappa < \infty$, *und* $Y_0 \in \mathbf{K}_0(\mathfrak{G})$, *so hat der Kern* K_Y *höchstens* $\varkappa$ *negative Quadrate; ist diese Darstellung regulär, so hat* K_Y *genau* $\varkappa$ *negative Quadrate.*

Wir beweisen nur die zweite Aussage; der einfachere Beweis des ersten Teiles sei dem Leser überlassen. Für den Kern K_Y besteht die Identität

*d.h., es gilt $Y_0^*(z)\, Y_0(z) \leqq I \quad (z \in \mathcal{E})$

$$K_Y(z,\zeta) = K_{Y_0}(z,\zeta) + Y^*_{n+1}(\zeta) K_{B_n^{-1}}(z,\zeta) Y_{n+1}(z) + \dots$$

$$\dots + Y_2^*(\zeta) K_{B_1^{-1}}(z,\zeta) Y_2(z),$$

und durch elementare Rechnung überzeugt man sich von der Richtigkeit der Beziehung

$$K_{B_j^{-1}}(z,\zeta) = \sum_{k=1}^{k_j} P'_{jk} \frac{1 - \left(\frac{1-\bar{\zeta}\alpha_j}{\bar{\zeta}-\bar{\alpha}_j}\right)^k \left(\frac{1-z\bar{\alpha}_j}{z-\alpha_j}\right)^k}{1-z\bar{\zeta}}$$

wobei wir $P'_{jk_j} = P_{jk_j}$ und $P'_{jk} = P_{j,k+1} - P_{jk}$, $k = 1,2,\dots,k_j-1$, gesetzt haben. Da der skalare Kern

$$\frac{1 - \left(\frac{1-\bar{\zeta}\alpha_j}{\bar{\zeta}-\bar{\alpha}_j}\right)^k \left(\frac{1-z\bar{\alpha}_j}{z-\alpha_j}\right)^k}{1-z\bar{\zeta}},$$

negativ definit ist mit genau k negativen Quadraten, ergibt sich die Behauptung z.B. durch Betrachtung von Ausdrücken der Gestalt

$$\iint [K_Y(z,\zeta) f(z), f(\zeta)]\, dz\, d\bar{\zeta}$$

mit geeigneten, im Inneren des Einheitskreises holomorphen Funktionen f mit Werten in $\mathfrak{Y}$; dabei wird längs eines Kreises integriert, der ganz in $\mathfrak{E}$ liegt und alle Punkte $\alpha_1, \alpha_2, \dots, \alpha_n$ umschliesst.

Aus Lemma 3.5 folgt, dass sich bei Division einer Funktion $Y_0 \in \mathbf{K}_0(\mathfrak{Y})$ durch ein Blaschke-Potapov-Produkt B von endlicher Ordnung eine Funktion $Y = B^{-1} Y_0$ ergibt, deren zugehöriger Kern K_Y endlich viele negative Quadrate hat. Der folgende Satz besagt nun, dass auf diese Weise sogar jede Funktion Y mit der angegebenen Eigenschaft erhalten wird.

Satz 3.2. *Es sei* $\mathfrak{G}$ *ein Hilbertraum und* Y *eine in einer Umgebung* $\mathfrak{U}_0 \subset \mathfrak{E}$ *eines Punktes* z_0, $|z_0| < 1$, *definierte und holomorphe Funktion mit Werten in* $[\mathfrak{G}, \mathfrak{G}]$. *Dann und nur dann hat der Kern*

$$K_Y(z,\zeta) = \frac{I - Y^*(\zeta)\,Y(z)}{1 - z\bar{\zeta}}, \qquad z, \zeta \in \mathfrak{U}_0,$$

genau $\varkappa$, $0 < \varkappa < \infty$, *negative Quadrate, wenn* Y *eine reguläre Darstellung* (3.14) *mit einem Blaschke-Potapov-Produkt* B *der Ordnung* $\varkappa$ *und einer in* $\mathfrak{E}$ *kontrahierenden und holomorphen Operatorfunktion* Y_0 *gestattet.*

Beweis. Ohne Beschränkung der Allgemeinheit setzen wir $z_0 = 0$ voraus. Der Kern K_Y habe genau $\varkappa$ negative Quadrate. Wir gehen über zur Funktion $X(z) = zY(z)$. Dann gilt $X(0) = 0$, und der Kern K_X hat ebenfalls genau $\varkappa$ negative Quadrate. Letzteres folgt aus den Identitäten

$$\frac{I - X^*(\zeta)\,X(z)}{1 - z\bar{\zeta}} = \frac{I - Y^*(\zeta)\,I(z)}{1 - z\bar{\zeta}} + Y^*(\zeta)\,Y(z),$$

$$\frac{I - X^*(\zeta)\,X(z)}{1 - z\bar{\zeta}} = \bar{\zeta} z \,\frac{I - Y^*(\zeta)\,Y(z)}{1 - z\bar{\zeta}} - I.$$

Gemäss Satz 3.1 existiert ein einfacher π-isometrischer Operator $V \in \mathfrak{N}(\Pi_\varkappa; \mathfrak{G})$, so dass $X(z) = X_V(z)$ gilt für alle Punkte z der betrachteten Umgebung von $z_0 = 0$. Wir setzen X durch diese Gleichung zu einer in $\mathfrak{E}$ meromorphen Funktion fort, die wir ebenfalls mit X bezeichnen. Diese hat auf Grund von Lemma 3.3 höchstens eine endliche Anzahl von Polen $\alpha_1, \alpha_2, \dots, \alpha_n$ ($\alpha_j \neq \alpha_{j'}$ für $j \neq j'$; $j, j' = 1, 2, \dots, n$) in $\mathfrak{E}$, die alle ungleich Null sind.

In der Umgebung eines solchen Poles α_1 gestattet X die Entwicklung

$$X(z) = \frac{A_{-k_1}}{(z - \alpha_1)^{k_1}} + \dots + \frac{A_{-1}}{z - \alpha_1} + A(z)$$

mit endlichdimensionalen Operatoren $A_{-1}, \dots, A_{-k_1}$; $A_{-k_1} \neq 0$ und einer bei $z = \alpha_1$ holomorphen Funktion A . Wir bezeichnen mit P_{1k_1} die orthogonale Projektion auf den Wertebereich $\mathcal{R}(A_{-k_1})$ und setzen $Q_{1k_1} = I - P_{1k_1}$. Dann hat die Funktion

$$\left(\left(\frac{z-\alpha_1}{1-z\alpha_1}\right)^{k_1} P_{1k_1} + Q_{1k_1}\right) X(z)$$

bei $z = \alpha_1$ einen Pol der Ordnung $\leqq k_1 - 1$ oder sie ist bei $z = \alpha_1$ holomorph. Im ersten Fall liegt der Wertebereich aller bei $z = \alpha_1$ singulären Glieder ihrer Laurententwicklung um $z = \alpha_1$ in $\mathcal{R}(Q_{1k_1})$ und durch evtl. Wiederholung dieser Betrachtungen wir eine Funktion

$$B_1(z) = \prod_{k=1}^{k_1} \left(\frac{z-\alpha_1}{1-\bar{\alpha}_1 z} P_{1k} + Q_{1k}\right),$$

deren Projektionen P_{1k} die Eigenschaft (3.15) haben und für die

$$X_2(z) = B_1(z) X(z)$$

bei $z = \alpha_1$ holomorph ist. Anschliessend spalten wir in der gleichen Weise von $X_2(z)$ einen Faktor $B_2(z)$ ab, der zu einem evtl. auftretenden Pol α_2 von $X_2(z)$ gehört, usw. Nach n Schritten ergibt sich eine Funktion

$$X_0(z) = B_n(z) \cdots B_1(z) X_z ,$$

die im Inneren des Einheitskreises holomorph ist. Da der Kern K_{X_0} höchstens endlich viele negative Quadrate hat, ist diese Anzahl auf Grund von Lemma 3.4 gleich Null, d.h., der Kern K_{X_0} ist positiv definit. Damit haben wir für X die Darstellung

$$X(z) = B_1^{-1}(z) B_2^{-1}(z) \dots B_n^{-1}(z) X_0(z)$$

erhalten, und aus der Konstruktion von B_j ersieht man, dass diese sogar regulär ist. Aus der Verallgemeinerung des Schwarzschen Lemmas für operatorwertige Funktionen (vgl. [14], Satz I.1) folgt, dass die Funktion $Y_0(z) = \frac{X_0(z)}{z}$ in $\mathfrak{E}$ kontrahierend ist, also gestattet Y eine reguläre Darstellung (3.14). Schliesslich ist auf Grund von Lemma 3.5 die Ordnung des Blaschke-Potapov-Produktes $B(z) = \overset{\curvearrowleft}{\prod_{j=1}^{n}} B_j(z)$ gleich $\varkappa$.

Umgekehrt wurde bereits in Lemma 3.5 gezeigt, dass der Kern K_Y genau $\varkappa$ negative Quadrate hat, wenn Y eine reguläre Darstellung (3.14) mit einem Blaschke-Potapov-Produkt der Ordnung $\varkappa$ gestattet.

Folgerung 3.3. *Ist* $\mathfrak{G}$ *ein Hilbertraum,* Y *eine in einer Umgebung* $\mathfrak{U}_0 \subset \mathfrak{E}$ *des Punktes* z_0, $|z_0| < 1$, *holomorphe Funktion mit Werten in* $[\mathfrak{G}, \mathfrak{G}]$, *so hat der Kern* K_Y *genau dann* $\varkappa$ *negative Quadrate, wenn dies für den Kern*

$$\bar{K}_{Y*}(z,\zeta) = \frac{I - Y(\bar{\zeta})Y^*(\bar{z})}{1 - z\bar{\zeta}} \qquad (\bar{z}, \bar{\zeta} \in \mathfrak{U}_0)$$

zutrifft.

Hat nämlich K_Y genau $\varkappa$ negative Quadrate, so gestattet Y eine Darstellung (3.14). Daraus ergibt sich leicht

$$\bar{K}_{Y*}(z,\zeta) = \bar{K}_{(B^{-1})^*}(z,\bar{\zeta}) + B^{-1}(\bar{\zeta})\bar{K}_{Y_0*}(z,\zeta)(B^{-1})^*(\bar{z}),$$

also hat $\bar{K}_{Y*}$ höchstens $\varkappa$ negative Quadrate; diese Anzahl sei $\varkappa'$. Dann gestattet aber Y^* gemäss Satz 3.2 eine Darstellung

$$Y^*(\bar{z}) = \tilde{B}^{-1}(z)\tilde{Y}_0(z)$$

mit einem Blaschke-Potapov-Produkt $\tilde{B}$ der Ordnung $\varkappa'$. Daraus folgt wie oben, dass der Kern K_Y höchstens $\varkappa'$ negative Quadrate hat. Andererseits ist diese Anzahl gleich $\varkappa$, also muss $\varkappa' = \varkappa$ gelten. Schliesslich vermerken wir noch eine aus den Sätzen 3.1 und 3.2 folgende Kennzeichnung der charakteristischen

Funktion eines einfachen π -isometrischen Operators $V \in \mathfrak{N}[\Pi_\varkappa ; \mathfrak{G}]$ mit positiv definiten Defekträumen.

Folgerung 3.4. *Es sei* $\mathfrak{G}$ *ein Hilbertraum. Eine Funktion* $X \in \mathbf{K}(\mathfrak{G})$ *ist genau dann die charakteristische Funktion eines einfachen* π *-isometrischen Operators* $V \in \mathfrak{N}(\Pi_\varkappa ; \mathfrak{G}]$, *wenn gilt:*

1) $X(0) = 0$;

2) X *gestattet eine reguläre Darstellung (3.14) mit einem Blaschke-Potapov-Produkt der Ordnung* $\varkappa$.

§ 4. VERALLGEMEINERTE RESOLVENTEN EINES π -ISOMETRISCHEN OPERATORS

1. Es sei jetzt V wieder ein Operator der Klasse $\mathfrak{N}(\Pi_\varkappa ; \mathfrak{G})$, d.h., V ist ein π -isometrischer Operator im $\Pi_\varkappa$ -Raum $\Pi_\varkappa$, dessen Definitionsbereich $\mathfrak{D}(V)$ und Wertebereich $\mathfrak{R}(V)$ abgeschlossene Teilräume sind, auf denen das π -Skalarprodukt nicht entartet, und die Dimensionen von $\mathfrak{G} = \mathfrak{D}(V)^{[\perp]}$ und $\mathfrak{R}(V)^{[\perp]}$ stimmen überein. Wir wählen eine im folgenden festgehaltene π -unitäre Erweiterung $\mathcal{U}$ von V in $\Pi_\varkappa$ und bilden mit dieser für $z \,\overline{\in}\, \sigma(\mathcal{U})$ wie in § 3.2 die Operatoren

$$\Gamma_z = \mathcal{U}(\mathcal{U} - zI)^{-1}\Gamma_0 , \quad \Gamma_\infty = 0 , \tag{4.1}$$

wenn Γ_0 die π-orthogonale Projektion von $\Pi_\varkappa$ auf $\mathfrak{G}$ bezeichnet. Der Operator Γ_z bildet den Teilraum $\mathfrak{G}$ eineindeutig und in beiden Richtungen stetig auf $\mathfrak{N}_z = \mathfrak{R}_{\bar z}^{[\perp]}$ ab, $\mathfrak{R}_z = (zV - I)^{-1}\mathfrak{D}(V)$. Wir bemerken, dass in Verallgemeinerung von (4.1) sogar für beliebige $z, z' \,\overline{\in}\, \sigma(\mathcal{U})$ die Beziehung

$$\Gamma_z = (\mathcal{U} - z'I)(\mathcal{U} - zI)^{-1}\Gamma_{z'}$$

besteht. Daraus ergeben sich leicht die Gleichungen

$$\Gamma_{\frac{1}{\bar z}} - \Gamma_{\bar \zeta} = (1 - z\bar\zeta)\frac{1}{z}\mathcal{U}^+(I - \zeta\mathcal{U}^+)^{-1}\Gamma_{\frac{1}{\bar z}} , \tag{4.2}$$

$$\Gamma_z^+ - \Gamma_{\frac{1}{\bar z}}^+ = -(1-z\bar\zeta)\frac{1}{\bar\zeta}\Gamma_{\frac{1}{\bar\zeta}}^+ \mathfrak{U}(1-z\mathfrak{U})^{-1}. \tag{4.3}$$

Der π-adjungierte Operator Γ_z^+ bildet den Raum $\Pi_\varkappa$ und damit auch $\mathfrak{N}_z$ in $\mathfrak{G}$ ab. Für $z \bar\in \mathfrak{D}_V$ führen wir noch die partiellen Inversen $\Gamma_z^{(-1)}$ und $\Gamma_z^{+(-1)}$ von $\mathfrak{N}_z$ auf $\mathfrak{G}$ bzw. $\mathfrak{G}$ auf $\mathfrak{N}_z$ ein. Schliesslich bezeichne X die mit der π-unitären Erweiterung $\mathfrak{U}$ von V gebildete charakteristische Funktion des Operators V, d.h.

$$X(z) = z\Gamma_0 \mathfrak{U}^+(I - zT^+)^{-1}\Gamma_0, \quad \frac{1}{z} \in \varrho(T^+).$$

Wir betrachten eine π-unitäre Erweiterung $\tilde{\mathfrak{U}}$ von V in einem $\Pi_\varkappa$-Raum $\tilde\Pi_\varkappa \supset \Pi_\varkappa$, der dieselbe Anzahl negativer Quadrate wie der Ausgangsraum hat, d.h., auf den Teilraum $\tilde\Pi_\varkappa [-] \Pi_\varkappa$ sei das π-Skalarprodukt positiv definit. Mit $\tilde P$ bezeichnen wir die π-orthogonale Projektion von $\tilde\Pi_\varkappa$ auf $\Pi_\varkappa$. Die für $z \in \varrho(\tilde{\mathfrak{U}}^+)$ definierte Funktion

$$R_z = \tilde P(\tilde I - z\tilde{\mathfrak{U}})^{-1}$$

mit Werten in $[\Pi_\varkappa, \Pi_\varkappa]$ heisse die von der Erweiterung $\tilde{\mathfrak{U}}$ erzeugte *verallgemeinerte Resolvente* des Operators V. Ist dabei $\tilde\Pi_\varkappa = \Pi_\varkappa$, d.h., $\tilde{\mathfrak{U}}$ eine π-unitäre Erweiterung im Ausgangsraum, so nennen wir die verallgemeinerte Resolvente *orthogonal.*

Eine verallgemeinerte Resolvente R_z ist im Inneren und Äusseren des Einheitskreises meromorph und hat dort höchstens eine endliche Anzahl von Polen. Wir überlegen uns, dass sie für $z \in \mathcal{C}$ mit Ausnahme höchstens abzählbar vieler isolierter Punkte eine Inverse $R_z^{-1} \in [\Pi_\varkappa, \Pi_\varkappa]$ besitzt.

Zu diesem Zweck wählen wir in $\tilde\Pi_\varkappa$ einen bezüglich eines positiv definiten Skalarproduktes der Form (1.2) unitären Operator $\tilde W$, der sich von $\tilde{\mathfrak{U}}$ nur durch einen endlichdimensionalen Operator unterscheidet; dass ein solcher existiert, folgt z.B. leicht aus der allgemeinen Darstellung (3.6) von [15] eines π-unitären Operators. Dann ist mit

$$\tilde{S}_z = (\tilde{I} - z\tilde{W})^{-1}, \qquad S_z = \tilde{P}\tilde{S}_z$$

der Operator $D_z = R_z - S_z$ für jedes z mit $|z| < 1$, $z \bar{\in} \sigma(\tilde{U}^+)$, endlichdimensional, und S_z hat eine Inverse in $[\Pi_\varkappa, \Pi_\varkappa]$. Letzteres ergibt sich wie in [10] aus der Beziehung

$$\mathrm{Re}(\tilde{S}_z h, h) = \frac{1}{2}\{(1-|z|^2)\|\tilde{S}_z^* h\|^2 + \|h\|^2\}, \qquad h \in \Pi_\varkappa,$$

denn für $|z| < 1$ erhält man damit

$$\|h\|^2 \leqq 2\mathrm{Re}(\tilde{S}_z h, h) \leqq 2\|S_z h\| \cdot \|h\|,$$

und der Wertebereich von S_z ist der ganze Raum $\Pi_\varkappa$, da aus $(S_z f, g) = 0$ für alle $f \in \Pi_\varkappa$ sofort $0 = (\tilde{S}_z g, g) = \mathrm{Re}(\tilde{S}_z g, g) \geqq \frac{1}{2}\|g\|^2$, d.h. $g = 0$ folgt. Die Behauptung erhalten wir jetzt aus der Gleichung $R_z = (I + D_z S_z^{-1}) S_z$, denn nach einem bekannten Satz der Störungstheorie ([16], Satz I.5.1) hat der Operator $I + D_z S_z^{-1}$ auf Grund von $D_0 = 0$ für $z \in \mathcal{E}$ mit Ausnahme höchstens abzählbar vieler isolierter Punkte eine Inverse.

2. Ist $\mathfrak{G}$ wieder ein $\Pi_\varkappa$-Raum mit dem π-Skalarprodukt [. , .], so bezeichnen wir im folgenden mit $\mathfrak{K}(\mathfrak{G})$ die Menge derjenigen Funktionen aus $\mathbf{K}_0(\mathfrak{G})$, die bei $z = 0$ holomorph sind. Gemäss Folgerung 3.2 sind diese meromorph in $\mathcal{E}$, und Lemma 3.1 besagt, dass es genau die in $\mathcal{E}_X$ π-kontrahierenden Funktionen X sind. Ist das π-Skalarprodukt auf $\mathfrak{G}$ insbesondere positiv definit, d.h. $\mathfrak{G}$ ein Π_0-Raum, so ist eine solche Funktion X sogar holomorph in $\mathcal{E}$; ist das π-Skalarprodukt auf $\mathfrak{G}$ negativ definit, so gilt

$$|[X(z)f, X(z)f]| \geqq |[f, f]|, \qquad f \in \mathfrak{G}, \qquad z \in \mathcal{E}_X.$$

Hauptergebnis dieses Paragraphen sind die nachfolgenden Sätze 4.1 und 4.2. Dabei setzen wir stets $|z| < 1$ voraus; eine Erweiterung dieser Aussagen auf das Gebiet $|z| > 1$ ergibt sich aus den Gleichungen

$$(I - zU)^{-1} = I - (I - \tfrac{1}{z}U^+)^{-1}, \quad R_z = I - R^+_{\frac{1}{\bar{z}}}.$$

Satz 4.1. *Die Gesamtheit aller verallgemeinerten Resolventen* R_z *des* π *-isometrischen Operators* V *ist gegeben durch die Beziehung*

(4.4) $$R_z = (I - z\mathfrak{U})^{-1} + \Gamma_{\frac{1}{\bar z}} P(z) \Gamma_{\bar z}^{+}, \quad z \in \varrho(\tilde{\mathfrak{U}}^{+}) \cap \varrho(\mathfrak{U}^{+}), \ |z| < 1,$$

mit

$$P(z) = (I - \mathcal{E}(z))(I - X^{+}(\bar z)\mathcal{E}(z))^{-1}(I - X(\bar z)),$$

wobei $\mathcal{E}$ *die Klasse* $\mathfrak{K}(\mathfrak{g})$ *durchläuft. Die verallgemeinerte Resolvente* R_z *ist genau dann orthogonal, wenn* $\mathcal{E}$ *ein von* z *unabhängiger* π *-unitärer Operator in* $\mathfrak{g}$ *ist.*

Der Satz 4.1 hängt sehr eng zusammen mit dem folgenden

Satz 4.2. *Durchläuft in der Beziehung*

(4.5) $$\Gamma_0(\tilde I + z\tilde{\mathfrak{U}})(\tilde I - z\tilde{\mathfrak{U}})^{-1}\Gamma_0 = (I + X^{+}(\bar z)\mathcal{E}(z))(I - X^{+}(\bar z)\mathcal{E}(z))^{-1}$$

der Operator $\tilde{\mathfrak{U}}$ *die Gesamtheit aller* π *-unitären Erweiterungen von* V *mit eventuellem Austritt in einen Oberraum* $\tilde{\Pi}_\varkappa \supset \Pi_\varkappa$ *, so durchläuft* $\mathcal{E}$ *die Klasse* $\mathfrak{K}(\mathfrak{g})$*. Die in* (4.4) *und* (4.5) *auftretenden Funktionen* $\mathcal{E}$ *stimmen überein.*

Dabei besteht die Beziehung (4.5) für alle Punkte $z \in \mathfrak{L}$, für die sämtliche auftretenden Operatoren definiert sind. Den Operator Γ_0 haben wir erweitert auf $\tilde{\Pi}_\varkappa$ durch die Festsetzung $\Gamma_0 f = 0$ für, $f \in \tilde{\Pi}_\varkappa [-] \Pi_\varkappa$. Gleichung (4.5) ist offensichtlich äquivalent mit der Beziehung

(4.6) $$\Gamma_0 R_z \Gamma_0 = (I - X^{+}(\bar z)\mathcal{E}(z))^{-1}$$

wenn R_z wieder die von $\tilde{\mathfrak{U}}$ erzeugte verallgemeinerte Resolvente des Operators V bezeichnet.

Wir beweisen die Sätze 4.1 und 4.2, indem wir zunächst für eine beliebige π -unitäre Erweiterung $\tilde{\mathfrak{U}}$ von V die Gleichung (4.5) herleiten, damit die Darstellung (4.4) erhalten und anschliessend zeigen, dass umgekehrt für $\mathcal{E} \in \mathfrak{K}(\mathfrak{g})$ die rechte Seite von (4.4) eine verallgemeinerte Resolvente von V definiert, für die ihrerseits die Beziehung (4.6) und damit auch (4.5) besteht.

3. Es sei $\tilde{\mathfrak{U}}$ eine π-unitäre Erweiterung von V, $R_z = \tilde{P}(\tilde{I} - z\tilde{\mathfrak{U}})^{-1}$ die von ihr erzeugte verallgemeinerte Resolvente. Wir betrachten die in $\mathfrak{C}$ mit Ausnahme höchstens abzählbar vieler isolierter Punkte definierte Funktion W:

$$W(z) = \frac{1}{z}(I - R_z^{-1}) \in [\Pi_\varkappa, \Pi_\varkappa].$$

Ihre im folgenden benötigten Eigenschaften fassen wir zusammen in dem

Lemma 4.1. *Die Funktion* W *gehört zur Klasse* $\mathfrak{G}(\Pi_\varkappa)$ *und es gilt*

(4.7) $$W(z) \supset V, \quad W^+(\bar{z}) \supset V^{-1} \quad \textit{und} \quad \Gamma_0 W^+(\bar{z}) = W^+(\bar{z})P,$$

wenn P *die* π*-orthogonale Projektion auf* $\mathfrak{R}(V)^{[\perp]}$ *bezeichnet.*

Beweis. Aus $R_0 = I$ und der Holomorphie von R_z^{-1} in einer Umgebung von $z = 0$ ergibt sich die Holomorphie von $W(z)$ bei $z = 0$ Weiter gilt

$$K_W(z,\zeta) = \frac{1}{1 - z\bar{\zeta}}\left\{I - \frac{1}{z\bar{\zeta}}(I - (R_\zeta^{-1})^+)(I - R_z^{-1})\right\} =$$

$$= \frac{1}{(1 - z\bar{\zeta})z\bar{\zeta}}(R_\zeta^{-1})^+\{-(1 - z\bar{\zeta})R_\zeta^+ R_z + R_z + R_\zeta^+ - I\}R_z^{-1} =$$

$$= \frac{1}{z\bar{\zeta}}(R_\zeta^{-1})^+ \tilde{P}(\tilde{I} - \bar{\zeta}\tilde{\mathfrak{U}}^+)^{-1}(\tilde{I} - \tilde{P})(\tilde{I} - z\tilde{\mathfrak{U}})^{-1}\tilde{P}R_z^{-1},$$

woraus wegen $[(\tilde{I} - \tilde{P})\tilde{h}, \tilde{h}] \geqq 0$ $(\tilde{h} \in \tilde{\Pi}_\varkappa)$ folgt, dass der $\Pi_\varkappa$-Kern K_W nichtnegativ ist. Auf Grund von Lemma 3.1 erhalten wir daraus $W \in \mathfrak{G}(\Pi_\varkappa)$.

Man sieht ohne Schwierigkeit, dass $W(z)$ eine Erweiterung von V und ebenso $W^+(\bar{z})$ eine Erweiterung von V^{-1} ist. Dann folgt aber für $f \in \Pi_\varkappa$, $h \in \mathfrak{D}(V)$:

$$[W^+(z)Pf, h] = [Pf, Vh] = 0,$$

d.h. $(I-\Gamma_0)W^+(z)P = 0$ und ebenso $\Gamma_0 W^+(z)(I-P) = 0$. Daraus ergibt sich die letzte Gleichung von (4.7).

4. Zum Beweis der Beziehung (4.5) (vgl. den Beweis von (3.8)) gehen wir aus von den Gleichungen $W^+(\bar{z})-T^+ = W^+(\bar{z})P$ und

$$(I-zT^+)^{-1}-(I-zW^+(\bar{z}))^{-1} = -z(I-zW^+(\bar{z}))^{-1}W^+(\bar{z})P(I-zT^+)^{-1} .$$

Durch Multiplikation mit Γ_0 von rechts und mit $z\Gamma_0\mathfrak{U}^+$ von links ergibt sich

$$X(z)-z\Gamma_0\mathfrak{U}^+(I-zW^+(\bar{z}))^{-1}\Gamma_0 = -z\Gamma_0\mathfrak{U}^+(I-zW^+(\bar{z}))^{-1}W^+(\bar{z})\mathfrak{U}\Gamma_0X(z) ,$$

also $\mathfrak{U}X(z) = zP(I-zW^+(\bar{z}))^{-1}\Gamma_0(I-\Gamma_0W^+(\bar{z})\mathfrak{U}\Gamma_0X(z))$, woraus mit $\mathcal{E}(z) = \Gamma_0\mathfrak{U}^+W(z)\Gamma_0$ bei Beachtung von (4.7) leicht

$$\mathcal{E}^+(\bar{z})X(z) = z\Gamma_0W^+(\bar{z})(I-zW^+(\bar{z}))^{-1}\Gamma_0(I-\mathcal{E}^+(\bar{z})X(z))$$

folgt. Nun ist weiter $zW^+(\bar{z})(I-zW^+(\bar{z}))^{-1} = R^+_{\bar{z}}-I = z\widetilde{P}\widetilde{\mathfrak{U}}^+(I-z\widetilde{\mathfrak{U}}^+)^{-1} = z\widetilde{P}(\widetilde{\mathfrak{U}}-z\widetilde{I})^{-1}$, also erhalten wir

$$\mathcal{E}^+(\bar{z})X(z)(I-\mathcal{E}^+(\bar{z})X(z))^{-1} = z\Gamma_0(\widetilde{\mathfrak{U}}-z\widetilde{I})^{-1}\Gamma_0 .$$

Diese Gleichung besteht zunächst für hinreichend kleine z ; durch analytische Fortsetzung folgt ihre Richtigkeit für alle Punkte $z\in\mathfrak{G}$, für die nur sämtliche auftretenden Operatoren sinnvoll sind. Die Funktion $\mathcal{E}$ gehört zur Klasse $\mathfrak{K}(\mathfrak{G})$, denn auf Grund von Lemma 4.1 ist sie bei $z=0$ holomorph und der $\mathfrak{G}$ -Kern

$$K_{\mathcal{E}}(z,\zeta) = \frac{I-\mathcal{E}^+(\zeta)\mathcal{E}(z)}{1-z\bar{\zeta}} = \Gamma_0K_W(z,\zeta)\Gamma_0$$

positiv definit. Damit ist die Beziehung (4.5) bewiesen.

Wir bemerken, dass gemäss der Definition der Funktion $\mathcal{E}$ zur Erweiterung $\mathfrak{U}$ die Operatorfunktion $\mathcal{E}(z) = I$ gehört, d.h. es gilt

(4.8) $$\Gamma_0(I+z\mathcal{U})(I-z\mathcal{U})^{-1}\Gamma_0 = (I+X^+(\bar{z}))(I-X^+(\bar{z}))^{-1}.\ *$$

Aus dieser Beziehung folgert man noch die später benötigten Gleichungen

(4.9) $$\Gamma_0\Gamma_{\frac{1}{z}} = \Gamma_0\left(I-\frac{1}{z}\mathcal{U}^+\right)^{-1}\Gamma_0 = -X^+(\bar{z})(I-X^+(\bar{z}))^{-1},$$

(4.10) $$\Gamma_{\bar{z}}^+\Gamma_0 = \Gamma_0(I-z\mathcal{U})^{-1}\Gamma_0 = (I-X^+(\bar{z}))^{-1},$$

(4.11) $$\Gamma_{\frac{1}{\zeta}}^+\Gamma_{\frac{1}{z}} = z\bar{\zeta}(I-X(\bar{\zeta}))^{-1}\frac{I-X(\bar{\zeta})X^+(z)}{1-z\bar{\zeta}}(I-X^+(\bar{z}))^{-1}.$$

Als nächstes zeigen wir, dass aus (4.6) die Darstellung (4.4) folgt. Dazu überzeugt man sich zuerst ohne Schwierigkeit von der Beziehung

$$R_z f-(I-z\mathcal{U})^{-1}f \begin{cases} =0 & f\in\mathcal{R}_z \\ \in\mathfrak{N}_{\frac{1}{z}} & f\in\mathfrak{N}_{\bar{z}}. \end{cases}$$

Für $z\in(\varrho(\mathcal{U}^+)\cap\varrho(\tilde{\mathcal{U}}^+))\setminus\mathfrak{D}_V$, bildet folglich der Operator

$$P(z) = \Gamma_{\frac{1}{z}}^{(-1)}\left(R_z-(I-z\mathcal{U})^{-1}\right)\Gamma_{\bar{z}}^{(-1)+}$$

den Raum $\mathfrak{G}$ in sich ab. Wir schreiben diese Gleichung in der Form

$$R_z = (I-z\mathcal{U})^{-1}+\Gamma_{\frac{1}{z}}P(z)\Gamma_{\bar{z}}^+.$$

* Das ist nichts anderes als Gleichung (3.8).

Nach Multiplikation von rechts und links mit Γ_0 ergibt sich

$$\Gamma_0 R_z \Gamma_0 = \Gamma_0 (I - z\mathfrak{U})^{-1} \Gamma_0 + \Gamma_0 \Gamma_{\frac{1}{z}} P(z) \Gamma^+_{\bar{z}} \Gamma_0$$

und daraus auf Grund (4.6), (4.9) und (4.10)

$$P(z) = (I - \mathcal{E}(z))(I - X^+(\bar{z})\mathcal{E}(z))^{-1}(I - X^+(\bar{z})).$$

Damit ist die Darstellung (4.4) bewiesen.

5. Es sei jetzt eine Funktion $\mathcal{E} \in \mathfrak{K}(\mathfrak{G})$ gegeben. $\mathfrak{M}_0$ bezeichne eine zur reellen Achse symmetrisch gelegene Umgebung des Punktes $z = 0$, die ganz zu $\varrho(\mathfrak{U})$ gehört, in der $\mathcal{E}$ holomorph ist und $(I - X^+(\bar{z})\mathcal{E}(z))^{-1} \in \in [\mathfrak{G}, \mathfrak{G}]$ existiert. Die mit $\mathcal{E}(z)$ gebildete rechte Seite von (4.4) bezeichnen wir mit $S(z)$:

$$S(z) = (I - z\mathfrak{U})^{-1} + \Gamma_{\frac{1}{z}} P(z) \Gamma^+_{\bar{z}},$$

insbesondere ist also $S(0) = I$. Wir zeigen, dass $S(z)$ eine verallgemeinerte Resolvente des Operators V ist.

Zu diesem Zweck extrapolieren wir die Funktion $S(z)$ durch die Festsetzung $S(\frac{1}{z}) = I - S^+(\bar{z})$ auf die Menge $\mathfrak{M}_0^* = \{z \mid \frac{1}{\bar{z}} \in \mathfrak{M}_0\}$ und definieren auf $\mathfrak{M} = \mathfrak{M}_0 \cup \mathfrak{M}_0^*$ den $\Pi_\varkappa$-Kern

$$K(z,\zeta) = \frac{S(z) + S^+(\zeta) - I}{1 - z\bar{\zeta}}$$

Dann gilt

(4.12) $$K(z,\zeta) = S^+(\zeta) S(z) + \Delta(z,\zeta),$$

wobei wir $\Pi_\varkappa$-Kern Δ die folgende Gestalt hat $(z,\zeta \in \mathfrak{M}_0)$:

(4.13) $$\Delta(z,\zeta) = \Gamma_{\bar\zeta} B_1^+(\zeta) \frac{I-\mathcal{E}^+(\zeta)\mathcal{E}(z)}{1-z\bar\zeta} B_1(z) \Gamma_{\bar z}^+ ,$$

(4.14) $$\Delta(z,\frac{1}{\zeta}) = \Gamma_{\frac{1}{\zeta}} B_2^+(\zeta) \frac{\mathcal{E}(z)-\mathcal{E}(\bar\zeta)}{z-\bar\zeta} B_1(z) \Gamma_{\bar z}^+ ,$$

(4.15) $$\Delta(\frac{1}{z},\frac{1}{\zeta}) = \Gamma_{\frac{1}{\bar\zeta}} B_2^+(\zeta) \frac{I-\mathcal{E}(\bar\zeta)\mathcal{E}^+(\bar z)}{1-z\bar\zeta} B_2(z) \Gamma_{\frac{1}{\bar z}}^+$$

mit $B_1(z) = z(I-X^+(\bar z)\mathcal{E}(z))^{-1}(I-X^+(\bar z))$, $B_2(z) =$
$= (I-X(z)\mathcal{E}^+(\bar z))^{-1}(I-X(z))$.

Gleichung (4.13) ergibt sich z.B. folgendermassen:

$$\Delta(z,\zeta) = \frac{1}{1-z\bar\zeta}\{(\Gamma_{\frac{1}{z}} - \Gamma_{\bar\zeta}) P(z) \Gamma_{\bar z}^+ + \Gamma_{\bar\zeta}(P(z)+P^+(\zeta))\Gamma_{\bar z}^+ -$$

$$- \Gamma_{\bar\zeta} P^+(\zeta)(\Gamma_{\bar z}^+ - \Gamma_{\frac{1}{\zeta}}^+)\} - \Gamma_{\bar\zeta} P^+(\zeta)\Gamma_{\frac{1}{\zeta}}^+(I-z\mathcal{U})^{-1} -$$

$$- (I-\bar\zeta\mathcal{U}^+)^{-1}\Gamma_{\frac{1}{z}} P(z)\Gamma_{\bar z}^+ - \Gamma_{\bar\zeta} P^+(\zeta)\Gamma_{\frac{1}{\zeta}}^+ \Gamma_{\frac{1}{z}} P(z)\Gamma_{\bar z}^+ =$$

$$= \Gamma_{\bar\zeta}\{\frac{z\bar\zeta}{1-z\bar\zeta}(P(z)+P^+(\zeta)) - P^+(\zeta)\Gamma_{\frac{1}{\zeta}}^+ \Gamma_{\frac{1}{z}} P(z)\}\Gamma_{\bar z}^+ =$$

$$= \frac{z\bar{\zeta}}{1-z\bar{\zeta}} \Gamma_{\bar{\zeta}} \{ P(z) + P^{+}(\zeta) - P^{+}(\zeta)(I - X(\bar{\zeta}))^{-1} \cdot$$

$$\cdot (I - X(\bar{\zeta}) X^{+}(\bar{z}))(I - X^{+}(\bar{z}))^{-1} P(z) \} \Gamma_{\bar{z}}^{+} =$$

$$= z\bar{\zeta} \Gamma_{\bar{\zeta}} (I - X(\bar{\zeta}))(I - \mathcal{E}^{+}(\zeta) X(\bar{\zeta}))^{-1} \cdot$$

$$\cdot \frac{I - \mathcal{E}^{+}(\zeta)\mathcal{E}(z)}{1 - z\bar{\zeta}} (I - X^{+}(\bar{z})\mathcal{E}(z))^{-1} (I - X^{+}(\bar{z})) \Gamma_{\bar{z}}^{+} ;$$

dabei haben wir die Beziehungen (4.2). (4.3) und (4.11) benutzt. Aus (4.13) – (4.15) und der Ungleichung (siehe (3.2))

$$\left| \sum_{j,k} \left[\frac{\mathcal{E}(z_j) - \mathcal{E}(\bar{\zeta}_k)}{z_j - \bar{\zeta}_k} f_j, g_k \right] \right|^2 \leqq$$

$$\leqq \sum_{j,k} \left[\frac{I - \mathcal{E}^{+}(z_k)\mathcal{E}(z_j)}{1 - z_j \bar{z}_k} f_j, f_k \right] \sum_{j,k} \left[\frac{I - \mathcal{E}(\bar{\zeta}_k)\mathcal{E}^{+}(\bar{\zeta}_j)}{1 - \zeta_j \bar{\zeta}_k} g_j, g_k \right]$$

folgert man jetzt leicht, dass der Kern $\Delta(z,\zeta)$ in $\mathfrak{N}$ positiv definit ist.

Wir konstruieren den Raum $\widetilde{\Pi}_{\varkappa}$, in dem die π-unitäre Erweiterung $\widetilde{U}$ von V wirkt. Zu diesem Zweck betrachten wir die lineare Menge $\widetilde{\mathcal{L}}$ aller formalen Summen $\tilde{f} = \sum_{z \in \mathfrak{V}} \varepsilon_z f_z$, $f_z \in \Pi_\varkappa$, wobei nur endlich viele f_z Nullelement verschieden sein sollen, und definieren zwischen zwei solchen Elementen $\tilde{f}$ und $\tilde{g} = \sum_{\zeta \in \mathfrak{V}} \varepsilon_\zeta g_\zeta \in \tilde{\mathcal{L}}$ das π-Skalarprodukt durch die Gleichung

(4.16) $$[\tilde{f}, \tilde{g}] = \sum_{z, \zeta \in \mathfrak{V}} [K(z,\zeta) f_z, g_\zeta] .$$

Dann gilt auf Grund (4.12)

$$[\tilde{f}, \tilde{g}] = [\sum_z S(z) f_z, \sum_\zeta S(\zeta) g_\zeta] + \sum_{z,\zeta} [\Delta(z,\zeta) f_z, g_\zeta],$$

somit hat das π-Skalarprodukt (4.16) auf $\tilde{\mathcal{L}}$ höchstens $\varkappa$ negative Quadrate. Andererseits vermittelt die Zuordnung $f \to \tilde{f} = \varepsilon_0 f$ eine eineindeutige π-isometrische Beziehung zwischen dem Ausgangsraum $\Pi_\varkappa$ und einem Teilraum von $\tilde{\mathcal{L}}$, also hat (4.16) auf $\tilde{\mathcal{L}}$ genau $\varkappa$ negative Quadrate. Wir betten $\tilde{\mathcal{L}}$ kanonisch in einen $\Pi_\varkappa$-Raum $\tilde{\Pi}_\varkappa$ ein; dieser ist der gesuchte Oberraum von $\Pi_\varkappa$. Im folgenden identifizieren wir noch $f \in \Pi_\varkappa$ mit dem ihm zugeordneten Element $\varepsilon_0 f \in \tilde{\Pi}_\varkappa$ und die Elemente aus $\tilde{\mathcal{L}}$ mit ihren Bildern in $\tilde{\Pi}_\varkappa$.

Auf der Menge $\tilde{\mathcal{L}}_0$ aller $\tilde{f} = \sum_{z \in \mathfrak{V}} \varepsilon_z f_z \in \tilde{\mathcal{L}}$ mit der Eigenschaft $f_0 = 0$ wird ein linearer Operator $\tilde{U}$ erklärt durch die Festsetzung

$$\tilde{U} \varepsilon_z f = \frac{\varepsilon_z f - f}{z}, \qquad z \neq 0 .$$

Offensichtlich ist $\tilde{\mathcal{L}}_0$ ein dichter Teil von $\tilde{\Pi}_\varkappa$. Aus den leicht nachweisbaren Beziehungen

$$[\varepsilon_z f, \varepsilon_z f] \to 0 \text{ und } [\varepsilon_z f, \varepsilon_\zeta g] \to 0 \text{ für } z \to \infty, \quad f, g \in \Pi_\varkappa,$$

folgt $\varepsilon_z f \to 0$ in $\tilde{\Pi}_\varkappa$ für $z \to \infty$. Deshalb liegt auch der Wertebereich von $\tilde{U}$ dicht in $\tilde{\Pi}_\varkappa$.

Der Operator $\tilde{\mathfrak{U}}$ ist π-isometrisch:

$$[\tilde{\mathfrak{U}} \sum_{\substack{z\in\mathfrak{A}\\ z\neq 0}} \varepsilon_z f_z, \tilde{\mathfrak{U}} \sum_{\substack{\zeta\in\mathfrak{A}\\ \zeta\neq 0}} \varepsilon_\zeta g_\zeta] = \sum_{z,\zeta} [\frac{\varepsilon_z f_z - f_z}{z}, \frac{\varepsilon_\zeta g_\zeta - g_\zeta}{\zeta}] =$$

$$= \sum_{z,\zeta} [\frac{S(z)+S^+(\zeta)-I}{1-z\bar{\zeta}} f_z, g_\zeta] = [\sum_z \varepsilon_z f_z, \sum_\zeta \varepsilon_\zeta g_\zeta] ;$$

nach [12], ist er folglich auch beschränkt, also durch Stetigkeit zu einem π-unitären Operator in $\tilde{\Pi}_\varkappa$ fortsetzbar. Für $f \in \vartheta(V)$ gilt weiter $\lim_{z\to 0} \varepsilon_z f = Vf$. Um das zu sehen, beachten wir die Beziehungen $\lim_{z\to 0} \Gamma_z^+ f = 0$ sowie

$$\frac{S(z)-I}{z} f \longrightarrow Vf, \qquad \frac{1}{z}\Delta(z,\zeta)f \longrightarrow 0 \qquad (z\to 0,\ f\in\vartheta(V)).$$

Aus ihnen folgt für $f\in\vartheta(V)$ und $z \longrightarrow 0$:

$$[\tilde{\mathfrak{U}}\varepsilon_z f - Vf, \varepsilon_\zeta g] =$$

$$= [S^+(\zeta)(\frac{S(z)-I}{z} - V)f, g] + \frac{1}{z}[\Delta(z,\zeta)f, g] \longrightarrow 0,$$

$$[\tilde{\mathfrak{U}}\varepsilon_z f - Vf, \tilde{\mathfrak{U}}\varepsilon_z f - Vf] =$$

$$= [K(z,z)f,f] - 2\,\mathrm{Re}[\frac{S(z)-I}{z} - f, Vf] + [Vf, Vf] \longrightarrow 0,$$

also $\tilde{\mathfrak{U}}\varepsilon_z f \longrightarrow Vf$.

Die für $f, g \in \Pi_\varkappa$ bestehende Beziehung

$$[(\tilde{I} - z\tilde{\mathfrak{U}})^{-1} f, g] = [\varepsilon_z f, g] = [S(z)f, g]$$

besagt schliesslich, dass $S(z)$ die von der Erweiterung $\tilde{\mathfrak{U}}$ erzeugte verallgemeinerte Resolvente des Operators V ist.

Aus (4.4) ergibt sich auch die Beziehung (4.6) und damit (4.5), denn es ist wegen (4.8), (4.9) und (4.10)

$$\Gamma_0(\tilde{I}-z\tilde{\mathfrak{U}})^{-1}\Gamma_0 = \Gamma_0(I-z\mathfrak{U})^{-1}\Gamma_0+\Gamma_0\Gamma_{\frac{1}{z}}P(z)\Gamma^{+}_{\bar{z}}\Gamma_0 = (I-X^{+}(\bar{z})\mathcal{E}(z))^{-1}.$$

Für eine orthogonale verallgemeinerte Resolvente von V, d.h. für eine π-unitäre Erweiterung $\tilde{\mathfrak{U}}$ von V im Ausgangsraum $\Pi_\varkappa$, ist $W(z) = \tilde{\mathfrak{U}}$ also auch $\mathcal{E}(z)$ ein von z unabhängiger π-unitärer Operator. Andererseits verschwindet für solches $\mathcal{E}$ der Kern Δ auf Grund der Beziehungen (4.13) – (4.15), d.h., es ist $\tilde{\Pi}_\varkappa = \Pi_\varkappa$ für den oben konstruierten Raum $\tilde{\Pi}_\varkappa$. Damit sind die Sätze 4.1 und 4.2 bewiesen.

LITERATUR

[1] M.G. Kreĭn – H. Langer, *Über die Defekträume und die verallgemeinerten Resolventen eines hermiteschen Operators im Raume* $\Pi_\varkappa$ [russisch].

[2] M.S. Livšic, Über eine Klasse linearer Operatoren im Hilbertraum, *Mat. Sbornik,* 19 (61) (1946), 239-262 [russisch].

[3] M.S. Livšic, Isometrische Operatoren mit gleichen Defektzahlen, *Mat. Sbornik,* 26 (68) (150), 247-264 [russisch].

[4] M.G. Kreĭn – H. Langer, Analytische Eigenschaften der Schmidtschen Paare eines Hankelschen Operators und die verallgemeinerte Aufgabe von Schur-Takagi, *Mat. Sbornik,* (im Druck) [russisch].

[5] B. Sz.-Nagy – A. Korányi, Operatortheoretische Behandlung und Verallgemeinerung eines Problemkreises in der komplexen Funktionentheorie, *Acta Math.,* 100 (1958), 171-202.

[6] A.V. Kužel', Spektralanalyse quasi-unitärer Operatoren im Raume mit indefiniter Metrik, *Teor. Funkciĭ Funkcional Anal. i Priložen.,* 4 (1967), 3-27 [russisch].

[7] V.P. Potapov, Multiplikative Struktur J -kontrahierender Matrixfunktionen, *Trudy Moskov, Mat. Obšč.,* 4 (1955) 125-236 [russisch].

[8] Ju.P. Ginzburg, Das Maximumprinzip für J -kontrahierende Operatorfunktionen und einige Anwendungen, *Isv. Vyss. Učebn. Zaved. Matematika,* I (32) (1963), 42-53 [russisch].

[9] Š.N. Saakjan, Zur Theorie der Resolventen eines symmetrischen Operators und unendlichen Defektzahlen, *Dokl. Akad. Nauk. Arm. SSR,* XLI, 4 (1965), 193-197 [russisch].

[10] M.E. Čumakin, Verallgemeinerte Resolventen isometrischer Operatoren, *Sibirsk. Mat. Ž.,* VIII, 4 (1967), 876-892 [russisch].

[11] A.G. Gibson, Triples of Operator-Valued Functions related to the Unit-Circle, *Pacific J. Math,* 28, 3 (1969), 504-531.

[12] I.S. Iohvidov – M.G. Kreĭn, Spektraltheorie der Operatoren im Raume mit indefiniter Metrik. I, *Trudy Moskov. Mat. Obšč.*, 5 (1956) 367-431; II, ebenda, 8 (1959), 413-496 [russisch].

[13] M.G. Kreĭn – Ju. L. Šmul'jan, Über Plus-Operatoren im Raume mit indefiniter Metrik. *Mat. Issled.*, I, 1 (1966) 131-161 [russisch].

[14] Ju.P. Ginzburg, Über multiplikative Darstellungen J -kontrahierender Operatorfunktionen. I, *Mat. Issled.*, II, 1 (1967), 52-83 [russisch].

[15] M.G. Kreĭn – ju.L. Šmul'jan, Über gebrochen lineare Abbildungen mit operatorwertigen Koeffizienten, *Mat. Issled.*, II, 3 (1967), 64-96 [russisch].

[16] I.Z. Gohberg – M.G. Kreĭn, *Einführung in die Theorie der linearen nichtselbstadjungierten Operatoren im Hilbertraum* (Moskau, 1965) [russisch].

COLLOQUIA MATHEMATICA SOCIETATIS JÁNOS BOLYAI
5. HILBERT SPACE OPERATORS, TIHANY (HUNGARY), 1970

Products of subgroups and projective multipliers

G. W. MACKEY

1. INTRODUCTION **

Let G be a separable locally compact group and let $x \longrightarrow U_x$ be a mapping from G into the unitary operator in some separable Hilbert space $\mathcal{H}(U)$. We shall say that $x \longrightarrow U_x$ is a *projective unitary representation* (or *unitary ray representation*) of G if the following conditions hold

(i) $U_e = I$ where e is the identity of G.

(ii) $x \longrightarrow (U_x(\varphi), \psi)$ is a Borel function of x for all φ and ψ in $\mathcal{H}(U)$.

(iii) There exists a Borel function σ from $G \times G$ to the complex numbers of modulus one such that

$$U_{xy} = \sigma(x, y) U_x \times U_y$$

* The author is a fellow of the John Simon Guggenheim Memorial Foundation. This article was written while he was a visitor at the Institut des Hautes Etudes Scientifiques.

** The first two thirds of this introduction is an exposition of known material designed for members of the conference who are not specialists in group representation theory.

for all x and y in G . The function σ is called the *projective multiplier* or just the *multiplier* for the projective representation and is clearly uniquely determined by U . When $\sigma \equiv 1$, $x \to U_x$ is well known to be a strongly continuous unitary representation of G .

Using the associative law one verifies easily that any multiplier σ satisfies the identity

$$(*) \qquad \sigma(x,y)\sigma(xy,z) \equiv \sigma(x,yz)\sigma(y,z)$$

characteristic of two cocycles in homological algebra. In addition it follows from (i) that σ is *normalized* in the sense that $\sigma(x,e) \equiv \sigma(e,x) = 1$. Conversely it may be shown (see [4]) that every Borel function σ from $G \times G$ to $|z| = 1$ which satisfies (*) and the normalization condition is in fact the multiplier for some projective unitary representation of G .

A unitary operator in a Hilbert space $\mathcal{H}$ defines an automorphism of the ortho complemented lattice $\mathcal{L}(\mathcal{H})$ of all closed subspaces of $\mathcal{H}$, and two unitary operators define the same automorphism if and only if one is a constant multiple of the other. Thus every projective unitary representation U of G defines a homomorphism $\tilde{U}$ of G into the group of all automorphisms of the ortho complemented lattice $\mathcal{L}(\mathcal{H}(U))$ and U is uniquely determined by $\tilde{U}$ up to multiplication by a Borel function ρ from G to the complex numbers of modulus unity. Now if U is a projective unitary representation with multiplier σ and $U'_x = \rho(x)U_x$ then U' is obviously a projective unitary representation with multiplier σ' where

$$(**) \qquad \sigma'(x,y) = \frac{\sigma(x,y)\rho(xy)}{\rho(x)\rho(y)} .$$

Thus σ' and σ are associated with the same homomorphism $\tilde{U}$ if and only if they are connected by a ρ as in (**). In the language of homological algebra this is just the condition that σ' and σ should be *cohomologous.* Hence the homomorphisms $\tilde{U}$ divide naturally into classes according to the cohomology class of the σ associated with U and it is only in the case in which every normalized σ satisfying (*) is of the form $x,y \to \dfrac{\rho(xy)}{\rho(x)\rho(y)}$ that every U is defined by an ordinary unitary representation of G .

In applications of group theory to quantum mechanics one encounters group representations by way of homomorphisms of a group into the group of automorphisms of some $\mathcal{L}(\mathcal{H})$. Thus where G admits multipliers σ which are not cohomologous to 1 one must be prepared to deal with projective unitary representations of G as well as ordinary ones. (Actually unless every element of G is a square one may have to admit anti-unitary operators as well as unitary ones). Projective unitary representations arise also when one attempts to relate the unitary representations of a separable locally compact group G to those of a normal subgroup N. In classifying the different representations of G associated with a given family of representations of N one finds oneself reduced to classifying the projective representations of a certain closed subgroup of G/N. Thus in a complete theory one has to consider the projective representation along with the ordinary ones. For further details see [4].

For any fixed choice of σ one may develop a theory of the projective unitary representatives with multiplier σ (σ representations which is closely parallel to the theory of ordinary unitary representations [4]). One even has a theory of induced σ representations. To a considerable extent this theory may be deduced from the theory of ordinary representations by generalising a construction of Schur. One introduces an auxiliary group G' having G as quotient and shows that there is a natural one to one correspondence between the σ representations of G and certain ordinary representations of G'. However there are several instances both in the theory and in the applications where thinking in terms of σ representation seems to have advantages over working with the related ordinary representation of G'. See for example sections 14 and 15 of [5] and the author's review (MR 29, 2324) of a paper of Weil as well as the introduction to [4].

In view of the foregoing one is interested in determining all possible cohomology classes of multipliers σ for each separable locally compact group G – in the language of homological algebra this is the problem of determining the cohomology group $H^2(G,K)$ where K is the circle group and where we restrict ourselves to Borel cocycles. An obvious approach to this problem is to attempt to relate the multipliers σ for G to those of some of its subgroups and so proceed inductively to reduce the problem to that for groups small enough to be handled directly.

An easy theorem which contributes to this program is proved in [4]. It asserts that if $G = G_1 \times G_2$ then the most general multiplier σ for G is cohomologous to one constructed as follows. Chose multipliers σ_1 and σ_2 for G_1 and G_2 respectively and let g be an arbitrary continuous function from $G_1 \times G_2$ to the complex numbers of modulus one which is multiplicate in each variable. Then

$x_1, y_1, x_2, y_2 \longrightarrow \sigma_1(x_1, x_2)\sigma(y_1, y_2)g(x_1, y_2)$ is the desired multiplier for $G_1 \times G_2$. From this result it follows more or less immediately that $H^2(G_1 \times G_2, K)$ is isomorphic to $H^2(G,K) \times H^2(G,K) \times A$ where A is the group of all continuous homomorphisms of G_1 into $\hat{G}_2$. Here $\hat{G}_2$ is the group of all one dimensional characters of G_2 i.e. the group of all continuous homomorphisms of G_2 into K. We see in particular that $G_1 \times G_2$ can have non trivial projective multiplier even when G_1 and G_2 do not. For example if G_1 is commutative and $G_2 = \hat{G}_1$ we may take $g(x_1, \chi_2) = \chi_2(x_1)$, $\sigma_1 = \sigma_2 = 1$. The corresponding multiplier σ for $G_1 \times G_2 = G_1 \times \hat{G}_1$ is never trivial but is identically one on both G_1 and G_2. It is well known that $H^2(G_1, K)$ reduces to the identity whenever G_1 is the additive group of the real line. However in this case $\hat{G}_1 \simeq G_1$ so the additive group of the plane $\simeq G_1 \times \hat{G}_1$ has non trivial multipliers.

Projective multipliers of the form $\sigma(x_1, \chi_1, x_2, \chi_2) = \chi_2(x_1)$ where $G = G_1 \times \hat{G}_1$ are expecially interesting. Although G is commutative whenever G_1 is, the σ representation theory of G is strikingly different from the ordinary representation theory of a commutative group. There is (to within equivalence) a unique irreducible σ representation and its dimension is equal to the order of G. This fact, as pointed out in [4], is equivalent to the generalized Stone–von Neumann theorem [3] concerning the uniqueness of solutions of the Heisenberg commutation rules and its specialization to the additive groups of p-adic vector spaces plays a key role in [9].

The theorem about the multipliers for $G_1 \times G_2$ proved in [4] is actually obtained there as a corollary of a more general theorem about the multipliers for a "semi direct product" $G = N \circledS H$. We recall that G is said to be a *semi direct product* of the closed subgroups N and H if N is normal and every element of G is uniquely of the form nh where $n \in N$ and $h \in H$. Each element $h \in H$ defines an automorphism of N via the formula $n \to hnh^{-1}$ and the mapping from H to authomorphism of N is a group homomorphism. Conversely given the groups N and H and the homomorphism $h \longrightarrow \alpha_h$ from H to the group of automorphisms of N we may reconstruct G.

The theorem proved in [4] about multipliers for semi-direct products is like that described above for direct products in that every multiplier is shown to be cohomologous to a product $\sigma_1(n_1, n_2)\sigma_2(h_1, h_2)g(n_1, h_2)$ where σ_1 and σ_2 are multipliers for N and H respectively and g is a function from $N \times H$ to K.

It differs in that σ_1 is no longer arbitrary and in that as a function of n, g satisfies a more complicated identity than being multiplicative. The condition in σ_1 is that its transforms under the automorphisms α_h must be cohomologous to σ_1 and the identity that g must satisfy as a function of n is

$$\frac{\sigma_1(\alpha_h(n_1),\alpha_h(n_2))}{\sigma_1(n_1,n_2)} = \frac{g(n_1,n_2,h)}{g(n_1,h)g(n_2,h)} .$$

Actually this identity expresses the α_h invariance of the cohomology class of σ_1 and reduces to multiplicativity when σ_1 itself is α_h invariant.

The principal result of this note is that the semi direct product theorem just described has a generalization to the case in which neither subgroup is normal. We assume only that our separable locally compact group G has two closed subgroups T and H such that every $x \in G$ may be written uniquely in the form $x = th$ where $t \in T$ and $h \in H$ and show how to describe the most general projective multiplier for G in terms of those for T and H and a complex valued function from $T \times H$ to K. An important class of examples is that in which G is a semi simple Lie group with finite center, H is a maximal compact subgroup and T is the solvable subgroup occuring in the Iwasawa decomposition.

For other results on the problem of determining $H^2(G,K)$ where G is a separable locally compact group the reader is referred to the survey of Parthasarathy [8] and to papers of Kleppner and C.C. Moore [2], [7].

2. GROUPS WHICH ARE PRODUCTS OF SUBGROUPS

Let G be a group and let T and H be subgroups of G such that every $x \in G$ is uniquely of the form th where $t \in T$, $h \in H$. Then in particular for every $t \in T$ and every $h \in H$ the product ht must have the form $t_1 h_1$ where t_1 and h_1 are members of T and H respectively. Thus we have well defined mappings ϑ and η from $T \times H$ to T and H such that

$$ht = \vartheta(t,h)\,\eta(t,h) \quad \text{for all} \quad h \in H \quad \text{and} \quad t \in T .$$

Clearly we can reconstruct G once we are given T and H as abstract groups and the two functions ϑ and η. Indeed G will be isomorphic to the set of all pairs t,h with $t \in T$, $h \in H$ where multiplication is defined by the rule

(1) $$(t_1,h_1)(t_2,h_2) = t_1\vartheta(t_2,h_1),\eta(t_2,h_1)h_2 .$$

If G is a topological group such that $t,h \to th$ is a homeomorphism of $T\times H$ with G then ϑ and η will be continuous functions and G will be determined as a topological group by giving the quadruple T, H, ϑ, η.

The following question now suggest itself. Given groups T and H and functions ϑ and η from $T\times H$ to T and H respectively under what conditions on ϑ and η will it be true that the set $T\times H$ is a group under the composition law defined by (1). The answer which can be given to this question takes an especially transparent form if we change our notation and point of view as follows. For each fixed $h\in H$, $t\to\vartheta(t,h)$ is a mapping of T into T. Let us denote the map of t by h by $h[t]$. So that $\vartheta(t,h)\equiv h[t]$. In other words let us regard ϑ as a map from H to maps of T into T instead of a map from $T\times H$ to T. Similarly let us regard η as a map from T to maps of H into H and write $\eta(t,h)$ as $[h]t$. The reason for writing the argument on the left in this case will be apparent in what follows.

Theorem 1.* *Let* T *and* H *be groups whose identity elements we denote by* e *and let* $h[t]$ *and* $[h]t$ *be defined for all* $h\in H$, $t\in T$ *so that* $h[t]\in T$ $[h]t\in H$. *Define a multiplication in the set* $T\times H$ *by setting* $(t_1,h_1)\circ(t_2,h_2)= = t_1\circ h_1[t_2],\ [h_1]t_2\circ h_2$. *Then* $T\times H$ *is a group with respect to this multiplication having* e, e *as identity element if and only if the following conditions are satisfied for all* $t_1, t_2, t\in T$ *and all* $h_1, h_2, h\in H$

(1) $$h[e] = e,\ [e]t = e$$

(2) $$e[t] = t \quad \text{and} \quad h_1\circ h_2[t] = h_1[h_2[t]]$$

(3) $$[h]e = h \quad \text{and} \quad [h]t_1\circ t_2 = [[h]t_1]t_2$$

(4) $$h[t_1\circ t_2] = h[t_1]\circ[h]t_1[t_2]$$

(5) $$[h_1\circ h_2]t = [h_1]h_2[t]\circ[h_2]t .$$

*The idea that such a theorem might exist was suggested to the author by Greenberg's paper [1]. Greenberg is interested in showing that a product group $G = TH$ must be a semi direct product when T is the Poincaré group and certain other special conditions are satisfied. In his proof he makes it clear that associativity leads to manageable identities. However, he does not formulate a general theorem or organize and interpret the identities as we have done. Greenberg's theorem is an easy corollary of our theorem 1 but as Michel has shown in [6] it can be proved directly by a short argument.

Proof. By definition $(e,e)\circ(t,h) = e\circ e[t],[e]t\circ h$ and

$$(t,h)\circ(e,e) = t\circ h[e],\ [h]e\circ e\,.$$

Thus e,e acts as a left and right dentity if and only if for all $t\in T,\ h\in H$ we have

$$e[t] = t = t\circ h[e] \quad\text{and}\quad [e]t\circ h = h = [h]e\,.$$

That is if and only if for all $t\in T,\ h\in H$ we have

(A) $$h[e] = e,\ [e]t = e$$

(B) $$e[t] = t,\ [h]e = h\,.$$

Again by definition

$$((t_1,h_1)\circ(t_2,h_2)\circ(t_3,h_3)) = (t_1\circ h_1[h_2],[h_1]t_2\circ h_2)\circ(t_3,h_3) =$$
$$= t_1\circ h_1[t_2]\circ([h_1]t_2\circ h_2)[t_3],([[h_1]t_2\circ h_2]t_3)\circ h_3$$

and

$$(t_1,h_1)\circ((t_2,h_2)\circ(t_3,h_3)) = (t_1,h_1)\circ(t_2\circ h_2[t_3],[h_2]t_3\circ h_3)=$$
$$= t_1\circ h_1[t_2\circ h_2[t_3]],([h_1](t_2\circ h_2[t_3]))\circ([h_2]t_3\circ h_3)\,.$$

Equating corresponding components and cancelling t_1 from one equation and h_3 from the other we see that our composition law is associative if and only if the two following identities hold for all $t_2,t_3\in T,\ h_1,h_2\in H$

(C) $$h_1[t_2]\circ([h_1]t_2\circ h_2)[t_3] = h_1[t_2\circ h_2[t_3]]$$

(D) $$([[h_1]t_2\circ h_2]t_3 = [h_1](t_2\circ h_2[t_3]))\circ[h_2]t_3$$

Setting $h_2 = e$ and using (A), (C) becomes

(C') $$h_1[t_2]\circ([h_1]t_2)[t_3] = h_1[t_2\circ t_3]$$

and using (C'), (C) may be written in the form

$$h_1[t_2]\circ([h_1]t_2\circ h_2)[t_3] = h_1[t_2]\circ([h_1]t_2)[h_2[t_3]]$$

which reduces to

(C'') $$(h_1\circ h_2)[t_3] = h_1[h_2[t_3]]$$

Thus in the presence of (A), (C) holds if and only if (C') and (C") both hold.

Similarly setting $t_2 = e$ and using (B), (D) becomes

(D') $$[h_1 \circ h_2]t_3 = [h_1](h_2[t_3])) \circ [h_2]t_3$$

and using (D'), (D) may be written in the form

$$[[h_1]t_2](h_2[t_3]] \circ [h_2]t_3) = [h_1](t_2 \circ h_2[t_3]) \circ [h_2]t_3$$

which reduces to

(D") $$[[h_1]t_2]t_3 = [h_1]t_2t_3.$$

Thus in the presence of (B), (D) holds if and only if (D') and (D") hold.

We have now shown that our operation is associative and has e,e as a left and right unit if and only if (A), (B), (C') (C") (D') and (D") all hold. But these six conditions are precisely (1), (2), (3), (4), (5) rearranged and with a slightly changed notation. To complete the proof of the theorem we observe that $t_1 \circ h_1[t_2] = e$ and $[h_1]t_2 \circ h_2 = e$ can always be solved for h_1 and t_1 given h_1 and t_2 and vice versa, so that

Remark (1). Conditions (2) and (3) say that our binary operation make T into a left H spaces and H into a right T space. Conditions (4) and (5) state that the operations on T and H defined by the elements of H and T are "almost" multiplication preserving and fail to be so only to the extent that the second group also acts on the first. In particular if h is fixed under the T action then $t \to h(t)$ is an automorphism and similarly when the roles of T and H are interchanged. Condition (1) is actually a consequence of condition (4) and (5) and is listed separately for convenience and because its truth is worth soluting out. Thus theorem 1 may be summerized as follows. To find the most general group G which is a product of subgroups T and H as indicated above make T into a left H space and H into a right T space so that the operators of the actions are almost automorphisms in the sense described by (4) and (5). Then set $G = T \times H$ and define

$$(t_1, h_1) \circ (t_2, h_2) = t_1 \circ h_1[t_2], [h_1]t_2 \circ h_2.$$

Remark (2). The semi-direct product case is that in which either the H space or the T space is trivial. Then one of (4) and (5) becomes vacuous and the other asserts that the operators are automorphisms.

3. MULTIPLIERS FOR GROUPS WHICH ARE PRODUCTS OF SUBGROUPS

Theorem 2. *Let* G *be a separable locally compact group and let* T *and* H *be closed subgroups such that* $t,h \to th$ *is a homeomorphism of* $T \times H$ *on* G. *Let* $ht = h[t][h]t$ *where* $h[t] \in T$, $[h]t \in H$ *so that the functions* $h,t \to h[t]$ *and* $h,t \to [h]t$ *satisfy the conditions listed in theorem 1. Then every normalized projective multiplier* v *for* G *is cohomologous to another such* v' *which may be uniquely represented in the following form*

$$(*) \qquad v'(t_1h_1, t_2h_2) = \sigma(t_1, h_1[t_2])\,\omega([h_1]t_2, h_2)\,g(t_2, h_1)$$

where σ *and* ω *are normalized projective multipliers for* T *and* H, *respectively,* g *is a Borel function from* $T \times H$ *to the complex numbers of modulus one which is* 1 *on* $T \times e$ *and* $e \times H$ *and* σ, ω *and* g *satisfy the following identities* $(h, h_1, h_2 \in H,$ $t, t_1, t_2 \in T)$:

$$\text{(A)} \qquad \frac{\sigma(h[t_1], ([h]t_1)[t_2])}{\sigma(t_1, t_2)} \equiv \frac{g(t_1 \circ t_2, h)}{g(t_1, h)\,g(t_2, [h]t_1)}$$

$$\text{(B)} \qquad \frac{\omega([h_1](h_2[t]), [h_2]t)}{\omega(h_1, h_2)} \equiv \frac{g(t, h_1 \circ h_2)}{g(t, h_2)\,g(h_2[t], h_1)} .$$

Conversely if σ *and* ω *are normalized projective multipliers for* T *and* H *respectively and* g *is a Borel function from* $T \times H$ *to the complex numbers of moduls one which are in* $T \times e$ *and* $e \times H$ *and* σ, ω *and* g *satisfy* (A) *and* (B) *then the function* v' *defined by* (*) *is a normalized projective multiplier for* G.

Proof. Let V be any v representation of G. Then for all $t \in T$, $h \in H$ we have $V_{th} = v(t,h)V_tV_h$. Let $V'_{th} = \dfrac{V_h}{v(t,h)}$. Then V' is a v' representation of G where v' is cohomologous to v. Note that $V'_t = \dfrac{t}{v(t,e)} = V_t$ and $V'_h = \dfrac{V_h}{v(e,h)} = V_h$. Thus $V'_{th} = \dfrac{V_{th}}{v(t,h)} = V_tV_h = V'_tV'_h = A_tB_h$ where A and B denote the restrictions of V' to T and H respectively. Let σ and ω denote the restrictions of v' to T and H. Then $V'_{(t_1h_1)(t_2h_2)} =$

$$= v'(t_1h_1,t_2h_2)A_{t_1}B_{h_1}A_{t_2}B_{h_2} = V'_{t_1h_1[t_2][h_1]t_2h_2} = A_{t_1h_1[t_2]}B_{[h_1]t_2h_2} =$$

$$= \sigma(t_1,h_1[t_2])\,\omega([h_1]t_2,h_2)A_{t_1}A_{h_1[t_2]}B_{[h_1]t_2}B_{h_2}$$

for all $t_1,t_2 \in H$, $h_1,h_2 \in H$. Hence

$$v'(t_1h_1,t_2h_2)B_{h_1}A_{t_2} = \sigma(t_1,h_1[t_1])\,\omega([h_1]t_2,h_2)A_{h_1[t_2]}B_{[h_1]t_2}.$$

Thus

$$A_{h_1[t_2]}B_{[h_1]t_2}A_{t_2}^{-1}B_{h_1}^{-1} = \frac{v'(t_1h_1,t_2h_2)}{\sigma(t_1,h_1[t_1])\,\omega([h_1]t_2,h_2)}\,I,$$

where I is the identity. Hence the function on the right depends only on h_1 and t_2 In other words there exists a unique Borel function g from $T\times H$ to the complex numbers of modulus one such that

$$v'(t_1h_1,t_2h_2) = \sigma(t_1,h_1[t_2])\,\omega([h_1]t_2,h_2)\,g(t_2,h_1).$$

If we let $t_1 = h_1 = e$ we get $v'(h_1,t_2) \equiv g(t_2,h_1)$ from which it follows that $g(e,h) \equiv g(t,e) \equiv 1$.

Now let σ and ω be arbitrary multipliers for T and H respectively and let g be an arbitrary Borel function from $T\times H$ to the complex numbers of modulus one such that $g(t,e) = g(e,h) = 1$ for all $t\in T$, $h\in H$. Let us define v on $G\times G$ by the equation

$$v'(t_1h_1,t_2h_2) = \sigma(t_1,h_1[t_2])\,\omega([h_1]t_2,h_2)\,g(t_2,h_1).$$

Then applying the definition we see that v' is a normalized projective multiplier for G if and only if the following identity holds for all $t_1,t_2,t_3 \in T$ and all $h_1,h_2,h_3 \in H$

$$\sigma(t_1,h_1[t_2])\,\omega([h_1]t_2,h_2)\,g(t_2h_1)\,\sigma(t_1\circ h_1[t_2],([h_1]t_2\circ h_2)[t_3])\cdot$$

$$\cdot\,\omega([[h_1]t_2\circ h_2]t_3,h_3)\,g(t_3,[h_1]t_2\circ h_2) \equiv$$

$$\equiv \sigma(t_1,h_1[t_2\circ h_2[t_3]])\,\omega([h_1](t_2\circ h_2[t_3],[h_2]t_3\circ h_3)\cdot$$

$$\cdot\,g(t_2\circ h_2[t_3],h_2)\,\sigma(t_2,h_2[t_3])\,\omega([h_2]t_3,h_3)\,g(t_3,h_2).$$

If we set $h_2 = h_3 = t_1 = e$ this identity reduces to

$$g(t_2,h_1)\sigma(h_1[t_2],[h_1]t_2[h_3])g(t_3,[h_1]t_2) \equiv \sigma(t_2,t_3)g(t_2 \circ t_3,h_1),$$

which is clearly equivalent to (A) in the statement of our theorem.

Similarly if we set $t_1 = t_2 = h_3 = e$ it reduces to

$$\omega(h_1,h_2)g(t_3,h_1 \circ h_2) \equiv \omega([h_1](h_2[t_3],[h_2]t_3))g(h_2[t_3],h_1)g(t_3,h_2)$$

which is clearly equivalent to (B). Thus (A) and (B) are certainly necessary conditions.

Conversely assuming (A) and (B) to hold the left and right sides of our identity become respectively

$$\sigma(t_1,h_1[t_2])\sigma(t_1 \circ h_1[t_2],([h_1]t_2 \circ h_2)[t_3])$$

$$\frac{\omega([h_1]t_2,h_2)}{\omega([h_1]t_2,h_2)}\ \omega([[h_1]t_2 \circ h_2]t_3,h_3)\omega([h_1](t_2 \circ h_2[t_3]),[h_2]t_3)$$

$$g(t_3,h_2)g(h_2[t_3],[h_1]t_2)g(t_2,h_1)$$

and

$$\sigma(t_1,h_1[t_2 \circ h_2[t_3]])\ \frac{\sigma(t_2,h_2[t_3])}{\sigma(t_2,h_2[t_3])}\ \sigma(h_1[t_2],([h_1]t_2)[h_2[t_3]])$$

$$\omega([h_1](t_2 \circ h_2[t_3],[h_2]t_3 \circ h_3)\,\omega([h_2]t_3,h_3)$$

$$g(t_2,h_1)g(h_2[t_3],[h_1]t_2)g(t_3,h_2).$$

Now the g factors in both extensions are identical. Moreover the product of the two σ factors on the left is identical, with the product of the two non cancelling σ factors on the right by the identity defining a projective multiplier. Since a corresponding statement is true for the ω factors we see that conditions A and B are sufficient as well as necessary and the proof of the theorem is complete.

Theorem 4. *The projective multiplier defined by* σ, ω *and* g *as described in the statement of theorem 3 is trivial if and only if there exist Borel functions* a *and* b *from* A *and* B *to the complex numbers of modulus one such that*

$$\sigma(t_1,t_2) \equiv \frac{a(t_1t_2)}{a(t_1)a(t_2)} \qquad \omega(t_1,t_2) = \frac{b(t_1t_2)}{b(t_1)b(t_2)}$$

and

$$g(t,h) \equiv \frac{a(h[t])}{a(t)} \frac{b(h[t])}{b(h)} .$$

In particular if σ *and* ω *are both identically one then the projective multiplier defined by* g *is trivial if and only if there exist one dimensional characters* a *and* b *of* T *and* H *respectively such that*

$$g(t,h) \equiv \frac{a(h[t])}{a(t)} \frac{b(h[t])}{b(h)} .$$

Proof. By definition the multiplier in question is trivial if and only if there exists a Borel function ρ from $T \times H$ to the complex numbers of modulus one such that

$$(*) \quad \sigma(t_1,h_1[t_2])\omega([h_1]t_2,h_2)g(t_2,h_1) \equiv \frac{\rho(t_1 \circ h_1[t_2],[h_1]t_2 \circ h_2)}{\rho(t_1,h_1)\rho(t_2,h_2)}$$

In the special case in which $t_2 = h_1 = e$ this reduces to

$$1 \equiv \frac{\rho(t_2,h_2)}{\rho(t_1,e)\rho(e,h_2)} .$$

Thus $\rho(t,h) \equiv a(t)b(h)$ where $a(t) \equiv \rho(t,e)$ and $b(h) = \rho(e,h)$. On the other hand when $h_1 = h_2 = e$ (*) reduces to

$$\sigma(t_1,t_2) \equiv \frac{\rho(t_1 \circ t_2,e)}{\rho(t_1,e)\rho(t_2,e)} = \frac{a(t_1 \circ t_2)}{a(t_1)a(t_2)}$$

and when $t_1 = t_2 = e$ it reduces to

$$\omega(h_1,h_2) \equiv \frac{b(t_1 \circ t_2)}{b(t_1)b(t_2)} .$$

Substituting there three necessary conditions in (*) it becomes

$$\frac{a(t_1 \circ h_1[t_2])}{a(t_1)a(h_1[t_2])} \frac{b([h_1]t_2 \circ h_2)}{b([h_1]t_2)b(h_2)} g(t_2,h_1) \equiv \frac{a(t_1 \circ h_1[t_2])b([h_1]t_2 \circ h_2)}{a(t_1)b(h_1)a(t_2)b(h_2)}$$

which simplifies to

$$g(t_2,h_1) = \frac{a(h_1[t_2])}{a(t_2)} \frac{b([h_1]t_2)}{b(h_1)}$$

The truth of the theorem is now obvious.

BIBLIOGRAPHY

[1] O.W. Greenberg, Coupling of internal and space time symmetries, *Phys. Rev.*, 135 (1964), B 1447-B 1450.

[2] A. Kleppner, Multipliers on abelian groups, *Math. Ann.*, 158 (1958), 11-34.

[3] G. W. Mackey, On a theorem of Stone and von Neumann, *Duke Math. J.*, (1949)

[4] ——————— Unitary representations of group extensions. I, *Acta. Math.*, 99 (1958), 265-311.

[5] ——————— Induced representatives of locally compact groups and applications. Pages 132 to 166 of Functional analysis and related fields. *Proceedings of the Conference in honor of M.H. Stone*, University of Chicago 1968.

[6] L. Michel, Relations between internal symmetry and relativistic invariance, *Phys. Rev.*, 137 (1965), B 405- B 408.

[7] C.C. Moore, Extensions and low dimensional cohomology theory of locally compact groups. I, *Trans. Amer. Math. Soc.*, 113 (1964), 40-63.

[8] K.R. Parthasararthy, *Multipliers on locally compact groups* (Lecture Notes in Mathematics, vol. 93 (Springer Verlag, 1969).

[9] A. Weil, Sur certains groupes d'opérateurs unitaires, *Acta Math.*, 111 (1964), 143-211.

COLLOQUIA MATHEMATICA SOCIETATIS JÁNOS BOLYAI
5. HILBERT SPACE OPERATORS. TIHANY (HUNGARY), 1970

Remarks on eigenpackets of self-adjoint operators

P. MASANI

1. PURPOSE OF THIS PAPER

In his book on differential operators [8, §10.4] G. Hellwig has discussed the idea of an eigenpacket of a self-adjoint operator, the utility of which has been emphasized by F. Rellich in his lectures [19', §15] and the origins of which he traces back to Hellinger [7, §5]. The purpose of this paper is to recast this idea in terms of vector- and operator-valued measures, and to suggest that the resulting widened concept is adequate for the rigorization of the theory of pseudoeigenfunction expansions of self-adjoint operators over L_2 spaces. In this approach one does not leave the given L_2 space in favor of a larger space by rigging or other devices, and this is in many ways an advantage. We shall indicate that every self-adjoint operator H from a Hilbert space $\mathcal{H}$ to $\mathcal{H}$ possesses a full system of eigenpackets, i.e. $\mathcal{H}$-valued, orthogonally scattered measures, or equivalently a single quasi-isometric measure, in terms of which any vector in $\mathcal{H}$ can be "expanded" as an infinite integral (§5). When $\mathcal{H}$ is an L_2 space and iH a differential operator D thereon, our integral expansion can be con-

#This work was supported by the National Science Foundation, U.S.A.

strued as a rigorous expression of the expansion in terms of the pseudo-eigenfunctions of D. We shall also suggest a way to render precise the notion of "pseudo-eigenfunction" (§6).

The vector- and operator-measure theoretic foundation on which our approach is based is not as yet generally familiar. We shall therefore devote §2 to a brief description of this foundation.# In §§3-5 we shall reformulate the Hellwig idea in this measure-theoretic framework. We shall state theorems, but defer their proofs to a later and longer paper. The more tentative §6 is devoted to the explication of the notion of "pseudo-eigenfunction" of a s.a. operator on an L_2 space.

It is necessary to show first that our framework is relevant to the issue at hand. This initial analysis involves certain vector-measures and integrals, arising from recent work on stochastic processes, the theory of which is still in embryonic form. As we are not concerned with such measures and integration except in a transitory way, we have placed their discussion in an Appendix.

2. QUASI-ISOMETRIC AND ORTHOGONALLY SCATTERED MEASURES

It is best to depart from the actual development and broach the subject by means of a concept introduced by Halmos in the early sixties:

2.1 Def. (Halmos [6, p.102]) A subspace W of a Hilbert space $\mathcal{H}$ is said to *wander* with respect to a continuous linear operator T on $\mathcal{H}$ to $\mathcal{H}$, iff. $T^m(W) \perp T^n(W)$ wherever m, n are distinct positive integers.

The concept defined concerns the relation between W and the operator T, but we could say equally well that it concerns the relation between W and the sequence $(T^n)_1^\infty$. Now the nth term of this sequence could well have been different from the power T^n, and this suggests replacing T^n by an arbitrary continuous, linear operator T^n in Def. 2.1. Next, a measure theorist would observe that the operator-valued sequence $(T^n)_1^\infty$ prescribes an atomic operator-valued measure $T(\cdot)$ defined on the δ-ring## $\mathcal{R}$ of finite subsets $A, B, \dots$ of the set $\mathbf{N}_+$ of positive integers, such that $T(A)(W) \perp T(B)(W)$ when $A \parallel B$###. He

At the Conference several participants expressed an interest in seeing such a brief description of the relevant vector- and operator-measure theory in print.

A δ-ring is by definition a ring closed under countable intersections.

We write $A \parallel B$ to mean "A & B are disjoint."

would also notice that both the atomicity of $T(\cdot)$ and the finiteness of sets in $\mathcal{R}$ are fortuitous as far as the "wandering" idea is concerned. He might thus be led to generalize Def.2.1 as follows:

2.2 Def. Let $\mathfrak{X}$ and $\mathcal{H}$ be complex Hilbert spaces, and let $T(\cdot)$ be a $\mathfrak{X}$-to-$\mathcal{H}$, continuous, linear operator-valued measure on a δ-ring $\mathcal{R}$ over an arbitrary space Λ, which is countably additive (c.a.) in the strong operator topology. Then a subspace W of $\mathfrak{X}$ is said to *wander under* $T(\cdot)$, iff.

$$A, B \in \mathcal{R} \ \& \ A \parallel B \implies T(A)(W) \perp T(B)(W) \text{ in } \mathcal{H}.$$

Now let the subspace W of $\mathfrak{X}$ wander under the measure $T(\cdot)$, let# $T_0(\cdot) \underset{d}{=} \mathrm{Rstr}._W T(\cdot)$, let $*$ denote the adjoint of operators from W to $\mathcal{H}$ and write $M(\cdot) \underset{d}{=} T_0(\cdot)^* T_0(\cdot)$. Then it is easy to verify that

$$\text{(2.3)} \qquad \forall A, B \in \mathcal{R}, \qquad T_0(B)^* T_0(A) = M(A \cap B),$$

where $M(A \cap B)$ is a non-negative hermitian operator on W to W, and the measure $M(\cdot)$ is c.a. on $\mathcal{R}$ under the strong operator topology. It is also easy to see that for an $A \in \mathcal{R}$ for which $M(A)$ is invertible, $T(A)\{\sqrt{M(A)}\}^{-1}$ is an isometry on W to $\mathcal{H}$, cf. [17, 8.6]. We therefore call a measure $T_0(\cdot)$ subject to (2.3) *quasi-isometric*. More fully, when (2.3) holds we say that $T_0(\cdot)$ *is a* W*-to-*$\mathcal{H}$*, countably additive, quasi-isometric (c.a.q.i.) measure on the* δ*-ring* $\mathcal{R}$ *having the non-negative hermitian measure* $M(\cdot)$. The entire italicized expression we then abbreviate by writing " $T_0(\cdot)$ *is a* W*-to-*$\mathcal{H}$ *, c.a.q.i. measure over* $(\Lambda, \mathcal{R}, M)$." A good deal can be done with such measures as we shall now indicate.

Let $T(\cdot)$ be a W-to-$\mathcal{H}$, c.a.q.i. measure over $(\Lambda, \mathcal{R}, M)$##. Associated with $T(\cdot)$ in a natural way are a subspace-valued set-function $\mathfrak{M}_T(\cdot)$ and a projection-valued set-function $Q_T(\cdot)$ defined on the affiliate σ-algebra:

$$\text{(2.4)} \qquad \mathcal{R}^{loc} \underset{d}{=} \{E : \ E \subseteq \Lambda \ \& \ \forall A \in \mathcal{R}, \ E \cap A \in \mathcal{R}\}$$

of "locally $\mathcal{R}$-measurable" sets, by the equations

$\mathrm{Rstr}._S F$ means the restriction of the function F to the domain S.
The symbol $\underset{d}{=}$ means "equals by definition".

We now drop the subscript from T_0 since it is no longer relevant to think of our measure as being obtained by restricting to a space W operators defined on a larger space $\mathfrak{X}$.

$$(2.5)^{\#}\quad \begin{cases} \mathfrak{M}_T(E) \underset{d}{=} \mathfrak{S}\{T(A)(W):\ A \in \mathcal{R}\ \&\ A \subseteq E\} \subseteq \mathcal{H} . \\ Q_T(E) \underset{d}{=} \text{ the projection on } \mathfrak{M}_T(\Lambda) \text{ onto } \mathfrak{M}_T(E). \end{cases}$$

It follows easily that $\mathfrak{M}_T(\cdot)$ is a subspace-valued, c.a. orthogonally scattered ## measure on $\mathcal{R}^{loc}$, and $Q_T(\cdot)$ a spectral measure on $\mathcal{R}^{loc}$ for the Hilbert space $\mathfrak{M}_T(\Lambda)$, cf. [17, 10.25]. We call $\mathfrak{M}_T(\cdot)$ the *spatial measure of* $T(\cdot)$, $Q_T(\cdot)$ the *spatial spectral measure of* $T(\cdot)$, and $\mathfrak{M}_T(\Lambda)$ the *subspace of* T. We call the measure $T(\cdot)$ *basic,* in case $\mathfrak{M}_T(\Lambda) = \mathcal{H}$.

Also associated with $T(\cdot)$ in a natural way is the space

$$(2.6)\qquad \mathcal{L}_{2,W} \underset{d}{=} L_2(\Lambda, \mathcal{R}, M;\ W)$$

of function Φ on Λ to W which are "square-integrable" with respect to the W-to-W, non-negative hermitian measure $M(\cdot)$, in the sense that $\int_\Lambda \Phi(\lambda)^* M(d\lambda)\Phi(\lambda)$, which is a non-negative real number, exists. For finite dimensional W such integrals are familiar from multivariate prediction and filtering, and their theory is well understood. It is known for instance that the more general complex-valued integral $\int_\Lambda \Psi(\lambda)^* M(d\lambda)\Phi(\lambda)$ can be defined simply to be the Bochner integral

$$(2.7)\qquad \int_\Lambda \Psi(\lambda)^* \frac{dM}{d\tau}(\lambda)\Phi(\lambda)\tau(d\lambda), \quad \text{where } \tau(\cdot) = \operatorname{trace} M(\cdot) .$$

For finite dimensional W it is also known that $\mathcal{L}_{2,W}$ is a Hilbert space under the inner product

$$(2.8)\qquad ((\Phi, \Psi)) \underset{d}{=} \int_\Lambda \Psi(\lambda)^* M(d\lambda)\Phi(\lambda) \in \mathbf{C} ,$$

and the $\mathcal{R}$-simple functions on Λ to W are everywhere dense in $\mathcal{L}_{2,W}$, cf. e.g. Rosenberg [22]. For infinite dimensional W the last two assertions about $\mathcal{L}_{2,W}$ may fail for certain $M(\cdot)$, cf. e.g. Kuroda [11, p.71]. We call a measure $M(\cdot)$ *adequate* in case they hold, cf. [17, 9.9]. In all applications up to now $M(\cdot)$ has turned out to be adequate . This is so, of course, in the important special case in which $M(\cdot) = \mu(\cdot)I_W$, $\mu(\cdot)$ being a real- or complex-valued, c.a. measure on $\mathcal{R}$.###

\# $\mathfrak{S}(X) \underset{d}{=}$ the closed linear subspace spanned by the set $X \subseteq \mathcal{H}$.

\## so-called, because $\mathfrak{M}_T(A) \perp \mathfrak{M}_T(B)$ when $A \parallel B$.

\### The question of adequacy is studied by Mandrekar and Salehi in [12] .

Now let $T(\cdot)$ be a W-to-W, c.a.q.i. measure over $(\Lambda, \mathcal{R}, M)$, where $M(\cdot)$ is *adequate*. Then we can define $\forall \Phi \in \mathcal{L}_{2,W}$ an integral $\int_\Lambda T(d\lambda)\Phi(\lambda) \in \mathcal{H}$ in a straightforward way by exploiting (2.3), and so bypassing many difficulties besetting the general theory of operator-valued measures and integrals, cf. [17, 10.4, 10.6].# The nice properties of this integral are summed up in the following theorem [17, 10.8, 10.9, 10.25(c)].

2.9 Isomorphism Thm. (c.a.q.i.) *Let* T *be a* W*-to-*W, *c.a.q.i. measure over* $(\Lambda, \mathcal{R}, M)$ *where* M *is adequate. Then the correspondence* $\sum_T : \Phi \rightarrow \int_\Lambda T(d\lambda)\Phi(\lambda)$ *is a unitary operator on the Hilbert space* $\mathcal{L}_{2,W} \underset{d}{=} L_2(\Lambda, \mathcal{R}, M; W)$ *onto the subspace* $\mathfrak{M}_T(\Lambda)$ *of* $\mathcal{H}$. *For all* $B \in \mathcal{R}^{loc}$, *this* $\sum_T$ *transforms the operation* M_{χ_B} *of multiplication by* χ_B *on* $\mathcal{L}_{2,W}$ *into the spatial spectral projection* $Q_T(B)$ *on* $\mathfrak{M}_T(\Lambda)$.##

Now consider our W-to-W, c.a.q.i. measure T over $(\Lambda, \mathcal{R}, M)$ in the special case $W = \mathbf{C}$, the complex number field. Each $M(A)$ is now a non-negative hermitian operator on $\mathbf{C}$ to $\mathbf{C}$, and may therefore be identified with a non-negative real number $\mu(A)$. Each $T(A)$ is a continuous linear operator on $\mathbf{C}$ to $\mathcal{H}$, and so may be identified with a vector $\xi(A)$ in $\mathcal{H}$. Our equation (2.3) relating T and M reduces to one relating ξ and μ, viz.

(2.10) $$\forall A, B \in \mathcal{R}, \qquad (\xi(A), \xi(B)) = \mu(A \cap B).$$

Our structure thus reduces to a non-negative, c.a. measure $\mu(\cdot)$ on $\mathcal{R}$, along with an $\mathcal{H}$-valued, countable additive, orthogonally scattered ### (c.a.o.s.) measure $\xi(\cdot)$ on $\mathcal{R}$ satisfying (2.10), or as we shall briefly say to an $\mathcal{H}$*-valued, c.a.o.s. measure* ξ *over* $(\Lambda, \mathcal{R}, \mu)$. Now the theory of such vector measures is more widely known, cf. [14; 17, §2], and under a stochastic interpretation its origins can be traced back to Wiener's fundamental work [23] of 1923 on the Brownian motion ####. Bounded

Think of this integral as follows. When the "infinitesmal" W-to-$\mathcal{H}$ operator $T(d\lambda)$ acts on the vector $\Phi(\lambda)$ in W, we get a tiny vector $T(d\lambda)\{\Phi(\lambda)\}$ in $\mathcal{H}$. Our integral is the vector-sum of a very large number of these tiny vectors.

χ_B denotes the indicator function of the set B.

so-called, because $\xi(A) \perp \xi(B)$ when $A \| B$.

In the stochastic interpretation $\mathcal{H} = L_2(\Omega, \mathcal{F}, P)$, where $(\Omega, \mathcal{F}, P)$ is a probability space. The $\mathcal{H}$-valued $\xi(\cdot)$ is then called a random measure and the corresponding integration is called stochastic integration, cf. [4, p.426]. Wiener [23] was the first to define and use such measures and integration. For him $\mathcal{R}$ was the δ-ring of Borel subsets of $\mathbf{R}$ of finite Lebesgue measure, and ξ was defined by the equation

$$\xi(a, b] = x_b - x_a,$$

where $(x_t(\cdot), \; t \in \mathbf{R})$ is the Brownian motion SP. ($\mathbf{R}$ denotes the real number field.)

measures of this type are also familiar in operator theory in the guise $E(\cdot)x$, where $x \in \mathcal{H}$ and $E(\cdot)$ is a spectral measure for $\mathcal{H}$.

The important class of $\mathcal{H}$-valued, c.a.o.s. measures is thus buried in the class of W-to-W, c.a.q.i. operator-valued measures. Concepts and results of c.a.o.s. measure theory can all be derived from corresponding concepts and results of c.a.q.i. measure theory by considering the special case $W = \mathbf{C}$. Thus when $W = \mathbf{C}$, the spatial and spatial spectral measures of T, cf. (2.5), become spatial and spatial spectral measures of ξ, defined by

$$(2.11)\quad \begin{cases} \forall E \in \mathcal{R}^{\text{loc}}, \ \mathfrak{M}_\xi(E) \underset{d}{=} \mathfrak{S}\{\xi(A) : A \in \mathcal{R} \ \& \ A \subseteq E\} \subseteq \mathcal{H}, \\ Q_\xi(E) \underset{d}{=} \text{the projection on } \mathfrak{M}_\xi(\Lambda) \text{ onto } \mathfrak{M}_\xi(E). \end{cases}$$

Also the integral $\int_\Lambda T(d\lambda)\Phi(\lambda)$ gives way to an $\mathcal{H}$-valued integral $\int_\Lambda \phi(\lambda)\,\xi(d\lambda)$, where $\phi \in L_2(\Lambda, \mathcal{R}, \mu; \mathbf{C})$. Our Isomorphism Thm. 2.9 immediately yields the corresponding theorem for the latter integration, cf. [17, 2.3], viz.

2.12 Isomorphism Thm. (c.a.o.s.) *Let* ξ *be a* $\mathcal{H}$*-valued, c.a.o.s. measure over* $(\Lambda, \mathcal{R}, \mu)$. *Then the correspondence* $\sum_\xi : \phi(\cdot) \longrightarrow \int_\Lambda \phi(\lambda)\,\xi(d\lambda)$ *is a unitary operator on* $\mathcal{L}_2 \underset{d}{=} L_2(\Lambda, \mathcal{R}, \mu; \mathbf{C})$ *onto the subspace* $\mathfrak{M}_\xi(\Lambda)$ *of* $\mathcal{H}$. *For all* $B \in \mathcal{R}^{\text{loc}}$, *this* $\sum_\xi$ *transforms the operation* M_{χ_B} *of multiplication by* χ_B *on* $\mathcal{L}_2$ *into the spatial projection* $Q_\xi(B)$ *on* $\mathfrak{M}_\xi(\Lambda)$.

Even when $W \neq \mathbf{C}$ a close relationship persists between our W-to-W, c.a.q.i. measure $T(\cdot)$ over $(\Lambda, \mathcal{R}, M)$ and $\mathcal{H}$-valued, c.a.o.s. measures. For let $\forall w, w' \in W$, $\xi_w(\cdot) \underset{d}{=} T(\cdot)w$ and $\mu_{ww'}(\cdot) \underset{d}{=} (M(\cdot)w, w')$. Then the set-functions $\xi_w(\cdot)$, $\mu_{ww'}(\cdot)$, $\mu_{ww}(\cdot)$ are easily seen to be $\mathcal{H}$-valued, complex-valued, and non-negative real-valued, c.a. measures on $\mathcal{R}$, respectively. Moreover, by (2.3),

$$(2.13)\quad \forall A, B \in \mathcal{R}, \quad (\xi_w(A), \xi_{w'}(B)) = \mu_{ww'}(A \cap B).$$

In other words, $\{\xi_w(\cdot),\ w \in W\}$ is a *biorthogonal family*# of $\mathcal{H}$-valued, c.a.o.s. measures on $\mathcal{R}$.

In [17, §§3 & 10B] we have shown that every isometry on an L_2 space, whether of scalar or vector valued functions, to an arbitrary Hilbert space $\mathcal{H}$ falls

cf. 3.5 and 3.6 below.

within the ambit of our Thms. 2.12 or 2.9, i.e. can be recovered explicity by an integration with respect to c.a.o.s. or c.a.q.i. measures. We have also shown the effectiveness of this explicit approach by exhibiting the space W and the measures μ, M, ξ and $T(\cdot)$ for several important representation and transform theorems [17, §4-6, §13-15; 18; 13; 15; 16]. Moreover the writer feels that failure to date in other instances stems from his own mathematical limitations. It is therefore worth seeing if this approach can cope with the difficulties besetting the theory of pseudo-eigenfunction expansions for s.a. operators.

3. EIGENPACKETS OF SYMMETRIC LINEAR OPERATORS

In this section

(3.1)
- (i) $\mathfrak{X}$ is a (complex) Banach space
- (ii) $\mathcal{H}$ is a (complex) Hilbert space
- (iii) $\mathcal{B}$ is the σ-algebra of Borel subsets of $\mathbf{R}$
- (iv) $\mathcal{B}_0$ is the δ-ring of all bounded Borel subsets of $\mathbf{R}$
- (v) $\mathcal{R}$ is a δ-ring over $\mathbf{R} \ni \mathcal{B}_0 \subseteq \mathcal{R} \subseteq \mathcal{B}$.

The Def. 3.2 we are about to give involves an $\mathfrak{X}$-valued, c.a. measure ξ on a δ-ring $\mathcal{R}$ satisfying (3.1) (v). Such measures ξ are not of the type mentioned in §2. They are discussed briefly in the Appendix, since they play only a transitory role in this paper. To understand Def. 3.2 it suffices to know only a few properties of ξ. Firstly, the total variation measure $|\xi|(\cdot)$ of $\xi(\cdot)$ will in general be infinite even of sets A in $\mathcal{B}_0$ but the semi-variation $s_\xi(\cdot)$ of $\xi(\cdot)$ is finite-valued on $\mathcal{R}$ though possibly unbounded on $\mathcal{R}$, cf. A.2. Consequently it is possible to define the integral of a complex-valued, $\mathcal{R}$ measurable function ϕ on $\mathbf{R}$ with respect to ξ in a standard way, and to show that it exists when, for instance, ϕ is bounded, cf. A.6.

Secondly, it follows from (3.1) (iv) (v) that every bounded interval $(a,b] \in \in \mathcal{R}$. We can therefore associate with an $\mathfrak{X}$-valued measure ξ defined on $\mathcal{R}$ a two-parameter family of curves $x(\cdot)$ in $\mathfrak{X}$:

$$x(\lambda) \underset{d}{=} \begin{cases} x_0 + \xi(\lambda_0, \lambda], & \lambda \geq \lambda_0 \\ x_0 - \xi(\lambda, \lambda_0], & \lambda < \lambda_0 \end{cases}$$

where $\lambda_0 \in \mathbf{R}$ and $x_0 \in \mathcal{H}$ are arbitrary. It is easy to convert the integral of a continuous function ϕ on $\mathbf{R}$ to $\mathbf{C}$ with respect to ξ into a Riemann–Stieltjes type integral of ϕ with respect to $x(\cdot)$, cf. A.8.

We shall now state our definition, and then give its motivation.

3.2. Def. Let H be a linear operator (not necessarily closed or continuous from a Banach space $\mathcal{X}$ to itself. We say that $\xi(\cdot)$ is an *eigenpacket of* H, iff. $\xi(\cdot)$ is an $\mathcal{X}$-valued, c.a. measure on a δ-ring $\mathcal{R}$ satisfying (3.1) (v) and such that

$$\forall A \in \mathcal{R}, \quad \xi(A) \in \mathfrak{S}_H \ \& \ H\{\xi(A)\} = \int_A \lambda \xi(d\lambda).$$

One is led to this definition by the following heuristic considerations. Let $\mathcal{X}$ be a function-space such as $L_p(\mathbf{R})$, and H an operator such as differentiation, which makes sense even outside $\mathcal{X}$. Suppose that for all λ in some subinterval Λ of $\mathbf{R}$, we can find an x_λ in some space Y satisfying

(1) $$H(x_\lambda) = \lambda x_\lambda .^{\#}$$

We may then call x_λ a *pseudo-eigenvector* of H and call the integral

(2) $$\xi(A) \underset{d}{=} \int_A x_\lambda d\lambda, \quad A \subseteq \Lambda$$

a *pseudo-eigenpacket* of H an A. If it so happens that $\forall A \subseteq \Lambda, \xi(A) \in \mathcal{X}$, then it would be reasonable to omit the second prefix "pseudo". Now by (1) and (2)

(3) $$H\{\xi(A)\} = \int_A H(x_\lambda) d\lambda = \int_A \lambda x_\lambda d\lambda .$$

If on the basis of (2) we replace the "eigendifferential" $x_\lambda d\lambda$ by $\xi(d\lambda)$, then (3) becomes

(4) $$H\{\xi(A)\} = \int_A \lambda \xi(d\lambda) .$$

Although the assumptions underlying (1) and the inference of (3) and (4) from (1) are questionable, the equation (4) by itself makes sense and may be utilized to define an eigenpacket as in 3.2.

For instance, for $\mathcal{X} = L_2(\mathbf{R})$ and $H = -iD$, we find on letting $x_\lambda(t) \underset{d}{=} e^{it\lambda}$, that $H(x_\lambda) = \lambda x_\lambda$. Here the $x_\lambda \notin L_2(\mathbf{R}) \underset{d}{=} \mathcal{X}$, but $x_\lambda \in L_\infty(\mathbf{R}) \underset{d}{=} Y$.

It is largely a matter of convenience whether we apply the term "eigen-packet" to the set-function ξ as in 3.2, or to the curves $\Phi_{\lambda_0}(\cdot) \underset{d}{=} \xi(\lambda_0, \cdot]$ in $\mathfrak{X}$ obtainable from ξ by the specification of a point of origin λ_0 as Hellwig does [8, p.162].# Hellwig requires, however, that the $\Phi_{\lambda_0}(\cdot)$ be continuous on $\mathbf{R}$, i.e. that ξ be purely non-atomic on $\mathcal{R}$, cf. A.4, and only to this extent our concepts of eigenpacket do not really tally.

From the equality of integrals with respect to ξ and RS integrals with respect to the associated curves $x(\cdot)$, the formula for integration by parts and the decomposability of ξ into atomic and non-atomic parts, cf. A.8–A.10 & A.5, we get the following trivialities:

3.3 Triv. *Let* (i) ξ *be an eigenpacket of* H *with domain* $\mathcal{R}$,

$$\text{(ii)} \quad x(\lambda) \underset{d}{=} \begin{cases} x_0 + \xi(0,\lambda], & \lambda \geq 0, \\ x_0 - \xi(\lambda, 0], & \lambda < 0, \end{cases} \qquad x_0 \in \mathcal{H}.$$

Then $\forall a, b \in \mathbf{R} \ni a \leq b$, *we have* $\xi(a,b] \in \mathfrak{S}_H$ *and*

$$H\{\xi(a,b]\} = \int_a^b \lambda \, dx(\lambda) = bx(b) - ax(a) - \int_a^b x(\lambda)\, d\lambda.$$

3.4 Triv. *Let* (i) ξ *be an eigenpacket of* H *with domain* $\mathcal{R}$, (ii) ξ_c, ξ_d *be the purely non-atomic and atomic parts of* ξ. *Then*

(a) ξ_c, ξ_d *are eigenpackets of* H *with the same domain* $\mathcal{R}$;

(b) $\forall A \in \mathcal{R}, \quad H\{\xi_d(A)\} = \sum_{\lambda \in A \cap \mathfrak{S}_\xi} \lambda \xi\{\lambda\},$

where $\mathfrak{S}_\xi \underset{d}{=} \{\lambda : \lambda \in \mathbf{R} \ \& \ \xi\{\lambda\} \neq 0\}$; ##

(c) $\mathfrak{S}_\xi \subseteq \sigma_d(H) \underset{d}{=}$ *the discrete spectrum of* H; *more fully, every* $\lambda \in \mathfrak{S}_\xi$ *is an eigenvalue of* H, *a corresponding eigenvector being* $\xi\{\lambda\}$.

We cannot say much more about the eigenpacket of linear operators from an arbitrary Banach space $\mathfrak{X}$ to $\mathfrak{X}$. But this situation improves when $\mathfrak{X}$ is a *Hilbert space* $\mathcal{H}$ and H is a *symmetric operator* from $\mathcal{H}$ to $\mathcal{H}$. To see this we

We have consistently felt that formulations involving set-functions are clearer and more readily generalizable than those resulting from point-functions.

Since, cf. A.4, $\mathfrak{S}_\xi$ is countable, the sum on the RHS of the first equality in (b) makes sense.

must first refer to one or two notions from the theory of $\mathcal{H}$ -valued measures:

3.5 Def.[#] A set $\mathcal{F}$ of $\mathcal{H}$ -valued, c.a. measures on an arbitrary δ-ring $\mathcal{R}$ is called a *biorthogonal family,* iff.

$$A, B \in \mathcal{R} \;\&\; A \| B \Rightarrow \forall \xi, \eta \in \mathcal{F}, \quad \xi(A) \perp \eta(B).$$

3.6 Lma. *Let* $\mathcal{F}$ *be a biorthogonal family of* $\mathcal{H}$ *-valued, c.a. measures on an arbitrary* δ *-ring* $\mathcal{R}$ *, Then*

(a) $\forall \xi, \eta \in \mathcal{F}, \exists$ *a unique complex-valued, c.a. measure* $\mu_{\xi\eta}$ *on* $\mathcal{R}$ *such that*

$$\forall A, B \in \mathcal{R}; \quad (\xi(A), \eta(B)) = \mu_{\xi\eta}(A \cap B).$$

(b) $\forall \xi \in \mathcal{F}, \xi(\cdot)$ *is a c.a.o.s. measure on* $\mathcal{R}$ *, and the measure* $\mu_{\xi\xi}(\cdot) = |\xi(\cdot)|^2$ *is non-negative.*

(c) $\sqrt{\mu_{\xi\xi}(\cdot)} = |\xi(\cdot)|$ *is the semi-variation* $s_\xi(\cdot)$ *of* ξ.

$\mu_{\xi\eta}$ is called the *(cross-)covariance measure of* ξ *with* η , and $\mu_{\xi\xi}$ the *(auto-)covariance measure of* ξ or the *non-negative measure of* ξ .These measures are of course finite on $\mathcal{R}$ but neither need be bounded on $\mathcal{R}$.

We can now state our first important theorem, which is essentially equivalent to the combined Thms. 1, 2 of Hellwig [8, pp.163,164]:

3.7 Thm. *Let* H *be a symmetric operator from* $\mathcal{H}$ *to* $\mathcal{H}$*, and* $\mathcal{R}$ *be any* δ *-ring over* $\mathbf{R}$ *satisfying* (3.1) (v) . *Then the eigenpackets of* H *having the domain* $\mathcal{R}$ *form a biorthogonal family of* $\mathcal{H}$ *-valued measures.*

Some comments on the proof of this theorem are in order. We split our packets ξ and η into ξ_c, ξ_d and η_c, η_d à la 3.4. That ξ_d, η_d are biorthogonal then follows from the fact that distinct eigenvalues of H have orthogonal eigenvectors. Also, each of ξ_d, η_d is biorthogonal to ξ_c, η_c as shown by Hellwig [8, p.163]. Finally, as Hellwig has shown [8, p.164], ξ_c is biorthogonal to η_c . Putting the pieces together we conclude that ξ is biorthogonal to η .[##]

[#] Def. 3.5 is a measure-theoretic extension of the more familiar definition of a biorthogonal family of sequences $(x_k)_1^\infty, (y_k)_1^\infty, \ldots$ in $\mathcal{H}$, cf. [20, p.208] .

[##] A more direct proof would be to show that, since H is symmetric, the measure M defined on $\mathcal{R} \times \mathcal{R}$ by

$$M(A \times B) \underset{d}{=} (\xi(A), \eta(B)), \quad A, B \in \mathcal{R}$$

is carried on the diagonal of $\mathcal{R} \times \mathcal{R}$, and hence $M(A \times B) = \mu(A \cap B)$. Unfortunately, certain lacunae remain in completing this argument.

3.8 Cor. *Every eigenpacket* ξ *of a symmetric operator from* $\mathcal{H}$ *to* $\mathcal{H}$ *is a* $\mathcal{H}$*-valued, c.a.o.s. measure.*

This is immediate from Thm. 3.7 and Lma. 3.6(b). In the study of such packets we may accordingly put the theory outlined in §2 to full use. An immediate result is the following relation between the approximate or generalized spectrum of H and the spectrum of the non-negative measure $\mu_{\xi\xi}$ of an eigenpacket ξ of H :

3.9 Cor. *Let* (i) ξ *be an eigenpacket of a symmetric operator* H *from* $\mathcal{H}$ *to* $\mathcal{H}$,

(ii) $\sigma(\mu_{\xi\xi})$ *be the spectrum of* $\mu_{\xi\xi}$, *i.e.*

$$\sigma(\mu_{\xi\xi}) \underset{d}{=} \{\lambda : \lambda \in \mathbf{R} \ \& \ \forall \varepsilon > 0, \ \mu_{\xi\xi}(\lambda-\varepsilon, \lambda+\varepsilon) > 0\}$$

(iii) $\sigma_g(H) \underset{d}{=} \{\lambda :$ λ *is a generalized eigenvalue* # *of* H$\}$.

Then $\sigma(\mu_{\xi\xi}) \subseteq \sigma_g(H) \subseteq \sigma(H)$.

The reverse inclusion fails in general, but can be secured for self-adjoint H by careful selection of ξ as we shall see in 4.10 below.

The eigenpackets of a non-self-adjoint, symmetric operator H will not of course suffice for its complete spectral analysis, since there is a residual spectrum $\sigma_r(H) \subseteq \mathbf{C} \setminus \mathbf{R}$. For this reason we shall deal only with self-adjoint operators from here on.

4. EIGENPACKETS OF SELF-ADJOINT OPERATORS

In this section

(4.1)
- (i) $\mathcal{H}, \mathcal{B}, \mathcal{B}_0, \mathcal{R}$ are as in (3.1),
- (ii) $H \underset{d}{=} \int_{\mathbf{R}} \lambda E(d\lambda)$ is a s.a. operator from $\mathcal{H}$ to $\mathcal{H}$, $E(\cdot)$ being its spectral measure on $\mathcal{R}$,
- (iii) $\forall x, y \in \mathcal{H} \ \& \ \forall B \in \mathcal{B}, \ \mu_{xy}(B) \underset{d}{=} (E(B)x, E(B)y)$,
- (iv) $\forall x \in \mathcal{H}, \ \mathcal{S}_x \underset{d}{=} \{E(B)x : B \in \mathcal{B}\} \underset{d}{=}$ the E-cyclic subspace due to x .

We call λ a generalized (or approximate) eigenvalue of $\mathcal{H}$, iff. $\exists$ a sequence $(x_n)_1^\infty$ such that

$$x_n \in H, \ |x_n| = 1 \ \& \ Hx_n - \lambda x_n \to 0.$$

Before continuing our discussion of eigenpackets, we must recall a few facts regarding the measures $E(\cdot)$ and μ_{xx}.

First, $\forall x \in \mathcal{H}$ we define the equivalence class:

$[\mu_{xx}] \underset{d}{=} \{\nu : \nu$ is a non-negative $(\leq \infty)$ c.a. Borel measure on $\mathcal{B}, \nu \approx \mu_{xx}\}$.#

From the Hellinger-Hahn theory we know that $\exists \alpha \in \mathcal{H}$ such that $\forall x \in \mathcal{H}$, $\mu_{xx} \ll \mu_{\alpha\alpha}$.## Any such α is said to be of *maximal measure-type with respect to* E, and the equivalence class $[\mu_{\alpha\alpha}]$ is called the *(maximal-) measure-type of* $E(\cdot)$. The *spectrum of* $E(\cdot)$ is defined by

$$\sigma(E) \underset{d}{=} \{\lambda : \lambda \in \mathbf{R} \ \& \ \forall \delta > 0,\ E(\lambda-\delta, \lambda+\delta) \neq 0\}.$$

The spectrum $\sigma(\nu)$ of a non-negative measure ν on $\mathcal{B}$ is defined similarly. These spectra are related to the spectrum $\sigma(H)$ of H by

$$\sigma(H) = \sigma(E) = \sigma(\nu),\ \forall \nu \text{ in the measure-type of } E. \tag{4.2}$$

Next, for any $x \in \mathcal{H}$ and any Borel measurable function ϕ on $\mathbf{R}$ to $\mathbf{C}$ let us write

$$\text{(4.3)} \quad \begin{cases} \mathcal{B}_{\phi,x} \underset{d}{=} \{B : B \in \mathcal{B} \ \& \ \phi(\cdot)\chi_B(\cdot) \in L_2(\mathbf{R}, \mathcal{B}, \mu_{xx}; \mathbf{C})\}, \\ \mathcal{R}_{\phi,x} \underset{d}{=} \mathcal{B}_{\phi I,x} \quad \text{where} \quad I(\lambda) \underset{d}{=} \lambda,\ \forall \lambda \in \mathbf{R}, \\ \mathcal{R}_x \underset{d}{=} \mathcal{R}_{1,x} = \mathcal{B}_{I,x}. \end{cases}$$

In other words, $\mathcal{B}_{\phi,x}, \mathcal{R}_{\phi,x}, \mathcal{R}_x$ are the families of Borel subsets B of $\mathbf{R}$ for which

$$\int_B |\phi(\lambda)|^2 \mu_{xx}(d\lambda) < \infty, \quad \int_B \lambda^2 |\phi(\lambda)|^2 \mu_{xx}(d\lambda) < \infty, \quad \int_B \lambda^2 \mu_{xx}(d\lambda) < \infty,$$

respectively. Obviously, $\mathcal{B}_{\phi,x}, \mathcal{R}_{\phi,x}, \mathcal{R}_x$ are δ subrings of $\mathcal{B}$, and

By a Borel measure ν we mean one: $\nu(B) < \infty$, $\forall B \in \mathcal{B}_0$; "$\nu \approx \mu$" means: ν & μ are mutually absolutely continuous.

"$\nu \ll \mu$" means: ν is absolutely continuous with respect to μ.

$$(4.4)\quad \begin{cases} \text{(a)} & \mathcal{B}_0 \subseteq \mathcal{R}_x \subseteq \mathcal{B} \\ \text{(b)} & \mathcal{B}_0 \subseteq \mathcal{B}_{\phi,x}{}^{\#} \Longrightarrow \mathcal{B}_0 \subseteq \mathcal{R}_{\phi,x} \subseteq \mathcal{B}_{\phi,x} \subseteq \mathcal{B}. \end{cases}$$

We now return to eigenpackets. Our first job is to show that every s.a. operator H has such a packet. This is utterly obvious from 4.4 (a):

4.5 Triv.## $\forall x \in \mathcal{H}$, $\xi_x(\cdot) \underset{d}{=} \text{Rstr.}_{\mathcal{R}_x} E(\cdot)x$ *is a bounded eigenpacket of* H *on the* δ*-ring* $\mathcal{R}_x$, *and* $\mathcal{M}_{\xi_x}(\mathbf{R}) = \mathcal{S}_x$.

From these bounded eigenpackets we shall now get others, not necessarily bounded, by indefinite integration. Again, for any $x \in \mathcal{H}$ and any Borel measurable function ϕ on $\mathbf{R}$ to $\mathbf{C}$ let us write

$$(4.6)\quad \begin{cases} \nu_{\phi,x}(B) \underset{d}{=} \int_B |\phi(\lambda)|^2 \mu_{xx}(d\lambda), & B \in \mathcal{B}_{\phi,x} \\ \eta_{\phi,x}(B) \underset{d}{=} \int_B \phi(\lambda) E(d\lambda) x, & B \in \mathcal{B}_{\phi,x}. \end{cases}$$

Then, cf. [14, p.84, 5.17],

(4.7) $\quad \eta_{\phi,x}$ is a $\mathcal{H}$-valued, c.a.o.s. measure over $(\mathbf{R}, \mathcal{B}_{\phi,x}, \nu_{\phi,x})$,

and from (4.4) (b) we get:

4.8 Thm. *Let* (i) $x \in \mathcal{H}$, (ii) ϕ *be a Borel measurable function on* $\mathbf{R}$ *to* $\mathbf{C} \ni \mathcal{B}_0 \subseteq \mathcal{B}_{\phi,x}$###. *Then* $\zeta_{\phi,x} \underset{d}{=} \text{Rstr.}_{\mathcal{R}_{\phi,x}} \eta_{\phi,x}$ *is an eigenpacket of* H *on the* δ*-ring* $\mathcal{R}_{\phi,x}$.

4.9 Remarks. 1. The condition 4.8 (ii) will of course be fulfilled by any ϕ in $L_2(\mathbf{R}, \mathcal{B}, \mu_{xx}; \mathbf{C})$. For such a ϕ, $\mathcal{B}_{\phi,x} = \mathcal{B}$ and $\eta_{\phi,x}$ will be bounded on $\mathcal{B}$; consequently, the eigenpacket $\zeta_{\phi,x}$ will be bounded on $\mathcal{R}_{\phi,x}$, in fact

This inclusion simply means that ϕ is L_2 with respect to μ_{xx} on all bounded Borel subsets of $\mathbf{R}$.

Hellwig [8] has a result corresponding to this in his Thm. 1 on p. 166.

Concerning the meaning of this inclusion see the last-but-one footnote.

$$|\zeta_{\phi,x}(B)|^2 \leq \int_{\mathbf{R}} |\phi(x)|^2 \mu_{xx}(d\lambda), \quad B \in \mathcal{R}_{\phi,x}.$$

2. Let $x \in \mathcal{H}$ and ν be any non-negative ($\leq \infty$) c.a. *Borel measure* # on $\mathcal{B}$ such that $\nu << \mu_{xx}$. Then we can find an eigenpacket ζ of H defined on a δ-ring $\mathcal{R}$ and such that in the notation of 3.6, $\mu_{xx} \subseteq \nu$. For this we need only take $\mathcal{R} = \mathcal{R}_{\phi,x}$, and $\zeta = \zeta_{\phi,x}$, where $\phi = \sqrt{(d\nu/d\mu_{xx}}$, noting that $\mathcal{B}_0 \subseteq \mathcal{B}_{\phi,x}$ since ν is a Borel measure; thus

$$\zeta(B) \underset{d}{=} \int_B \sqrt{\{\frac{d\nu}{d\mu_{xx}}(\lambda)\}} E(d\lambda)x, \quad B \in \mathcal{R}.$$

3. In particular for any ν in the maximal measure-type of $E(\cdot)$, there exists an eigenpacket ζ of H defined on a δ-ring $\mathcal{R}$ such that $\mu_{\zeta\zeta} \subseteq \nu$.

For eigenpackets ζ of H of the type mentioned in 4.9 (3) we can sharpen the inclusion relations given in Cor. 3.9. This follows at once from (4.2) and the fact that when $\mu_{\zeta\zeta} \subseteq \nu$, we have $\sigma(\mu_{\zeta\zeta}) = \sigma(\nu)$. Thus:

4.10 Cor.## *Let* ν *be any measure in the maximal measure-type of* E, *and* ζ *be any eigenpacket of* H *such that* $\mu_{\zeta\zeta} \subseteq \nu$. *Then* $\sigma(\mu_{\zeta\zeta}) = \sigma(\mathrm{H})$.

The results 4.5, 4.8 show that every s.a. operator has a plentiful supply of eigenpackets, both bounded and unbounded. All these packets are of course c.a.o.s. measures, cf. Cor. 3.8. We shall now show, conversely, that every $\mathcal{H}$-valued, c.a.o.s. measure ξ on a δ-ring $\mathcal{R}$ satisfying 3.1 (v) is, after suitable restriction, the eigenpacket of a s.a. operator H_ξ from $\mathcal{H}$ to $\mathcal{H}$. Observing that

(4.11) *if* $\mathcal{R}$ *satisfies* (3.1) (v), *then* $\mathcal{R}^{loc} = \mathcal{B}$,

we define H_ξ to be the s.a. operator from $\mathfrak{M}_\xi(\mathbf{R})$ to $\mathfrak{M}_\xi(\mathbf{R})$ whose spectral measure is the spatial spectral measure $Q_\xi(\cdot)$ of ξ on $\mathcal{R}^{loc}$, i.e. on $\mathcal{B}$; thus

(4.12) $$\mathrm{H}_\xi \underset{d}{=} \int_{\mathbf{R}} \lambda Q_\xi(d\lambda), \quad \text{cf. 2.11.}$$

i.e. a measure ν such that $\nu(B) < \infty$, $\forall B \in \mathcal{B}_0$.

This result corresponds in a way to Hellwig's Thm.3, p.169 in [8].

We call H_ξ the *s.a. operator associated with* ξ .

Before stating our theorem we must recall that under the isomorphism Σ_ξ appearing in Thm. 2.12, the projection $Q_\xi(B)$, $B \in \mathcal{B}$, corresponds to multiplication by χ_B on $\mathcal{L}_2$, i.e.

(4.13) $$Q_\xi(B) = \Sigma_\xi \cdot M_{\chi_B} \cdot \Sigma_\xi^{-1}, \quad B \in \mathcal{B},$$

and must record certain consequences of this result:

4.14. Lma. *Let* (i) ξ *be an* $\mathcal{H}$ *-valued, c.a.o.s. measure over* $(\mathbf{R}, \mathcal{R}, \mu)$ *where* $\mathcal{R}$ *is a* δ *-ring satisfying* 3.1 (v), (ii) Σ_ξ *be the isomorphism* $\phi \to \int_{\mathbf{R}} \phi(\lambda) \cdot \xi(d\lambda)$ *on* $\mathcal{L}_2 =_d L_2(\mathbf{R}, \mathcal{R}, \mu; \mathbf{C})$ *onto* $\mathcal{M}_\xi(\mathbf{R}) \subseteq \mathcal{H}$, *cf. Th. 2.12,* (iii) $\forall x \in \mathcal{M}_\xi(\mathbf{R})$,
$$\phi(\cdot, x) =_d \Sigma_\xi^{-1}(x) \in \mathcal{L}_2 .$$
Then

(a) $\forall x \in \mathcal{M}_\xi(\mathbf{R})$ & $\forall B \in \mathcal{B}$, $Q_\xi(B)x = \int_B \phi(\lambda, x)\xi(d\lambda)$;

(b) $\forall A \in \mathcal{R}$ & $\forall B \in \mathcal{B}$, $Q_\xi(B)\{\xi(A)\} = \xi(A \cap B)$;

(c) Q_ξ *has total multiplicity* 1 *; a vector* x *in* $\mathcal{M}_\xi(\mathbf{R})$ *is* Q_ξ*-cyclic, iff.* $\phi(\cdot, x)$ *vanishes almost nowhere* (μ) *on* $\mathbf{R}$;

(d) μ *is in the maximal measure-type of* $Q_\xi(\cdot)$ *, and hence* $\forall Q_\xi$*-cyclic* $x \in \mathcal{M}_\xi(\mathbf{R})$, $|Q_\xi(\cdot)x| \approx \mu$.

Our theorem now reads as follows:

4.15 Thm. *Let* (i) ξ *be an* $\mathcal{H}$ *-valued, c.a.o.s. measure over* $(\mathbf{R}, \mathcal{R}, \mu)$, *where* $\mathcal{R}$ *is a* δ *-ring over* $\mathbf{R}$ *satisfying* 3.1 (v), (ii) $\mathcal{B}_\xi =_d \{B : B \in \mathcal{R}$ & $I(\cdot)\chi_B(\cdot) \in L_2(\mathbf{R}, \mathcal{R}, \mu_{\xi\xi}; \mathbf{C})\}$.
Then

(a) $\mathcal{B}_\xi$ *is a* δ^**-ring* $\ni \mathcal{B}_0 \subseteq \mathcal{B}_\xi \subseteq \mathcal{R} \subseteq \mathcal{B}$;

(b) $\xi_1 =_d \text{Rstr.}_{\mathcal{B}_\xi} \xi$ *is an eigenpacket of the s.a. operator* H_ξ *associated with* ξ *, cf.* (4.12), *on the* δ *-ring* $\mathcal{B}_\xi$;

(c) $\forall n \in \mathbf{N}_+$ & $\forall B \in \mathcal{B}_0$, $\xi(B) \in \mathcal{D}_{H_\xi^n}$ & $H_\xi^n\{\xi(B)\} = \int_B \lambda^n \xi(d\lambda)$.

The result 4.15 (c) says that for any c.a.o.s. measure ξ on $\mathcal{R}$, the values $\xi(B)$ of ξ on bounded Borel subsets B of $\mathbf{R}$ are in a sense "infinitely smooth" with respect to the associated s.a. operator H_ξ. The following example should make the meaning of this clear:

4.16 Example. Let $\mathcal{H} \underset{d}{=} L_2(\mathbf{R})$,

$$\mathcal{R} \underset{d}{=} \{B;\ B \in \mathcal{B}\ \&\ \text{Leb.}\, B < \infty\}$$

$$\forall B \in \mathcal{R},\quad \xi_B(t) \underset{d}{=} \int_B e^{it\lambda}\, d\lambda .$$

Then, cf. [17, (6.7)] or [16],

ξ *is an* $L_2(\mathbf{R})$*-basic, c.a.o.s. measure over* $(\mathbf{R}, \mathcal{R}, \text{Leb.})$. It is easy to verify that the associated s.a. operator H_ξ is given by

$$iH_\xi = D = \text{differentiation on } L_2(\mathbf{R}).$$

By 4.15 (c), $\forall B \in \mathcal{B}_0$, $\xi_B(\cdot)$ is infinitely differentiable and all its derivatives are in $L_2(\mathbf{R})$.

Now let as in (4.1) H be any s.a. operator from $\mathcal{H}$ to $\mathcal{H}$, and ξ be an eigenpacket of H on a δ-ring $\mathcal{R}$. Then of course $\xi_1 = \text{Rstr.}_{\mathcal{B}_\xi}\xi$ is an eigenpacket of H on the δ-ring $\mathcal{B}_\xi$. But by Thm. 4.15, ξ_1 is also an eigenpacket of the s.a. operator H_ξ from $\mathcal{M}_\xi(\mathbf{R})$ to $\mathcal{M}_\xi(\mathbf{R})$ associated with ξ. Moreover by 4.14 (c), the multiplicity of H_ξ is one. This suggests that $H_\xi \subseteq H$. This is indeed the case as our next theorem attests:

4.17 Thm. *Let* ξ *be any eigenpacket of the s.a. operator* H *in* (4.1). *Then*

$$H_\xi = \text{Rstr.}_{\mathcal{M}_\xi(\mathbf{R})}H, \qquad Q_\xi(\cdot) = \text{Rstr.}_{\mathcal{M}_\xi(\mathbf{R})}E(\cdot) .$$

Combining this theorem with the results 4.14, and 4.15 on Q_ξ and H_ξ we get two interesting corollaries:

4.18 Cor. *Let* (i) ξ *be any eigenpacket of the s.a. operator* H *in* (4.1) *on a* δ*-ring* $\mathcal{R}$, (ii) $\forall x \in \mathcal{M}_\xi(\mathbf{R})$, $\phi(\cdot, x) = \Sigma_\xi^{-1}(\phi) \in L_2(\mathbf{R}, \mathcal{R}, \mu_{\xi\xi}; \mathbf{C})$, *where* Σ_ξ *is the isomorphism given in Thm. 2.12.*

Then

(a) $\forall x \in \mathcal{M}_\xi(\mathbf{R})$ & $\forall B \in \mathcal{B}$, $E(B)(x) = \int_B \phi(\lambda, x)\xi(d\lambda)$;

(b)# $\forall A \in \mathcal{R}$ & $\forall B \in \mathcal{B}$, $E(B)\{\xi(A)\} = \xi(A \cap B)$;

(c) $\forall n \in \mathbf{N}_+$ & $\forall B \in \mathcal{B}_0$, $\xi(B) \in \mathcal{D}_{H^n}$ & $H^n\{\xi(B)\} = \int_B \lambda^n \xi(d\lambda)$.

Cor. 4.18 (c) tells us that for any eigenpacket ξ of H the values $\xi(B)$ of ξ on bounded Borel subsets B of $\mathbf{R}$ are "infinitely smooth" with respect to H in the sense indicated after 4.15. For instance, if ξ is *any* eigenpacket of the s.a. operator iD on $L_2(\mathbf{R})$, then for all $B \in \mathcal{B}_0$, the function $\xi_B(\cdot)$ will be infinitely differentiable and all its derivatives will be in $L_2(\mathbf{R})$.

4.19 Cor. *Let the multiplicity of the s.a. operator* H *in* (4.1) *be* 1. *Then* H *possesses an* $\mathcal{H}$*-basic eigenpacket* ξ, *i.e. one such that* $\mathcal{M}_\xi(\mathbf{R}) = \mathcal{H}$.

5. EIGENPACKET EXPANSIONS

Let $\mathcal{H}, \mathcal{B}, \mathcal{B}_0, \mathcal{R}$ be as in (3.1) and let the self-adjoint operator $H = \int_{\mathbf{R}} \lambda E(d\lambda)$, the measures μ_{xy} and the subspaces $\mathcal{S}_x$ be as in (4.1) (ii)-(iv).

Suppose first that the multiplicity of $E(\cdot)$ is 1. Then by Cor. 4.19 H has one and therefore many $\mathcal{H}$*-basic* eigenpackets. Some of these packets may possess a desirable or convenient attribute, such as for instance having Lebesgue measure as non-negative measure. Select in any convenient way one such packet ξ and let $\mathcal{R}$ denote its domain. Recall, cf. 3.2, that $\mathcal{R}$ is a δ-ring such that $\mathcal{B}_0 \subseteq \mathcal{R} \subseteq \mathcal{B}$, and by Cor. 3.8

$$\xi \textit{ is a } \mathcal{H}\textit{-basic, c.a.o.s. measure over } (\mathbf{R}, \mathcal{R}, \mu_{\xi\xi}).$$

Hence by the Isomorphism Thm. 2.12, $\forall x \in \mathcal{H}$, $\exists \phi(\cdot, x) \in \mathcal{L}_2 \overset{=}{_d} L_2(\mathbf{R}, \mathcal{R}, \mu_{\xi\xi}; \mathbf{C})$ such that

$$x = \int \phi(\lambda, x)\xi(d\lambda). \tag{5.1}$$

Thus *every* x *in* $\mathcal{H}$ *has an "expansion" in terms of the eigenpacket* ξ *of* H.

Next let the multiplicity of $E(\cdot)$ be q where $1 \leq q \leq \aleph_0$. Then by

A special case of this result appears in Hellwig's treatment [8] in Thm. 2, p.167.

the Hellinger-Hahn Thm. [17, 14.7], $\exists$ a sequence $(\beta_j)_{j\in J}$ of vectors β_j in $\mathcal{H}$ such that

(5.2)
$$\begin{cases} \mathcal{H} = \sum_{j\in J} \mathcal{S}_{\beta_j}, \quad \mathcal{S}_{\beta_j} \perp \mathcal{S}_{\beta_k}, \quad |\beta_j| = 1, \quad j,k \in J \ \&\ j \neq k \\ \textit{and letting } \mu_j \underset{d}{=} \mu_{\beta_j\beta_j}, \textit{ we have} \\ \qquad \mu_j(\cdot) = \mu_1(\Lambda_j \cap \cdot)/\mu_1(\Lambda_j) \\ \textit{where } \Lambda_{j+1} \subseteq \Lambda_j \subseteq \Lambda_1 \subseteq \mathbf{R} \ \&\ \mu_1(\lambda_j) > 0 . \end{cases}$$

Now each $\mathcal{S}_{\beta_j}$ reduces H as well as the projections $E(B)$, $B \in \mathcal{B}$. Hence $\text{Rstr.}_{\mathcal{S}_{\beta_j}} E(\cdot)$ is a spectral measure for the Hilbert space $\mathcal{S}_{\beta_j}$. Since the last is the spectral measure of the s.a. operator $H_j \underset{d}{=} \text{Rstr.}_{\mathcal{S}_{\beta_j}} H$, it follows from Cor. 4.19 that H_j possesses an $\mathcal{S}_{\beta_j}$-basic eigenpacket. Obviously this packet is also an eigenpacket of H itself. In short, $\forall j \in J$, H has one and therefore many $\mathcal{S}_{\beta_j}$-basic eigenpackets. For each j in J, select an $\mathcal{S}_{\beta_j}$-basic eigenpacket ξ_j in any convenient way, and let $\mathcal{R}_j$ be its domain. Then, cf. 3.2, $\mathcal{R}_j$ is a δ-ring such that $\mathcal{B}_0 \subseteq \mathcal{R} \subseteq \mathcal{B}$. Now define

(5.3) $$\mathcal{R} \underset{d}{=} \bigcap_{j\in J} \mathcal{R}_j, \quad \eta_j(\cdot) \underset{d}{=} \text{Rstr.}_{\mathcal{R}}\, \xi_j(\cdot), \quad \nu_j(\cdot) \underset{d}{=} |\eta_j(\cdot)|^2 .$$

It follows readily that $\mathcal{R}$ is a δ-ring such that $\mathcal{B}_0 \subseteq \mathcal{R} \subseteq \mathcal{B}$, and each $\eta_j(\cdot)$ is a $\mathcal{H}$-valued, c.a.o.s. measure over $(\mathbf{R}, \mathcal{R}, \nu_j)$. Moreover, since $\mathcal{R} \subseteq \mathcal{R}_j \subseteq \mathcal{B} = \sigma(\mathcal{R})$, we get from [14, 3.3], $\mathfrak{M}_{\eta_j}(\mathbf{R}) = \mathfrak{M}_{\xi_j}(\mathbf{R}) = \mathcal{S}_{\beta_j}$. In short

(1) $\forall j \in J$, η_j *is a* $\mathcal{S}_{\beta_j}$*-basic eigenpacket of* H on $\mathcal{R}$.

We claim that *the* η_j, $j \in J$, *constitute a full system of eigenpackets of* H.

To justify this claim, let $x \in \mathcal{H}$. Then by (5.2)

$$x = \sum_{j\in J} x_j, \text{ where } x_j \in \mathcal{S}_{\beta_j} \ \&\ x_j \perp x_k \quad \text{for } j \neq k.$$

But by (1) and the Isomorphism Thm. 2.12, $\forall j \in J,\ x_j = \int_{\mathbf{R}} \phi_j(\lambda, x)\eta_j(d\lambda)$, where $\phi_j(\cdot, x) \in L_2(\mathbf{R}, \mathcal{R}, \nu_j; \mathbf{C})$. Hence

$$(5.4) \qquad x = \sum_{j \in J} \int_{\mathbf{R}} \phi_j(\lambda, x)\eta_j(d\lambda).$$

Thus *every* x *in* $\mathcal{H}$ *has an "expansion" in terms of the eigenpackets* η_j, $j \in J$. In this sense, the family $\{\eta_j, j \in J\}$ is a full system of eigenpackets of H.

It is possible to replace the full system of eigenpackets η_j, $j \in J$, defined in (5.3) by a single c.a.q.i. "operator-valued eigenpacket" $T(\cdot)$ of H. To see this let the β_j be as in (5.2), and let

$$(5.5) \qquad \begin{cases} W \underset{d}{=} \mathfrak{S}\{\beta_j : j \in J\} \\ \forall j \in J,\ P_j \underset{d}{=} \text{the projection on } W \text{ onto the } 1\text{-dimensional space } \mathfrak{S}\{\beta_j\}. \end{cases}$$

Let $\mathcal{R}, \eta_j, \nu_j$ be as in (5.3), and define the set-functions $M(\cdot)$ and $T(\cdot)$ by

$$(5.6) \qquad \begin{cases} \forall A \in \mathcal{R}, \qquad M(A) = \sum_{j \in J} \nu_j(A) P_j \\ T(A)(w) \underset{d}{=} \sum_{j \in J} (w, \beta_j)\eta_j(A), \qquad w \in W. \end{cases}$$

Then we can show that

(2) $M(\cdot)$ *is an adequate,* W*-to-*W, *non-negative hermitian operator-valued measure on* $\mathcal{R}$,

and

(3) $T(\cdot)$ *is a* W*-to-*W *basic, c.a.q.i. measure over* $(\mathbf{R}, \mathcal{R}, M)$.

The measure $T(\cdot)$ resembles an eigenpacket in that it satisfies an equation similar to that in Def. 3.2, viz.

$$(4) \qquad \forall A \in \mathcal{R}, \qquad H \cdot T(A) = \int_A \lambda T(d\lambda).$$

where the last integral is defined as in [17, 16.2]. We may therefore call $T(\cdot)$ an *"operator-valued eigenpacket"* of H.

When the spectral multiplicity of H exceeds 1, this single operator-valued eigenpacket $T(\cdot)$ can be used instead of the full system of (vector-valued) eigenpackets η_j, $j \in J$ for expanding vectors in $\mathcal{H}$. In fact, the adequate measure $M(\cdot)$, cf. (2), provides us with a complete representation space

$$\mathcal{L}_{2,W} \underset{d}{=} L_2(\mathbf{R}, \mathcal{R}, M;\, W).$$

Also, letting

$$(5) \qquad \forall x \in \mathcal{H} \ \& \ \forall \lambda \in \quad , \qquad \Phi(\lambda, x) \underset{d}{=} \sum_{j \in J} \phi_j(\lambda, x)\beta_j \in W,$$

where the ϕ_j are as in (5.4), it follows that $\Phi(\cdot, x) \in \mathcal{L}_{2,W}$ and from (5.4) that $x = \int_{\mathbf{R}} T(d\lambda)\Phi(\lambda, x)$. We may sum up in the following result:

5.7 Thm. *When the spectral multiplicity of the s.a. operator* H *from* $\mathcal{H}$ *to* $\mathcal{H}$ *exceeds* 1, *let* β_j, $j \in J$ *and* W *be as in* (5.2) *and* (5.5), *the* δ*-ring* $\mathcal{R}$ *be as in* (5.3), *and the operator-valued measures* $M(\cdot)$ *and* $T(\cdot)$ *be as in* (5.6) *Then*

(a) T *is a* W*-to-*$\mathcal{H}$ *basic, c.a.q.i. measure over* $(\mathbf{R}, \mathcal{R}, M)$, *such that* $\forall A \in \mathcal{R}$, $H \cdot T(A) = \int_A \lambda T(d\lambda)$;

(b) $\forall x \in \mathcal{H}$, $\exists \Phi(\cdot, x) \in L_2(\mathbf{R}, \mathcal{R}, M;\, W)$ *such that*

$$x = \int_{\mathbf{R}} T(d\lambda)\Phi(\lambda, x).$$

The expansion (5.4), which of course subsumes (5.1) in case $q = 1$, may be regarded as a rigorization of the nonrigorous idea of "expansion of x in terms of the pseudo-eigenvectors of H". Alternatively, we may regard the expansion given in 5.7 (b) as being such a rigorization. These rigorizations, in which pseudo integrals with respect to scalar measures are replaced by genuine integrals with respect to $\mathcal{H}$-valued or operator-valued measures, seem to us to be as effective as those obtained by devices such as rigging. They have the advantage of not making us move from our Hilbert space $\mathcal{H}$ to some non-unique, larger space. Moreover, they are conceived in measure-theoretic terms, which have proved fruitful in the theory of stochastic processes. Thus they secure a certain unification that is missing in the other approaches.

6. THE EXISTENCE OF PSEUDO-EIGENFUNCTIONS

In §5 we deduced the existence of a full system of eigenpackets of a s.a. operator H from its spectral resolution. Our appeal to this resolution was only for purposes of proof. In practice it is often easier to find the eigenpackets of H than to find its spectral measure $E(\cdot)$, as Hellwig has remarked [8, p.171].

For instance, for certain s.a. differential operators H on L_2 spaces it may be possible to solve the eigenvalue problem:

$$H(\Psi) = \lambda\Psi, \qquad \lambda \in \mathbf{R},$$

obtaining thereby solutions $\Psi(\cdot,\lambda)$ which though not in L_2 are such that the associated functions $\xi_B(\cdot)$, where

$$\xi_B(x) \underset{d}{=} \int_B \Psi(x,\lambda)\,\mu(d\lambda),$$

are in L_2 for sets B in some appropriate δ-ring $\mathcal{R}$ such as the $\mathcal{B}_0$ of 3.1 (iv), and the L_2-valued, set-function ξ is an eigenpacket of H. In short, we may be able to get the eigenpacket ξ of H by integrating with respect to the variable λ, on sets in $\mathcal{R}$, a kernel $\Psi(x,\lambda)$, where each $\Psi(\cdot,\lambda)$ is an easily obtained pseudo-eigenfunction of H. If this eigenpacket is L_2-basic, it can be used for expansion purposes as in (5.1). If it is not L_2-basic, then a full family of packets may be found by integrating kernels in the manner just indicated, and these used for expansion purposes as in (5.4) or 5.7 (b).

To discuss the idea of pseudo-eigenfunction more precisely, let H be a s.a. operator on the L_2-space $\mathcal{H} \underset{d}{=} L_2(X,\mathcal{F},\mu)$ where μ is a non-negative, σ-finite, c.a. measure on a σ-algebra $\mathcal{F}$ over a set X. Let ξ be an eigenpacket of H defined on a δ-ring $\mathcal{R}$. $\mathcal{R}$ will of course satisfy condition (3.1) (v). Now define

$$(6.1)\qquad \begin{cases} \mathcal{F}_\mu \underset{d}{=} \{A : A \in \mathcal{F} \ \&\ \mu(A) < \infty\} \\ \forall A \times B \in \mathcal{F}_\mu \times \mathcal{R}, \qquad M_\xi(A \times B) \underset{d}{=} (\chi_A, \xi_B)_{\mathcal{H}}. \end{cases}$$

Then $M_\xi(\cdot)$ is a complex-valued set-function on the δ-ring $\mathcal{F}_\mu \times \mathcal{R}$ over $X \times \mathbf{R}$.

Now suppose that the values ξ_B, $B \in \mathcal{R}$, of this eigenpacket ξ are obtainable by integrating over B with respect to $\hat{\mu}$ a complex-valued kernel $\Psi(\cdot,\cdot)$

defined on $X \times \mathbf{R}$:

(1) $$\xi_B(x) = \int_B \Psi(x,\lambda)\hat{\mu}(d\lambda), \quad x \in X,$$

where $\hat{\mu} \underset{d}{=} \mu_{\xi\xi}$. Then it follows from (6.1) that

(2) $$\forall A \times B \in \mathcal{F}_\mu \times \mathcal{R}, \quad M_\xi(A \times B) = \iint_{A\times B} \overline{\Psi(x,\lambda)}(\mu \times \hat{\mu})\{d(x,\lambda)\}.$$

Thus, the set-function M_ξ will be a c.a. measure on the δ-ring $\mathcal{F}_\mu \times \mathcal{R}$, and $M_\xi << \mu \times \hat{\mu}$; furthermore

(6.2) $$\overline{\Psi} = \frac{dM_\xi}{d(\mu \times \hat{\mu})}, \text{ a.e. } \mu \times \hat{\mu}.$$

This brings up the following question:

Question. *What conditions must a s.a. operator* H *on the space* $\mathcal{H} \underset{d}{=} L_2(X, \mathcal{F}, \mu)$ *satisfy in order that the complex-valued set-function* M_ξ *defined on the* δ*-ring* $\mathcal{F}_\mu \times \mathcal{R}$ *over* $X \times \mathbf{R}$ *by* (6.1), *in which* ξ *is an eigenpacket of* H *be a c.a. measure, absolutely continuous with respect to the product measure* $\mu \times \hat{\mu}$ *?*

No good answer to this question seems to be known. But for all H's for which the answer is affirmative, it would be reasonable to call each function $\Psi(\cdot,\lambda)$ where $\lambda \in \mathbf{R}$ and $\Psi(\cdot,\cdot)$ is given by (6.2), a *pseudo-eigenfunction of* H .

Our vector-measure-theoretic treatment thus suggest our defining a "pseudo-eigenfunction" of a s.a. operator H on an L_2 space as a function obtained by slicing a Radon-Nikodym derivative of a certain measure $\bar{M}_\xi$ constructed from an eigenpacket ξ of H . The measure $\bar{M}_\xi$ plays a fundamental role in L_2-transform theory by its occurrence in Bochner's Duality Thm., cf. [17, 2.9], [16, 2.16] or [14, 9.6-9.8]. We now see that the consideration of a RN derivative of this very measure leads to a new explication of the notion of pseudo-eigenfunction. It would seem that this framework is well suited for the study of kernels in eigenfunction expansions, and that its bearing on the important work of Mautner [19], and Bade and Schwartz [1] on the existence of such kernels deserves scrutiny.

APPENDIX. UNBOUNDED VECTOR MEASURES ON δ-RINGS

Our Def. 3.2 of eigenpacket involves a c.a. measure ξ having values in a Banach space $\mathcal{X}$, which in general is unbounded and defined only on a δ-ring. Such measures fall outside the scope of the Bartle-Dunford-Schwartz theory, which is restricted to bounded vector-valued measures defined on σ-algebras [5, Ch.IV, § 10]. They are not covered by Dinculeanu's treatment [3], which is concerned primarily with measures whose total variation-measures are finite. In typical situations the total variation-measure of our measure ξ will be infinite on *all* non-void open sets.

Interestingly enough, the measure and integration theory needed for Def. 3.2 originated in attempts to extend the theorems of Khinchine, Kolmogorov and Wiener on stationary stochastic proccesses to non-stationary ones. The pioneering paper [2] is by Cramer, who however assumes that the variation measure is finite. It was Rosanov [21, §1] who seems to have freed the Cramer theory from this restriction by introducing a new non-additive "variation", which in today's parlance is known as the *semi-variation*, cf. [5, p.310] or [3, p.51]. We shall now discuss the parts of this theory which are needed in §3 above.

Let

(A.1)
- (i) $\mathcal{X}, \mathcal{B}, \mathcal{B}_0, \mathcal{R}$ *be as in* (3.1),
- (ii) ξ *be a* $\mathcal{X}$*-valued, c.a. measure on* $\mathcal{R}$,
- (iii) $s_\xi(\cdot)$ *be the semi-variation of* ξ, cf. [5, p.310].

It follows, cf. [3, p.56, Prop. 7], *that* ξ *is a locally bounded measure, and*

(A.2) $\quad s_\xi(\cdot)$ *is a finite (but generally unbounded) set-function on* $\mathcal{R}$.

Since $\mathcal{B}_0 \subseteq \mathcal{R}$, the following definition, where $x_0 \in \mathcal{X}$ and $\lambda_0 \in \mathbf{R}$ makes sense:

$$x(\lambda) \underset{d}{=} \begin{cases} x_0 + \xi(\lambda_0, \lambda] & \lambda \geq \lambda_0, \\ x_0 - \xi(\lambda, \lambda_0] & \lambda < \lambda_0. \end{cases} \tag{A.3}$$

The properties of this curve $x(\cdot)$ in $\mathcal{X}$ and its connection to the $\mathcal{X}$-valued measure ξ are given by the following:

A.4 Triv. (a) $\forall a,b \in \mathbf{R} \ni a \le b$, $\xi(a,b] = x(b) - x(a)$.

(b) $x(\cdot)$ *has only simple discontinuities on* $\mathbf{R}$, *i.e.* $\forall \lambda \in \mathbf{R}$, *the limits* $x(\lambda-)$ & $x(\lambda+)$ *exist.*

(c) $x(\cdot)$ *is right continuous on* $\mathbf{R}$, *i.e.* $\forall \lambda \in \mathbf{R}$, $x(\lambda+) = $ $= x(\lambda)$ & $\xi\{\lambda\} = x(\lambda) - x(\lambda-)$.

(d) *The set* $D_\xi \underset{d}{=} \{\lambda: \lambda \in \mathbf{R} \ \& \ \xi\{\lambda\} \neq 0\}$ *is countable, and* D_ξ & $\mathbf{R} \setminus D_\xi \in \mathcal{B}$.

From A.4 we easily deduce the decomposability of ξ into purely atomic and purely non-atomic parts:

A.5 Lma. *Let* $\forall A \in \mathcal{R}$, $\xi_c(A) \underset{d}{=} \xi(A \setminus D_\xi)$, $\xi_d(A) \underset{d}{=} \xi(A \cap D_\xi)$. *Then* (a) ξ_c, ξ_d *are* $\mathcal{X}$ *-valued, c.a. measures on* $\mathcal{R}$ *;*

(b) ξ_c *is purely non-atomic, i.e.* $\forall \lambda \in \mathbf{R}$, $\xi_c(\lambda) = 0$;

(c) ξ_d *is purely atomic with carrier* D_ξ, *cf.* A.4(d), *and*

$$\forall A \in \mathcal{R}, \quad \xi_d(A) = \sum_{\lambda \in A} \xi\{\lambda\} ;$$

(d)
$$\xi(\cdot) = \xi_c(\cdot) + \xi_d(\cdot).$$

We now turn to integration. The definition of $\int_{\mathbf{R}} \phi(\lambda)\xi(d\lambda)$ is given first for an $\mathcal{R}$-simple function ϕ on $\mathbf{R}$ to $\mathbf{C}$, and then extended to functions ϕ in a vector-space $L_1(\mathbf{R}, \mathcal{B}, \xi; \mathbf{C})$, so that the integral becomes a linear functional on the latter. For the details we refer to [21, p.274-]. A result we need is as follows:

A.6 Thm. *Let* ϕ *be a complex-valued, bounded* $\mathcal{R}$ *measurable function on* $\mathbf{R}$, *and* S *be its support, (so that* $S \in \mathcal{R}$*). Then* $\phi \in L_1(\mathbf{R}, \mathcal{B}, \xi; \mathbf{C})$ *and*

$$\left| \int_{\mathbf{R}} \phi(\lambda)\xi(d\lambda) \right|_{\mathcal{X}} \le |\phi|_\infty \cdot s_\xi(S).$$

This immediately yields the corollary:

A.7 Cor. *Let* ϕ *be continuous on* $\mathbf{R}$ *to* $\mathbf{C}$ *and* $B \in \mathcal{B}_0$ *. Then*

$$\int_B \phi(\lambda)\xi(d\lambda) \ \textit{exist}, \ \& \ \left| \int_B \phi(\lambda)\xi(d\lambda) \right| \le \sup_{\lambda \in B} |f(\lambda)| \cdot s_\xi(B).$$

We consider next the conversion of the integral of a continuous complex-valued function ϕ on $\mathbf{R}$ with respect to the measure ξ into a *Riemann-Stieltjes (RS) integral* of ϕ with respect to an associated curve $x(\cdot)$ defined as in (A.3). For a study of vector-valued RS integrals see [9, pp.62-67]. Our theorem reads as follows:

A.8 Thm. *Let* ϕ *be continuous on* $\mathbf{R}$ *to* $\mathbf{C}$, $a, b \in \mathbf{R}$ *and* $a < b$. *Then*

$$(\mathrm{RS}) \int_a^b \phi(\lambda)\, dx(\lambda) \ \textit{exist} \ \& = \int_{(a,b]} \phi(\lambda)\, \xi(d\lambda).$$

The proof of the existence of the RS integral on the left resembles that of the corresponding more familiar result for functions $x(\cdot)$ of bounded vaiation – we replace the variation by the semi-variation, which is bounded on $[a, b]$, cf. (A.2).

Combining A.8 with a known result on RS integration by parts [9, Thm. 3.3.1], we easily get:

A.9 Cor. *Let* ϕ *be continuous on* $\mathbf{R}$ *to* $\mathbf{C}$, $a, b \in \mathbf{R}$ *and* $a < b$. *Then*

$$(\mathrm{RS}) \int_a^b x(\lambda)\, d\phi(\lambda) \ \textit{exist} \ \& = \phi(b)x(b) - \phi(a)x(a) - \int_a^b \phi(\lambda)\, dx(\lambda).$$

Finally, taking $\phi(\lambda) = \lambda$, $\lambda \in \mathbf{R}$, we get a result used by Hellwig [8, p.163] viz.

$$\text{(A.10)} \quad (\mathrm{RS}) \int_a^b \lambda\, dx(\lambda) \ \textit{exist} \ \& = bx(b) - ax(a) - \int_a^b x(\lambda)\, d\lambda.$$

This result holds even when $x(\cdot)$ is discontinuous, cf. A.4.

REFERENCES

[1] W.G. Bade and J. Schwartz, On abstract eigenfunction expansions, *Proc. Nat. Acad. Sci. U.S.A.*, 42 (1956), 519-525.

[2] H. Cramer, A contribution to the theory of stochastic processes, *Proc. Second Berkeley Sympos. Math. Statist. and Prob. (1950)*, Berkeley, Calif., 1951, pp. 329-339.

[3] N. Dinculeanu, *Vector measures* (Oxford, 1967).

[4] J.L. Doob, *Stochastic Processes* (New York, 1953).

[5] N. Dunford and J. Schwartz, *Linear operators.* vol. I (New York, 1958).

[6] P.R. Halmos, Shifts of Hilbert spaces, *J. Reine Angew. Math.*, 208 (1961), 102-112.

[7] E. Hellinger, Neue Begrundung der Theorie quadratischer Formen von unendlich vielen Veränderlichen, *J. Reine Angew. Math.*, 136 (1909), 210-271.

[8] G. Hellwig, *Differential operators of mathematical physics. An introduction* (Berlin, 1964); English transl. (Reading, Mass., 1954).

[9] E. Hille and R.S. Phillips, *Functional analysis and semigroups* (Providence, R. I., 1957).

[10] A.N. Kolmogorov, Stationary sequences in Hilbert space, *Bull. Math. Univ. Moscow* 2 (1941), 1-40. (English transl.)

[11] S.T. Kuroda, An abstract stationary approach to perturbation of continuous spectra and scattering theory, *J. Analyse Math.*, 20 (1967), 57-117.

[12] V. Mandrekar and H. Salehi, The square-integrability of operator-valued functions with respect to a non-negative operator valued mesure, *J. Math. Mech.*, 20 (1970-71), 545-564.

[13] P. Masani, Isometric flows on Hilbert space, *Bull. Amer. Math. Soc.*, 68 (1962), 624-632.

[14] ———— Orthogonally scattered measures, *Advances in Math.*, 2 (1968), 61-117. (Originally, *Technical Report* 38, Mathematics Research Center, University of Wisconsin, 1967.)

[15] ———— On the representation theorem of scattering, *Bull. Amer. Math. Soc.*, 74 (1968), 618-624.

[16] ———— Explicit form for the Fourier-Plancherel transform over locally compact abelian groups, in *Abstract spaces and approximation* (P.L. Butzer and B. Sz.–Nagy Editors) (Basel, 1969), pp. 162-182.

[17] ———— Quasi-isometric measures and their applications, *Bull. Amer. Mat. Soc.*, 76 (1970), 427-528.

[18] P. Masani and J. Robertson, The time-domain analysis of continuous parameter weakly stationary stochastic processes, *Pacific J. Math.*, 12 (1962), 1362-1378.

[19] F.I. Mautner, On eigenfunction expansions, *Proc. Nat. Acad. Sci. U.S.A.*, 39 (1953), 49-53.

[19'] F. Rellich, *Spectral theory of a second order ordinary differential operator* (Lecture Notes, NYU, 1950-51).

[20] F. Riesz and B. Sz.–Nagy, *Functional analysis* (Budapest, 1953); English transl. (New York, 1955).

[21] Ju.A. Bozanov, Spectral analysis of abstract functions, *Teor. Verojatnost. i Primenen.*, 4 (1959), 291-310.

[22] M. Rosenberg, The square-integrability of matrix-valued functions with respect to a non-negative Hermitian measure, *Duke Math J.*, 31 (1964), 291-298.

[23] N. Wiener, Differential space, *J. Math. and Phys.*, 2 (1923), 131-174.

COLLOQUIA MATHEMATICA SOCIETATIS JÁNOS BOLYAI
5. HILBERT SPACE OPERATORS, TIHANY (HUNGARY), 1970

The determining function method in the treatment of commutator systems

J. D. PINCUS

INTRODUCTION

In a previous paper [1] we introduced the notion of the Determining Function of a commutator pair and studied certain of its properties. In the present paper we extend this definition somewhat and study further properties of commutator pairs in the case where the commutator is not necessarily positive.

Our object of study is the pair of bounded self-adjoint operators $\{U, V\}$ defined on a separable Hilbert space $\mathcal{H}$ and satisfying the commutator relation

$$VU - UV = \frac{1}{\pi i} C \tag{1}$$

where C is trace class and where, for the most part, we assume that U has an absolutely continuous spectral measure. *

* A parallel development to that of the present paper can be largely carried through without the assumption that U or V is bounded if the commutator identity is interpreted to mean that

$$(V-\ell)^{-1}(U-z)^{-1} - (U-z)^{-1}(V-\ell)^{-1} = \frac{1}{\pi i}(V-\ell)^{-1}(U-z)^{-1} - (U-z)^{-1}(V-\ell)^{-1}$$

The determining function of the pair $\{\mathcal{U}, \mathcal{V}\}$ is an operator valued function defined on an auxilliary Hilbert space. It characterizes the pair up to unitary equivalence and makes it possible to compute the unitary invariants of $\mathcal{V}$ given those of $\mathcal{U}$.

It also allows us to find the complete generalized eigenfunction expansion associated with the absolutely continuous part of $\mathcal{V}$ if the spectral representation of $\mathcal{U}$ is known.

The major result of this theory is the connection that is established between the Determining Function of the commutator pair and a homogeneous Riemann-Hilbert barrier relation associated with the pair.

We begin by stating the following hypothesis which, unless otherwise noted, shall be imposed for the remainder of this paper.

Hypothesis. $\mathcal{U}$ and $\mathcal{V}$ are bounded self-adjoint operators defined on a separable Hilbert space, $\mathcal{H}$. We assume that $\mathcal{U}$ has a purely absolutely continuous spectral measure *, and that $\mathcal{V}\mathcal{U} - \mathcal{U}\mathcal{V} = \frac{1}{\pi i} C$ where C is trace class on $\mathcal{H}$.** We further assume that the smallest invariant subspace for both $\mathcal{U}$ and $\mathcal{V}$ which contains the range of C is $\mathcal{H}$ itself.

Let h be a fixed ℓ_2 space whose dimension is equal to the maximum of the dimension of the range of C and the spectral multiplicity of $\mathcal{U}$.

Let the spectral representation of C be given by $C = \sum (\operatorname{sgn} c_n)\lambda_n^2 \varphi_n(\cdot, \varphi_n)$ where $\{\varphi_n\}$ denotes the orthonormalized set of eigenvectors of C, completed if necessary if C has a null space, and $c_n = (\operatorname{sgn} c_n)\lambda_n^2$ is the n th eigenvalue of C.

(We use throughout the notation $\frac{c_n}{|c_n|} = \operatorname{Sgn} c_n$ if $c_n = 0$ and $\operatorname{Sgn} c_n = 0$ if $c_n = 0$).

Let $\{\vartheta_j\}$ be a fixed complete orthonormal set in h. Define $k^*: h \to \mathcal{H}$ by setting $k^*\vartheta_j = \lambda_j \varphi_j$, and $k: \mathcal{H} \to h$ by setting $k\varphi_j = \lambda_j \vartheta_j$. We further define an (indefinite metric) operator $\operatorname{Sgn} C$ on h by setting $\operatorname{Sgn} C \vartheta_n = \operatorname{Sgn} c_n \vartheta_n$. Finally, let us introduce the operator $J: h \to h$ defined by setting $J\vartheta_n = \varepsilon_n \vartheta_n$,

* This condition can be somewhat relaxed.

** See (2) for a relaxation of the same class condition when C is positive.

where $\varepsilon_n = 1$ if $\mathrm{Sgn}\, c_n = 1$, $\varepsilon_n = i$ if $\mathrm{Sgn}\, c_n = -1$ and $\varepsilon_n = 0$ if $\mathrm{Sgn}\, c_n = 0$.

Definition. The determining function of the pair $\{U, V\}$ is defined as a map from h into h by setting

$$(2) \qquad E(\ell, z) = 1 + \frac{1}{\pi i} Jk(V-\ell)^{-1}(U-Z)^{-1}k^*J \quad \text{for all} \quad \ell, z$$

not in $\sigma(V)$, $\sigma(U)$, where 1 is the identity on h.

It is at once obvious that the closure of the range of k is an invariant subspace for $E(\ell,z)$, and we will call this subspace of h, $\not{h}$, and unless we specify the contrary we will consider the operator $E(\ell,z)$ defined above to be restricted to $\not{h}$.

Theorem 1. *Suppose* $\{U', V'\}$ *is another pair satisfying our hypothesis, with* $V'U' - U'V' = \frac{1}{\pi i} C'$. *Then* $T' \equiv U' + iV'$ *is unitarily equivalent to* $T \equiv U + iV$ *if and only if* $E'(\ell,z)$ *is unitarily equivalent on* $\not{h}'$ *to* $E(\ell,z)$ *on* $\not{h}$.

Note that this theorem asserts that if $T'T'^* - T'^*T' = C'$ and $TT^* - T^*T = C$ where C and C' belong to trace class, then $T \cong T' \Leftrightarrow E(\ell,z) \cong E'(\ell,z)$.

Proof. Simple multiplication and use of the relations

$$(3) \qquad \begin{aligned} &\frac{1}{\pi i}(V-\ell)^{-1}(U-Z)^{-1}C(U-Z)^{-1}(V-\ell)^{-1} = \\ &= (V-\ell)^{-1}(U-Z)^{-1} - (U-Z)^{-1}(V-\ell)^{-1} = \\ &= \frac{1}{\pi i}(U-Z)^{-1}(V-\ell)^{-1}C(V-\ell)^{-1}(U-Z)^{-1} \end{aligned}$$

show that the following relations are equivalent to (2)

$$(4a) \qquad Jk(V-Y)^{-1}(U-X)^{-1} = E(Y,X)Jk(U-X)^{-1}(V-Y)^{-1}$$

$$(4b) \qquad (U-X)^{-1}(V-Y)^{-1}k^*J^* = (V-Y)^{-1}(U-X)^{-1}k^*J^*E^*(Y,X).$$

Thus

$$(5) \quad (U'-X)^{-1}(V'-Y)^{-1}k'^*J'^* = (V'-Y)^{-1}(U'-X)^{-1}k'^*J'^*E'^*(X,Y).$$

By assumption there is a unitary operator S with $SU'S^{-1} = U$ and $SV'S^{-1} = V$. We now wish to show that $E'(x,y)$ is unitarily equivalent to $E(x,y)$ for all non-real x, y – this last unitary equivalence being realized by an operator that is x, y independent.

Relations (4) and (5) allow us to conclude that

(6) $$S(U'-X)^{-1}S^{-1}S(V'-Y)^{-1}S^{-1}S\,k'^*J'^* =$$
$$= S(V'-Y)^{-1}S^{-1}S(U'-X)^{-1}S^{-1}Sk'^*J'^*E'^*(X,Y).$$

Let $(k'^*)^{-1}$ denote the psuedo inverse of k'^* and multiply both sides of (6) by $(J'^*)^{-1}(k'^*)^{-1}S^{-1}k^*$ on the right. We note that the range of $S^{-1}k^*$ coincides with the range of k'^*.

Hence, since $k'^*(k'^*)^{-1}|_{R(k'^*)} = 1$, we can conclude that

(7) $$(U-x)^{-1}(V-y)^{-1}k^*J^* =$$
$$= (V-y)^{-1}(U-x)^{-1}Sk'^*J'^*E'^*(x,y)J'^{*-1}(k'^*)^{-1}S^{-1}k^*J^* =$$
$$= (V-y)^{-1}(U-x)^{-1}k^*J^*E^*(x,y).$$

Thus

(8) $$Sk'^*J'^*E'^*(x,y)\cdot J'^{-1}(k'^*)^{-1}S^{-1}k^*J^* = k^*J^*E^*(x,y).$$

Let $(k^*)^{-1}$ denote the psuedo inverse of k^* and multiply both sides of the above equation by $J(k^*)^{-1}$ on the left.

Now we assert that

(9) $$J(k^*)^{-1}k^*J^*E^*(x,y) = E^*(x,y)$$

because the range of $E^*(x,y)$ is the range of k which is certainly included within the range of $(k^*)^{-1}k^*$, and $(k^*)^{-1}k^*|_{R[(k^*)^{-1}k^*]} = 1$.

We have therefore been able to conclude that

(10) $$E^*(x,y) = J(k^*)^{-1}Sk'^*J'^*E'^*(x,y)J'(k'^*)^{-1}S^{-1}k^*J^*$$

our proof will be finished when we show that $J(k^*)^{-1}Sk'^*J'^*$ is unitary.

Since $J'^*\vartheta_n' = \varepsilon_n^*\vartheta_n'$, $k'^*\vartheta_n' = \lambda_n\varphi_n'$, $S\varphi_n' = \varphi_n'$, $(k^*)^{-1}\varphi_n = \frac{1}{\lambda_n}\vartheta_n$, and $\vartheta_n = \varepsilon_n\vartheta_n$ we see at once that $J(k^*)^{-1}Sk'^*J'^*\vartheta_{n'} = \vartheta_n$, $n = 1, 2, \ldots$ and we are done.

The converse is considerably more difficult to demonstrate.

We begin with a consequence of the hypothesis that the smallest invariant subspace for both U and V which contains the range of C is $\mathcal{H}$.

Lemma. *If* $Jk(V-\ell)^{-1}(U-z)^{-1}f = 0$ *for all* (ℓ, z) *outside* $\sigma(V)\times\sigma(U)$, *then* $f = 0$.

Proof. For arbitrary $g \in \mathcal{H}$, form

$$
\begin{aligned}
(Jk(V-\ell)^{-1}(U-z)^{-1}f,\ J^*k(V-\bar{\ell})^{-1}(U-\bar{z})^{-1}g) &= \\
&= ((U-z)^{-1}(V-\ell)^{-1}C(V-\ell)^{-1}(U-z)^{-1}f, g) = \\
&= \pi i((U-z)^{-1}(V-\ell)^{-1}f - (V-\ell)^{-1}(U-z)^{-1}f, g).
\end{aligned}
$$

By hypothesis this is zero. Hence

$$(U-z)^{-1}(V-\ell)^{-1}f = (V-\ell)^{-1}(U-z)^{-1}f.$$

Now consider the linear manifold formed from the finite linear combinations of the form

$$\sum a_{kj}(V-\ell_k)^{-1}(U-z_j)^{-1}f.$$

Call the closure of this linear manifold N_f.

N_f is invariant under both V and U because

$$(V-q)^{-1}a_{kj}(V-\ell_k)^{-1}(U-z_j)^{-1}f \in N_f$$

by the resolvent identity, and

$$
\begin{aligned}
(U-z)^{-1}a_{kj}(V-\ell_k)^{-1}(U-z_j)^{-1}f &= a_{kj}(U-z)^{-1}(U-z_j)^{-1}(V-\ell_k)^{-1} = \\
&= \frac{a_{kj}}{z-z_j}(U-z)^{-1}(V-\ell_k)^{-1}f - \frac{a_{kj}}{z-z_j}(U-z_j)^{-1}(V-\ell_k)^{-1}f =
\end{aligned}
$$

$$= \frac{a_{kj}}{z-z_j}(V-\ell_k)^{-1}(U-z)^{-1}f - \frac{a_{kj}}{z-z_j}(V-\ell_k)^{-1}(U-z_j)^{-1}f$$

is also in N_f.

But all of N_f is in the null space of k . So $CN_f = 0$.

Hence N_f is a closed subspace of $\mathcal{H}$ invariant for both V and U on which V und U commute.

If we assume – as we have done – that there are no subspaces satisfying the above condition, we must conclude that $N_f = \{0\}$ and hence $f = 0$.

Now we outline the construction of a special reproducing kernel Hilbert space $\mathcal{H}(E)$ associated with the determining function (this construction is discussed in greater detail in the appendix.)

With vectors $f \in \mathcal{H}$ we associate h-valued analytic functions $F(z_1,z_2)$ defined for $(z_1,z_2) \notin (\sigma(V),\sigma(U))$ by means of the formula

$$F(z_1,z_2) = Jk(V-z_1)^{-1}(U-z_2)^{-1}f = \mathcal{J}f .$$

The resulting linear space of h-valued analytic functions of two complex variables endowed with a scalar product derived from the reproducing kernel

$$k(w_1,w_2;z_1,z_2) =$$

$$= \frac{i}{\pi}\left\{\frac{E(z_1,z_2)\,\mathrm{Sgn}\,CE^*(w_1,\bar{z}_2) - E(z_1,\bar{w}_2)\,\mathrm{Sgn}\,CE^*(w_1,w_2)}{(z_1-\bar{w}_1)(z_2-\bar{w}_2)}\right\}$$

is called $\mathcal{H}(E)$.

It is easy to show that $k(w_1,w_2;z_1,z_2)$ is a reproducing kernel in the sense that for any $q \in h$ and non-real pair (w_1,w_2) $k(w_1,w_2;z_1,z_2)$ belongs to $\mathcal{H}(E)$ as a function of z_1 and z_2 and

$$\langle F(w_1,w_2),q\rangle_h = \langle F(t_1,t_2),k(w_1,w_2;t_1,t_2)q\rangle_{\mathcal{H}(E)}$$

holds for every $F(z_1,z_2) \in \mathcal{H}(E)$.

The lemma proved above shows that the map $\mathcal{J}: \mathcal{H} \to \mathcal{H}(E)$ is in fact an isometric isomorphism.

In the appendix we show that if

$$(F, z_1, z_2) = \mathcal{J}f, \text{ then } \frac{F(z_1, z_2) - F(z_1, w_2)}{z_2 - w_2} = \mathcal{J}(U - w_2)^{-1} f \text{ and}$$

$$\frac{F(z_1, z_2) - E(z_1, z_2)E(w_1, z_2)^{-1}F(w_1, z_2)}{z_1 - w_1} = \mathcal{J}(V - w_1)^{-1} f.$$

The main result proved in the appendix is that if $E'(\ell, z) = \mathcal{G}^{-1}E(\ell, z)\mathcal{G}$ then $E'(\ell, z)$ is the deterinining function of a commutator pair $\{U, V\}$ satisfying our hypothesis. Accordingly, since $E' = \mathcal{G}^{-1}E\mathcal{G}$, we can conclude that $k' = \mathcal{G}^{-1}k\mathcal{G}$, where k' is the reproducing kernel associated with $\{U', V'\}$. Hence $\mathcal{G}$ induces an isometric isomorphism between $\mathcal{H}(E')$ and $\mathcal{H}(E)$. We can write $\mathcal{G}^{-1}F(z_1, z_2) = F'(z_1, z_2)$, and then note that

$$\mathcal{G}^{-1}\mathcal{J}(U - w_2)^{-1}\mathcal{J}^{-1}\mathcal{G}F'(z_1, z_2) = \mathcal{G}^{-1}\left\{\frac{F(z_1, z_2) - F(z_1, w_2)}{z_2 - w_2}\right\} =$$

$$= \frac{F'(z_1, z_2) - F'(z_1, w_2)}{z_2 - w_2} = \mathcal{J}'(U' - w_2)^{-1}\mathcal{J}'^{-1}F'(z_1, z_2)$$

while

$$\mathcal{G}^{-1}\mathcal{J}(V - w_1)^{-1}\mathcal{J}^{-1}\mathcal{G}F'(z_1, z_2) =$$

$$= \mathcal{G}^{-1}\frac{F(z_1, z_2) - \mathcal{G}^{-1}E(z_1, z_2)\mathcal{G}\mathcal{G}^{-1}E(w_1, z_2)^{-1}\mathcal{G}\mathcal{G}^{-1}F(w_1, z_2)}{z_1 - w_1} =$$

$$= \mathcal{J}'(V' - w_1)^{-1}\mathcal{J}'^{-1}F(z_1, z_2).$$

Since these relations hold for arbitrary vectors $F' \in \mathcal{H}(E')$, we can conclude that

$$(\mathcal{U}' - w_2)^{-1} = (\mathcal{J}'^{-1}\mathcal{G}^{-1}\mathcal{J})(\mathcal{U} - w_2)^{-1}(\mathcal{J}^{-1}\mathcal{G}\mathcal{J}')$$
$$(V' - w_1)^{-1} = (\mathcal{J}'^{-1}\mathcal{G}^{-1}\mathcal{J})(V - w_1)^{-1}(\mathcal{J}^{-1}\mathcal{G}\mathcal{J}') .$$

Our conclusion that $\mathcal{U}'$ and $\mathcal{U}, V'$ and V are simultaneously unitarily equivalent follows at once.

The theorem above may be put into the language of algebras by asserting that the operator algebras $\mathfrak{A}$ and $\mathfrak{A}'$ generated by $\{\mathcal{U}, V\}$ and $\{\mathcal{U}', V'\}$ are unitarily equivalent if and only if the corresponding determining functions are unitarily equivalent.

Definition. We call the algebra $\mathfrak{A}$ generated by $\{\mathcal{U}, V\}$ reducible if there is a direct sum decomposition of the space $\mathcal{H} = \sum \oplus \mathcal{H}_i$ such that $\mathcal{H}_i$ is invariant for both V and $\mathcal{U}$ for each i . If we denote the restrictions to $\mathcal{H}_i$ of $V, \mathcal{U}$ and C by $V_i, \mathcal{U}_i$, and C_i , then we will have $V_i \mathcal{U}_i - \mathcal{U}_i V_i = \frac{1}{\pi i} C_i$.

Definition. If the C_i above have one dimensional range then we say that $\mathfrak{A}$ is completely reducible.

The following theorem is trivial.

Theorem 2. *$\mathfrak{A}$ is completely reducible if and only if there exists a unitary map $\mathcal{R}$: $h \to h$ such that $\mathcal{R}E(x,y)\mathcal{R}^{-1}$ is diagonal, i.e. has a matrix representation of the form $e_j(x,y)\delta_{ij}$ relative to some complete orthonormal set in h .*

Proof. We assume that $E(x,y)$ has the indicated diagonal form relative to the $\{\vartheta_n\}$ basis and we show that $\mathfrak{A}$ is completely reducible.

Let the closure of the finite linear combinations of the form

$$\sum_{i,k} (\mathcal{U} - x_i)^{-1}(V - y_k)^{-1} k^* J^* \vartheta_n$$

be called A_n.

Lemma 1. *Each A_n is invariant, under both V and $\mathcal{U}$ and A_n is orthogonal to A'_n if $n \neq n'$.*

Lemma 2. *If the restrictions to A_n of V and $\mathcal{U}$ are called respectively V_n and $\mathcal{U}_n$ then $V_n\mathcal{U}_n - \mathcal{U}_n V_n = \frac{1}{\pi i} C_n$, where $C_n = (\operatorname{Sgn}\lambda_n)\lambda_n^2 \varphi_n(\cdot, \varphi_n)$.*

Proof of lemma 2.

$$(10)\quad C(U-x)^{-1}(V-y)^{-1}k^*J^*\vartheta_n =$$

$$= C(V-y)^{-1}(U-x)^{-1}k^*J^*E^*(x,y)\vartheta_n =$$

$$= k^*\,\mathrm{Sgn}\,CK(V-y)^{-1}(U-x)^{-1}k^*J^*e_n^*(x,y)\vartheta_n =$$

$$= k^*JJk(V-y)^{-1}(U-x)^{-1}k^*J^*e_n^*(x,y)\vartheta_n =$$

$$= k^*J(\pi i)\,e_n^*(x,y)(E(x,y)-1)\,\mathrm{Sgn}\,C\vartheta_n =$$

$$= k^*J(\pi i)\,e_n^*(x,y)(e_n(x,y)-1)\,\mathrm{Sgn}\,C\vartheta_n =$$

$$= \varepsilon_n^*(\pi i)\lambda_n e_n^*(x,y)(e_n(x,y)-1)\varphi_n .$$

But

$$\mathrm{Sgn}(\lambda_n)\lambda_n^2\varphi_n((U-x)^{-1}(V-y)^{-1}k^*J^*\vartheta_n,\varphi_n) =$$

$$= \lambda_n\varphi_n((U-x)^{-1}(V-y)^{-1}k^*J^*\vartheta_n,k^*J^2\vartheta_n) =$$

$$= \lambda_n\varphi_n(J^*k(U-x)^{-1}(V-y)^{-1}k^*J^*\vartheta_n,J\vartheta_n) =$$

$$= \lambda_n\varphi_n(J^*k(V-y)^{-1}(U-x)^{-1}k^*J^*E^*(x,y)\vartheta_n,J\vartheta_n) =$$

$$= \lambda_n\varphi_n e_n^*(x,y)(\pi i)(\mathrm{Sgn}\,C(E(x,y)-1)\,\mathrm{Sgn}\,C\vartheta_n,J\vartheta_n) =$$

$$= \lambda_n\varphi_n e_n^*(x,y)(\pi i)\,\varepsilon_n^*(e_n(x,y)-1) .$$

Thus $CA_n = C_nA_n$.

Proof of lemma 1. The fact that A_n is invariant under both V and U is the assertion of lemma 2.1 of [1] without essential change.

The fact that A_n is orthogonal to A'_n if $n \neq n'$ follows from a tedious reduction of

$$((U-x')^{-1} V-y')^{-1} k^* J^* \vartheta_k , (U-x)^{-1}(V-y)^{-1} k^* J^* \vartheta_j)$$

making use of the identity (3).

Theorem 3. *If $\mathfrak{U}$ is completely reducible, the eigenvalues of $E(x,y)$ have the form*

$$e_j(x,y) = \exp\left\{\pm \frac{1}{2\pi i} \int_{\sigma(U)} \int_{\sigma(V)} g_j(v,u) \frac{dv}{v-x} \frac{du}{u-y}\right\} \tag{11}$$

where $g_j(v,u)$ is the characteristic function of some subset of $\sigma(U) \times \sigma(V)$ and the $+$ sign is taken if $c_j \geq 0$ and the minus sign is taken if $c_j \leq 0$.

This theorem follows at once from theorem 2 and theorem 7.1 of [1].

Theorem 4. *If $\mathfrak{U}$ is completely reducible both V_i and U_i have purely absolutely continuous spectrum and the spectral multiplicity of $\xi \in \sigma(V_i)$ is obtained in the following way:*

Let $\Lambda_\xi^{(i)} = \{\lambda \,\rfloor\, g_i(\xi,\lambda) = 1\}$. If $\Lambda_\xi^{(i)}$ is the union of n-disjoint intervals then the spectral multiplicity of ξ is n; otherwise it is infinite. Similarly, the spectral multiplicity of $\mu \in \sigma(U_i)$ is computed by constructing $\Omega_\mu^{(i)} = \{\nu \,\rfloor\, g_i(\nu,\mu) = 1\}$ and is equal to M if $\Omega_\mu^{(i)}$ is the union of M disjoint intervals; otherwise it is infinite.

This theorem follows at once from theorem 2 and theorem 1.1 of [1].

We now turn to various fundamental identities satisfied by the determining function

Lemma 3.

$$E^*(\bar{x},\bar{y})\, \mathrm{Sgn}\, C E(x,y) = \mathrm{Sgn}\, C. \tag{12}$$

The proof is immediate

Lemma 4. *If we define*

$$\varepsilon(\ell,z) = 1 + \frac{1}{\pi i} J^* k (V-\ell)^{-1}(U-z)^{-1} k^* J^* , \tag{13}$$

then $\operatorname{Sgn} CE(\ell, Z) = \varepsilon(\ell, Z) \operatorname{Sgn} C$.

The proof is immediate.

Theorem 5.

$$\frac{1}{\pi i}(J^* k(U-X)^{-1}(V-\ell)^{-1}(U-Y)^{-1} k^* J) = \operatorname{Sgn} C \cdot \frac{1}{x-y}(1-\varepsilon^*(\bar{\ell}, \bar{x}) E(\ell, y)). \tag{14}$$

The proof is straightforward * and we have the following corollary

$$\frac{1}{\pi}(J^* k(U-\bar{Y})^{-1}(V-\ell)^{-1}(V-\bar{\ell})^{-1}(U-Y)^{-1} k^* J) = \frac{1}{(\ell-\bar{\ell})(y-\bar{y})}(E^*(\bar{\ell}, y)(\operatorname{Sgn} CE(\ell, y) - E^*(\ell, y) \operatorname{Sgn} CE(\bar{\ell}, y)) \geq 0. \tag{15}$$

A more general result is contained in the following theorem, whose proof is also quite trivial.

Theorem 6. *The following two equations are valid:*

$$i(E(\ell_1, Z_2) \operatorname{Sgn} CE^*(\ell_2, \bar{Z}_2) - E(\ell_1, \bar{Z}_1) \operatorname{Sgn} CE^*(\ell_2, Z_1)) = \pi(\ell_1 - \bar{\ell}_2)(Z_2 - \bar{Z}_1) J k (V-\ell_1)^{-1}(U-Z_2)^{-1}(U-\bar{Z}_1)^{-1}(V-\bar{\ell}_2)^{-1} k^* J^* ; \tag{16}$$

$$i(E^*(\ell_2, \bar{Z}_2) \operatorname{Sgn} CE(\bar{\ell}_2, \bar{Z}_1) - E^*(\bar{\ell}_1, \bar{Z}_2) \operatorname{Sgn} CE(\ell_1, \bar{Z}_1)) = \pi(\ell_1 - \bar{\ell}_2)(Z_2 - \bar{Z}_1) J^* k (U-Z_2)^{-1}(V-\ell_1)^{-1}(V-\bar{\ell}_2)^{-1}(U-Z_1)^{-1} k^* J^*. \tag{17}$$

The proof is an immediate computation.

We may summarize by noting that we have shown that the determining function satisfies the following relations

* A minor modification of the corresponding proof in [1] , p. 238.

(18)

$a)\quad E^*(\bar{\ell},\bar{Z})\,\mathrm{Sgn}\,C E(\ell,Z) = \mathrm{Sgn}\,C$

$\beta)\quad \dfrac{i}{(\ell-\bar{\ell})(y-\bar{y})}\left[E^*(\bar{\ell},y)\,\mathrm{Sgn}\,CE(\ell,y) - E^*(\ell,y)\,\mathrm{Sgn}\,CE(\bar{\ell},y)\right] \geq 0$

$\gamma)\quad \| E(\ell,Z)-1 \| = O\left| \dfrac{1}{\mathrm{Im}\,\ell\ \mathrm{Im}\,z} \right|$ as $|\mathrm{Im}\,\ell\ \mathrm{Im}\,z| \longrightarrow \infty$

$\delta)\quad E(\ell,z)-1 \in C_1$

$\beta')\quad i\left\{\dfrac{E^*(\ell_2,\bar{Z}_2)\,\mathrm{Sgn}\,CE(\bar{\ell}_2,\bar{Z}_1) - E^*(\bar{\ell}_1,\bar{Z}_2)\,\mathrm{Sgn}\,CE(\ell_1,\bar{Z}_1)}{(\ell_1-\bar{\ell}_2)(Z_2-\bar{Z}_1)}\right\} \geq 0$

$\beta'')\quad i\left\{\dfrac{E(\ell_1,Z_2)\,\mathrm{Sgn}\,CE^*(\ell_2,\bar{Z}_2) - E(\ell_1,\bar{Z}_1)\,\mathrm{Sgn}\,CE^*(\ell_2,Z_1)}{(\ell_1-\bar{\ell}_2)(Z_2-\bar{Z}_1)}\right\} \geq 0 .$

We will prove $\beta'')$ only . Let

$$K(\ell_2,Z_1;\ell_1,Z_2) =$$

$$= \frac{i}{\pi}\,\frac{E(\ell_1,Z_2)\,\mathrm{Sgn}\,CE^*(\ell_2,\bar{Z}_2) - E(\ell_1,\bar{Z}_1)\,\mathrm{Sgn}\,CE^*(\ell_2,Z_1)}{(\ell_1-\bar{\ell}_2)(Z_2-\bar{Z}_1)} .$$

Let (Z_i,ℓ_i) be a finite set of non-real pairs, and let (C_i) be a set of vectors in Then, by (16)

$$(19)\quad \sum_{i,j} \bar{C}_j k(\ell_j,Z_i;Z_j,\ell_i) =$$

$$= \sum_{i,j} \bar{C}_j Jk(V-\ell_j)^{-1}(U-Z_j)^{-1}(U-\bar{Z}_i)^{-1}(V-\bar{\ell}_i)^{-1}k^*J^*C_i =$$

$$= \left\langle \sum_i (U-Z_i)^{-1}(V-\ell_i)^{-1}k^*J^*C_i,\ \sum_j (U-Z_j)^{-1}(V-\ell_j)^{-1}k^*J^*C_i \right\rangle \geq 0 .$$

These relations enable us to characterize the determining function analytically. Thus, in the appendix to this paper, we sketch the proof of

Theorem 7.*,** *An* h *-operator valued analytic function* $E(\ell,Z)$ *satisfies* α), β''), γ) *and* δ) *if and only if it is the determining function of a commutator pair.*

There are several other analytic characterizations of the determining function which bring out the fundamental connection between determining functions and factorization problems of the Riemann – Hilbert type. However these characterizations involve conditions on the determining function when one of its variable is real.

Similarly, using the next theorems we can give a much more efficient calculation of the commutator pair given the determining function than that presented in the appendix.

Theorem 8. V *is unitarily equivalent to the singular integral operator given as*

$$Lx(\lambda) = A(\lambda)x(\lambda) + \frac{1}{\pi i}\int_{\sigma(V)} \frac{k^*(\lambda)\,\mathrm{Sgn}\,C\,k(\mu)}{\mu-\lambda}\,x(\mu)\,d\mu \tag{20}$$

on a certain direct sum Hilbert space H *where* $x(\lambda) = \{x_1(\lambda),\ldots,x_p(\lambda) \in h \text{ a.a. } \lambda\}$ *and* H *consists of those* $x(\lambda)$ *with* $\|x\|_H^2 = \int_{\sigma(V)} \sum_{i=1}^{p} |x_i(\lambda)|^2\, d\lambda < \infty$.

and $A(\lambda)$ *is a bounded symmetric operator on* H, *which can be taken to be diagonal, and* $k(\lambda)$ *is a bounded operator on* H *which is Hilbert-Schmidt on* h *for almost all* λ. *Both* $A(\lambda)$ *and* $k(\lambda)$ *are measurable operator-valued functions of* λ *which are essentially bounded as functions of* λ.

We will sketch a proof of this theorem.

Let $\hat{H}$ be a minimal direct sum decomposition of $\mathcal{H}$ into invariant subspaces of U, $\mathcal{H}_i$ each generated by a cyclic vector. Choose the canonical isometry, Q from $\hat{H}$ to the appropriate direct sum of L_2 spaces, H, associated with the minimal decomposition so that

* See footnote in the appendix.

** See J. Pincus and J. Rovnyak [2] for use for the condition β).

$$Qf = \{g_1(\lambda), \dots, g_n(\lambda)\} = g(\lambda)$$

$$QUf = \{\lambda g_1(\lambda), \dots, \lambda g_n(\lambda)\}$$

where n is the total spectral multiplicity of U.

Set $Q_i f = g_i(\lambda)$, and let $\{\varphi_j\}_1^\infty$ denote the (completed) orthonormal set of eigenvectors of C, and define, for each λ, the square matrix $k^*(\lambda)$ with i,j element $(k^*(\lambda))_{ij} = \lambda_j Q_i \varphi_j$.

If we define m = dimension (range of C) then i and j range from 1 to $p = max(n,m)$ and the matrix $(k^*(\lambda))_{ij}$ is filled out with zeros where necessary, i.e. we take $\lambda_{m+1}, \dots, \lambda_p$ all to be zero and if the spectral multiplicity of λ is less than m we have $Q_i \varphi_j(\lambda) = 0$ for $i > m$.

Then $QCf = QCQ^{-1}g = \int_{\sigma(V)} k^*(\lambda) \operatorname{Sgn} C k(\mu) g(\mu) d\mu$.

Furthermore, $\sum_{i,j} \int_{\sigma(V)} |Q_j(\lambda_\ell \varphi_\ell)(\mu)|^2 d\mu = \sum |\lambda_\ell|^2 < \infty$ and hence $\sum_{i,j} \int |Q_j(\lambda_\ell \varphi_\ell)(\mu)| < \infty$ a.a. μ so that $k(\mu)$ is a Hilbert Schmidt operator in ℓ_2^p.

Lemma (a).* $\|E(\Delta)CE(\Delta)\| \le \pi \|V\|(b-a)$ *where* Δ *denotes the interval* (a,b) *and* $E(\Delta)$ *is the resolution of the identity corresponding to* U.

For,
$$\frac{1}{\pi} E(\Delta)CE(\Delta) = \int_\lambda \lambda dE_\lambda VE(\Delta) - E(\Delta)V \int_\Delta \lambda dE\lambda =$$
$$= \int_\Delta (\lambda-\alpha) dE_\lambda VE(\Delta) - E(\Delta)V \int_\Delta (\lambda - a) dE_\lambda$$

for any complex number α. Choose α to be the midpoint of Δ, and note that $\|E(\Delta)CE(\Delta)\| \le 2\pi \|\int_\Delta (\lambda-\alpha) dE_\lambda\| \, \|V\|$. But

$$\|\int_\Delta (\lambda-\alpha) dE_\lambda\| = \sup_{\lambda \in \Delta} |\lambda - \alpha| \le \frac{1}{2}(b-a).$$

* Estimates of this type have been extensively utilized by C. Putnam..

Lemma (b).

$$\|C\| = \int \|k(\lambda)\|^2_{\ell_2^p} \, d\lambda .$$

Proof. We note that $\|C\| = \lambda_\tau^2$ where $c_\tau = (\mathrm{Sgn}\, c_\tau)\lambda_\tau^2$ is the largest eigenvalue of C. Let the corresponding eigenvector of QCQ^{-1} be called $\hat{\varphi}_\tau$, i.e. $\hat{\varphi}_\tau = Q\varphi_\tau$. Then

$$\|k^*(\lambda)e_\tau\|^2_{\ell_2^p} = \langle k^*(\lambda)e_\tau, k^*(\lambda)e_\tau\rangle_{\ell_2^p} = \lambda_\tau^2 \langle \varphi_\tau(\lambda), \varphi_\tau(\lambda)\rangle_{\ell_2^p}$$

and thus $\int \|k^*(\lambda)e_\tau\|^2_{\ell_2^p} d\lambda = \|C\|$. However,

$$\int_{\sigma(V)} \|k^*(\lambda)e_\tau\|^2_{\ell_2^p} d\lambda = \int_{\sigma(V)} \|k^*(\lambda)\|^2_{\ell_2^p} d\lambda = \int_{\sigma(V)} \|k(\lambda)\|^2_{\ell_2} d\lambda .$$

Hence $\|E(\Delta)CE(\Delta)\| = \int_{\sigma(V)\cap\Delta} \|k(\lambda)\|^2_{\ell_2} d\lambda .$

Lemma (c). $\|k(\lambda)\|^2_{\ell_2}$ *is essentially bounded.*

Proof. Take $\Delta_m = [\alpha_m, \beta_m]$, then the fundamental theorem of the calculus implies that for almost every t_0

$$\lim_{m\to\infty} \frac{1}{\beta_m - \alpha_m} \int_{\alpha_m}^{\beta_m} \|k(t)\|^2_{\ell_2} dt = \|k(t_0)\|^2_{\ell_2}$$

provided that $\alpha_m < t_0 < \beta_m$ and $\lim_{m\to\infty} \alpha_m = \lim_{m\to\infty} \beta_m = t_0$. The preceeding lemmas then establish our assertion.

We can now complete the proof of the theorem.

We can now assert that the operator R defined by

$$Rx(\lambda) = \frac{1}{\pi i} \int_{\sigma(V)} \frac{k^*(\lambda)\,\mathrm{Sgn}\, C k(\mu)}{\mu - \lambda} x(\mu)\, d\mu$$ is bounded, since the

Hilbert transform is bounded. R satisfies

$$[TM-MT]x(\lambda) = \frac{1}{\pi i} \int_{\sigma(V)} k^*(\lambda)\, \mathrm{Sgn}\, Ck(\mu)x(\mu)\,d\mu$$

where $Mx(\lambda) = \lambda x(\lambda)$.

Hence if R' is another bounded operator satisfying this commutator equation then $R' - R$ will commute with M – and hence must be a bounded Borel function of λ, which we can call $A(\lambda)$.

Lemma. *The determining function of the pair* $\{M, L\}$ *on* H *has the form*

$$E(\ell, z)_{\{M,L\}} = 1 + \frac{1}{\pi i} \int_{\sigma(V)} Jk(\lambda)(L-\ell)^{-1}(M-z)^{-1}K^*(\lambda)J\,d\lambda. \tag{22}$$

The proof of this lemma is almost exactly the same as that of the corresponding result in [1].

Our fundamental result is the following theorem.

Theorem 9. $E(\ell, Z)_{\{M,L\}}$ *is the unique solution of the operator Homogeneous Riemann – Hilbert problem*

$$E(\ell, \lambda + io) = E(\ell, \lambda - io)\, \frac{1 + Jk(\lambda)(A(\lambda)-\ell)^{-1}k^*(\lambda)J}{1 - Jk(\lambda)(A(\lambda)-\ell)^{-1}k^*(\lambda)J} \tag{23}$$

$$E(\ell, z) \to 1 \text{ as } |\mathrm{Im}\,\ell\ \mathrm{Im}\,z| \to \infty$$

i.e. the unique sectionally holomorphic function which satisfies these two conditions.

The proof of this theorem is substantially the same as that of the corresponding result in [1]; however, there is one additional complication. It is necessary to show that all the partial indices associated with the factorization are zero. (See [4] for definition.) This fact may be proved in two ways: First by using results of [4] concerning the stability of partial indices and then letting $\ell \to \infty$, or, more directly, by noting that if

$$\frac{1 + Jk(\lambda)(A(\lambda)-\ell)^{-1}k^*(\lambda)J}{1 - Jk(\lambda)(A(\lambda)-\ell)^{-1}k^*(\lambda)J} = N_-(\ell, \lambda - io)^{-1}\left(\frac{\lambda - i}{\lambda + i}\right)^r N_+(\ell, \lambda + io) \tag{24}$$

where $r=(r_1, r_2, \ldots, r_n)$, and N_- and N_+ denote functions analytic in the lower and upper half planes of their second variable. Then if z is real but not in $\sigma(\mathcal{U})$

$$(25) \qquad N_-(\ell,z) = \left(\frac{z-i}{z+i}\right)^r N_+(\ell,z),$$

and we can set $\Phi^{-1}(\ell,z) = (z+i)^{-r} N_+(\ell,z) = (z-i)^{-r} N_-(\ell,z)$ and use a slightly modified version of the proof of theorem (3.1) in [1] to conclude that, in the notation of [1],

$$(26) \qquad \frac{N_+(\ell,\lambda+io)}{(\lambda+i)^r}\left[F^+ - (M-z)^{-1}G^+\right] = \frac{N_-(\ell,\lambda-io)}{(\lambda-i)^r}\left[F^- - (M-z)^{-1}G^-\right]$$

and hence by the argument in [1] that $E(\ell,z) = \Phi^{-1}(z)$. But since

$$\|\Phi^{-1}(\ell,z)\| = O\left(\frac{1}{z^r}\right) \text{ as } z \to \infty \text{ and } E(\ell,z) \to 1 \text{ as } z \to \infty,$$

we can conclude that $r = (0,0,\ldots,0)$.

We wish also to take this opportunity to note that the restriction in theorem 3.2 of [1] that λ range over a bounded set (i.e. that the operator $\mathcal{U}$ be bounded) is completely unnecessary. Thus, we can argue exactly as in [1] to prove that

$$(27) \quad -Jk(\lambda)(L-\omega)^{-1}(M-z)^{-1}k^*(\lambda)J = \frac{(M-Z)^{-1}}{2}\left[\varphi(\lambda+io) - \varphi(\lambda-io)\right]\varphi(z)^{-1}$$

and then, instead of invoking Cauchy's theorem, as was done in [1], argue that this last expression is

$$-(M-z)^{-1}Jk(\lambda)(L-\omega)^{-1}k^*(\lambda)J\varphi(z)^{-1}.$$

Hence, by equation (4b)

$$(28) \qquad Jk(\lambda)(L-\omega)^{-1}(M-z)^{-1}k^*(\lambda)J\varphi(z)\,\mathrm{Sgn}\,C =$$

$$= Jk(\lambda)(M-z)^{-1}(L-\omega)^{-1}k^*(\lambda)J^* =$$

$$= Jk(\lambda)(L-\omega)^{-1}(M-z)^{-1}k^*(\lambda)J^*E^*(\bar{\omega},\bar{z}).$$

From which it is easy to conclude that $\operatorname{Sgn} C\,\varphi(\omega,z)\operatorname{Sgn} C = E^*(\bar{\omega},\bar{z})$ and thus by equation (12) $E(\omega,z) = \varphi^{-1}(\omega,z)$.

Theorem 10. *There exists a real valued summable function* $G(\nu,\lambda)$ *such that*

(29)
$$\iint G(v,\lambda)\,dv\,d\lambda = 2\,\mathrm{Trace}\,C,$$

and

(30)
$$\text{determinant}\ \frac{1+Jk(\lambda)(A(\lambda)-\ell)^{-1}k^*(\lambda)J}{1-Jk(\lambda)(A(\lambda)-\ell)^{-1}k^*(\lambda)J} = e^{\int G(v,\lambda)\frac{dv}{v-\ell}}.$$

Proof.

(31)
$$\det(1+Jk(\lambda)(A(\lambda)-\ell)^{-1}k^*(\lambda)J) = $$
$$= \det(1+(A(\lambda)-\ell)^{-1}k^*(\lambda)\operatorname{Sgn} C\,k(\lambda)).$$

Since $\det(1+RS) = \det(1+SR)$ if RS and SR are trace class and R is completely continuous and S is bounded [3].

By a result of I.M. Livšic [5] which we use in a version due to M.G. Kreĭn [6] [7]:

(32)
$$\det(1+R(\ell)T) = e^{\int \gamma(v)\frac{dv}{v-\ell}}$$

where $\int \gamma(v)\,dv = \mathrm{Trace}\,T$ and $\int |\gamma(v)|\,dv < \infty$, provided $R(\ell)$ is the resolvent of a self-adjoint operator and T is both self-adjoint and Trace class.

A simple calculation shows that

(33)
$$\mathrm{Tr}\int Jk(\lambda)k^*(\lambda)J\,d\lambda = \mathrm{Tr}\int k(\lambda)\operatorname{Sgn} C\,k^*(\lambda)\,d\lambda = \mathrm{Tr}\,C,$$

and hence

(34)
$$2\int \mathrm{Tr}\,k(\lambda)\operatorname{Sgn} C\,k^*(\lambda)\,d\lambda = \int d\lambda \int G(v,\lambda)\,dv.^{*}$$

* We note that $G(v,\lambda)$ is obtained via the Plemelj formulae completely explicity from the argument of $\det \dfrac{1+Jk(\lambda)(A(\lambda)-\ell)^{-1}k^*(\lambda)J}{1-Jk(\lambda)(A(\lambda)-\ell)^{-1}k^*(\lambda)J}$.

Theorem 11. *

$$(35)\qquad \det E(\ell,z) = e^{\frac{1}{2\pi i}\iint G(v,u)\frac{dv}{v-\ell}\,\frac{du}{u-z}}$$

Proof.

$$(36)\qquad \log\det\left[\frac{1+Jk(\lambda)(A(\lambda)-\ell)^{-1}k^*(\lambda)J}{1-Jk(\lambda)(A(\lambda)-\ell)^{-1}k^*(\lambda)J}\right] =$$

$$= \log\det E(\ell,\lambda+io) - \log\det E(\ell,\lambda-io)$$

and $E(\ell,z)\to 1$ as $z\to\infty$. Hence $\log\det E(\ell,z) =$

$$= \frac{1}{2\pi i}\iint G(v,u)\,\frac{dv}{v-\ell}\,\frac{du}{u-z}\;.$$

Theorem 12. *There exists a unique positive trace class (on* h *) valued one parameter family of measures defined on the Borel sets of the real axis* $dR_\lambda(\cdot)$ *such that*

$$(37)\qquad \frac{1+Jk(\lambda)(A(\lambda)-\ell)^{-1}k^*(\lambda)J}{1-Jk(\lambda)(A(\lambda)-\ell)^{-1}k^*(\lambda)J} = 1+\operatorname{Sgn}C\int_{\sigma(V)}\frac{dR_\lambda(v)}{v-\ell}\;.$$

Proof. Let $\int\frac{dS_\lambda(v)}{v-\ell} = k(\lambda)(A(\lambda)-\ell)^{-1}k^*(\lambda)$. Then $[\operatorname{Sgn}C-\int\frac{dS_\lambda(v)}{v-\ell}]^{-1}$ has its imaginary part positive in the upper half plane, and hence there exists a positive measure $dr_\lambda(v)$ such that

$$\left[\operatorname{Sgn}C-\int\frac{dS_\lambda(v)}{v-\ell}\right]^{-1} = \operatorname{Sgn}C+\int\frac{dr_\lambda(v)}{v-\ell} = E_2(\ell).$$

* If the λ interval were unbounded we would have the factor $\frac{\mu z+1}{\mu-z}\,\frac{1}{\mu^2+1}$ instead of $\frac{1}{\mu-z}$.

But $E_2(\ell)^{-1} = \operatorname{Sgn} C - \int \frac{dS_\lambda(v)}{v-\ell}$, so if we set

$\operatorname{Sgn} C + \int \frac{dS_\lambda(v)}{v-\ell} = E_1(\ell)$ we must have $\operatorname{Sgn} C - E_2(\ell)^{-1} = E_1(\ell) - \operatorname{Sgn} C$

or $E_1(\ell)\, E_2(\ell) = (2\operatorname{Sgn} C - E_2(\ell)^{-1})\, E_2(\ell) =$

$$= 2\operatorname{Sgn} C E_2(\ell) - 1 = 2\operatorname{Sgn} C\left[\operatorname{Sgn} C + \int \frac{dr_\lambda(v)}{v-\ell}\right]^{-1} =$$

$$= 1 + 2\operatorname{Sgn} C \int \frac{dr_\lambda(v)}{v-\ell}$$

The indicated result follows.

This theorem combined with the fundamental theorem shows that the determining function is prescribed once the measure $dr_\lambda(\cdot)$ is specified.

Corresponding to this result (in V), we have the following (U) result.

Theorem 13. *There exists a unique one parameter family of* h *-valued positive operator measures defined on the Borel subsets of the real line* $dM_\xi(\cdot)$ *such that* $dM_\xi(\cdot)$ *is trace class on* h *and*

$$\operatorname{Sgn} C E^*(\xi - io, \bar{Z}) \operatorname{Sgn} C E(\xi - io, Z) \operatorname{Sgn} C = \tag{38}$$

$$= \operatorname{Sgn} C + \int \frac{dM_\xi(\mu)}{\mu - z} .$$

We have seen that (Theorem 6)

$$((L-\ell)^{-1} - (L-\bar{\ell})^{-1})\left(\frac{K^*(\lambda)J}{\lambda - y}\alpha, \frac{K^*(\lambda)J}{\lambda - y}\beta\right) =$$

$$= \frac{\pi i}{(\ell-\bar{\ell})(y-\bar{y})}\left[E^*(\bar{\ell},y)\operatorname{Sgn} C E(\ell,y) - E^*(\ell,y)\operatorname{Sgn} C E(\bar{\ell},y)\right]$$

and we define

(39) $$P_\zeta(x,\bar{y}) \equiv \frac{1}{2}\frac{1}{x-y}\,[E^*(\zeta-io,\bar{x})\,\mathrm{Sgn}C\,E(\zeta+io,y) -$$

$$- E^*(\zeta+io,\bar{x})\,\mathrm{Sgn}C\,E(\zeta-io,y)]$$

whenever the latter limit exists.

Then

(40) $$\varepsilon(\zeta-io,y)P_\zeta^*(x,y)\varepsilon^*(\zeta-io,x) =$$

$$= \frac{1}{2}\frac{1}{\bar{x}-y}\{\mathrm{Sgn}C\,E(\zeta-io,\bar{x})\varepsilon^*(\zeta-io,x) - \varepsilon(\zeta-io,y)E^*(\zeta-io,\bar{y})\mathrm{Sgn}C\} =$$

$$= \frac{1}{2}\frac{1}{\bar{x}-y}\{\mathrm{Sgn}C\,E(\zeta-io,\bar{x})\,\mathrm{Sgn}C\,E^*(\zeta-io,x)\,\mathrm{Sgn}C -$$

$$- \mathrm{Sgn}C\,E(\zeta-io,y)\mathrm{Sgn}C\,E^*(\zeta-io,\bar{y})\mathrm{Sgn}C\} .$$

Now set $y = x$, then

(41) $$\{\mathrm{Sgn}C\,E(\zeta-io,\bar{x})\,\mathrm{Sgn}C\,E^*(\zeta-io,x)\,\mathrm{Sgn}C]^* =$$

$$= \mathrm{Sgn}C\,E(\zeta-io,x)\,\mathrm{Sgn}C\,E^*(\zeta-io,\bar{x})\,\mathrm{Sgn}C$$

and

(42) $$0 \leq \frac{1}{\mathrm{Im}\,x}\,\mathrm{Im}(\mathrm{Sgn}C\,E(\zeta-io,\bar{x})\,\mathrm{Sgn}C\,E^*(\zeta-io,x)\mathrm{Sgn}C).$$

Thus, there exists a measure $dM_\zeta(\cdot)$ with

(43) $$\mathrm{Sgn}C\,E(\zeta-io,\bar{x})\,\mathrm{Sgn}C\,E^*(\zeta-io,x)\,\mathrm{Sgn}C =$$

$$= \mathrm{Sgn}C + \int \frac{dM_\zeta(\mu)}{\mu-\bar{x}} .$$

Theorem 14.

$$P_\xi(x,y) = \frac{1}{2}\,\varepsilon^{-1}(\xi - io, x) \int \frac{dM_\xi(\mu)}{(\mu - x)(\mu - \bar{y})}\, \varepsilon^{*-1}(\xi - io, y). \tag{44}$$

Proof.

$$\varepsilon(\xi - io, y)\, P_\xi^*(x,y)\, \varepsilon^*(\xi - io, x) =$$

$$= \frac{1}{2}\,\frac{1}{\bar{x} - y}\left\{ \mathrm{Sgn}\, C + \int \frac{dM_\xi(\lambda)}{\mu - \bar{x}} - \mathrm{Sgn}\, C - \int \frac{dM_\xi(\lambda)}{\mu - y} \right\} =$$

$$= \frac{1}{2} \int \frac{dM_\xi(\lambda)}{(\mu - \bar{x})(\mu - y)}.$$

It now follows exactly as in [1] that $dM_\xi(\cdot)$ is absolutely continuous with respect to the measure $d(\mathrm{Tr}\, M_\xi(\cdot))$. Call the resulting Radon-Nikodym derivative $M'_\xi(\mu)$ and let the j th eigenvalue of $M'_\xi(\mu)$ be denoted by $\lambda_j(\xi,\mu)$, each eigenvalue appearing in this enumeration according to its multiplicity, in such a way that

$$0 \le \ldots \le \lambda_3(\xi,\mu) \le \lambda_2(\xi,\mu) \le \lambda_1(\xi,\mu) \le 1.$$

Define, for each Borel set Δ of R, the Scalar measures

$$\mu_\xi^{(j)}(\Delta) \equiv \int_\Delta \lambda_j(\xi,\mu)\, d(\mathrm{Tr}\, M_\xi(\mu)), \tag{45}$$

and let $H_j(\xi)$ denote the space of complex-valued function on $\sigma(U)$, which are square summable with respect to $dM_\xi^{(j)}(\cdot)$.

Theorem 15. *Let* $m(\xi) = \sum_j \dim H_j(\xi)$. *Then* $m(\xi)$ *is the (von-Neumann) spectral multiplicity function for the absolutely continuous part of* V.

The proof is exactly the same as the proof of Theorem (1.4) in (1).

We would like to note that we could now write out a generalized eigenfunction expansion together with the corresponding Parsevel relations for the absolutely continuous part of V, i.e. the partial isometries which occur in the proof of this theorem have the generalized eigenfunctions as kernels when we express them as integral operators and there is no difficulty in writing them down explicitly (in terms of the determining function).

In a large number of special cases the foregoing reduction of the spectral theory of L to the calculation of the determining function enables us, as explicitly as can be expected, to diagonalize the absolutely continuous part of L .

As noted in [1], N.P. Vekua [8] has shown how $E(\ell, z)$ may be constructed explicitly if $A(\lambda)$ and $K(\lambda)$ are matrices with rational entries – or piecewise rational; while for Hölder continuous $A(\lambda)$ and $K(\lambda)$ the classical reduction due to J. Plemelj [10] of the homogeneous Riemann – Hilbert problem to a Fredholm equation holds.

For numerical purposes there are now several techniques available, and in particular the method of G. Mandshewidze [9] seems effective for the determination of the generalized eigenfunctions of L by means of the theorems given here.

If, however, less information is desired – such as the spectrum alone – then it is often possible to calculate this unitary invariant from the coefficients without solving for the determining function.

Theorem 16. *If* $C \geq 0$, *then*

$$\begin{cases} \xi \in \sigma(v) \Leftrightarrow \int G(\xi, \mu)\, d\mu \neq 0 \\ \mu \in \sigma(u) \Leftrightarrow \int G(\xi, \mu)\, d\xi \neq 0 . \end{cases}$$

Without the assumption that C is positive we can conclude less.

Theorem 17.

$$\begin{cases} \int G(\xi, \mu)\, d\mu \neq 0 \Rightarrow \xi \in \sigma(v) \\ \int G(\xi, \mu)\, d\xi \neq 0 \Rightarrow \mu \in \sigma(u) . \end{cases}$$

These results follow from a contour integration argument which we now sketch.

$$\det E(\ell,z) = e^{\frac{1}{2\pi i}\iint G(\nu,\mu)\frac{d\nu}{\nu-\ell}\frac{d\mu}{\mu-z}} = e^{\operatorname{Tr}\log E(\ell,z)} .$$

But this means that

(46) $$\operatorname{Tr}\log E(\ell,z) = \frac{1}{2\pi i}\iint G(\nu,\mu)\frac{d\nu}{\nu-\ell}\frac{d\mu}{\mu-z} .$$

For large z, $\log E(\ell,z) \sim \frac{JK}{\pi i}(V-\ell)^{-1}(U-z)^{-1}k^*J+\dots$. Thus if we integrate around the spectrum of V and take residues at infinity we will get

(47) $$\begin{aligned}\int \operatorname{Tr}\log E(\ell,z)\,dz &= 2\operatorname{Tr}(JK(V-\ell)^{-1}K^*J) = \\ &= 2\operatorname{Tr}(\operatorname{Sgn}C\,K(V-\ell)^{-1}K^*) = \\ &= \int\frac{dv}{v-\ell}\int G(v,u)\,du .\end{aligned}$$

On the other hand $\det E(\ell,z)$ has another representation obtained from the characterization of $E(\ell,z)$ as the solution of the Hilbert problem of Th. 9; namely,

(48) $$\det E(\ell,z) = e^{\frac{1}{2\pi i}\int\log\det\left(1+\int\frac{dM_v(u)}{u-z}\operatorname{Sgn}C\right)\frac{dv}{v-\ell}} .$$

Thus, we may use the Plemelj formulae to conclude that

(49) $$\int G(v,u)\frac{dv}{v-\ell} = \log\det\left[1+\operatorname{Sgn}C\int\frac{dR_\lambda(v)}{v-\ell}\right] ,$$

(50) $$\int G(v,u)\frac{du}{u-z} = \log\det\left[1+\int\frac{dM_v(u)}{u-z}\operatorname{Sgn}C\right] .$$

But

(51) $$\log\det\left[1+\int\frac{dM_v(u)}{u-z}\operatorname{Sgn}C\right] \sim \operatorname{Tr}\left[\int\frac{dM_v(u)}{u-z}\operatorname{Sgn}C+\dots\right]$$

for large z, thus by contour integration and evaluation of the residue at infinity we may conclude that the following three relations are valid.

Theorem 18.

(52) $$\mathrm{Tr}\left[\mathrm{Sgn}\, C \int dM_v(u)\right] = \int G(v,u)\,du\,,$$

(53) $$\mathrm{Tr}\left[\mathrm{Sgn}\, C \int dR_\lambda(v)\right] = \int G(v,\lambda)\,dv\,,$$

(54) $$\mathrm{Tr}\left[\mathrm{Sgn}\, C k(V-\ell)^{-1}k^*\right] = \frac{1}{2}\int \frac{dv}{v-\ell}\int G(v,u)\,du\,.$$

This theorem gives rise to a test for a number p to be an eigenvalue of V.

Theorem 19. *If* $f(v) = \int G(v,u)\,du$ *has a* δ *-function component at* $v = p$, *i.e. is the derivative of a singular measure with mass at* p, *then* p *is an eigenvalue of* V.

Theorem 20. *For fixed sufficiently large* ℓ, *i.e.* $\ell > \|V\|$, $JE^*(\ell,\bar{x})J^*$, *considered as a function of* x, *is a characteristic operator function in the sense of M.S. Livšic.*

Proof. By Theorem (5) we have

$$\frac{1}{\pi}\,\frac{x-\bar{x}}{i}\left\{J^*K(U-x)^{-1}(V-\ell)^{-1}(U-\bar{x})^{-1}K^*J\right\} = \mathrm{Sgn}\, C - E^*(\bar{\ell},\bar{x})\,\mathrm{Sgn}\, C E(\ell,\bar{x})$$

for $\ell > \|V\|$, $(V-\ell)^{-1}$ is a negative operator. Thus

α) $$(JE^*(\ell,\bar{x})J^*)\,\mathrm{Sgn}\, C\,(JE(\ell,\bar{x})J^*) \geq \mathrm{Sgn}\, C$$

when $\mathrm{Im}\, x \geq 0$.

β) $$(JE^*(\ell,\bar{x})J^*)\,\mathrm{Sgn}\, C\,(JE(\ell,\bar{x})J^*) = \mathrm{Sgn}\, C$$

when $\mathrm{Im}\, x = 0$.

Furthermore, for $|x|$ sufficiently large

$$E^*(\bar{\ell},\bar{x}) = 1 - \frac{1}{\pi i} J^*K(U-x)^{-1}(V-\ell)^{-1}K^*J^* \sim 1 + i\,\frac{J^*TJ}{x} + \dots$$

where $T \equiv -\frac{1}{\pi} k(V-\ell)^{-1}k^*$ is a non-negative Hermitian operator. Hence

γ) $$JE^*(\bar{\ell},\bar{x})J^* \sim 1 + i\frac{T}{x}\,\mathrm{Sgn}\,C + \ldots$$

Let $N\,\mathrm{Sgn}\,C = -\mathrm{Sgn}\,C$, then we have shown that

α') $$(JE^*(\ell,\bar{x})J^*)N\,\mathrm{Sgn}\,C(JE(\ell,\bar{x})J^*) \leq N\,\mathrm{Sgn}\,C.$$

By a result of Brodskiĭ [16], generalising a result of Livšic and Brodskiĭ, we can conclude that

$$JE^*(\bar{\ell},\bar{z})J^* = \overset{\curvearrowleft}{\int_{\sigma(U)}} e^{\frac{1}{2\pi i}\left(\int \frac{dT_\lambda(V)}{V-\ell} N\,\mathrm{Sgn}\,C\right)\frac{d\mu}{\mu - z}} .$$

where $dT_\lambda(V)$ is a positive trace class on h valued measure.

Hence

$$E(\ell,z) = J^* \overset{\curvearrowright}{\int_{\sigma(U)}} e^{\frac{1}{2\pi i}\left(\mathrm{Sgn}\,C\int \frac{dT_\lambda(V)}{V-\ell}\right)\frac{d\mu}{\mu - z}} J.$$

We know that

$$\det E(\ell,z) = e^{\frac{1}{2\pi i}\iint G(\nu,\mu)\frac{d\nu}{\nu-\ell}\frac{d\mu}{\mu-z}}$$

and $\int G(\nu,\mu)\frac{d\nu}{\nu-\ell} = \mathrm{trace}\left[\mathrm{Sgn}\,C\int \frac{dT_\mu(\nu)}{\nu-\ell}\right.$.

Let $T'_\lambda(\nu)$ denote the Radon-Nikodym derivative of $dT_\lambda(\nu)$ with respect to its own trace. Then

$$\mathrm{Trace}\left[\mathrm{Sgn}\,C\int \frac{dT_\mu(\nu)}{\nu-\ell}\right] = \int \frac{\mathrm{Tr}[\mathrm{Sgn}\,C\,T'_\mu(\nu)]}{\nu-\ell}\, d\mathrm{Tr}[T_\mu(\nu).$$

Hence we can conclude that $d\mathrm{Tr}[T_\lambda(\nu)]$ is absolutely continuous with respect to Lebesgue measure. In fact

$$G(\nu,\mu) = \mathrm{Tr}[\mathrm{Sgn}\,C\,T'_\mu(\nu)]\frac{d\mathrm{Tr}[T_\mu(\nu)]}{d\nu}.$$

We have sketched a demonstration of

Theorem 21. *There exists a positive trace class-valued on* h *function,* $B(\nu,\mu)$, *such that*

$$E(\ell,z) = J^* \overset{\curvearrowright}{\int_{\sigma(U)}} e^{\frac{1}{2\pi i}(\operatorname{Sgn} C \int B(\nu,\mu)\frac{d\nu}{\nu-\ell}\frac{d\mu}{\mu-z}} J .$$

In a similar way we can deduce the dual version of Theorem 21 in which the ordered product is taken over the spectrum of V.

These results and their connections with other investigations will be discussed elsewhere.

Here we note only that if $B(\nu,\mu)$ permits a block diagonalization independent of ν and μ then the algebra generated by U and V reduces.

Closing remarks. 1) The present results are perhaps applicable to perturbation theory, and relations between the S-matrix of the perturbed operator $H = H_0 + V$ where V is trace class will be studied in another paper. Of course the most interesting results here correspond to the fact that we may, in principle, discuss a much wider class of perturbations.

Our major interest is that the present theory seems to be applicable to the treatment of certain quantum mechanical systems which "almost" admit a symmetry.

2) Certain non-normal operators, for example those with $\frac{A-A^*}{i} \geq 0$ in case they have absolutely continuous spectrum (Livšic [11], Sahnovič [12]) can be described in terms of determining functions. It is known [11] that in this case A is unitarily equivalent to $\tilde{A}$ given as

$$Af(x) = xf(x) + i\int_a^x \beta(x)\beta(t)f(t)\,dt ,$$

where $f \in L^2 m(a,b)$ and $\beta(x) \geq 0$ with $\int_a^b \operatorname{Tr} \beta^2(x)\,dx < \infty$.
But $\tilde{A}$ may be written in the form $\tilde{U} + i\tilde{V}$ where

$$\tilde{U}f(x) = xf(x) + \frac{1}{2}\left(\int_a^x \beta(x)\beta(t)f(t)\,dt - \int_x^b \beta(x)\beta(t)f(t)\,dt \right)$$

$$\tilde{V}f = \int_a^b \beta(x)\beta(t)f(t)\,dt .$$

3) Singular integral operators on the unit circle corresponding to the case where, say, U is unitary are also treated immediately by the present results. See [13] for the case of a commutator with one dimensional range.

4) *The support of* $G(\nu,\mu)$ *and the spectrum of* $U + iV$. We give below some preliminary arguments leading to the *conjecture* that $\sigma(U+iV)$ is the essential support of $G(\nu,\mu)$, when the commutator C is semidefinite.

We consider the case where: $\sigma(U) = [a,b]$ and U is of simple multiplicity, C has one dimensional range, and V corresponds (as in Theorem [8]) to a singular integral operator

$$A(\lambda)x(\lambda) + \frac{1}{\pi i}\int_a^b \frac{k^*(\lambda)k(\mu)}{\mu-\lambda}\,x(\mu)\,d\mu$$

with continuous coefficients $A(\lambda)$ and $k(\lambda)$.

In this case we know [17] that the essential spectrum of $U + iV$ is the periphery of a certain curvilinear quadrilateral which consists of the two curves

$$\begin{aligned} A_1B_1 &: (\lambda + iA(\lambda)) + i|k(\lambda)|^2 \\ A_2B_2 &: (\lambda + iA(\lambda)) - i|k(\lambda)|^2 \end{aligned} \qquad (\lambda \in \sigma(U) = (a,b))$$

together with the rectilinear segments

$$A_1 = (a + iA(a)) + i|k(a)|^2$$

$$B_1 = (b + iA(b)) - i|k(b)|^2 .$$

On the other hand we know that the Weyl spectrum of $U+iV \equiv N$, $w(N)$ which is defined to be $w(N) = \cap\, \sigma(N+K)$ where the intersection is taken over all compact operators K, is characterized as follows [18]:

$$w(N) = \{\ell \mid N-\ell \notin \mathfrak{F}\} \cup \{\ell \mid N-\ell \in \mathfrak{F} \text{ and } i(N-\ell) \neq 0\}$$

where $\mathfrak{F}$ denotes the Fredholm operators and

$$i(N) = \dim N(N) - \dim R(N)^{\perp}.$$

The set $\{\ell \mid N-\ell \notin \mathfrak{F}\}$ is just the curvilinear quadrilateral discussed above, and we recall a theorem of L. Coburn [19] generalizing a classical result of H. Weyl:

If $N^*N \geq NN^*$, then $w(N)$ consists precisely of all points in $\sigma(N)$ except the isolated eigenvalues of finite multiplicity.

Furthermore, we note that the curvilinear quadrilateral above is just the topological boundary of the determining set since

$$G(\nu,\mu) = \frac{1}{\pi} \arg \frac{A(\mu)-\nu-io-|k(\mu)|^2}{A(\mu)-\nu-io+|k(\mu)|^2}.$$

But, by the classical Noether characterization of index in terms of winding number [20], the set

$$\{\lambda \mid N-\lambda \in \mathfrak{F} \text{ and } i(N-\lambda) \neq 0\}$$

is the interior of the quadrilateral $A_1B_1A_2B_2$.

Similar arguments extend to the case where $\mathcal{U}$ has finite spectral multiplicity, but the general situation in which the coefficients may not be smooth remains open.

APPENDIX

Theorem.* *An* h *-operator valued analytic function* $E(\ell,z)$ *satisfies* α), β''), γ) *and* δ) *if and only if it is the determining function of a commutator pair.*

* In a special situation this theorem was discovered by the present author and more generally, but still assuming $C \geq 0$, by L. deBranges (private letter) as a follow up to conversations held some time before. I have also benefitted from conversations with J. Rovnyak which were devoted to the case $C \geq 0$.

We have seen that the conditions $a) - \delta)$ are satisfied by the determining function of a commutator pair. We now show how it is possible given a function $E(\ell,z)$ satisfying these conditions to construct a pair of operators H_1 and H_2 defined on a Hilbert space $\mathcal{H}(E)$, associated with $E(\ell,z)$ so that $E(\ell,z)$ will be the determining function of the pair H_1 and H_2.

From (β") it follows immediately in a standard way that $k(\omega_1,\omega_2;z_1,z_2)$ is the reproducing kernel of a Hilbert space $\mathcal{H}(E)$ of h-valued analytic functions $F(z_1,z_2)$, which are defined when z_1 and z_2 are not real, such that for any vector $q \in h$ and non-real numbers ω_1 and ω_2 $k(\omega_1,\omega_2;z_1,z_2)$ belongs to $\mathcal{H}(E)$ as a function of z_1 and z_2 and

$$\langle F(\omega_1,\omega_2), q\rangle_h = \langle F(t_1,t_2), k(\omega_1,\omega_2;t_1,t_2)q\rangle_{\mathcal{H}(E)}$$

holds for every $F(z_1,z_2)$ in $\mathcal{H}(E)$.

Lemma. *The map* $\vartheta: \mathcal{H} \to \mathcal{H}(E)$ *given by*

$$\vartheta: f \to F(z_1,z_2) = JK(V-z)^{-1}(U-z_2)^{-1}f$$

is a partial isometry of $\mathcal{H}$ *onto* $\mathcal{H}(E)$ *whose kernel is the largest closed subspace of* $\mathcal{H}$ *which reduces both* U *and* V *and on which* U *and* V *commute.*

The kernel is easily computed from the fact that $F(z_1,z_2) = 0$ for all z_1 and z_2 implies that $f = 0$ provided there is no invariant subspace of the pair U and V on which they commute. (Under this assumption, which we adopt now for simplicity, ϑ is one to one.)

The set $\mathcal{L}$ of functions $F(z_1,z_2)$ obtained as the image of $\mathcal{H}$ under the transformation ϑ is made into a Hilbert space in the norm which makes the one to one correspondence with $\mathcal{H}$ isometric.

Thus

$$\langle F,G\rangle_{\mathcal{L}} \equiv (f,g)_{\mathcal{H}} = \langle F(z_1,z_2),\ k(\omega_1,\omega_2;z_1,z_2)q\rangle_{\mathcal{H}(E)} =$$

$$= \langle F(\omega_1,\omega_2),q\rangle_h = (f,(U-\bar{\omega}_2)^{-1}(V-\bar{\omega}_1)^{-1}k^*J^*q)_{\mathcal{H}}$$

if $g = (U-\bar{\omega}_2)^{-1}(V-\bar{\omega}_1)^{-1}k^*J^*q$, and

$$G(z_1,z_2) = K(\omega_1,\omega_2;z_1,z_2)q,$$

Lemma. *Let the map* $\mathcal{U}(\omega_2)\colon \mathcal{H}(E) \to \mathcal{H}(E)$ *be defined by*

$$\mathcal{U}(\omega_2)G(z_1,z_2) \equiv G(z_1,z_2) + (\omega_2 - \bar{\omega}_2)\frac{G(z_1,z_2) - G(z_1,\omega_2)}{z_2 - \omega_2},$$

for a non-real ω_2. Then $\mathcal{U}(\omega_2)$ extends by continuity and linearity to an everywhere defined isometry in $\mathcal{H}(E)$ whose adjoint is $\mathcal{U}(\bar{\omega}_2)$; moreover, $\mathcal{U}(\omega_2)$ is unitary.

The proof consists in first verifying that

$$\frac{[K(\alpha_1,\alpha_2;z_1,z_2) - K(\alpha_1,\alpha_2;z_1,\omega_2)]}{z_2 - \omega_2} = \frac{[K(\alpha_1,\alpha_2;z_1,z_2) - K(\alpha_1,\omega_2;z_1,z_2)]}{\bar{\alpha}_2 - \omega_2}$$

and then verifying by an additional tedious calculation, directly from the definitions, that

$$(\mathcal{U}(\omega_2)F, \mathcal{U}(\omega_2)G)_{\mathcal{H}(E)} = (F,G)_{\mathcal{H}(E)}.$$

By the usual theory of Cayley transforms it can now be shown that there is a unique self-adjoint operator H_2 in $\mathcal{H}(E)$ such that

$$\mathcal{U}(\omega_2) = (H_2 - \omega_2)(H_2 - \omega_2)^{-1}.$$

Then by the definition of $\mathcal{U}(\omega_2)$ we have

$$(H_2 - \omega_2)^{-1}F(z_1,z_2) = \frac{F(z_1,z_2) - F(z_1,\omega_2)}{z_2 - \omega_2}.$$

A similar calculation shows that there is a unique self-adjoint transformation H_1 in $\mathcal{H}(E)$ such that

$$(H_1 - \omega_1)^{-1}F(z_1,z_2) = \frac{[F(z_1,z_2) - E(z_1,z_2)E(\omega_1,z_2)^{-1}F(\omega_1,z_2)]}{z_1 - \omega_1}$$

We do not wish to go through all of the verification of this assertion here, but we will show how these formulae arise when it is known that $E(\ell,z)$ is the determining function of a pair

In this case, if we let $g = (\mathcal{U} - \omega_2)^{-1}f$ where f is any element of $\mathcal{H}$, and set $F(z_1,z_2)$ and $G(z_1,z_2)$ to be the images of f and g under ϑ in $\mathcal{H}(E)$, we will have

$$G(z_1,z_2) = Jk(V-z_1)^{-1}(U-z_2)^{-1}(U-\omega_2)^{-1}f =$$

$$= Jk(V-z_1)^{-1}[(U-z_2)^{-1} - (U-\omega_2)^{-1}]f/(z_2-\omega_2) =$$

$$= (F(z_1,z_2) - F(z_1,\omega_2))]/(z_2-\omega_2) .$$

If we now let $g=(V-\omega_1)^{-1}f$ and let $F(z_1,z_2)=\vartheta f$, $(z_1,z_2)=\vartheta g$, then we can use (4a) to conclude that

$$G(z_1,z_2) = Jk(V-z_1)^{-1}(U-z_2)^{-1}(V-\omega_1)^{-1}f =$$

$$= E(z_1,z_2)Jk(U-z_2)^{-1}(V-z_1)^{-1}(V-\omega_1)^{-1}f =$$

$$= E(z_1,z_2)Jk(U-z_2)^{-1}[(V-z_1)^{-1} - (V-\omega_1)^{-1}]f/(z_1-\omega_1) =$$

$$= [Jk(V-z_1)^{-1}(U-z_2)^{-1} - E(z_1,z_2)E(\omega_1,z_2)^{-1}Jk(V-\omega_1)^{-1}(U-z_2)^{-1}]f/(z_1-\omega_1) =$$

$$= [F(z_1,z_2) - E(z_1,z_2)E(\omega_1,z_2)^{-1}F(\omega_1,z_2)]/(z_1-\omega_1) .$$

Define $k_{\mathcal{H}(E)}$: $\mathcal{H}(E) \longrightarrow h$ by setting

$$k_{\mathcal{H}(E)}G = J^*(\mathrm{Sgn}C - \mathrm{Sgn}CE^*(\omega_1,\omega_2))q$$

when $G(z_1,z_2)=k(\omega_1,\omega_2;z_1,z_2)q$ (and extending by linearity). The adjoint of the resulting transformation maps h into $\mathcal{H}(E)$ and is given by

$$k^*_{\mathcal{H}(E)}q = [\mathrm{Sgn}C - E(\omega_1,\omega_2)\mathrm{Sgn}C]Jq .$$

Lemma $k^*_{\mathcal{H}(E)}$ *defines a continuous transformation from* h *into* $\mathcal{H}(E)$ *and the commutator of* H_1 *and* H_2 *is* $k^*_{\mathcal{H}(E)}\mathrm{Sgn}Ck_{\mathcal{H}(E)}$; *furthermore,* $k^*_{\mathcal{H}(E)}\mathrm{Sgn}Ck_{\mathcal{H}(E)}$ *is trace class on* $\mathcal{H}(E)$ *and* $E(z_1,z_2) = 1+\frac{1}{i}Jk_{\mathcal{H}(E)}(H_1-z_1)^{-1}(H_2-z_2)^{-1}k_{\mathcal{H}(E)}J$.

The boundedness of $k^*_{\mathcal{H}(E)}$ is established by noting first that

$$\lim_{v_1,v_2\to\infty} -iv_1v_2k(iv_1,iv_2,z_1,z_2)q = (\mathrm{Sgn}C - E(z_1,z_2)\mathrm{Sgn}C)q$$

in the sense of pointwise convergence with respect to the norm topology in h , and then noting that

$$\lim_{v_1,v_2\to\infty} \| v_1 v_2 k(iv_1, iv_2 t_1, t_2) q \| < \infty$$

so that $(\operatorname{Sgn} C - E(z_1,z_2)\operatorname{Sgn} C)q$ belongs to $\mathcal{H}(E)$.

We omit the tedious verification that the commutator of H_1 and H_2 is as stated, and we show that

$$E(\omega_1,\omega_2) = 1 + \frac{1}{i} J k_{\mathcal{H}(E)} (H_1-\omega_1)^{-1}(H_2-\omega_2)^{-1} k^*_{\mathcal{H}(E)} J .$$

Indeed, for any vector q , we have

$$iJk_{\mathcal{H}(E)}(H_1-\omega_1)^{-1}(H_2-\omega_2)^{-1} k^*_{\mathcal{H}(E)} Jq =$$

$$= iJk_{\mathcal{H}(E)}(H_1-\omega_1)^{-1}(H_2-\omega_2)^{-1}[1-E(z_1,z_2)]\operatorname{Sgn} C J^2 q =$$

$$= iJk_{\mathcal{H}(E)}(H_1-\omega_1)^{-1} \frac{E(z_1,z_2)-E(z_1,\omega_2)}{z_2-\omega_2} q =$$

$$= -iJk_{\mathcal{H}(E)} \frac{E(z_1,z_2)\operatorname{Sgn} CE^*(\omega_1,z_2) - E(z_1,\omega_2)\operatorname{Sgn} CE^*(\omega_1,\omega_2)}{(z_1-\omega_1)(z_2-\omega_2)} \operatorname{Sgn} CE(\omega_1,\omega_2) q =$$

$$= -Jk_{\mathcal{H}(E)} k(\omega_1,\omega_2,z_1,z_2)\operatorname{Sgn} CE(\omega_1,\omega_2) q =$$

$$= -JJ^*(\operatorname{Sgn} C - \operatorname{Sgn} CE^*(\omega_1,\omega_2))\operatorname{Sgn} CE(\omega_1,\omega_2) q = -(E(\omega_1,\omega_2)-1)q .$$

Note added in proof. The author has now proved that the conjecture made in this paper about the spectrum of $T = U + iV$ is correct.

In "The spectrum of seminormal operators" (*Proc. Nat. Acad. Sci. U.S.A.* (1970)) it is proved that if $T = U + iV$ is completely non-normal and $i(VU - UV)$ semidefinite and of trace class then $\sigma(T) = D(U,V)$, where $D(U,V)$ is the closure of the support of any almost everywhere determined version of the function $G(v,u)$ defined in equation (30).

REFERENCES

[1] J.D. Pincus, Commutators and Systems of Singular Integral Equations. I, *Acta Math.*, 121 (1968), 219-249.

[2] J.D. Pincus and J. Rovnyak, A Spectral Theory for some unbounded self-adjoint singular integral operators, *Amer. J. of Math.*, XCI (1969), 619-636.

[3] I.C. Gohberg and M.G. Kreĭn, Theory of linear non-self adjoint operators, *A. M. S. Transl.*, 1969.

[4] I.C. Gohberg and M.G. Kreĭn, Systems of integral equations, *A. M. S. Transl.*, Sec. 2, 14, 217-287.

[5] I.M. Livšic, Some problems of the Dynamic Theory of non-ideal crystal Lattices, Suppl. to vol. III, ser. X, *Nuovo Cinento* (1956), 716-734.

[6] M.G. Kreĭn, Perturbation determinants and a formula for the traces of unitary and self-adjoint operators, *Dokl. Akad. Nauk. SSSR*, 144 (1962), 268-271.

[7] M.G. Kreĭn, On the trace formula in perturbation theory, *Mat. Sbornik*, 33 (75) (1953), 597-626.

[8] N.P. Vekua, *Systems of singular integral equations*, (Groningen, 1967).

[9] G.F. Mandshewidze, see reference [5] and [6] under Mandshewidze listed by N.I. Muskhelishvili for these references.

[10] N.I. Muskhelishvili, *Singulare Integral gleichungen* (Berlin 1965).

[11] M.S. Livšic, On spectral decomposition of a linear non-self adjoint operator, *A.M.S. Transl.*, (2) 5 (1957), 67-114.

[12] L.A. Sahnovič, On dissipative operators with absolutely continuous spectrum. *Trans. Moscow Math. Society*, 19 (1968).

[13] J.D. Pincus, Singular Integral operators on the unit circle, *Bull. Amer. Math Soc.*, 73 (1967), 195-199.

[14] M.S. Brodskiĭ and M.S. Livšic, Spectral analysis of non-self adjoint operators and intermediate systems, *A. M. S. Transl., ser.* 2, V. 13, p. 265.

[15] J.D. Pincus, Commutators and Systems of Singular Integral Equations. II (in preparation).

[16] M.S. Brodskiĭ, A multiplicative representation of certain analytic operator functions, *Soviet Math. Dokl.*, 2 (1961), 695-698.

[17] R. Denčev, Spectrum of singular integrals on domains with a boundary, *Soviet Math. Dokl.*, 10 (1969), 773-775.

[18] M. Schecter, Invariance of the essential spectrum, *Bull. Amer. Math. Soc.*, 71 (1965), 365-367.

[19] L. Coburn, Weyl's theorem for non-normal operators, *Michigan Math. J.* 13 (1966), 285-288.

[20] D.P. Rolewicz and S. Rolewicz, *Equations in linear spaces* (Warsaw 1968).

COLLOQUIA MATHEMATICA SOCIETATIS JÁNOS BOLYAI
5. HILBERT SPACE OPERATORS. TIHANY (HUNGARY). 1970

Problems on invariant subspaces and operator algebras

P. ROSENTHAL

1. INTRODUCTION

Von Neumann algebras are determined by their reducing subspaces; i.e., if $\mathfrak{A}$ and $\mathfrak{L}$ are von Neumann algebras which have the same family of reducing subspaces, then $\mathfrak{A} = \mathfrak{L}$. More generally, if $\mathfrak{A}$ is a von Neumann algebra, and if every reducing subspace of $\mathfrak{A}$ reduces the operator A, then $A \in \mathfrak{A}$.

In the past few years there has been work by a number of people investigating the circumstances under which similar results hold for invariant subspaces and weakly closed (not self-adjoint) operator algebras. My work on these problems has been in collaboration with Heydar Radjavi. Below we summarize the results that have been obtained and mention a number of unsolved problems. We consider transitive, Hermitian, reflexive and triangular operator algebras, and some questions relating to commutants of weakly closed algebras.

2. TRANSITIVE OPERATOR ALGEBRAS

An algebra of bounded linear operators on a Hilbert space $\mathcal{H}$ is *transitive* if the only subspaces which are invariant under all the operators in the algebra are $\{0\}$ and $\mathcal{H}$. A very well-known unsolved problem, which might be called the *transitive algebra problem,* asks: if $\mathfrak{A}$ is a weakly closed transitive operator algebra on $\mathcal{H}$ must $\mathfrak{A}$ be $\mathcal{B}(\mathcal{H})$, the algebra of all operators on $\mathcal{H}$? Burnside's Theorem [10, p. 276] states that the answer is yes if $\mathcal{H}$ is finite-dimensional. Arveson initiated the study of transitive algebras on Hilbert space in his elegant paper [1], in which he proved that the answer to the transitive algebra problem is affirmative if it is assumed that $\mathfrak{A}$ contains a m.a.s.a. (maximal abelian self-adjoint algebra) or that $\mathfrak{A}$ contains the unilateral shift. An affirmative answer has also been obtained in the cases where $\mathfrak{A}$ contains a Donoghue operator [12], or any non-zero finite-rank operator [12], or a unilateral shift of finite-multiplicity [13], or a Hermitian operator of finite uniform multiplicity [7].

Since an affirmative answer to the transitive algebra problem would be a much stronger result than an affirmative answer to the question of whether or not every operator has an invariant subspace, it is very likely that there does exist a transitive algebra other than $\mathcal{B}(\mathcal{H})$. Nordgren [13] has observed that the general transitive algebra problem is equivalent to the problem with the additional hypothesis that $\mathfrak{A}$ contains a unilateral shift of infinite-multiplicity. Similar considerations show that it would be sufficient to prove the result with the additional hypothesis that $\mathfrak{A}$ contains a non-trivial algebraic operator, or a non-trivial projection. Also, an affirmative answer to the problem on non-separable spaces would imply an affirmative answer on separable spaces, (see the remarks in the introduction to [16]), unlike the invariant subspace problem.

The basic technique that has been used to obtain the above special cases of the transitive algebra problem is Arveson's reduction to the problem of proving that certain densely defined operators commuting with $\mathfrak{A}$ are multiples of the identity, (see also the Theorem of [12]). The assumption that $\mathfrak{A}$ contains a certain operator is then used in getting information about densely defined operators commuting with $\mathfrak{A}$. Other special cases where one could hope to use this technique include the case where $\mathfrak{A}$ contains any weighted shift; (the result that the commutant of a weighted shift is the weakly closed algebra generated by it [24] might be useful here). Another case, suggested by Eric Nordgren, might be where $\mathfrak{A}$ contains the Volterra operator (or, more generally, the compressions S of the unilateral shift studied by Sarason [23]).

The case where $\mathfrak{U}$ contains a non-zero compact operator must be very difficult – its corollaries would include a very powerful theorem on existence of invariant subspaces (see the remarks in [12]).

Theorems giving sufficient conditions that a transitive algebra be $\mathcal{B}(\mathcal{H})$ can have interesting consequences. For example, it follows from Arveson's Theorem on algebras containing a m.a.s.a. that to any operator A there corresponds a compact Hermitian operator K such that the weakly closed algebra generated by A and K is $\mathcal{B}(\mathcal{H})$, (see [15]). Also, it is an immediate corollary of Nordgren's Theorem on algebras containing finite-multiplicity shifts that every operator that can be written as an $n \times n$ matrix whose entries are analytic Toeplitz operators has a non-trivial hyperinvariant subspace.

3. HERMITIAN OPERATOR ALGEBRAS

An algebra of operators is said to be *Hermitian* if it is weakly closed, contains the identity, and has the property that every invariant subspace is reducing (i.e., whenever $\mathfrak{M}$ is invariant under the algebra so is $\mathfrak{M}^{\perp}$). A question raised in [16], which might be called the ***generalized transitive algebra problem,*** asks: is every Hermitian operator algebra self-adjoint? An affirmative answer to this question would immediately imply an affirmative solution to the transitive algebra problem. It is shown in [16] that every Hermitian algebra of operators on a finite-dimensional space is self-adjoint, and that the answer to the generalized transitive algebra problem is yes under the additional hypothesis that the algebra contains a m.a.s.a. Earlier, Sarason [22] showed that a Hermitian algebra of normal operators is self-adjoint. There are a number of open problems about Hermitian operator algebras which seem tractable. Must a Hermitian algebra containing a finite-multiplicity unilateral shift, or a Hermitian operator of finite uniform multiplicity, be self-adjoint? Must the restriction of a Hermitian algebra to a reducing subspace be weakly closed?

It is likely, of course, that the answer to the generalized transitive algebra problem is negative in general. A counter-example might be a very good beginning on a counter-example to the transitive algebra problem, and perhaps also on a counter-example to the invariant subspace problem. The recent very interesting result of Dyer and Porcelli [8] can be re-phrased (using the above-mentioned result of [22]): every operator has a non-trivial invariant subspace if and only if every singly generated Hermitian operator algebra is self-adjoint.

There is an algebraic problem related to the generalized transitive algebra problem which seems interesting. It is well known that Burnside's Theorem in the finite-dimensional case has an infinite-dimensional analogue: if $\mathfrak{U}$ is an algebra of operators that has no invariant linear manifolds then $\mathfrak{U}$ is strictly dense (see [17]). In particular, a weakly closed algebra of operators which has no invariant linear manifolds is $\mathcal{B}(\mathcal{H})$. Now the algebraic version of the generalized transitive algebra problem would be: if $\mathfrak{U}$ is an algebra of operators, and $\mathfrak{U}^* = \{A^*: A \in \mathfrak{U}\}$, and if every linear manifold invariant under $\mathfrak{U}$ is invariant under $\mathfrak{U}^*$, must the strict closure of $\mathfrak{U}$ contain $\mathfrak{U}^*$? In particular, is a Hermitian operator algebra which has the property that every invariant linear manifold is invariant under $\mathfrak{U}^*$ necessarily self-adjoint? The answer must be yes.

4. REFLEXIVE OPERATOR ALGEBRAS

An operator algebra $\mathfrak{U}$ is reflexive if it is determined by its invariant subspaces as von Neumann algebras are; i.e., if $\mathfrak{U} = \{ B : B$ leaves invariant all the invariant subspaces of $\mathfrak{U} \}$. Obviously every reflexive algebra is weakly closed and contains the identity operator. Ringrose ([18], [19]) has studied properties of reflexive algebras whose invariant subspaces are totally ordered (he called such algebras *nest algebras*). The transitive algebra problem can be re-phrased: is every weakly closed transitive algebra reflexive? The generalized transitive algebra problem is equivalent to the question: is every Hermitian algebra reflexive?

There are several known sufficient conditions that an algebra be reflexive. The first theorem along these lines was Sarason's elegant result [22] that every weakly closed algebra of normal operators which contains the identity is reflexive. Sarason [22] also showed that the weakly closed algebra generated by any analytic Toeplitz operator is reflexive.

Arveson's Theorem on transitive algebras containing a m.a.s.a. was generalized in [14] to the theorem: every weakly closed operator algebra which contains a m.a.s.a., and whose lattice of invariant subspaces is totally ordered, is reflexive. This result gives a concrete description of the operators which have given chains of invariant subspaces. For example, if $\mathcal{H} = \mathcal{L}^2(0,1)$, and if $\mathfrak{M}_\alpha = \{f \in \mathcal{H} : f = 0$ a.e. on $[0,\alpha]\}$, then $\{A \in \mathcal{B}(\mathcal{H}) : A\mathfrak{M}_\alpha \subset \mathfrak{M}_\alpha$ for all $\alpha\}$ is the weakly closed algebra generated by the Volterra operator and multiplication by the independent variable.

None of the other known results in the transitive case generalize to the reflexive totally ordered case in the same way as Arveson's result does. For example, if $\mathfrak{A}$ is the set of all operators which have lower triangular matrices (with respect to a given o.n. basis) that are constant on the main diagonal (i.e., $(Ae_i,e_i) = (Ae_j,e_j)$ whenever e_i and e_j are basis vectors), then it is easily verified that the invariant subspaces of $\mathfrak{A}$ are exactly those exhibited by the triangular form. Thus $\mathfrak{A}$ has a discrete totally ordered lattice of invariant subspaces and is not reflexive (*every* lower triangular matrix leaves invariant all of $\mathfrak{A}$'s invariant subspaces). Note that $\mathfrak{A}$ contains the unilateral shift and many finite-rank operators. Also $\mathfrak{A}^*$ (which is not reflexive since $\mathfrak{A}$ isn't), contains all Donoghue operators. For another example of a non-reflexive algebra let $\mathcal{L}$ denote the algebra of all operators of the form $\begin{pmatrix} A & B \\ 0 & A \end{pmatrix}$ such that $A, B \in \mathcal{B}(\mathcal{H})$. Then the only non-trivial invariant subspace of $\mathcal{L}$ is $\mathcal{H} \oplus \{0\}$ and $\mathcal{L}$ is therefore not reflexive. The algebra $\mathcal{L}$ contains *all* normal operators of even uniform multiplicity, all unilateral shifts of even multiplicity, many finite-rank operators, etc.

Radjavi and I conjecture that a more general theorem is true in the case where $\mathfrak{A}$ contains a m.a.s.a.; namely, we feel that every weakly closed algebra containing a m.a.s.a. must be reflexive. Other cases of this result are known in addition to the Hermitian and totally ordered cases discussed so far. In [14] it is observed that the totally ordered case immediately implies the result in the case where the lattice of invariant subspaces is a direct product of totally ordered lattices, and in [3] it is shown that the result is true in certain other cases (essentially atomic ones). It is hard to believe that all these special cases could be true and yet the general conjecture false, but we have been unable to prove it. The general conjecture vould have some interesting consequences (cf. the remarks in [16]).

One problem that has apparently not been seriously studied is the question of which reflexive algebras are finitely-generated (this general question, of course, includes the famous unsolved problem of whether or not every von Neumann algebra on a separable space is singly generated). In particular, is every reflexive algebra with totally ordered invariant subspace lattice on a separable space finitely-generated? It seems reasonable to guess that every such algebra is generated by two operators. To prove this it would suffice to show: given any chain of subspaces whose projections are in a given m.a.s.a. $\mathcal{R}$ (on a separable Hilbert space) there exists an operator A such that the invariant subspaces of A whose projections are in $\mathcal{R}$ are exactly those in the given chain.

The general question of which operator algebras are reflexive is interesting, and there are undoubtedly many other theorems to be proved. In particular there is little known about the question: for which operators A is the weakly closed algebra generated by A reflexive? In addition to the results which follow from Sarason's theorems discussed above it is known that the algebras generated by inflations [14] and by isometries [4] are reflexive. The operators on finite-dimensional spaces which generate reflexive algebras have been determined ([5]).

5. TRIANGULAR OPERATOR ALGEBRAS

Kadison and Singer [11] defined a *triangular operator algebra* $\mathfrak{A}$ as an operator algebra with the property that $\mathfrak{A} \cap \mathfrak{A}^*$ is a m.a.s.a. The idea is that triangular operator algebras should have a relation to invariant subspaces analogous to the relationship between triangular matrices and invariant subspaces in the finite-dimensional case. A triangular algebra is *hyperreducible* if the projections onto its invariant subspaces generate $\mathfrak{A} \cap \mathfrak{A}^*$. Kadison and Singer [11] gave an analysis of hiperreducible triangular algebras.

There is an interesting problem raised in [11] which is still unsolved: is every operator a member of at least one hyperreducible maximal triangular algebra? Equivalently, is it true that given any operator A there exists a chain $\{P_\alpha\}$ of projections onto invariant subspaces of A such that $\{P_\alpha\}$ generates a m.a.s.a.? An affirmative answer to this question, like the transitive algebra and generalized transitive algebra problems, would give a result much stronger than the existence of invariant subspaces. Thus a counter-example is expected. Such a counter-example would be very interesting. (Note: as is well-known, the assertion added in proof to [11] to the effect that the paper [6] leads to a counter-example is erroneous). It is not known whether or not the result is true for compact operators, in spite of the existence of invariant subspaces, although a partial result has been obtained by Erdos [9] (using techniques of [18],[19]) in this case.

Ringrose [19] observed that a reflexive algebra is triangular if and only if it is hyperreducible. This, together with the theorem of [14] referred to in section 4 above, gives the result that every weakly closed maximal triangular algebra is hiperreducible ([21]). The general conjecture made in section 4 above would imply, in the same way, the result that every weakly closed triangular algebra is hyperreducible. This question seems interesting, and it is very likely that it could be proved without proving the conjecture of section 4.

6. COMMUTANTS

A von Neumann algebra $\mathfrak{A}$ is equal to its double commutant: $\mathfrak{A} = \mathfrak{A}''$. The general question of determining which (non-self-adjoint) operator algebras satisfy $\mathfrak{A} = \mathfrak{A}''$ is obviously very difficult, even in the finite-dimensional case. In the finite-dimensional case it is well known [10] that every singly-generated algebra is equal to its double commutant. This is not true in the infinite-dimensional case: as R.G. Douglas has pointed out, the algebra generated by the bilateral shift is a counter-example. Nonetheless it would be interesting to have sufficient conditions that $\mathfrak{A} = \mathfrak{A}''$. In particular, if $\mathfrak{A}$ is a Hermitian operator algebra must $\mathfrak{A} = \mathfrak{A}''$?

It might also be interesting to study those algebras which are neither reflexive nor equal to their double commutants, but which satisfy $\mathfrak{A} = \{B : B\mathfrak{M} \subset \mathfrak{M}$ whenever $\mathfrak{M}$ invariant under $\mathfrak{A}\} \cap \mathfrak{A}''$. The algebra $\mathfrak{A}$ given as an example in section 4 above is a weakly closed algebra which does not satisfy this relation. It seems possible that every commutative weakly closed algebra containing I satisfies this relation (I don't even know if this is true in the finite-dimensional case). In particular does every singly generated algebra satisfy it; i.e., if A is any operator, $B \in \{A\}''$ and every invariant subspace of A is invariant under B, must B be in the weakly closed algebra generated by A? A weaker result that this, which is still unproved, would be the statement that every operator of the form $A \oplus A$ generates a reflexive algebra. A stronger result, conjectured (and left unproved) independently by D. Sarason and myself, is: if $B \in \{A\}'$ and every invariant subspace of A is invariant under B, then B is in the weakly closed algebra generated by A. This is known to be true in the finite-dimensional case ([2]). This result would imply (using Corollary 1 of [20]), that the commutant of a unicellular operator is the weakly closed algebra generated by the operator, which would be a very nice theorem by itself. The algebraic analogue of this question seems to be unknown; namely, if $AB = BA$ and every invariant linear manifold of A is invariant under B, must B be in the strict closure of the set of polynomials in A?

REFERENCES

[1] W.B. Arveson, A density theorem for operator algebras, *Duke Math. J.*, 34 (1967), 635-647.

[2] L. Brickman and P.A. Fillmore, The invariant subspace lattice of a linear transformation, *Can. J. Math.*, 19 (1967), 810-822.

[3] Ch. Davis, H. Radjavi and P. Rosenthal, On operator algebras and invariant subspaces, *Can. J. Math.*, 21 (1969), 1178-1181.

[4] J.A. Deddens, Every isometry is reflexive (to appear).

[5] J.A. Deddens and P.A. Fillmore, Reflexive linear transformations (to appear).

[6] W.F. Donoghue, The lattice of invariant subspaces of completely continuous quasi-nilpotent transformation, *Pac. J. Math.*, 7 (1957), 1031-1035.

[7] R.G. Douglas and C. Pearcy, Hyperinvariant subspaces and transitive algebras (to appear).

[8] J. Dyer and P. Porcelli, Concerning the invariant subspace problem, *Notices Amer. Math. Soc.*, 17 (August 1970), 788.

[9] J.A. Erdos, Some results on triangular operator algebras, *Amer. J. Math.*, 98 (1967), 85-92.

[10] N. Jacobson, *Lectures in Abstract Algebra.* II (Princeton, 1953).

[11] R.V. Kadison and I. Singer, Triangular Operator Algebras, *Amer. J. Math.*, 82 (1960), 227-259.

[12] E. Nordgren, H. Radjavi and P. Rosenthal, On density of transitive algebras, *Acta. Sci. Math.*, 30 (1969), 175-179.

[13] E. Nordgren, Transitive operator algebras, *J. Math. Anal. and Appl.* (to appear).

[14] H. Radjavi and P. Rosenthal, On invariant subspaces and reflexive algebras, *Amer. J. Math.*, 91 (1969), 683-692.

[15] H. Radjavi and P. Rosenthal, Matrices for operators and generators of $\mathfrak{B}(\mathcal{H})$, *J. London Math. Soc.* (to appear).

[16] H. Radjavi and P. Rosenthal, A sufficient condition that an operator algebra be self-adjoint (to appear).

[17] C.E. Rickart, *General Theory of Banach Algebras* (Princeton, 1960).

[18] J.R. Ringrose, Super-diagonal forms for compact linear operators, *Proc. London Math. Soc.*, (3) 12 (1962). 367-384.

[19] J.R. Ringrose, On some algebras of operators. II, *Proc. London Math. Soc.*, (3) 16 (1966), 385-402.

[20] P. Rosenthal, A note on unicellular operators, *Proc. Amer. Math. Soc.*, 19 (1968), 505-506.

[21] P. Rosenthal, Weakly closed maximal triangular algebras are hyperreducible, *Proc. Amer. Math. Soc.*, 24 (1970). 220.

[22] D.E. Sarason, Invariant subspaces and unstarred operator algebras, *Pac. J. Math.*, 17 (1966), 511-517.

[23] D.E. Sarason, Generalized interpolation in $\mathcal{H}^\infty$, *Trans. Amer. Math. Soc.*, 127 (1967), 179-203.

[24] A.L. Shields and L.J. Wallen, The commutants of certain Hilbert space operators (to appear).

COLLOQUIA MATHEMATICA SOCIETATIS JÁNOS BOLYAI
5. HILBERT SPACE OPERATORS. TIHANY (HUNGARY). 1970

On the structure of measurable linear operators

JU. A. ROZANOV

Let $(X, \mathfrak{U}, P)$ be a measurable linear space with a complete measure P on the σ-algebra of sets $\mathfrak{U}$ generated by a family U of linear functionals $u = (u, x)$ on X. We shall say that the linear operator S defined on a linear subspace $E \subseteq X$ of full measure $(SE \subseteq X)$ is weakly measurable, if for any $u \in U$ the function (u, Sx), $x \in X$, is measurable; S will be called measurable if, in addition, $S\mathfrak{U} \subseteq \mathfrak{U}$.

In what follows a description of all measurable linear operators S in X is given for the case of a gaussian measure P.

Recall that a gaussian measure is determined by the mean value

$$a(u) = \int_X (u, x) \, P(dx) , \qquad u \in U,$$

and the correlation function

$$B(u,v) = \int_X [(u,x) - a(u)][(v,x) - a(v)]\, P(dx)\,; \qquad u, v \in \mathcal{U}.$$

Without loss of generality one may assume that $a(u) \equiv 0$, $B(u,u) > 0$ for $u \neq 0$, and that the initial family $\mathcal{U}$ of linear functionals $u = (u,x)$, $x \in X$, is a linear space. On this space let us introduce the scalar product

(1) $$\langle u, v \rangle = B(u,v).$$

Consider (u, y) as a linear form in $u \in \mathcal{U}$ for fixed $y \in X$. We select the linear subspace $\mathfrak{I}$ of all elements y for which the linear functional (u,y), $u \in \mathcal{U}$, is continuous relative to the scalar product (1). We shall assume that the following reflexivity condition is satisfied: for any $v \in \mathcal{U}$ there can be found a $y \in \mathfrak{I}$ such that

(2) $$(u,y) = \langle u, v \rangle, \qquad u \in \mathcal{U}.$$

It should be noted that condition (2) is fulfilled for all known spaces X in which we are able to build up a gaussian measure explicitely with the help of the correlation function $B(u,v)$ (see [1]).

Example. Let X be a Hilbert space, $\mathcal{U} = X^*$, and

$$B(u,v) = (Bu, v)\,; \qquad u, v \in X \; (= X^*),$$

where B is a positive nuclear operator in X. Then $\mathfrak{I} = B^{\frac{1}{2}} X$.

In the general case relation (2) yields a mapping $\mathcal{U} \longrightarrow \mathfrak{I}$. We identify all elements $y \in \mathfrak{I}$ corresponding to the same element $v \in \mathcal{U}$. Introduce the scalar product

$$\langle y_1, y_2 \rangle = \langle v_1, v_2 \rangle,$$

where $y_i \longleftrightarrow v_i$, $i = 1, 2$. As a result, we have a Hilbert space $\mathfrak{I} \subseteq X$ constructed with the aid of the original measure P with the correlation function $B(u,v)$; $u, v \in \mathcal{U}$.

Observe that every linear subspace $E \subseteq X$, $P(E) > 0$, contains $\mathfrak{I}$, so that every measurable linear operator S is defined on the subspace $\mathfrak{I}$. It can be shown (see [1]) that the subspace $\mathfrak{I}$ is invariant relative to any measurable linear transformation S.

Furthermore, for each transformation of this kind the operator S is bounded in the Hilbert space $\mathfrak{I} \subseteq X$ (with the scalar product (2)). Let the adjoint operator (in $\mathfrak{J}$) be denoted by S^* .

The fundamental result (see [1]; cf. also [2]) can be stated as follows.

Theorem. *The linear operator* S *in* X *is measurable if and only if* $S\mathfrak{J} \subseteq \mathfrak{J}$, S *is a bounded operator in the Hilbert space* $\mathfrak{J}$ *having a bounded inverse operator* S^{-1} *, and the difference* $I - SS^*$ *is a Hilbert–Schmidt operator in* $\mathfrak{J}$.

BIBLIOGRAPHY

[1] Ju.A. Rozanov, Gaussian distributions of infinite dimension, *Trudy Mat. Inst. Akad. Nauk,* vol. 108.

[2] G.E. Šilov and Fan Dyk Tan', *Integral, measure and derivative on linear spaces* (Moscow, 1967).

COLLOQUIA MATHEMATICA SOCIETATIS JÁNOS BOLYAI
5. HILBERT SPACE OPERATORS. TIHANY (HUNGARY). 1970

On global type II_1 w^*-algebras

S. SAKAI

1. INTRODUCTION

A W^*-algebra whose center consists only of scalar multiples of the identity is called a factor. Otherwise it is called to be global.

Let $\mathfrak{M}$ be a factor, C a commutative W^*-algebra and let $\mathfrak{M} \bar{\otimes} C$ be the W^*-tensor product. Then the center of $\mathfrak{M} \bar{\otimes} C$ is $1 \otimes C$, where 1 is the identity of $\mathfrak{M}$. Therefore if C is not one-dimensional, $\mathfrak{M} \bar{\otimes} C$ is a global W^*-algebra. More generally let $\{\mathfrak{M}_\alpha\}_{\alpha \in \Pi}$ be a family of factors, $\{C_\alpha\}_{\alpha \in \Pi}$ a family of commutative W^*-algebras. Then a W^*-algebra $\sum_{\alpha \in \Pi} \oplus (\mathfrak{M}_\alpha \bar{\otimes} C_\alpha)$ is also a global W^*-algebra. We shall call such W^*-algebras trivial global W^*-algebras.

My talk is concerned with the problem whether there exists a nontrivial global type II_1 W^*-algebra. Henceforward we shall assume that W^*-algebras considered here are ones on separable Hilbert spaces.

Let $\mathfrak{N}$ be a global W^*-algebra. Then by the reduction theory of von Neumann, $\mathfrak{N}$ can be expressed uniquely as a direct integral of factors as follows:

$\mathcal{N} = \int_{\Omega} \mathcal{N}(t)\, d\mu(t)$. If all of the factors $\{\mathcal{N}(t) \mid t \in \Omega\}$ are $*$-isomorphic to a factor $\mathcal{N}_0$, $\mathcal{N}$ is $*$-isomorphic to $\mathcal{N}_0 \bar{\otimes} L^{\infty}(\Omega, \mu)$. Hence $\mathcal{N}$ is a trivial global W^*-algebra (cf. [1]).

More generally if there exists a sequence $\{\mathcal{N}_i\}$ of factors such that each of the factors $\{\mathcal{N}(t) \mid t \in \Omega\}$ is $*$-isomorphic to some of $\{\mathcal{N}_i\}$, then $\mathcal{N} = \sum_{i=1}^{\infty} \oplus (\mathcal{N}_i \bar{\otimes} C_i)$, where C_i is a commutative W^*-algebra. Hence $\mathcal{N}$ is again trivial. Therefore in order that there exist a non-trivial global W^*-algebra, it is necessary that there exist uncountably many non-isomorphic factors.

The converse is not trivial. To integrate an uncountable family of factors, we have to introduce a measure structure, to which the reduction theory is applicable, into the family.

We shall introduce a compact group structure into the uncountable family of type II_1-factors constructed in [2]. By using that structure, we shall show the existence of non-trivial global type II_1 W^*-algebras.

Since a part of the content of my talk will be published in the Journal of Functional Analysis, here we shall show an outline of the construction and raise some related problems.

2. OUTLINE OF THE CONSTRUCTION AND PROBLEMS

First of all, we shall explain some results in [2]. Suppose $G_1, G_2, \dots ;$ $H_1, H_2, \dots$ are two sequences of groups. We denote by $(G_1, G_2, \dots ; H_1, H_2, \dots)$ the group generated by the G_i's and the H_i's with the additional relations that H_i, H_j commute elementwise for $i \neq j$, and G_i, H_j commute elementwise for $i \leq j$.

Let $L_1 = (G_1, G_2, \dots ; H_1, H_2, \dots)$ with $G_i = Z$ and $H_i = Z$ for all i, where Z is the group of all integers. Define L_k inductively by $L_k = (G_1, G_2, \dots ;$ $H_1, H_2, \dots)$, where $G_i = Z$, $H_i \doteq L_{k-1}$ for all i.

Now let $\Pi_1 = (p_i)$ be a sequence of positive integers. Define $M_i(\Pi_1) = \sum_{j=1}^{i} \oplus L p_j$ for all i, if Π_1 is infinite; $M_i(\Pi_1) = \sum_{j=1}^{i} \oplus L p_j$ for all $i \leq n_0$ and $M_i(\Pi_1) = M_{n_0}(\Pi_1)$ for $i \geq n_0$, if $\Pi_1 = (p_1, p_2, \dots, p_{n_0})$

is finite. Define $G[\Pi_1] = (G_1, G_2, \ldots; M_1(\Pi_1), M_2(\Pi_1), \ldots)$ with $G_i = Z$ for all i.

Now let $\Pi_1 = (p_i)$ and $\Pi_2 = (q_i)$ be two sequences of positive integers. Suppose that $\Pi_1 \neq \Pi_2$ as a set – namely there exist a q_{i_0} such that $q_{i_0} \neq p_i$ for all i or $p_{i_0} \neq q_i$ for all i. Then, the type Π_1-factor $U(G[\Pi_1])$ is not $*$-isomorphic to the type Π_1-factor $U(G[\Pi_2])$, where $U(G[\Pi_1])$ is the W^*-algebra generated by the left regular representation of $G[\Pi_i]$ $(i = 1, 2)$ ([2]).

Now let Ω be the set of all infinite sequences (p_i) of positive integers such that $p_i = 1$ or i for $i = 1, 2, \ldots$.

Let $S = \{0,1\}$ be the cyclic compact group of order 2. Let $S_n = S$ for $n = 2, 3, \ldots$ and let $\Gamma = \prod_{n=2}^{\infty} S_n$ be the compact group obtained by the infinite direct product of $\{S_n\}$.

We shall identify an element Π of Ω with $G[\Pi]$. In the following considerations, we shall define a one-to-one correspondence between Ω and Γ. Let $\Pi = (p_i) \in \Omega$. Define $(\Pi)_n$ as follows: $(\Pi)_n = 1$ if $p_n = n$ and $(\Pi)_n = 0$ if $p_n = 1$ $(n = 2, 3, \ldots)$.

Then $((\Pi)_2, (\Pi)_3, \ldots)$ will define an element γ in Γ; define $\rho(\Pi) = \gamma$. Then the ρ is one-to-one.

By using the mapping ρ, we shall identify Ω with Γ; then $\Omega = \Gamma$ is a compact group.

Lemma 1. *Let* $\Delta_1 = (G_1, G_2, \ldots; H_1, H_2, \ldots)$ *and* $\Delta_2 = (G_1, G_2, \ldots; J_1, J_2, \ldots)$. *Suppose that there exists a homomorphism* φ_i *of* H_i *onto* J_i *for all* i. *Then we can define a homomorphism* φ *of* Δ_1 *onto* Δ_2 *such that* $\varphi =$ *the identity on* G_i *and* $\varphi = \varphi_i$ *on* H_i *for all* i, *where* G_i *and* H_i *are identified with the corresponding subgroups of* $(G_1, G_2, \ldots; H_1, H_2, \ldots)$.

Now let F_∞ be the free group of denumerable generators and let $\Lambda = (G_1, G_2, \ldots; H_1, H_2, \ldots)$ with $G_i = Z$ and $H_i = F_\infty$ for all i. Let $(r_1, r_2, \ldots, r_n)$ be a finite sequence of positive integers such that $r_i = 1$ or i for $i = 1, 2, \ldots, n$. Consider the group $\sum_{i=1}^{n} \oplus L_{r_i}$, then there exist a homomorphism

ξ of F_∞ onto $\sum_{i=1}^{n} \oplus L_{r_i}$. We shall pick up one homomorphism ξ and fix it; we denote this ξ by $\xi(r_1, r_2, \dots, r_n)$ and so ξ is a function of $(r_1, r_2, \dots, r_n)$. Let $\Pi \in \Omega$ with $\Pi = (p_1, p, \dots)$, then $G[\Pi] = (G_1, G_2, \dots; M_1(\Pi), M_2(\Pi), \dots)$; by Lemma 1, we can define a homomorphism $\xi(\Pi)$ of Λ onto $G[\Pi]$ such that $\xi(\Pi) =$ the identity on G_i and $\xi(\Pi) = \xi(p_1, p_2, \dots, p_n)$ on H_n for all n.

Let $R(\Lambda)$ be the group C^*-algebra of Λ. For $\lambda \in \Gamma \equiv \Omega$, we shall define a trace τ_λ on $R(\Lambda)$ as follows: Take the homomorphism $\xi(\lambda)$ of Λ onto $G[\lambda]$ and define $\tau_\lambda(g) = \delta_{\xi(\lambda)(g)}(e_\lambda)$, where $\delta_{\xi(\lambda)(g)}$ is the function on $G[\lambda]$ such that $\delta_{\xi(\lambda)(g)}(\ell) = 1$ for $\xi(\lambda)(g) = \ell$ and $\delta_{\xi(\lambda)(g)}(\ell) = 0$ for $\xi(\lambda)(g)$ ℓ with $\ell \in G[\lambda]$, and e_λ is the identity of $G[\lambda]$.

Then τ_λ is a central positive definite function on Λ such that $\tau_\lambda(e) = 1$; therefore it will define a unique trace on $R(\Lambda)$, denoted by the same notation τ_λ – namely, $\tau_\lambda(xy) = \tau_\lambda(yx)$, $\tau_\lambda(x^*x) \geq 0$ and $\tau_\lambda(1) = 1$ for $x, y \in R(\Lambda)$, where 1 is the identity of $R(\Lambda)$. Let $\{\pi_\lambda, \mathcal{H}_\lambda\}$ be the $*$-representation of $R(\Lambda)$ on a separable Hilbert space $\mathcal{H}_\lambda$ constructed via τ_λ. Then the weak closure of $\pi_\lambda(R(\Lambda))$ is $*$-isomorphic to $U(G[\lambda])$.

The mapping ξ: $\lambda \to \tau_\lambda$ of Γ into the state space $\mathfrak{S}$ of $R(\Lambda)$ is a one-to-ne continuous mapping.

The compactness of Γ implies that the ξ is homeomorphic.

Let $d\lambda$ be the Haar measure on Γ with the total mass 1; by using the ξ we can introduce a Radon measure ν on $\mathfrak{S}$ such that $d_\nu(\tau_\lambda) = d\lambda$.

Define a trace ψ on $R(\Lambda)$ such that $\psi(x) = \int_{\mathfrak{S}} \varphi(x) d_\nu(\varphi)$ for $x \in R(\Lambda)$.

Let $\{\pi_\psi, \mathcal{H}_\psi\}$ be the $*$-representation of $R(\Lambda)$ on a separable Hilbert space $\mathcal{H}_\psi$ constructed via ψ. Let $\overline{\pi_\psi(R(\Lambda))}$ be the weak closure of $\pi_\psi(R(\Lambda))$; then $\overline{\pi_\psi(R(\Lambda))}$ is a finite W^*-algebra and ψ can be uniquely extended to a normal faithful trace on $\overline{\pi_\psi(R(\Lambda))}$.

Then ν is the central measure of ψ; hence $\overline{\pi_\psi(R(\Lambda))} = \int_{\mathfrak{S}} \overline{\pi_\psi(R(\Lambda))} d_\nu(\varphi)$ is the central decomposition of $\overline{\pi_\psi(R(\Lambda))}$ and so the global type II_1 W^*-algebra $\overline{\pi_\psi(R(\Lambda))}$ is non-trivial.

Therefore we have

Theorem. *There exists a type* II_1 W^**-algebra* M *on a separable Hilbert space satisfying the following properties:*

(1) *The center of* M *is isomorphic to the* $L^\infty(\Gamma, d\lambda)$*; where* Γ *is an infinite compact group and* $d\lambda$ *is the Haar measure of* Γ.

(2) *For its central decomposition* $M = \int_\Gamma M(\lambda)\,d\lambda$, $M(\lambda_1)$ *is not* $*$*-isomorphic to* $M(\lambda_2)$ *for every two different* $\lambda_1, \lambda_2 \in \Gamma$.

Problem 1. Let $\mathfrak{N}$ be a global W^*-algebra on a separable Hilbert space. Find nice conditions under which $\mathfrak{N}$ can be written as $\mathfrak{M} \bar{\otimes} Z$ (Z, the center of $\mathfrak{N}$; $\mathfrak{M}$, a factor). By using those conditions, can we find directly a global W^*-algebra $\mathfrak{N}$ such that the reduction $\mathfrak{N} = \int_\Omega \mathfrak{N}(t)\,d\mu(t)$ satisfies that μ is a continuous measure and there exists a μ-measurable subset Ω_0 in Ω with $\mu(\Omega - \Omega_0) = 0$ such that for any two distinct t_1, t_2 in Ω_0 $\mathfrak{N}(t_1)$ is not * isomorphic to $\mathfrak{N}(t_2)$?

Problem 2. (Dixmier) In this paper, the existence of an uncountable family of type II_1-factors with a totally disconnected compact topology was shown. Can we have an uncountable family of type II_1-factors with a connected compact topology?

Remark. Problem 2 is true for type III-factors (R. Powers, *Ann. Math.*, 86 (1967), 138-171.)

REFERENCES

[1] J. Dixmier, *Les algèbres d'operateurs dans l'espace hilbertien* (2nd edition, 1969).

[2] S. Sakai, An uncountable number of II_1, II_∞-factors (to appear in the *Journal of Functional Analysis*).

COLLOQUIA MATHEMATICA SOCIETATIS JÁNOS BOLYAI
5. HILBERT SPACE OPERATORS. TIHANY (HUNGARY). 1970

Harnack inequalities for a functional calculus

I. SUCIU

1. Preliminaries. Let X be a compact Haudorff space and $C(X)$ the Banach algebra of all continuous complex-valued functions on X normed with the supnorm. In all that follows A will be a closed subspace of $C(X)$ which contains the constant functions and separates the points of X. In certain cases A will be supposed to be multiplicatively closed in $C(X)$ i.e. a *function algebra on* X.

Let H be a complex Hilbert space and $L(H)$ the Banach albegra of all bounded linear operators on H. $\mathcal{B}(A;H)$ will denote the vector space of all bounded linear maps of A into $L(H)$. If A is a function algebra on X, a *representation* of A on X is an algebra homomorphism φ of A into $L(H)$ such that $\varphi(1)=I$ (the identity operator on H) and $\|\varphi\| \leq 1$. Any representation π of $C(X)$ on H is given by a spectral measure on X with values in $L(H)$.

Let $\varphi \in \mathcal{B}(A;H)$. A triple $[K,V,\pi]$, where K is Hilbert space, V is a bounded linear operator from H into K, and π is a representation of $C(X)$ on K, is called a *spectral dilation* of φ if we have

(1.1) $$\varphi(f) = V^*\pi(f)V \qquad (f\in A).$$

Let us denote by $\mathcal{D}(A;H)$ the set of all $\varphi\in\mathcal{B}(A;H)$ which have spectral dilations. The spectral dilation $[K, V, \pi]$ of φ is called *minimal* if K is the closed linear span of all vectors $\pi(g)Vh$, $g\in C(X)$, $h\in H$. It is easy to see that if φ has a spectral dilation then it has a minimal one. In what follows all spectral dilation considered will be supposed to be minimal.

Let us remark that if $\varphi(1) = I$ then from (1.1) it results that $V^*V = I$, i.e. V is isometric. In this case H may be embedded in K and if we put $P = VV^*$ then P is the orthogonal projection of K onto VH. We have

$$V\varphi(f)h = P\pi(f)Vh \qquad (f\in A,\ h\in H).$$

An element $\mu\in\mathcal{B}(C(X);H)$ is *positive if* $\mu(g)$ is a positive operator on H for any positive function g in $C(X)$. We say that $\mu\in\mathcal{B}(C(X);H)$ is a *semi-spectral measure* attached to φ in $\mathcal{B}(A;H)$ if μ is positive and $\mu(f) = \varphi(f)$ for any $f\in A$. If $[K,V,\pi]$ is a spectral dilation of φ then μ defined by

(1.2) $$\mu(g) = V^*\pi(g)V \qquad (g\in C(X))$$

is a semi-spectral measure attached to φ.

Let n be a positive integer. M_n will denote the C^*-algebra of all $n\times n$ scalar matrices. $C(X)\otimes M_n$ denotes the algebra of all $n\times n$ matrices over $C(X)$ with the usual involution $(g_{ij})^* = (\bar{g}_{ij})$, where the bar means complex conjugate. For a matrix (g_{ij}) over $C(X)$ we put

$$\mathrm{Re}(g_{ij}) = \frac{1}{2}[(g_{ij}) + (g_{ij})^*].$$

The matrix (g_{ij}) over $C(X)$ is *positive* if there exists a matrix (u_{ij}) such that $(g_{ij}) = (u_{ij})^*(u_{ij})$.

Let $\varphi\in\mathcal{D}(A;H)$ and $[K,V,\pi]$ a spectral dilation of φ. Let (f_{ij}) be an $n\times n$ matrix over A such that the matrix $\mathrm{Re}(f_{ij})$ is positive, say $\mathrm{Re}(f_{ij}) = (u_{ij})^*(u_{ij})$. Then for each n-tuple $h_1,\dots,h_n$ of elements of H we have

$$2\,\mathrm{Re}\sum_{i,j}(\varphi(f_{ij})h_j, h_i) = 2\,\mathrm{Re}\sum_{i,j}(V^*\pi(f_{ij})Vh_j, h_i) =$$

$$= 2\,\mathrm{Re}\sum_{i,j}(\pi(f_{ij})Vh_j, Vh_i) = \sum_{i,j}(\pi(f_{ij})Vh_j, Vh_i) +$$

$$+ \sum_{i,j}\overline{(\pi(f_{ij})Vh_j, Vh_i)} = \sum_{i,j}(\pi(f_{ij})Vh_j, Vh_i) +$$

$$+ \sum_{i,j}(Vh_i, \pi(f_{ij})Vh_j) = \sum_{i,j}(\pi(f_{ij})Vh_j, Vh_i) +$$

$$+ \sum_{i,j}(\pi(\overline{f_{ij}})Vh_i, Vh_j) = \sum_{i,j}(\pi(f_{ij} + \overline{f_{ji}})Vh_j, Vh_i) =$$

$$= \sum_{i,j}\sum_{k}(\pi(u_{ki})^*\pi(u_{kj})h_j, h_i) = \sum_{k}\|\sum_{j}\pi(u_{kj})h_j\|^2 \geq 0.$$

We conclude that the $n \times n$ operator matrix $\mathrm{Re}(\varphi(f_{ij}))$ is positive.

We say that an element $\varphi \in \mathcal{B}(A;H)$ is *completely positive* if for each integer $n \geq 1$ and for each $n \times n$ matrix (f_{ij}) over A for which the matrix $\mathrm{Re}(f_{ij})$ is positive, the operator matrix $\mathrm{Re}(\varphi(f_{ij}))$ is positive.

If A is self-adjoint it is easy to see that φ is completely positive if and only if for each positive matrix (f_{ij}) over A the operator matrix $(\varphi(f_{ij}))$ is positive; this is the definition of complete positivity used by W.F. Stinespring in [6] and W.B. Arveson in [1].

Moreover, if $\varphi \in \mathcal{B}(A;H)$ is completely positive then it has a unique completely positive extension onto the uniform closure S of $A + \bar{A}$. Indeed, it is obvious that the bounded selfadjoint extension is unique. Since φ is completely positive we have, in particular, $\mathrm{Re}\,\varphi(f) \geq 0$ for any $f \in A$ with $\mathrm{Re} f \geq 0$, thus

(1.3) $$\|\mathrm{Re}\,\varphi(f)\| \leq \|\mathrm{Re} f\|\,\|\varphi(1)\| \qquad (f \in A).$$

Then we have

$$2\|\varphi(f) + \varphi(g)^*\| = \|\mathrm{Re}\,\varphi(f+g) - i\,\mathrm{Re}(-i(f-g)\| \leq$$

$$\leq \| \mathrm{Re}\,\varphi(f+g)\| + \| \mathrm{Re}\,\varphi(-i(f-g))\| \leq \|\varphi(1)\| (\| \mathrm{Re}(f+g)\| +$$

$$+ \| \mathrm{Re}(-i(f-g))\|) \leq 4 \|\varphi(1)\| \, \| f+\bar{g} \| .$$

Thus if we put

$$\tilde{\varphi}(f+\bar{g}) = \varphi(f) + \varphi(g)^* \qquad (f, g \in A),$$

we can extend $\tilde{\varphi}$ onto S. $\tilde{\varphi}$ is the bounded self-adjoint extension of φ onto S. It remains to show that $\tilde{\varphi}$ is completely positive. So let $(f_{ij} + \overline{g_{ij}})$ be a positive matrix over $A + \bar{A}$. We have $f_{ij} + g_{ij} = f_{ji} + g_{ji}$ and the matrix $\mathrm{Re}(f_{ij} + \overline{g_{ij}}) = \mathrm{Re}(f_{ij} + g_{ji})$ is positive. Then we have:

$$2 \sum_{i,j} (\tilde{\varphi}(f_{ij} + g_{ij}) h_j, h_i) = \sum_{i,j} (\tilde{\varphi}(f_{ij} + \overline{f_{ji}} + g_{ji} + \overline{g_{ij}}) h_j, h_i) =$$

$$= \sum_{i,j} (\varphi(f_{ij} + g_{ji}) h_j, h_i) + \sum_{i,j} (\varphi(f_{ji} + g_{ij})^* h_j, h_i) =$$

$$= \sum_{i,j} (\varphi(f_{ij} + g_{ji}) h_j, h_i) + \sum_{i,j} \overline{(\varphi(f_{ji} + g_{ij}) h_i, h_j)} =$$

$$= \sum_{i,j} (\varphi(f_{ij} + g_{ji}) h_j, h_i) + \sum_{i,j} \overline{(\varphi(f_{ij} + g_{ji}) h_j, h_i)} =$$

$$= 2\mathrm{Re} \sum_{i,j} (\varphi(f_{ij} + g_{ji}) h_j, h_i) \geq 0 .$$

Since $A \otimes M_n + \bar{A} \otimes M_n$ is dense in $S \otimes M_n$ we conclude that $\tilde{\varphi}$ is completely positive.

Let us summarize the above remarks together with the Naimark dilation theorem [5] and the Arveson extension theorem [1] (see also W.F. Stinespring [6]) in the following

Theorem 1. *Let* $\varphi \in \mathcal{B}(A;H)$. *The following assertions are equivalent:*

(i) $\varphi \in \mathcal{D}(A;H)$,
(ii) *there exists a semi-spectral measure* μ *attached to* φ,
(iii) φ *is completely positive.*

Proof. The implications (i) $\longrightarrow$ (ii) and (i) $\longrightarrow$ (iii) hold in view of the above considerations. The implication (ii) $\longrightarrow$ (i) is the Naimark dilation theorem, and the implication (iii) $\longrightarrow$ (ii) results, on account of the remark preceding theorem 1, from the Arveson extension theorem.

We say that $\varphi \in \mathcal{D}(A;H)$ has a unique spectral dilation if for any two spectral dilations $[K_1, V_1, \pi_1]$, $[K_2, V_2, \pi_2]$ of φ there exists a unitary operator $U: K_2 \to K_1$ such that

$$UV_2 = V_1 \quad \text{and} \quad U\pi_2(g) = \pi_1(g)U \qquad (g \in C(X)).$$

An element $\varphi \in \mathcal{D}(A;H)$ has a unique spectral dilation if and only if it has a unique extension to a semi-spectral measure on X. Indeed, suppose that φ has a unique spectral dilation and let μ_1, μ_2 be two semi-spectral measures attached to φ. Let $[K_1, V_1, \pi_1]$ be the spectral dilation of μ_1, and $[K_2, V_2, \pi_2]$ the spectral dilation of μ_2. It is clear that each of them is a spectral dilation of φ. Let $U: K_2 \longrightarrow K_1$ be such that

$$UV_2 = V_1 \quad \text{and} \quad U\pi_2(g) = \pi_1(g)U \qquad (g \in C(X)).$$

For any $g \in C(X)$ we have:

$$\mu_2(g) = V_2^*\pi_2(g)V_2 = V_2^*U^*\pi_1(g)UV_2 = V_1^*\pi_1(g)V_1 = \mu_1(g).$$

Thus $\mu_2 = \mu_1$. The converse assertion follows from the uniqueness part of the Naimark dilation theorem.

We say that a closed subspace M of H is doubly invariant with respect to φ if for any $f \in A$ we have $\varphi(f)M \subset M$, $\varphi(f)^*M \subset M$. If $\varphi \in \mathcal{D}(A;H)$ and M is a doubly invariant subspace with respect to φ then φ_M defined by $\varphi_M(f) = \varphi(f)|M$ $(f \in A)$ belongs to $\mathcal{D}(A;M)$. To see this, let $[K,V,\pi]$ be a spectral dilation of φ, K_M the closed linear span of the elements $\pi(g)V_M$, $g \in C(X)$, $m \in M$, and π_M the representation of $C(X)$ on K_M defined by $\pi_M(g) = \pi(g)|K_M$ for $g \in C(X)$. Then $[K_M, V_M, \pi_M]$ is a spectral dilation of φ_M.

Finally, we recall that if A is a Dirichlet algebra on X, i.e. if it has the property that each continuous real function on X can be uniformly approximated by real parts of functions of A , then any representation of A has a unique spectral dilation [3]. If A is a logmodular algebra on X , i.e. if A has the property that the set $\{u \in C(X):\ u = \log|f|, f, f^{-1} \in A\}$ is uniformly dense in the set of all continuous real-valued functions on X then the spectral dilation of the representation φ of A is unique provided it exists [4].

2. Harnack equivalence. For $\varphi_1, \varphi_2 \in \mathcal{B}(A;H)$ let us write $\varphi_1 \leq \varphi_2$ if $\varphi_2 - \varphi_1$ is completely positive. If $A = C(X)$ then $\varphi_1 \leq \varphi_2$ means just that for each positive function g $\varphi_2(g) - \varphi_1(g)$ is a positive operator on H .

Theorem 2. *Let* $\varphi_1, \varphi_2 \in \mathcal{B}(A;H)$. *The following assertions are equivalent:*

(i) *There exists a* c $(0<c<1)$ *such that*

$$c\varphi_2 \leq \varphi_1 \leq (1/c)\varphi_2 .$$

(ii) *There exist a* c $(0<c<1)$, *a semi-spectral measure* μ_1 *attached to* φ_1 , *and a semi-spectral measure* μ_2 *attached to* φ_2 , *such that*

$$c\mu_2 \leq \mu_1 \leq (1/c)\mu_2 .$$

(iii) *There exist spectral dilations* $[K_1, V_1, \pi_1]$ *and* $[K_2, V_2, \pi_2]$ *of* φ_1 *and* φ_2 , *and a bounded linear operator* $S:\ K_2 \to K_1$ *with bounded inverse, such that*

$$SV = V_1 \quad \text{and} \quad \pi_2(g) = S^{-1}\pi_1(g)S \qquad (g \in C(X)) .$$

Proof. (i) $\longrightarrow$ (ii). Let φ_1, φ_2 be as in (i). Since $\varphi_1 - c\varphi_2$ and $\varphi_2 - c\varphi_1$ are completely positive we can find semi-spectral measures m_1 and m_2 on X such that for each $f \in A$ we have:

$$\varphi_1(f) - c\varphi_2(f) = m_1(f), \qquad \varphi_2(f) - c\varphi_1(f) = m_2(f) .$$

Le us put

$$\mu_1 = (1-c^2)^{-1}(m_1 + cm_2), \qquad \mu_2 = (1-c^2)^{-1}(m_2 + cm_1) .$$

For each $f \in A$ we have

$$\varphi_1(f) = m_1(f) + c\varphi_2(f) = m_1(f) + c(c\varphi_1(f) + m_2(f)) .$$

Hence

$$\varphi_1(f) = (1-c^2)^{-1}(m_1(f) + cm_2(f)) = \mu_1(f) .$$

Similarly we obtain:

$$\varphi_2(f) = \mu_2(f) \qquad (f \in A).$$

It is clear that μ_1 and μ_2 are semi-spectral measures on X. We have:

$$c\mu_1 = (1-c^2)^{-1}(cm_1+c^2m_2) \leq (1-c^2)^{-1}(cm_1+m_2) = \mu_2$$

$$c\mu_2 = (1-c^2)^{-1}(cm_2+c^2m_1) \leq (1-c^2)^{-1}(cm_2+m_1) = \mu_1 .$$

The proof of (i) $\longrightarrow$ (ii) is complete. (ii) $\longrightarrow$ (iii). Let $[K_1,V_1,\pi_1]$ be as in (ii), $[K_1,V_1,\pi_1]$ a spectral dilation of μ_1 and $[K_2,V_2,\pi_2]$ a spectral dilation of μ_2. Then it is clear that $[K_1,V_1,\pi_1]$ is a spectral dilation of φ_1 and $[K_2,V_2,\pi_2]$ is a spectral dilation of φ_2.

Since $c\mu_2 \leq \mu_1$, so for each integer $n \geq 1$, each n-tuple $g_1,\dots,g_n$ of elements of $C(X)$ and each n-tuple $h_1,\dots,h_n$, of element of H we have:

$$c\|\sum_j \pi_2(g_j)Vh_j\|^2 = c(\sum_j \pi_2(g_j)V_2h_j, \sum_i \pi_2(g_i)V_2h_i) =$$

$$= c\sum_{i,j}(V_2^*\pi_2(\bar{g}_i g_j)V_2h_j, h_i) = c\sum_{i,j}(\mu_2(\bar{g}_i g_j)h_j, h_i) \leq$$

$$\leq \sum_{i,j}(\mu_1(\bar{g}_i g_j)h_j, h_i) = \sum_{i,j}(V_1^*\pi_1(\bar{g}_i g_j)V_1h_j, h_i) =$$

$$= \|\sum_j \pi_1(g_j)V_1h_j\|^2 .$$

Thus there exists a linear operator $S_1\colon K_1 \to K_2$ such that $\|S_1\| \leq 1/c$ and

$$S_1(\sum_j \pi_1(g_j)V_1h_j) = \sum_j \pi_2(g_j)V_2h_j .$$

In particular, we have

$$S_1V_1 = V_2 \quad \text{and} \quad S_1\pi_1(g) = \pi_2(g)S_1 \qquad (g \in C(X)) .$$

Similarly we can find $S_2 : K_2 \to K_1$ such that $\|S_2\| \le c^{-1}$ and
$S_2 V_2 = V_1$ and $S_2 \pi_2(g) = \pi_1(g) S_2$ $(g \in C(X))$.

For $g \in C(X)$ and $h \in H$, we have:

$$S_2 S_1 \pi_1(g) V_1 h = S_2 \pi_2(g) S_1 V_1 h = \pi_1(g) S_2 S_1 V_1 h =$$
$$= \pi_1(g) S_2 V_2 h = \pi_1(g) V_1 h .$$

Thus $S_2 S_1$ is the identity operator on K_1. In the same manner we can show that $S_1 S_2$ is the identity operator on K_2. Thus if we put $S = S_2$ then $S^{-1} = S_1$ and we have

$$SV_2 = V_1 \quad \text{and} \quad \pi_2(g) = S^{-1} \pi_1(g) S \qquad (g \in C(X)).$$

The implication (ii) $\longrightarrow$ (iii) is proved. (iii) $\longrightarrow$ (ii). Let $[K_1, V_1, \pi_1], [K_2, V_2, \pi_2]$ and S be as in (iii). Define $\mu_1, \mu_2 \in \mathcal{B}(C(X); H)$ by

$$\mu_1(g) = V_1^* \pi_1(g) V_1 \quad \text{and} \quad \mu_2(g) = V_2^* \pi_2(g) V_2 \qquad (g \in C(X)).$$

We have

$$\mu_2(g) = V_2^* \pi_2(g) V_2 = V_2^* S^{-1} \pi_1(g) S V_2 =$$
$$= V_1^* (S^{-1})^* S^{-1} \pi_1(g) V_1 .$$

Let $T = c(S^{-1})^* S^{-1}$, where c is a positive constant such that $\|T\| \le 1$. Then T is a positive operator on K_1, $T \le I$. For each $g \in C(X)$ we have:

$$T\pi_1(g) = c(S^{-1})^* S^{-1} \pi_1(g) = c(S^{-1})^* \pi_2(g) S^{-1} =$$
$$= c(\pi_2(\bar{g}) S^{-1})^* S^{-1} = c(S^{-1} \pi_1(\bar{g}))^* S^{-1} =$$
$$= \pi_1(g) c(S^{-1})^* S^{-1} = \pi_1(g) T .$$

Thus T commutes with $\pi_1(g)$ for each $g \in C(X)$. Let D be the positive square root of $I - T$. Then D commutes with $\pi_1(g)$ for every $g \in C(X)$. For every positive function g in $C(X)$ and for h in H we have:

$$((\mu_1(g) - c\mu_2(g)) h, h) = ((V_1^* \pi_1(g) V_1 - V_1^* T \pi_1(g) V_1) h, h) =$$
$$= (V_1^* (I - T) \pi_1(g) V_1 h, h) = (V_1 D^2 \pi_1(g) V_1 h, h) =$$

Thus $c\mu_2 \le \mu_1$. In the same manner we may find $c > 0$ such that $c\mu_1 \le \mu_2$. Thus $c\mu_2 \le \mu_1 \le (1/c)\mu_2$.

The implication (iii) $\longrightarrow$ (ii) is proved. Since the implication (ii) $\longrightarrow$ (i) is obvious, the proof of the theorem is complete.

Remark. In the proof of the implication (i) $\longrightarrow$ (ii) we used a similar argument to Bishop's proof of an analogous result for the complex homomorphisms of function algebras [2]. The proof of the equivalence (ii) $\longleftrightarrow$ (iii) was inspired to the author by the proof of Theorem 1.4.2 of Arveson's paper [1].

We say that $\varphi_1, \varphi_2 \in \mathcal{B}(A;H)$ are *Harnack equivalent* if either $\varphi_1 = \varphi_2$ or they satisfy one of the (equivalent) assertions of Theorem 2. The Harnack equivalence is an equivalence relation on $\mathcal{B}(A;H)$. The equivalence classes induced on $\mathcal{B}(A;H)$ by this equivalence relation will be called the *Harnack parts* of $B(A;H)$. Let us remark that if a Harnack part P of $\mathcal{B}(A;H)$ is not a single point then each $\varphi \in P$ has a spectral dilation.

Assertion (i) of Theorem 2 may be also expressed as follows: If φ_1, φ_2 belong to the same Harnack part of $\mathcal{B}(A;H)$ then there exists a c $(0 < c < 1)$ such that for any integer $n \ge 1$, any $n \times n$ matrix (f_{ij}) over A for which the matrix $\mathrm{Re}(f_{ij})$ is positive, and for any n-tuple $h_1 \dots h_n$ of elements of H we have:

$$c\,\mathrm{Re}\sum_{i,j}(\varphi_2(f_{ij})h_j, h_i) \le \mathrm{Re}\sum_{i,j}(\varphi_1(f_{ij})h_j, h_i) \le$$

$$\le (1/c)\,\mathrm{Re}\sum_{i,j}(\varphi_2(f_{ij})h_j, h_i).$$

These inequalities generalize the Harnack inequalities for positive harmonic functions in the complex plane.

3. Invariants. Let $\varphi \in D(A;H)$, $\varphi(1) = I$, and let $[K, V, \pi]$ be a spectral dilation of φ. Let us put

$$M_\pi = \{m \in H : \ \pi(g)Vm \in VH, \ g \in C(X)\};$$

M_π is a closed subspace of H. For any $m \in M_\pi$ and $g \in C(X)$ we have

$$VV^*\pi(g)Vm = \pi(g)Vm$$

because VV^* coincides with the identity on VH. M_π is a doubly-invariant subspace for φ. Indeed, if $m \in M$, $f \in A$ and $g \in C(X)$, we have:

$$\pi(g)V\varphi(f)m = \pi(g)VV^*\pi(f)Vm = \pi(g)\pi(f)Vm = \pi(fg)Vm \in VH ,$$

$$\pi(g)V\varphi(f)^*m = \pi(g)VV^*\pi(\bar{f})Vm = \pi(g)\pi(\bar{f})Vm = \pi(\bar{f}g)Vm \in VH .$$

Let us define $\mu \in B(C(X); M_\pi)$ by

$$\mu(g)m = V^*\pi(g)Vm \qquad (m \in M_\pi, g \in C(X)) .$$

We have $\mu(1) = I$, $\|\mu\| \leq 1$ and

$$\mu(g_1)\mu(g_2)m = V^*\pi(g_1)VV^*\pi(g_2)Vm =$$

$$= V^*\pi(g_1)\pi(g_2)Vm = V^*\pi(g_1g_2)Vm = \mu(g_1g_2)m .$$

Thus μ is a representation of $C(X)$ on M_π. We also have

$$\varphi(f)m = \mu(f)m \qquad (m \in M_\pi, f \in A) .$$

Theorem 3. *Let* $\varphi_1, \varphi_2 \in \mathcal{B}(A;H)$ $(\varphi_1 \neq \varphi_2)$ *belong to the same Harnack part. There exists spectral dilations* $[K_1, V_1, \pi_1]$ *and* $[K_2, V_2, \pi_2]$ *of* φ_1 *resp.* φ_2 *such that* $M_{\pi_1} = M_{\pi_2}$.

Proof. Since φ_1, φ_2 are Harnack equivalent and $\varphi_1 \neq \varphi_2$ so there exist spectral dilations $[K_1, V_1, \pi_1]$ and $[K_2, V_2, \pi_2]$ of φ_1 resp. φ_2 and a bounded linear operator $S: K_2 \longrightarrow K_1$ with bounded inverse, such that

$$SV_2 = V_1 \quad \text{and} \quad \pi_2(g) = S^{-1}\pi_1(g)S \qquad (g \in C(X)) .$$

Let $m \in M_\pi$, $g \in C(X)$ and $h \in H$ be such that $\pi_1(g)V_1m = V_1h$. We have

$$\pi_2(g)V_2m = S^{-1}\pi_1(g)SV_2m = S^{-1}\pi_1(g)V_1m = S^{-1}V_1h = V_2h .$$

Thus $M_{\pi_1} \subset M_{\pi_2}$. By symmetry, we have $M_{\pi_2} \subset M_{\pi_1}$. The proof of the theorem is complete.

In the sequel we will suppose that A has the uniqueness property for spectral dilations, i.e. for any complex Hilbert space H and any $\varphi \in \mathcal{D}(A;H)$ the spectral dilation of φ is unique. Let $\varphi \in \mathcal{D}(A;H)$ and let $[K, V, \pi]$ be its spectral dilation. We say that φ is *spectral* if it is the restriction to A of a representation of $C(X)$ on H. We have remarked that if $M = M_\pi$ then φ_M is spectral. Now we can show that M_π *is the largest subspace* M, *doubly-invariant with repsect to* φ, *for which* φ_M *is spectral.* Indeed, let a subspace M be doubly invariant with respect to φ and π_0 a representation of $C(X)$ on M such that $\varphi_M(f) = \pi_0(f)$, $f \in A$.

From the uniqueness property for the spectral dilation of A it results that for any $g \in C(X)$ and $m \in M$ we have

$$\pi_0(g)m = V_M^* \pi(g) Vm .$$

Applying this relation to any $m \in M$, $h \in M$ and $g, g_1 \in C(X)$ we obtain:

$$(VV_M^* \pi(g)Vm - \pi(g)Vm, \pi(g_1)Vh) =$$

$$= (V\pi_0(g)m - \pi(g)Vm, \pi(g_1)Vh) = (V\pi_0(g)m, \pi(g_1)Vh) -$$

$$- (\pi(g)Vm, \pi(g_1)Vh) = (V_M^* \pi(\bar{g}_1) V\pi_0(g)m, h) -$$

$$- (V_M^* \pi(\bar{g}_1)\pi(g)Vm, h) = (\pi_0(\bar{g}_1)\pi_0(g)m, h) -$$

$$- (V_M^* \pi(\bar{g}_1 g)Vm, h) = (\pi_0(\bar{g}_1 g)m, h) - (\pi_0(\bar{g}_1 g)m, h) = 0 .$$

Thus

$$\pi(g)Vm = VV_M^* \pi(g)Vm$$

for any $m \in M$ and $g \in C(X)$, i.e. $M \subset M_\pi$. We say that φ is *completely non-spectral* if for any doubly-invariant subspace M for which φ_M is spectral we have $M = \{0\}$. It is clear that

$$H = M_\pi \oplus M_\pi^\perp$$

is the *canonical* (necessarily unique) decomposition of φ into the spectral part and the completely non-spectral part. φ is spectral if and only if $M_\pi = H$; φ is completely non spectral if and only if $M = \{0\}$ (see [3]).

From the above remarks and Theorem 3 it follows

Theorem 4. *Suppose that* A *has the uniqueness property for spectral dilation and let* $\varphi_1, \varphi_2 \in \mathcal{D}(A;H)$ *belong to the same Harnack part. Then* φ_1 *and* φ_2 *have the same canonical decomposition. In particular,* φ_1 *is spectral (completely non-spectral) if and only if* φ_2 *has this property.*

Finally, let us remark that if a Harnack part P contains a spectral element φ then any element of P is unitarily equivalent to φ.

4. Application to contractions. Let X be the unit circle in the complex plane and A the algebra of all continuous functions on X which are uniform limits of polynomials on X. A is a Dirichlet algebra on X. Let H be a Hilbert space and T a contraction on H. Using the functional calculus of B. Sz.-Nagy – C. Foiaş [7] we can construct a representation φ_T of A on H such that $\varphi_T(z) = T$ (z means here the function f: $f(z) = z$ on X). Conversely, any representation of A on H may be obtained in this way from $T = \varphi(z)$. T is unitary (completely non-unitary) if and only if φ_T is spectral (completely non-spectral).

If U is a unitary dilation of T then T_1 is a spectral dilation of φ_T.

Let T_1, T_2 be contractions on H. We say that T_1 *is Harnack equivalent to* T_2 if φ_{T_1} is Harnack equivalent to φ_{T_2}. In this special case Theorem 2 has the following form.

Theorem 5. *Let* T_1, T_2 *be contractions on* H, *and* E_1, E_2 *the spectral measures attached to their unitary dilations* U_1, U_2, *respectively. Then the Harnack equivalence of* T_1 *and* T_2 *is equivalent to each of the following assertions:*

(i) *There exists a* c $(0 < c < 1)$ *such that*

$$cu(T_1) \leq u(T_2) \leq (1/c)\,u(T_1) \qquad (u \in \mathrm{Re}A,\ u \geq 0).$$

(ii) *There exists a* c $(0 < c < 1)$ *such that for any Borel subset* σ *of* X *and any* $h \in H$ *we have*

$$c(E_1(\sigma)h,h) \leq (E_2(\sigma)h,h) \leq (1/c)(E_1(\sigma)h,h).$$

Proof. It is sufficient to remark that if A is a Dirichlet algebra on X then complete positivity of φ is equivalent to positivity only.

From Theorem 4 it results that if T_1 is Harnack equivalent to T_2 then T_1 is unitary (completely non-unitary) if and only if T_2 has this property.

Corollary. *A contraction* T *is unitarily equivalent to a unitary operator* U *if and only if there exists a* c $(0 < c < 1)$ *such that*

$$cu(T) \leq u(U) \leq (1/c)\,u(T) \qquad (u \in \mathrm{Re}A,\ u \geq 0).$$

REFERENCES

[1] W.B. Arveson, Subalgebras of C^*-algebras, *Acta Math.*, 123 (1969), 141-224.

[2] E. Bishop, Representing measures for points in a uniform algebra, *Bull. Amer. Math. Soc.*, 70 (1964), 121-122.

[3] C. Foiaş – I. Suciu, Szegő measure and spectral theory in Hilbert spaces, *Rev. Roum. Math. pures et appl.*, XI (1966), 147-159.

[4] C. Foiaş – I. Suciu, On the operator representations of logmodular algebras, *Bull. Acad. Polon. Sci.*, XVI, (1968), 505-509.

[5] M.A. Naimark, Positive definite operator valued functions on a commutative group., *Bull. (Izvestia) Acad. Sci. USSR (ser. math.)*, 7 (1943), 237-244.

[6] W.F. Stinespring, Positive functions on C^*-algebras, *Proc. Amer. Math. Soc.*, 6 (1955), 211-216.

[7] B. Sz.-Nagy – C. Foiaş, *Analyse harmonique des opérateurs de l'espace de Hilbert* (Budapest, 1967).

COLLOQUIA MATHEMATICA SOCIETATIS JÁNOS BOLYAI
5. HILBERT SPACE OPERATORS. TIHANY (HUNGARY). 1970

Quasi-similarity of operators of class C_0

B. SZ.-NAGY

The class C_0 of operators on a Hilbert space $\mathfrak{H}$ consists of those (completely non-unitary) contractions T on $\mathfrak{H}$ for which there exists a non-zero function $u \in H^\infty$ on the open unit disc such that $u(T) = 0$. Among these functions u there exists then an inner one (i.e. with $|u(e^{it})| = 1$ almost everywhere) which is a divisor in H^∞ of all the others: this inner function is uniquely determined by T up to a constant factor of modulus 1; it is called the minimal function of T and denoted by m_T.

This class of operators was introduced in [1]; cf. also [2]. Its investigation deserves interest in particular because many facts known in linear algebra carry over to this class in a surprising manner. Note that for every operator T on a finite dimensional Hilbert space the operator αT belongs to C_0 if the numerical factor α is sufficiently small.

Let us mention in particular that for any given inner function m there exist operators T of class C_0 with $m_T = m$. Such is the operator $S(m)$ defined on the space

$$\mathfrak{H}(m) = H^2 \ominus mH^2$$

by

$$S(m)h = P_{\mathfrak{H}(m)}(e^{it}h) \qquad (h \in \mathfrak{H}(m)),$$

where H^2 denotes the Hardy-Hilbert space; cf. [2], Proposition III. 4. 3.

The aim of this lecture is to report on some of the recent results concerning the class C_0 of operators, obtained in collaboration with C. Foiaş in the papers [3]-[5]. These results show that the Jordan theory of matrices has a remarkable analog for these operators.

First we repeat some definitions.

A bounded linear operator X from a space into another is called a quasi-affinity if it has zero kernel and dense range. (Then X^* is also a quasi-affinity.) Two operators, say A and B, are *quasi-similar* if there exist quasi-affinities X and Y such that

$$AX = XB \qquad \text{and} \qquad BY = YA .$$

For any operator T on $\mathfrak{H}$ we define the *multiplicity* μ_T as the least cardinal number of a set of vectors in $\mathfrak{H}$ which, together with their transforms by $T, T^2, \ldots, T^n, \ldots,$ span the whole space $\mathfrak{H}$.

Thus, in particular, $\mu_T = 1$ means that there is a vector h in $\mathfrak{H}$ such that $h, Th, T^2h, \ldots$ together span $\mathfrak{H}$; that is, a *cyclic* vector.

Next we say what we mean by the *Jordan operator* $S(m_1, m_2, \ldots, m_K)$. Here $m_1, m_2, \ldots, m_K$ are non-constant inner functions, each of which is a divisor of the preceding one. The operator is defined as the orthogonal sum

$$S(m_1) \oplus S(m_2) \oplus \cdots \oplus S(m_K) .$$

Theorem A. *For every* T *of class* C_0 *we have* $\mu_T = \mu_{T^*}$. *If* $\mu_T < \infty$ *then* T *is quasi-similar to a uniquely determined Jordan operator* $S(m_1, m_2, \ldots, m_K)$

(the "Jordan model" of T *); here* $K = \mu_T$ *. Two operators of class* C_0 *are quasi-similar if and only if they have the same Jordan model.*

For the particular case $\mu_T = 1$ (i.e. for "multiplicity-free" operators) we have the following, more complete result:

Theorem B. *For an operator* T *of class* C_0 *the following conditions are equivalent:*

(i) $\mu_T = 1$;
(i*) $\mu_{T^*} = 1$;
(ii) T *is quasi-similar to* $S(m_T)$;
} special cases of Theorem A

(iii) *for every inner function* m *dividing* m_T *there is a unique invariant subspace* $\mathcal{L}$ *for* T *such that the minimal function of* $T|\mathcal{L}$ *equals* m *; in fact,* $\mathcal{L} = \ker m(T)$;

(iv) *for every proper invariant subspace* $\mathcal{L}$ *for* T *the minimal function of* $T|\mathcal{L}$ *is different from* m_T*;*

(v) *if* $\mathcal{L}_1$, $\mathcal{L}_2$ *are different invariant subspaces for* T *then* $T|\mathcal{L}_1$ *and* $T|\mathcal{L}_2$ *are not quasi-similar;*

(vi) *every operator* $A \in (T)'$ *is a "function" of* T *; in fact* $A = \varphi(T)$ *with* $\varphi \in N_T$.

[Here $(T)'$ denotes the "commutant" of T , i.e. the family of operators commuting with T . We are going to consider the "second commutant" $(T)''$ also: this consists of the operators commuting with the operators A in $(T)'$.]

Using Theorem A one is able to prove the following

Theorem C. *For every* T *of class* C_0 *and with* $\mu_T < \infty$, $(T)''$ *consists of "functions"* $\varphi(T)$ *of* T *, with* $\varphi \in N_T$.

In case $\mu_T = 1$, $(T)'$ is commutative (by B (vi). Conversely, if $(T)'$ is commutative for some operator T , then $(T)' = (T)''$ (note that $(T)'' \in (T)'$ for every operator). Thus, if $\mu_T < \infty$ then by C and B (vi) we have $\mu_T = 1$. Hence, *for* T *of class* C_0 *and with* $\mu_T < \infty$, $(T)'$ *is commutative if and only if* $\mu_T = 1$.

It is not known as vet whether the condition $\mu_T < \infty$ is essential.

The class N_T consists of (meromorphic) functions $\varphi = u/v$ on the open unit disc, where $u, v \in H^\infty$, $v(T)$ is (not necessarily boundedly) invertible, and $\varphi(T)$ is defined by $v(T)^{-1} \cdot u(T)$; cf. [2] (English edition), Chapter IV.

One of the steps in proving the above therems is to establish the following

Theorem D. *For any operator* T *of class* C_0 *on* $\mathfrak{H}$ *, there exists a vector* $h \in \mathfrak{H}$ *such that the restriction of* T *to the subspace spanned by* $h, Th, T^2h, \ldots$ *has the same minimal function as* T *itself. That is,* $u(T) = 0$ *if and only if* $u(T)h = 0$.

Originally, in [3], Theorem D has been proved under the restriction $\mu_T < \infty$, by means of an interesting arithmetical property of the set of inner functions. The general case was obtained in [5] using a Baire category argument.

In Theorems A and B, quasi-similarity cannot be replaced, in general, by similarity. A counter example is constructed in [3]. Choose two inner functions, say a and b , and consider the matrix valued function

$$\Theta = \frac{1}{\sqrt{2}} \begin{bmatrix} a & a \\ b & -b \end{bmatrix} ;$$

this is analytic on the unit disc, bounded in norm by 1 , and its boundary values $\Theta(e^{it})$ are a.e. unitary matrices. Thus Θ is a matrix valued inner function. Form the Hardy-Hilbert space H^2 of 2-vector valued functions and its subspace

$$\mathfrak{H}(\Theta) = H^2 \ominus \Theta H^2 .$$

Then

$$S(\Theta)h = P_{\mathfrak{H}(\Theta)}(e^{it}h) \qquad (h \in \mathfrak{H}(\Theta))$$

defines an operator on $\mathfrak{H}(\Theta)$, of class C_0 , whose minimal function is equal to the smallest inner multiple of the functions a and b . In case a and b have no non-constant inner divisor, then we can show that $S(\Theta)$ is multiplicity free, and hence quasi-similar to $S(m)$, where $m = m_{S(\Theta)} = ab$. But in order that $S(\Theta)$ be similar to $S(m)$, the following necessary condition should hold: existence of functions $x, y \in H^\infty$ satisfying the equation

$$ax + by = 1 .$$

Now this equation certainly cannot hold if we choose $a(\lambda) = \exp \frac{\lambda+1}{\lambda-1}$ and $b(\lambda) =$ a Blaschke product whose zeros are real and tend to 1.

The existence of an operator T which is quasi-similar, but not similar, to an operator $S(m)$, indicates that in any attempt to generalize the Jordan theory of matrices to the Hilbert space situation, one has to replace similarity by some weaker relation, such as quasi-similarity. Clearly, in finite dimensional space, the two notions coincide.

Let us also mention that in Theorems B and C the class N_T of functions cannot be avoided, that is, replaced by the class H^∞. Indeed, if $T = S(\Theta)$ is the multiplicity-free operator constructed above, then an operator B can be found (see [4]) which is in $(T)'$, but cannot be represented in the form $B = u(T)$ with some $u \in H^\infty$.

REFERENCES

B. Sz.-Nagy and C. Foiaş:

[1] Sur les contractions de l'espace de Hilbert. VII. Triangulations canoniques. Fonctions minimum, *Acta Sci. Math.*, 25 (1964), 12-37.

[2] *Analyse harmonique des opérateurs de l'espace de Hilbert* (Budapest, 1967; English edition, 1970).

[3] Opérateurs sans multiplicité, *Acta Sci. Math.*, 30 (1969), 1-18.

[4] Modèle de Jordan pour une classe d'opérateurs de l'espace de Hilbert, *ibidem*, 31 (1970), 91-115.

[5] Compléments à l'étude des opérateurs de classe C_0, *ibidem*, 31 (1970), 287-296.

COLLOQUIA MATHEMATICA SOCIETATIS JÁNOS BOLYAI
5. HILBERT SPACE OPERATORS. TIHANY (HUNGARY), 1970

Representations of von Neumann algebras by sheaves

S. TELEMAN

In a previous note (see [18]) we have shown that any finite von Neumann algebra $\mathfrak{A}$ is $*$-isomorphic to the $*$-algebra of all global sections in a soft sheaf of local $*$-algebras over $\mathbf{C}$, the basis of the sheaf being the maximal spectrum $\mathfrak{M}(\mathfrak{A})$ of the given algebra. This result is an immediate consequence of a theorem of representation by sheaves of the regular (harmonic) rings (see [16], theorems 6 and 10). Theorem 6 of [16] is a generalization (in a certain sense) of a theorem of J. Dauns and K.H. Hofmann, by which these authors give a representation by sheaves for biregular rings (see [1], theorem 1; [2], theorem XI). In order to obtain theorem 1 from [18] we had to require that the algebra be strongly semi-simple. This is always the case with finite von Neumann algebras (see [4], ch. III, § 5.2, prop. 2, cor. 3; [13], ch. II, §6, corollary of theorem 6.2), but not with any von Neumann algebra. For instance, if $\mathcal{H}$ is a Hilbert space of infinite dimension, the algebra $\mathcal{L}(\mathcal{H})$ of all linear continuous operators of $\mathcal{H}$ into itself is a von Neumann algebra which is not strongly semi-simple. Since the representation theorem given in [16] requires the strong semi-simplicity of the ring, we had to restrain ourselves in [18] to the case of finite von Neumann algebras.

On the other hand, it is well known that any von Neumenn algebra is harmonic (regular, in the terminology of [14], [15], [16]; completely regular, in the terminology of [12]; GS-algebra, in the terminology of [22]; see [18] for a proof of this fact). In [16] (see theorem 6.a) we have shown that for any harmonic ring R with unit element there exists a soft sheaf $\mathcal{R}$ (see [16], theorem 9; since R has the unit element, $\mathfrak{M}(R)$ is compact!) of local rings and an epimorphism

$$\varphi: R \longrightarrow \Gamma(\mathfrak{M}(R), \mathcal{R})$$

which turns out to be an isomorphism if, and only if, R is strongly semi-simple. Conversely, for any compact (Hausdorff) space and any soft sheaf $\mathcal{R}$ of local rings, having the space $\mathfrak{M}$ as basis, the ring $\Gamma(\mathfrak{M}, \mathcal{R})$ is harmonic (that is regular) and it is (strongly) semi-simple if, and only if, the sheaf $\mathcal{R}$ is semi-simple (see [16], theorem 12).

Therefore, we have the following problem: given a harmonic ring R, with a unit element, to construct a soft sheaf $\mathcal{R}$ of local rings, having the space $\mathfrak{M}(R)$ as basis (here $\mathfrak{M}(R)$ is the space of maximal (two-sided) ideals with the Stone-Jacobson--Zariski topology (that is the hull-kernel topology)), such that an isomorphism $\varphi: R \to \Gamma(\mathfrak{M}(R), \mathcal{R})$ exists. Theorem 10 from [16] gives the answer for the strongly semi--simple, harmonic rings, having the unit element, but we do not know the answer for the general case. Perhaps, the theory developed briefly by Chr. Mulvey in [9] would solve the general problem.

In this note we shall give the solution for the particular case of von Neumann algebras, by showing that for any von Neumann algebra $\mathfrak{A}$, a soft sheaf $\tilde{\mathfrak{A}}$ of local algebras may be constructed, having the compact (Hausdorff) space $\mathfrak{M}(\mathfrak{A})$ as basis, such that there exists an isomorphism $\varphi: \mathfrak{A} \to \Gamma(\mathfrak{M}(\mathfrak{A}) \tilde{\mathfrak{A}})$.

The constructions which we give are essentially due to R.S. Pierce (see [11], theorems 4.4 and 4.5), and J. Dauns and K.H. Hofmann (see [1]).

We conclude the paper by proving some properties of the given representation.

We wish to remind the reader that for von Neumann algebras satisfying some countability conditions there is a metric reduction theory (see [4], ch. II §3.3), whereas for finite von Neumann algebras there is an algebraic reduction theory (see [13], ch.III), which corresponds to the theory developed in [18]. The following general algebraic reduction theory allows the simultaneous consideration of a von Neumann algebra and of its commutant (if $\mathfrak{A}$ is finite, $\mathfrak{A}'$ may be not finite!). Since factors and local von Neumann algebras are the same thing, the following theory bears much resemblance to reduction theory, but it has weak points too, since the stalks of the sheaf that we construct are in general not C^*-algebras.

1. Let $\mathfrak{A}$ be a von Neumann algebra acting on a Hilbert space $\mathcal{H}$; let $\mathcal{Z}$ be its center (that is $\mathcal{Z} = \mathfrak{A} \cap \mathfrak{A}'$), $\mathfrak{M}(\mathfrak{A})$ the maximal spectrum of $\mathfrak{A}$, that is the set of maximal (two-sided) ideals of $\mathfrak{A}$ with the Stone-Jacobson-Zariski topology. It is well known that the mapping $M \to \mathcal{Z} \cap M$ determines a homeomorphism $\tau: \mathfrak{M}(\mathfrak{A}) \to \to \mathfrak{M}(\mathcal{Z})$ (see, for instance, [18], p. 143). It follows immediately that $\mathfrak{M}(\mathfrak{A})$ is a (hyper)-stonian space (see [3]), hence a totally disconnected space.

Here $\mathfrak{M}(\mathcal{Z})$ is the maximal spectrum of the commutative C^*-algebra $\mathcal{Z}$ and therefore its Stone-Jacobson-Zariski topology coincides with the Gelfand topology, whereas by the Gelfand representation the C^*-algebra $\mathcal{Z}$ is isomorphic to the algebra $C(\mathfrak{M}(\mathcal{Z}))$ of all complex continuous functions defined on $\mathfrak{M}(\mathcal{Z})$ (see [6], ch. I. § 8; [10], theorem 1, ch. III, § 16; [12], theorem 4.2.2).

Since the space $\mathfrak{M}(\mathfrak{A})$ is consequently a separated (Hausdorff) space, it follows that $\mathfrak{A}$ is a harmonic (regular, in the terminology of [14], [15], [16]) algebra.

For any $x \in \mathfrak{A}$ let $C(x)$ be its *central support:* it is the smallest central projection in $\mathfrak{A}$ such that $xC(x) = x$ (see [4], ch. I, § 1.4).

For any $x \in \mathfrak{A}$, let $s(x) \subset \mathfrak{M}(\mathfrak{A})$ be the *support* of x (see [14], § 4) and $s_0(x_0) \subset \mathfrak{M}(\mathcal{Z})$ the *support* of $x_0 \in \mathcal{Z}$. It is obvious that $s_0(x_0)$ defined as in [14], § 4, coincides with the support of the Gelfand transform $\hat{x}_0 \in C(\mathfrak{M}(\mathcal{Z}))$ of $x_0 \in \mathcal{Z}$.

Proposition 1. *For any central projection* $e \in \mathfrak{A}$ *we have*

$$\tau(s(e)) = s_0(e) .$$

Proof. Let $M \in \mathfrak{M}(\mathfrak{A})$ and $x \to x(M)$ be the canonical mapping of $\mathfrak{A}$ onto $\mathfrak{A}/M$. Then $e(M)$ is a central projection in $\mathfrak{A}/M$, which is a quasi-simple ring. It follows that $e(M) = 0$ or $= 1$ in $\mathfrak{A}/M$ and, therefore, $e \in M$ or $1-e \in M$ for any $M \in \mathfrak{M}(\mathfrak{A})$. It follows that $h(1-e) \cup h(e) = \mathfrak{M}(\mathfrak{A})$, where h is the symbol for the hull operation. If $M \in h(1-e) \cap h(e)$, then $1-e \in M$, $e \in M$ and consequently $1 \in M$, which is absurd. It follows that $h(1-e) \cap h(e) = \emptyset$, which shows that the set $h(e) \subset \mathfrak{M}(\mathfrak{A})$ is closed and open in $\mathfrak{M}(\mathfrak{A})$ for any central projection $e \in \mathfrak{A}$. It follows therefore that $s(e) = \complement h(e)$ for any central projection $e \in \mathfrak{A}$

(see [14], §4).

If $m \in \tau(s(e))$, then there exists a unique $M \in \mathfrak{M}(\mathfrak{A})$, such that $m = M \cap \mathfrak{Z}$ and $M \in (s(e))$. Consequently, $e \notin M$, whence $e \notin m$ and therefore $m \in s_0(e)$. Conversely, if $m \in s_0(e)$, we have $e \notin m$, whence $e \notin M$, which shows that $M \in s(e)$ It follows that $m = \tau(M) \in \tau(s(e))$ and the proof is ready.

Proposition 2. *Let* R *be a local ring and* $e \in R$ *a central idempotent. Then* $e = 0$ *or* $e = 1$.

Proof. We recall that by a *local* ring we mean a ring R, having a unit element $1 \neq 0$, and only one maximal (two-sided) ideal M. Because eR and $(1-e)R$ are two-sided ideals, we have either $R = eR$ or $eR \subset M$, and either $(1-e)R = R$ or $(1-e)R \subset M$. We cannot have $eR \subset M$ and $(1-e)R \subset M$ simultaneously, since then we would infer that $1 \in M$, and this is absurd. Consequently, we have either $eR = R$ or $(1-e)R = R$. In the first case there exists an element $f \in R$ such that $ef = 1$, and therefore $e = e^2 f = ef = 1$. In the second case we obtain $e = 0$, which ends the proof.

For any $M \in \mathfrak{M}(\mathfrak{A})$ let $P(M) = \{x \in \mathfrak{A} : C(x) \in M\}$. We have the following fundamental

Proposition 3. *For any* $M \in \mathfrak{M}(\mathfrak{A})$ *the set* $P(M)$ *is the smallest primary ideal contained in* M.

Proof. a) Let $x, y \in P(M)$. Then $C(x), C(y) \in M$, and therefore $C(x) + C(y) - C(x)C(y) \in M$.

From $(x-y)(C(x) + C(y) - C(x)C(y)) = x - y$ we infer immediately $C(x-y) \leq C(x) + C(y) - C(x)C(y)$ and therefore $C(x-y) = C(x-y) \cdot (C(x) + C(y) - C(x)C(y)) \in M$. It follows that $x - y \in P(M)$.

b) For any $z \in \mathfrak{A}$ we have $xzC(x) = xz$, and therefore $C(xz) \leq C(x)$. As before, we deduce that $xz, zx \in P(M)$ for any $x \in P(M)$, $z \in \mathfrak{A}$.

Thus we have shown that $P(M)$ is an ideal of $\mathfrak{A}$.

c) Let $x \in P(M)$. Then from $C(x) \in M$ and $x = C(x)x$ it follows that $x \in M$. Consequently we have $P(M) \subset M$ for any $M \in \mathfrak{M}(\mathfrak{A})$.

d) Let $P \subset M$ be a primary ideal (that is a two-sided ideal of $\mathfrak{A}$

contained in only one maximal ideal of $\mathfrak{A}$). Absurdly, let us suppose that $P(M)\not\subset P$. Then there exists $x_0 \in P(M)$ such that $x_0 \notin P$. It follows that $C(x_0) \notin P$, since if the contrary is true we would have $x_0 = C(x_0)x_0 \in P$. Since $\mathfrak{A}/P$ is a local ring, and the image γ of $C(x_0)$ in $\mathfrak{A}/P$ is a central idempotent, from $\gamma \neq 0$ and from proposition 2 we infer that $\gamma = 1$ in $\mathfrak{A}/P$. It follows that $1 - C(x_0) \in P$, and therefore $1 - C(x_0) \in M$. From $x_0 \in P(M)$ we have that $C(x_0) \in M$ and consequently $1 \in M$, which is absurd. Thus we have shown that for any primary ideal $P \subset M$ we have $P(M) \subset P$.

e) Let us show now that $P(M_0)$ is a primary ideal for any $M_0 \in \mathfrak{M}(\mathfrak{A})$. Indeed, if $M_1 \in \mathfrak{M}(\mathfrak{A})$, $P(M_0) \subset M_1$ and $M_1 \neq M_0$, then there exists a central projection $e \in \mathfrak{A}$, such that $e \in M_0$ and $e \notin M_1$ (see proposition 1). From $C(e) = e$ we have $e \in P(M_0)$, which is absurd. The proof is complete.

Corollary. *For any* $M \in \mathfrak{M}(\mathfrak{A})$ *the ring* $\mathfrak{A}/P(M)$ *is local.*

2. For any $x \in \mathfrak{A}$ we shall denote by $\hat{x}(M)$ the image of x in $\mathfrak{A}/P(M)$ by the canonical homomorphism. For any open set $U \subset \mathfrak{M}(\mathfrak{A})$ let

$$\hat{x}(U) = \{\hat{x}(M);\ M \in U\}.$$

We shall consider the disjoint union

$$\tilde{\mathfrak{A}} = \bigsqcup_{M \in \mathfrak{M}(\mathfrak{A})} \mathfrak{A}/P(M)$$

and the canonical projection

$$\pi : \tilde{\mathfrak{A}} \longrightarrow \mathfrak{M}(\mathfrak{A}).$$

Evidently, we have $\pi(\hat{x}(M)) = M$ for any $x \in \mathfrak{A}$ and $M \in \mathfrak{M}(\mathfrak{A})$.

Proposition 4. *For any* $x, y \in \mathfrak{A}$, *any open sets* $U, V \subset \mathfrak{M}(\mathfrak{A})$ *and any* $\xi_0 \in \hat{x}(U) \cap \hat{y}(V)$, *there exists an open set* $W \subset U \cap V$ *such that* $\pi(\xi_0) \in W$ *and* $\hat{x}(M) = \hat{y}(M)$ *for any* $M \in W$.

Proof. Let $M_0 = \pi(\xi_0)$. From $\xi_0 = \hat{x}(M_0) = \hat{y}(M_0)$ it follows that $x - y \in P(M_0)$ and therefore $C(x-y) \in M_0$. From proposition 1 we infer that there exists an open set $W \subset \mathfrak{M}(\mathfrak{A})$ such that:

$$M_0 \in W \subset U \cap V \quad \text{and} \quad C(x-y) \in M \quad \text{for any } M \in W.$$

We infer that $x-y \in P(M)$ for any $M \in W$ and therefore we have $\hat{x}(M) = \hat{y}(M)$, $M \in W$. The proof is complete.

Corollary. *The family of sets* $\hat{x}(U)$, $x \in \mathfrak{A}$, $U \subset \mathfrak{M}(\mathfrak{A})$ *open, forms a basis for the open sets of a topology on the set* $\tilde{\mathfrak{A}}$.

Proof. It is an immediate consequence of the fact $\tilde{\mathfrak{A}} = \bigcup \hat{x}(U)$, $x \in \mathfrak{A}$ $U \subset \mathfrak{M}(\mathfrak{A})$, open, and of the preceding proposition.

An immediate consequence of the way in which we introduced the topology on $\tilde{\mathfrak{A}}$ is the fact that π is a local homeomorphism, and therefore $\tilde{\mathfrak{A}}$ is a sheaf of local algebras, having the space $\mathfrak{M}(\mathfrak{A})$ as basis.

The continuity of the operations in $\tilde{\mathfrak{A}}$ follows immediately from the fact that for any $x \in \mathfrak{A}$ the mapping

$$\varphi(x) = \hat{x} : M \longrightarrow \hat{x}(M), \qquad M \in \mathfrak{M}(\mathfrak{A}),$$

is a continuous global section in $\tilde{\mathfrak{A}}$. In this way we obtain an algebra homomorphism

$$\varphi : \mathfrak{A} \longrightarrow \Gamma(\mathfrak{M}(\mathfrak{A}), \tilde{\mathfrak{A}}).$$

Proposition 5. φ *is an isomorphism.*

Proof. a) Let $x \in \mathfrak{A}$ be such that $\varphi(x) = 0$. Then $x \in P(M)$ for any $M \in \mathfrak{M}(\mathfrak{A})$ and therefore $C(x) \in M$ for any $M \in \mathfrak{M}(\mathfrak{A})$. It follows that $C(x) = 0$ and consequently, $x = 0$. Therefore φ is a monomorphism.

b) Let $\sigma \in \Gamma(\mathfrak{M}(\mathfrak{A}), \tilde{\mathfrak{A}})$ be a continuous section. For any $M_0 \in \mathfrak{M}(\mathfrak{A})$ there exist an open neighbourhood $U_{M_0} \subset \mathfrak{M}(\mathfrak{A})$ of M_0 and an element $x_{M_0} \in \mathfrak{A}$, such that

$$\sigma(M) = \hat{x}_{M_0}(M), \qquad M \in U_{M_0}.$$

Since $\mathfrak{M}(\mathfrak{A})$ is compact, totally disconnected, we may consider that we have a finite number of open and closed subsets $U_i \subset \mathfrak{M}(\mathfrak{A})$, $i = 1, 2, \ldots, n$, such that $\bigcup_{i=1}^{n} U_i = \mathfrak{M}(\mathfrak{A})$ and $U_i \cap U_j = \phi$ for $i \neq j$, and corresponding elements $x_i \in \mathfrak{A}$, $i = 1, 2, \ldots, n$, such that

$$\sigma(M) = \hat{x}_i(M) \quad \text{for} \quad M \in U_i .$$

Let e_i be the central projection of $\mathcal{A}$ which corresponds to the set U_i by the Gelfand isomorphism. Put $x_0 = \sum_{i=1}^{n} e_i x_i$. Then we have $\varphi(x_0)(M) = \sigma(M)$ for any $M \in \mathfrak{M}(\mathcal{A})$ and therefore $\varphi(x_0) = \sigma$. The proof is concluded.

Proposition 6. *The sheaf* $\tilde{\mathcal{A}}$ *is soft.*

Proof. It is an immediate consequence of the fact that the basis of the sheaf is compact and totally disconnected. Indeed, let $K \subset \mathfrak{M}(\mathcal{A})$ be any closed set and $\sigma: K \longrightarrow \tilde{\mathcal{A}}$ a continuous section. Because K has a fundamental system of paracompact neighbourhoods, from theorem 3.3.1 of [8], ch. II, it follows that σ may be extended to an open neighbourhood $U \supset K$, which we may suppose to be closed too. If σ_0 is the (continuous) extension of σ to U , we shall define $\sigma_1 \in \Gamma(\mathfrak{M}(\mathcal{A}), \tilde{\mathcal{A}})$ by putting $\sigma_1(M) = 0$ for $M \in \complement U$, and $\sigma_1(M) = \sigma_0(M)$, for $M \in U$. The proof is ended. We resume the results thus for obtained in

Theorem 1. *For any von Neumann algebra* $\mathcal{A}$ *there exists a soft sheaf* $\tilde{\mathcal{A}}$ *of local algebras, having the space* $\mathfrak{M}(\mathcal{A})$ *as basis, and an isomorphism*

$$\varphi: \mathcal{A} \longrightarrow \Gamma(\mathfrak{M}(\mathcal{A}), \tilde{\mathcal{A}}) .$$

3. Since the primary ideals $P(M)$, $M \in \mathfrak{M}(\mathcal{A})$, are in general not closed, not even in the norm topology of $\mathcal{A}$, on the algebras $\mathcal{A}/P(M)$ we have not a canonical structure of C^*-algebra (see proposition 24, corollary). Nevertheless, the stalks $\mathcal{A}/P(M)$, $M \in \mathfrak{M}(\mathcal{A})$ of the sheaf $\tilde{\mathcal{A}}$ have algebraic properties which bring them very close to factors.

First of all, they are local algebras. From the fact that $C(x) = C(x^*)$ for any $x \in \mathcal{A}$, we infer that $P(M) = P(M)^*$ for any $M \in \mathfrak{M}(\mathcal{A})$ (we have $M = M^*$, too) and therefore the algebras $\mathcal{A}/P(M)$ have a canonical involution.

Proposition 7. If $\alpha \in \mathcal{A}/P(M)$ *and* $\alpha^*\alpha = 0$, *then* $\alpha = 0$.

Proof. Let $a \in \alpha$. Then we have $a^*a \in P(M)$, and consequently, $C(a^*a) \in M$. From $C(a^*a) = C(a)$ we infer that $a \in P(M)$ and therefore $\alpha = 0$. The proof is complete.

We shall now show that the stalks $\mathcal{A}/P(M)$ are inductive limits of von Neumann algebras, for any $M \in \mathfrak{M}(\mathcal{A})$. Indeed, let S_M be the set of central projections $e \in \mathcal{A}$, such that $e \notin M$. It is obvious that $e, e' \in S_M \Rightarrow ee' \in S_M$

and therefore, with the natural order in the set of central projections, the set S_M is directed downward.

For any $e \in S_M$ the algebra $e\mathfrak{A} \subset \mathfrak{A}$ is a von Neumann algebra (see [4], ch. I, §2.1). For $e' \leq e$ $(e, e' \in S_M)$ we shall define a homomorphism of von Neumann algebras $\varphi^e_{e'}: e\mathfrak{A} \longrightarrow e'\mathfrak{A}$ by putting $\varphi^e_{e'}(x) = e'x$ for $x \in e\mathfrak{A}$. We get obviously an inductive system of algebras with involution. Let

$$\mathfrak{A}_M = \lim_{\substack{\longrightarrow \\ e \in S_M}} \text{ind}\,(e\mathfrak{A}) .$$

We have

Proposition 8. $\mathfrak{A}_M$ *is isomorphic to* $\mathfrak{A}/P(M)$.

Proof. For $e = 1 \in S_M$ the corresponding algebra of the inductive system is $1 \cdot \mathfrak{A} = \mathfrak{A}$. Therefore we have a canonical homomorphism $\Theta: \mathfrak{A} \longrightarrow \mathfrak{A}_M$, and for any $e \in S_M$ we have a canonical homomorphism $\Theta^e: e\mathfrak{A} \longrightarrow \mathfrak{A}_M$ such that $\Theta^e(ex) = \Theta(x)$ tor any $x \in \mathfrak{A}$. (We put $\Theta^1 = \Theta$.) It follows that Θ is an epimorphism. If $\Theta(x) = 0$ then there exists $e \in S_M$ such that $ex = 0$. It follows that $(1-e)x = x$ and therefore $C(x) \leq 1-e$. From $e \notin M$ we have $1-e \in M$ and consequently $C(x) \in M$, which shows that $x \in P(M)$. Conversely, if $x \in P(M)$ then for $e = 1 - C(x)$ we have $ex = 0$, $e \notin M$ which shows that $\Theta(x) = 0$. Thus we have shown that $P(M) = \ker \Theta$, and the proposition is proved.

If $e, e' \in S_M$ and $e \leq e'$, we have $e\mathfrak{A} \subset e'\mathfrak{A}$. For any $e \in S_M$ let us denote by $\sigma_e(x)$ the spectrum of $x \in e\mathfrak{A}$ in this algebra: it is the set of all complex numbers λ such that $(\lambda e - x)^{-1}$ does not exist in $e\mathfrak{A}$. It is obvious that for any $x \in e\mathfrak{A}$ and $e \leq e'$, $e, e' \in S_M$, we have $\sigma_e(x) \subset \sigma_{e'}(x)$.

From what we have already shown, for any $\xi \in \mathfrak{A}_M$ there exists $x \in \mathfrak{A}$, such that $\xi = \Theta(x)$. If $\sigma(\xi)$ is the spectrum of $\xi \in \mathfrak{A}_M$, we have $(M \in \mathfrak{M}(\mathfrak{A}))$

Proposition 9. *For any* $x \in \mathfrak{A}$ *we have* $\sigma(\Theta(x)) = \bigcap_{e \in S_M} \sigma_e(ex)$.

Proof. If $\lambda \in \sigma(\Theta(x))$, then $(\lambda 1_M - \Theta(x))^{-1}$ does not exist in $\mathfrak{A}_M$, where 1_M is the unit element of $\mathfrak{A}_M$ (which we identify with $\mathfrak{A}/P(M)$ by the isomorhism given in the proof of the preceding proposition). But then, for any $e \in S_M$ the element $(\lambda e - ex)^{-1}$ does not exist in $e\mathfrak{A}$, and therefore $\lambda \in \sigma_e(ex)$. If

$\lambda \notin \sigma(\Theta(x))$, then $\eta = (\lambda 1_M - \Theta(x))^{-1}$ exists in $\mathfrak{A}_M$ and therefore we have

$$\eta(\lambda 1_M - \Theta(x)) = (\lambda 1_M - \Theta(x))\eta = 1_M .$$

Let $y \in \mathfrak{A}$ be such that $\eta = \Theta(y)$. Then we have

$$y(\lambda 1 - x) - 1 \in P(M), \quad (\lambda 1 - x)y - 1 \in P(M).$$

It follows that there exist $e', e'' \in S_M$ such that

$$e'(y(\lambda 1 - x) - 1) = 0 \qquad e''((\lambda 1 - x)y - 1) = 0 .$$

For $e = e'e''$ we have $e \in S_M$ and

$$(ey)(\lambda e - ex) = e, \quad (\lambda e - ex)(ey) = e ,$$

which shows that $\lambda \notin \sigma_e(ex)$. We infer that $\lambda \notin \bigcap_{e \in S_M} \sigma_e(ex)$ and the proof is ended.

Corollary. *Every element* $\xi \in \mathfrak{A}_M$ *has a compact, non-void spectrum.*

Proof. It follows immediately from the fact that the family of compact, non-void sets $\sigma_e(ex)$ has the finite intersection property.

4. As in any $\mathbf{C}$-algebra with involution, we may define the *hermitian* elements of $\mathfrak{A}_M$ by the condition $\xi = \xi^*$, and the *normal* elements of $\mathfrak{A}_M$ by the condition $\xi\xi^* = \xi^*\xi$.

The results that we have obtained in the preceding paragraphs allow the definition of a functional calculus with germs of holomorphic functions, for any element of $\mathfrak{A}_M$, and that of a functional calculus with germs of continuous complex functions, for any normal element of $\mathfrak{A}_M$.

Lemma 1. a) *For any hermitian element* $\xi \in \mathfrak{A}_M$ *and for any* $e \in S_M$ *there exists a hermitian element* $x_e \in e\mathfrak{A}$ *such that* $\Theta^e(x_e) = \xi$.

b) *For any normal element* $\xi \in \mathfrak{A}_M$ *and for any* $e \in S_M$ *there exists a normal element* $x_e \in e\mathfrak{A}$ *such that* $\Theta^e(x_e) = \xi$.

Proof. a) Let $x \in \mathfrak{A}$ be such that $\Theta(x) = \xi$. Then we have $x - x^* \in P(M)$. Let $e_0 = 1 - C(x - x^*)$. We have $e_0 \in S_M$ and $e_0(x - x^*) = 0$. Consequently, if we put $x_e = ee_0x$, we have $x_e = x_e^*$ and $\Theta^e(x_e) = \xi$.

b) If $x \in \mathfrak{A}$ is such that $\Theta(x) = \xi$, we have $xx^* - x^*x \in P(M)$. Let $e_0 = 1 - C(xx^* - x^*x)$. As before, it follows that $e_0 \in S_M$ and $0 = e_0(xx^* - x^*x)$. For any $e \in S_M$ we put $x_e = ee_0x$. Then $x_e \in e\mathfrak{A}$ is normal, and $\Theta^e(x_e) = \xi$. The lemma is proved.

For any compact set $K \subset \mathbf{C}$, we shall denote by $\mathcal{C}(K)$ the $\mathbf{C}$-algebra of germs of complex continuous functions defined on neighbourhoods of K; by $\mathcal{O}(K)$ we shall denote the $\mathbf{C}$-algebra of germs of holomorphic functions defined on neighbourhoods of K. We have a canonical inclusion $\mathcal{O}(K) \subset \mathcal{C}(K)$.

Let $\xi \in \mathfrak{A}_M$ and let $\sigma(\xi)$ be its spectrum which, by the corollary of proposition 9, is compact and non-void. Let $1 \in \mathcal{O}(\sigma(\xi))$ be the unit element of this algebra (the germ of the constant function, equal to 1), and $\iota \in \mathcal{O}(\sigma(\xi))$ the germ of the identical function on $\mathbf{C}$. We have

Proposition 10. *There exists a homomorphism of* $\mathbf{C}$*-algebras* $\tau : \mathcal{O}(\sigma(\xi)) \to \mathfrak{A}_M$ *such that* $\tau(1) = 1_M$ *and* $\tau(\iota) = \xi$.

Proof. For any germ $\tilde{f} \in \mathcal{O}(\sigma(\xi))$ there exist an open set $U \supset \sigma(\xi)$ and a holomorphic function $f_U : U \to \mathbf{C}$, whose germ on $\sigma(\xi)$ is $\tilde{f}$.

From proposition 9 we infer that there exist an $e \in S_M$ such that $\sigma_e(ex) \subset U$, where $x \in \mathfrak{A}$ is chosen so that $\Theta(x) = \xi$.

It follows that the element $f_U(ex) \in e\mathfrak{A}$ is defined (by the functional calculus in the C^*-algebra $e\mathfrak{A}$). We shall put

$$\tau(\tilde{f}) = \Theta^e(f_U(ex)).$$

It is not difficult to prove that the mapping τ is thus correctly defined and that it is a homomorphism of $\mathbf{C}$-algebras having the required properties. The proposition is proved.

For any $\tilde{f} \in \mathcal{O}(\sigma(\xi))$ we shall write $\tilde{f}(\xi) = \tau(\tilde{f})$.

Proposition 11. *Let* $\xi \in \mathfrak{A}_M$ *be a normal element. Then there exists a homomorphism of* $\mathbf{C}$*-algebras with involution*

$$\tau : \mathcal{C}(\sigma(\xi)) \longrightarrow \mathfrak{A}_M,$$

such that $\tau(1) = 1_M$ *and* $\tau(\iota) = \xi$.

Proof. For any germ $\tilde{g} \in \mathcal{C}(\sigma(\xi))$ there exist an open set $\mathcal{U} \supset \sigma(\xi)$ and a continuous function $g_{\mathcal{U}}: \mathcal{U} \longrightarrow \mathbf{C}$, whose germ on $\sigma(\xi)$ is $\tilde{g}$.

From lemma 1 it follows that there exists a normal element $x \in \mathfrak{A}$ such that $\Theta(x) = \xi$. Then $ex \in e\mathfrak{A}$ is normal for any $e \in S_M$ and we have $\Theta^e(ex) = \xi$.

From proposition 9 it follows that there exists an element $e \in S_M$ such that $\sigma_e(ex) \subset \mathcal{U}$. Then $g_{\mathcal{U}}(ex) \in e\mathfrak{A}$ is defined (by the functional calculus in C^*-algebras (see [5] §1. 5)). We shall put

$$\tau(\tilde{g}) = \Theta^e(g_{\mathcal{U}}(ex)).$$

It is easy to prove that the mapping τ is thus correctly defined and that it is a homomorphism of $\mathbf{C}$-algebras with involution having the required properties. The proposition is proved.

For any $\tilde{g} \in \mathcal{C}(\sigma(\xi))$ we shall write $\tilde{g}(\xi) = \tau(\tilde{g})$.

In the algebras $\mathfrak{A}_M$, $M \in \mathfrak{M}(\mathfrak{A})$, one has an analogue of the spectral mapping theorem of N. Dunford. Indeed, we have

Proposition 12. *For any* $\xi \in \mathfrak{A}_M$, $M \in \mathfrak{M}(\mathfrak{A})$, *and* $\tilde{f} \in \mathcal{O}(\sigma(\xi))$, *we have* $\sigma(\tilde{f}(\xi)) = \tilde{f}(\sigma(\xi))$.

Proof. In the second member we have, in fact, the image of the set $\sigma(\xi)$ by any representative $f \in \tilde{f}$. Since this image does not depend on the choice of the representative, the notation is correct.

Let $\lambda \in \sigma(\tilde{f}(\xi))$ and, absurdly, suppose that $\lambda \notin \tilde{f}(\sigma(\xi))$. Let $f_{\mathcal{U}} \in \tilde{f}$ be a representative of the germ $\tilde{f}$, defined on the open set $\mathcal{U} \supset \sigma(\xi)$. Since $\lambda \notin f_{\mathcal{U}}(\sigma(\xi))$ there exists an open set $V \subset \mathcal{U}$ such that $V \supset \sigma(\xi)$ and $f_{\mathcal{U}}(z) \neq \lambda$ for any $z \in V$. We infer that the function $g_V = (\lambda 1 - f_{\mathcal{U}})^{-1}$ is holomorphic on V. Let $\tilde{g} \in \mathcal{O}(\sigma(\xi))$ be the corresponding germ from $\mathcal{O}(\sigma(\xi))$ and $\eta = \tilde{g}(\xi)$. We have

$$\eta(\lambda 1_M - \tilde{f}(\xi)) = (\lambda 1_M - \tilde{f}(\xi))\eta = 1_M,$$

and, consequently, $\lambda \notin \sigma(\tilde{f}(\xi))$, which is absurd.

If $\lambda \in \tilde{f}(\sigma(\xi))$, there exists a $\mu \in \sigma(\xi)$ such that $f_{\mathcal{U}}(\mu) = \lambda$, for any representative $f_{\mathcal{U}} \in \tilde{f}$. It follows that we can write $f_{\mathcal{U}} - \lambda 1 = (\iota - \mu 1) g_{\mathcal{U}}$, where $g_U: U \longrightarrow \mathbf{C}$ is a holomorphic function, and therefore

$$\tilde{f}(\xi) - \lambda 1_M = (\xi - \mu 1_M)\tilde{g}(\xi),$$

where $\tilde{g}$ is the germ of g_U on $\sigma(\xi)$ and hence we have $\lambda \in \sigma(\tilde{f}(\xi))$. The proof is ended.

In a similar fashion we may prove.

Proposition 13. *For any* $\xi \in \mathfrak{A}_M$ *we have* $\sigma(\xi^*) = \overline{\sigma(\xi)}$.

5. The set H_M of the hermitian elements of $\mathfrak{A}_M$ is obviously a real vector subspace, whereas the set P_M of the elements of $\mathfrak{A}_M$ having the form

$$\sum_{i=1}^{n} \xi_i^* \xi_i \qquad (\xi_i \in \mathfrak{A}_M;\ i = 1,2,\dots,n;\ n \in N)$$

is a cone, and we have $P_M \subset H_M$.

Proposition 14. *Every* $\eta \in P_M$ *can be uniquely written in the form* $\eta = \chi^2$, *where* $\chi \in P_M$.

Proof. Let $x_i \in \mathfrak{A}$ be such that $\Theta(x_i) = \xi_i$ $(i = 1,2,\dots,n)$, where $\eta = \sum_{i=1}^{n} \xi_i^* \xi_i$. Let $y = \sum_{i=1}^{n} x_i^* x_i$. Then y is a positive element of $\mathfrak{A}$, and therefore there exists only one positive element $h \in \mathfrak{A}$ such that $y = h^2$. We obtain $\eta = \Theta(y) = \chi^2$, where $\chi = \Theta(h)$. Evidently, we have $\chi \in P_M$ and thus the existence is proved.

Let now $\chi_1, \chi_2 \in P_M$ such that $\chi_1^2 = \chi_2^2$. From the preceding we deduce that there exist $\chi', \chi'' \in P_M$ such that $\chi'^2 = \chi_1$, $\chi''^2 = \chi_2$. From lemma 1 we infer that there exist hermitian elements $x', x'' \in \mathfrak{A}$ such that $\Theta(x') = \chi'$, $\Theta(x'') = \chi''$. It follows that $\Theta(x'^2) = \chi_1$, $\Theta(x''^2) = \chi_2$. By putting $x_1 = x'^2$, $x_2 = x''^2$ we infer that there exist *positive* elements $x_1, x_2 \in \mathfrak{A}$ such that $\Theta(x_1) = \chi_1$, $\Theta(x_2) = \chi_2$. From $\chi_1^2 = \chi_2^2$ we infer that $x_1^2 - x_2^2 \in P(M)$, and consequently, by putting $e = 1 - C(x_1^2 - x_2^2)$, we have $e \notin M$ and $e(x_1^2 - x_2^2) = 0$; hence $(ex_1)^2 = (ex_2)^2$. Since $ex_1, ex_2 \in \mathfrak{A}$ are positive, we infer that $ex_1 = ex_2$ and therefore $\chi_1 = \Theta(x_1) = \Theta(ex_1) = \Theta(ex_2) = \Theta(x_2) = \chi_2$, which concludes the proof.

Proposition 15. *The set* $P_M - 1_M$ *is convex and absorbant in* H_M.

Proof. Since the set P_M is a cone, it is convex and therefore $P_M - 1_M$ is convex too.

Let $\chi \in H_M$. Then in virtue of lemma 1, there exists a hermitian element $h \in \mathfrak{A}$ such that $\Theta(h) = \chi$. For any real number t such that $|t| < (1+\|h\|)^{-1}$, the operator $1 + th$ is positive in $\mathfrak{A}$, and consequently, $1_M + t\chi$ is positive in $\mathfrak{A}_M$. The proof is ended.

Proposition 16. *The cone* P_M *does not contain vector subspaces different from* $\{0\}$.

Proof. Let us suppose that there exists an element $\chi \in H_M$ such that $\chi \in P_M$ and $-\chi \in P_M$. Then there exists $h_1 \in \mathfrak{A}$, $h_1 \geq 0$, and $h_2 \in \mathfrak{A}$, $h_2 \geq 0$ such that $\Theta(h_1) = \chi$, $\Theta(h_2) = -\chi$. It follows that $\Theta(h_1 + h_2) = 0$ and therefore $h_1 + h_2 \in P_M$. We deduce that $C(h_1 + h_2) \in M$ and therefore we have $e = 1 - C(h_1 + h_2) \notin M$; whence $e \in S_M$. From $e(h_1 + h_2) = 0$ we obtain $eh_1 = eh_2 = 0$ and consequently $\chi = \Theta(eh_1) = 0$, which completes the proof.

Corollary. *The cone* P_H *induces an order relation on* H_M *by which this space becomes an ordered vector space.*

Later we shall see that, in general, the order relation induced on H_M by the cone P_H is not archimedian (see proposition 24).

6. Because a von Neumann algebra $\mathfrak{A}$ is given together with a Hilbert space $\mathcal{H}$, on which the elements of $\mathfrak{A}$ act as linear continuous operators ($\mathfrak{A} \subset \mathcal{L}(\mathcal{H})$), we have the problem of representing by a sheaf the space $\mathcal{H}$.

More generally, we may consider a representation $\varrho : \mathfrak{A} \to \mathcal{L}(\mathcal{K})$, where $\mathcal{K}$ is an arbitrary Hilbert space and ϱ is a homomorphism of C^*-algebras. Then $\mathcal{K}$ becomes a unitary left $\mathfrak{A}$-module, and so we get at the more general problem of representing unitary left $\mathfrak{A}$-modules.

For harmonic and (strongly) semi-simple algebras we have developed in [17] and [21] the general theory of the representation of modules by sheaves.

In this case, however, although a von Neumann algebra is always harmonic it is not always strongly semi-simple, and therefore the theory developed in [17] and [21] may not be applied.

Instead, we shall use some ideas of the general theory of P.S. Pierce (see [11]) and Chr. Mulvey (see [9]) and we shall apply them to the case of von Neumann algebras.

Let $\mathcal{K}$ be a unitary left $\mathcal{A}$-module. For any $M \in \mathfrak{M}(\mathcal{A})$ we shall consider the $\mathcal{A}$-submodule $P(M)\mathcal{K} \subset \mathcal{K}$ and the quotient $\mathcal{A}$-module $\mathcal{K}_M = \mathcal{K}/P(M)\mathcal{K}$. Let $\tilde{\mathcal{K}} = \bigsqcup_{M \in \mathfrak{M}(\mathcal{A})} \mathcal{K}_M$ be the disjoint union and $\pi: \mathcal{K} \to \mathfrak{M}(\mathcal{A})$ the canonical projection.

For any $a \in \mathcal{K}$ let $\hat{a}(M)$ be the canonical image of a in $\mathcal{K}_M$ and for any open set $U \subset \mathfrak{M}(\mathcal{A})$ let $\hat{a}(U) = \{\hat{a}(M);\ M \in U\}$.

Proposition 17. *Let $a, b \in \mathcal{K}$ and $U, V \subset \mathfrak{M}(\mathcal{A})$ be open sets. If $\alpha \in \hat{a}(U) \cap \hat{b}(V)$, then there exists an open neighbourhood W of $M_0 = \pi(\alpha)$ in $\mathfrak{M}(\mathcal{A})$ such that $\hat{a}(M) = \hat{b}(M)$ for any $M \in W$.*

Proof. From $\alpha = \hat{a}(M_0) = \hat{b}(M_0)$ we infer that $a - b \in P(M_0)\mathcal{K}$. It follows that there exist $x_i \in P(M_0)$ and $c_i \in \mathcal{K}$ $(i = 1, 2, \dots, n)$ such that $a - b = \sum_{i=1}^{n} x_i c_i$. From $x_i \in P(M_0)$ $(i = 1, 2, \dots, n)$ we deduce that there exists an open neighbourhood $W \subset U \cap V$ of M_0 in $\mathfrak{M}(\mathcal{A})$ such that $x_i \in P_M$ for any $M \in W$ and $i = 1, 2, \dots, n$. It follows that $\hat{a}(M) = \hat{b}(M)$ for any $M \in W$, and the proposition is proved.

Corollary. *The family of sets $\hat{a}(U)$, $a \in \mathcal{K}$, $U \subset \mathfrak{M}(\mathcal{A})$, open, forms a basis for the open sets of a topology on the set $\tilde{\mathcal{K}}$.*

Proof. It is an immediate consequence of the fact that $\tilde{\mathcal{K}} = \bigcup \hat{a}(U)$, $a \in \mathcal{K}$, $U \in \mathfrak{M}(\mathcal{A})$, open, and of the preceding proposition.

An immediate consequence of the way in which we introduced the topology on $\tilde{\mathcal{K}}$ is the fact that π is a local homeomorphism, and therefore $\tilde{\mathcal{K}}$ is a sheaf of complex vector spaces, having the space $\mathfrak{M}(\mathcal{A})$ as basis. (Moreover, $\tilde{\mathcal{K}}$ is a sheaf of $\mathcal{A}$-modules) The stalks of the sheaf $\tilde{\mathcal{K}}$ are the $\mathcal{A}$-modules $\mathcal{K}_M$, $M \in \mathfrak{M}(\mathcal{A})$.

For any $a \in \mathcal{K}$ the mapping

$$\rho(a) = \hat{a}: M \to \hat{a}(M), \qquad M \in \mathfrak{M}(\mathcal{A})$$

is a continuous global section in $\tilde{\mathcal{K}}$, whereas by the mapping $a \longrightarrow \hat{a}$ we define a homomorphism of $\mathfrak{A}$-modules:

$$\rho: \mathcal{K} \to \Gamma(\mathfrak{M}(\mathfrak{A}), \tilde{\mathcal{K}}) .$$

Proposition 18. ρ *is an isomorphism.*

Proof. a) Let $a \in \mathcal{K}$ be such that $\rho(a) = 0$. Then we have $\hat{a}(M) = 0$ for any $M \in \mathfrak{M}(\mathfrak{A})$ and therefore $a \in P(M)\mathcal{K}$, for any $M \in \mathfrak{M}(\mathfrak{A})$. It follows that for any $M_0 \in \mathfrak{M}(\mathfrak{A})$ there exist $a_i \in \mathcal{K}$, $x_i \in P(M_0)$ $(i = 1,2,\dots,n)$, and an open and closed neighbourhood U_{M_0} of M_0, such that $a = \sum_{i=1}^{n} x_i a_i$ and $ex_i = 0$ $(i = 1,2,\dots,n)$, where e is the central projection which corresponds to the characteristic function of U_{M_0} by the Gelfand isomorphism. Due to the fact that the space $\mathfrak{M}(\mathfrak{A})$ is compact and totally disconnected, we deduce that there exist central projections $e_j \in \mathfrak{A}$ $(j = 1,2,\dots,m)$ such that $\sum_{j=1}^{m} e_j = 1$, and elements $a_{ji} \in \mathcal{K}$, $x_{ji} \in \mathfrak{A}$ $(j = 1,2,\dots,m;\ i = 1,2,\dots,n(j))$ such that $a = \sum_{i=1}^{n(j)} x_{ji} a_{ji}$ $(j = 1,\dots,m)$ and $e_j x_{ji} = 0$ $(i = 1,2,\dots,n(j))$. It follows that

$$a = 1a = \sum_{j=1}^{m} e_j a = \sum_{j=1}^{m} e_j \sum_{i=1}^{n(j)} x_{ji} a_{ji} = 0 ,$$

and therefore ρ is a monomorphism.

b) Let $\sigma: \mathfrak{M}(\mathfrak{A}) \to \tilde{\mathcal{K}}$ be any global continuous section. Then for any $M_0 \in \mathfrak{M}(\mathfrak{A})$ there exist an open and closed set $U_{M_0} \ni M_0$ and an element $a_{M_0} \in \mathcal{K}$ such that $\hat{a}_{M_0}(M) = \sigma(M)$ for any $M \in U_{M_0}$. Again, by the compacteness and total disconnectedness of $\mathfrak{M}(\mathfrak{A})$, we deduce that there exists an open covering of $\mathfrak{M}(\mathfrak{A})$ by a finite number of (open and closed) sets $U_i \subset \mathfrak{M}(\mathfrak{A})$ $(i = 1,2,\dots,n)$ such that $\bigcup_{i=1}^{n} U_i = \mathfrak{M}(\mathfrak{A})$ and $U_i \cap U_j = \emptyset$ for $i \neq j$. Let e_i be the central projection of $\mathfrak{A}$ which corresponds to the characteristic function of U_i by the Gelfand isomorphism. Then we have $e_i \in P(M)$ for any $M \in \complement U_i$ $(i = 1,2,\dots,n)$. Also, there exist elements $a_i \in \mathcal{K}$ $(i = 1,2,\dots,n)$ such that $\hat{a}_i(M) = \sigma(M)$ for any $M \in U_i$. It follows that $\rho(\sum_{i=1}^{n} e_i a_i) = \sigma$, and thus ρ is an epimorphism by which the proof is ended.

Proposition 19. *$\tilde{\mathcal{K}}$ is a $\tilde{\mathfrak{A}}$-module.*

Proof. We define a structure of $\mathfrak{A}_M$-module on $\mathcal{K}_M$ by the equality

$$\varphi(x)(M)\,\rho(a)(M) = \rho(xa)(M),$$

where $M \in \mathfrak{M}(\mathfrak{A})$, $x \in \mathfrak{A}$, $a \in \mathcal{K}$. It is easy to show that the definition is correct. The proposition follows now immediately from the preceding equality.

Corollary. *The sheaf $\tilde{\mathcal{K}}$ is soft.*

Proof. It follows immediately from a general theorem (see [8], ch. II. theorem 3.7.1) or as for proposition 6.

Consequently, we have proved the following

Theorem 2. *For any unitary left $\mathfrak{A}$-module $\mathcal{K}$ there exist a soft $\mathfrak{A}$-module $\tilde{\mathcal{K}}$ and an isomorphism*

$$\rho:\ \mathcal{K} \to \Gamma(\mathfrak{M}(\mathfrak{A}), \tilde{\mathcal{K}}).$$

The following proposition shows that the stalk $\mathcal{K}_M$ of $\tilde{\mathcal{K}}$ is the localization of $\mathcal{K}$ in the ideal $P(M)$, $M \in \mathfrak{M}(\mathfrak{A})$.

Proposition 20. *For any $M \in \mathfrak{M}(\mathfrak{A})$ we have the isomorphism*

$$\varphi_M:\ \mathfrak{A}_M \underset{\mathfrak{A}}{\otimes} \mathcal{K} \simeq \mathcal{K}_M$$

given by $\sum_{i=1}^{n} \xi_i \otimes a_i \longrightarrow \sum_{i=1}^{n} \xi_i \hat{a}_i(M)$.

Proof. The application $(\xi, a) \to \xi\hat{a}(M)$ of $\mathfrak{A}_M \times \mathcal{K}$ in $\mathcal{K}_M$ is obviously $\mathbf{Z}$-bilinear and $\mathfrak{A}$-balanced. It follows that the induced linear mapping from the proposition is well defined. On the other hand, it is obvious that φ_M is an epimorphism.

Let now $\beta:\ \mathfrak{A}_M \times \mathcal{K} \to \mathcal{L}$ be a $\mathbf{Z}$-bilinear $\mathfrak{A}$-balanced mapping in the Abelian group $\mathcal{L}$. We shall define the $\mathbf{Z}$-linear mapping $\gamma:\ \mathcal{K}_M \to \mathcal{L}$ by putting

$$\gamma(\alpha) = \beta(1_M, a)\ \text{, where}\ \ a \in \alpha.$$

First of all, the definition is correct. Indeed, let us suppose that $a \in P(M)\mathcal{K}$. Then we have $a = \sum_{i=1}^{n} x_i a_i$, where $x_i \in P(M)$ and $a_i \in \mathcal{K}$ $(i = 1, 2, \dots, n)$. We have $\beta(1_M, a) = \sum_{i=1}^{n} \beta(1_M, x_i a_i) = \sum_{i=1}^{n} \beta(1_M x_i, a_i) = 0$, because from $x_i \in P(M)$ we have $\Theta(x_i) = 0$ and therefore $1_M \cdot x_i = 1_M \cdot \Theta(x_i) = 0$. Thus the mapping is correctly defined.

Now for any $(\xi, a) \in \mathfrak{A}_M \times \mathcal{K}$ and $x \in \xi$ we have $(b = xa)$ $(\gamma\varphi_M)(\xi \otimes a) = \gamma(\xi \hat{a}(M)) = \gamma(\hat{b}(M)) = \beta(1_M, xa) = \beta(1_M \cdot x, a) = \beta(\xi, a)$, and thus $\gamma\varphi_M \mu = \beta$, where $\mu: \mathfrak{A}_M \times \mathcal{K} \to \mathfrak{A}_M \otimes_{\mathfrak{A}} \mathcal{K}$ is the canonical $\mathbf{Z}$-bilinear mapping. the proposition is proved.

7. We shall define the *support* $\operatorname{supp} \mathcal{K}$ of an $\mathfrak{A}$-module $\mathcal{K}$ as being the support of the associated sheaf $\tilde{\mathcal{K}}$. Consequently, it is the set of points $M \in \mathfrak{M}(\mathfrak{A})$ such that for any open neighbourhood $U \ni M$ we have $\tilde{\mathcal{K}} | U \neq 0$. It follows that $\operatorname{supp} \mathcal{K} \subset \mathfrak{M}(\mathfrak{A})$ is a closed set for any (unitary, left) $\mathfrak{A}$-module $\mathcal{K}$.

Proposition 21. a) $\mathcal{K} = \{0\}$ *if and only if* $\operatorname{supp} \mathcal{K} = \emptyset$.
b) *If* $\mathcal{K}$ *is faithful, then* $\operatorname{supp} \mathcal{K} = \mathfrak{M}(\mathfrak{A})$.
c) *If* $\mathfrak{A}$ *is strongly semi-simple, and* $\operatorname{supp} \mathcal{K} = \mathfrak{M}(\mathfrak{A})$ *then* $\mathcal{K}$ *is faithful.*
d) *If* $\mathcal{K}$ *is a simple* $\mathfrak{A}$ *-module, then* $\operatorname{card} \operatorname{supp} \mathcal{K} = 1$.

Proof. a) If $\mathcal{K} = \{0\}$, then obviously $\mathcal{K}_M = \{0\}$ for any $M \in \mathfrak{M}(\mathfrak{A})$, and therefore $\operatorname{supp} \mathcal{K} = \emptyset$. Conversely, if $\mathcal{K} \neq \{0\}$ then there exists an $a \in \mathcal{K}$ such that $a \neq 0$, and therefore $\hat{a}(M) \neq 0$ $(M \in \mathfrak{M}(A))$. But then $\tilde{\mathcal{K}} \neq 0$, and consequently $\operatorname{supp} \mathcal{K} \neq 0$.

b) If there is a $M_0 \in \mathfrak{M}(\mathfrak{A}) \setminus \operatorname{supp} \mathcal{K}$, then there exists a central projection $e \in \mathfrak{A}$ such that $e \neq 0$ and $s(e) \subset \complement \operatorname{supp} \mathcal{K}$. But then we have $\hat{e}(M)\hat{a}(M) = 0$ for any $M \in \mathfrak{M}(\mathfrak{A})$ and $a \in \mathcal{K}$, because one or the other of the two factors is zero. It follows that $ea = 0$ for any $a \in \mathcal{K}$, because ϱ is a monomorphism, and thus $\mathcal{K}$ is not faithful.

c) If $\mathfrak{A}$ is strongly semi-simple and $\operatorname{supp} \mathcal{K} = \mathfrak{M}(\mathfrak{A})$, then we may apply theorem 2.b from [17] in order to obtain the desired implication.

d) If $\mathcal{K}$ is a simple $\mathfrak{A}$-module, then from a) we have $\operatorname{supp} \mathcal{K} \neq \emptyset$. If $M_0, M_1 \in \operatorname{supp} \mathcal{K}$, and $M_0 = M_1$, let $e \in \mathfrak{A}$ be a central projection such that $e \in M_0$ and $e \notin M_1$. Let us consider the $\mathfrak{A}$-modules $\mathcal{K}_0 = e\mathcal{K} \subset \mathcal{K}$ and $\mathcal{K}_1 = (1-e)\mathcal{K} \subset \mathcal{K}$. We have $\mathcal{K} = \mathcal{K}_0 \oplus \mathcal{K}_1$, $\mathcal{K}_0 \neq \{0\}$, $\mathcal{K}_1 \neq \{0\}$ and therefore $\mathcal{K}$ would not be simple. The proof is ended.

Example. Let $\mathfrak{A}$ be a von Neumann algebra and $\mathfrak{N}$ its strong radical

$(\mathfrak{N} = \cap M,\ M \in \mathfrak{M}(\mathfrak{A}))$. Then $\mathfrak{A}/\mathfrak{N}$ is an $\mathfrak{A}$ -module, which is faithful if and only if $\mathfrak{N} = \{0\}$. Since we have $\mathrm{supp}\,\mathfrak{A}/\mathfrak{N} = \mathfrak{M}(\mathfrak{A})$, it follows that the implication " $\mathrm{supp}\,\mathcal{X} = \mathfrak{M}(\mathfrak{A}) \Longrightarrow \mathcal{X}$ faithful" is true for any module $\mathcal{X}$ if, and only if $\mathfrak{A}$ is strongly semi-simple.

Proposition 22. *For any left maximal ideal $L \subset \mathfrak{A}$ there exists only one two-sided maximal ideal $M \in \mathfrak{M}(\mathfrak{A})$ such that $P(M) \subset L$.*

Proof. $\mathfrak{A}/L$ is a simple $\mathfrak{A}$ -module, and therefore we have $\mathrm{supp}\,\mathfrak{A}/L = \{M\}$, where $M \in \mathfrak{M}(\mathfrak{A})$. It follows that for any $M_0 \in \mathfrak{M}(\mathfrak{A})$, $M_0 \neq M$, there exists a central projection $e_0 \in M$ such that $e_0 \in S_{M_0}$ and $e_0\mathfrak{A} \subset I$, that is $e_0 \in L$. It follows that for any central projection $e \in \mathfrak{A}$ such that $e \in M$, we have $e \in L$, and consequently $x \in L$ for any $x \in P(M)$.

The proposition will be proved if we take into consideration the fact that for $M_1 \neq M_2$ we have $P(M_1) + P(M_2) = \mathfrak{A}$.

Proposition 23. *A primitive von Neumann algebra is a factor.*

Proof. Let $\mathfrak{A}$ be a primitive von Neumann algebra. Then there exists a complex vector space $\mathcal{X}$ and a $\mathbf{C}$ -algebra monomorphism $\mu : \mathfrak{A} \to \mathrm{End}_{\mathbf{C}}\mathcal{X}$, such that $\mathcal{X}$ is a simple $\mathfrak{A}$ -module (which is faithful, because μ is a monomorphism). From proposition 21 we infer that $\mathfrak{M}(\mathfrak{A}) = \mathrm{supp}\,\mathcal{X} = \{M_0\}$, which shows that $\mathfrak{A}$ is a local algebra, and hence a factor.

8. Unlike the case of von Neumann algebras, in which the order relation induced by the cone of positive elements in the real vector space of hermitian elements is archimedian, this property is in general no longer true for the stalks $\mathfrak{A}_M$, $M \in \mathfrak{M}(\mathfrak{A})$. In fact, we have

Proposition 24. *If $M \in \mathfrak{M}(\mathfrak{A})$ and if the order relation induced by the cone P_M in H_M is archimedian, then M is a P -point.*

Proof. Let $f\colon\ \mathfrak{M}(\mathfrak{A}) \to [0, +\infty)$ be a continuous function and $\tilde{f}$ its germ in M . By the Gelfand isomorphism, to f there corresponds a central hermitian operator $\tilde{f} \in \mathfrak{A}$, whose image in $\mathfrak{A}_M$ may be identified with $\tilde{f}$. Let us suppose that M is not a P -point. Then there exists a function f for which $f(M) = 0$ and $\tilde{f} \neq 0$.

From continuity consideration we obtain $n\tilde{f} \leq 1_M$ for any $n \in N$ and consequently H_M is not archimedian. End of the proof.

Corollary. *The vector spaces* H_M *, are archimedian for any* $M \in \mathfrak{M}(\mathfrak{A})$ *if and only if* $\mathfrak{A}$ *is isomorphic to a finite direct product of factors.*

Proof. If $\mathfrak{A}$ is isomorphic to a finite direct product of factors $\mathfrak{A} \cong \mathfrak{A}_1 \times \ldots$ $\ldots \times \mathfrak{A}_n$, then the space $\mathfrak{M}(\mathfrak{A})$ has n points and the corresponding stalks of the sheaf $\mathfrak{A}$ are isomorphic to the factors $\mathfrak{A}_i$.

Conversely, if the spaces H_M, $M \in \mathfrak{M}(\mathfrak{A})$ are all archimedian, then from proposition 24 if follows that all the points of $\mathfrak{M}(\mathfrak{A})$ are P-points. From a well-known theorem (see [7], ch. 4.4; [18], theorem 14) it follows that $\mathfrak{M}(\mathfrak{A})$ is finite, and therefore $\mathfrak{A}$ is isomorphic to a direct product of factors. End of the proof.

The preceding results justify a more profound study of the set of P-points in compact hyperstonian spaces.

9. We conclude by observing that, due to the natural homeomorphisms $\mathfrak{M}(\mathfrak{A}) \simeq \mathfrak{M}(\mathcal{Z}) \simeq \mathfrak{M}(\mathfrak{A}')$, the preceding theory allows the simultaneous algebraic reduction of $\mathfrak{A}$ and $\mathfrak{A}'$, which is not possible with the theory developed in [18], or [13], since the fact that $\mathfrak{A}$ is finite does not imply that $\mathfrak{A}'$ is finite.

BIBLIOGRAPHY

[1] J. Dauns – K.H. Hofmann, The representation of biregular rings by sheaves, *Math. Zeitschr.*, 91 (1966), 103-123.

[2] J. Dauns – K.H. Hofmann, Representation of rings by sections, *Memoirs of the Amer. Math. Soc.*, 83 (1968), 1-180.

[3] J. Dixmier, Sur certains espaces considérés par M.H. Stone, *Sum. Bras. Math.*, 11 (1951), 185-202.

[4] J. Dixmier, *Les algèbres d'opérateurs dans l'espace hilbertien (Algèbres de von Neumann)* (Paris, 1957).

[5] J. Dixmier, *Les* C^**-algèbres et leurs représentations* (Paris, 1969).

[6] I.M. Gel'fand – D.A. Raĭkov – G.E. Šilov, *Commutative normed rings* (Moscow, 1960). In Russian.

[7] L. Gillman – M. Jerison, *Rings of continuous functions* (Princeton, Toronto, London, New York, 1960).

[8] R. Godement, *Topologie algébrique et théorie des faisceaux* (Paris, 1958).

[9] Chr. Mulvey, Représentation des produits sous-directs d'anneaux par espaces annelés, *C.R. Acad. Sci. Paris*, (1970), 564-567.

[10] M.A. Naǐmark, *Normed rings* (Moscow, 1959). In Russian.

[11] R.S. Pierce, Modules over commutative regular rings, *Memoirs Amer. Math. Soc.*, 70 (1967).

[12] Ch. Rickart, *General theory of Banach algebras* (Princeton, Toronto, London, New York, 1960).

[13] S. Sakai, *The Theory of* W^**-algebras* (Yale University Press, 1962).

[14] S. Teleman, Analyse harmonique dans les algèbres regulières, *Rev. roum. math. pures appl.*, XIII 750, (1968), 691-750.

[15] S. Teleman, La représentation des anneaux tauberiens discrets par des faisceaux, *Rev. roum. math. pures appl.* XIV, (1969), 249-264.

[16] S. Teleman, La représentation des anneaux réguliers par les faisceaux, *Rev. roum. math. pures appl.*, XIV (1969), 703-717.

[17] S. Teleman, Représentation par faisceaux des modules sur les anneaux harmoniques, *C.R. Acad. Sci. Paris,* 269 (1969), 753-756.

[18] S. Teleman, La représentation des algèbres de von Neumann finies par faisceaux, *Rev. roum. math. pures appl.*, XV, (1970), 143-151.

[19] S. Teleman, Sur les anneaux reguliers , *Rev. roum. math. pures appl.*, XV. (1970), 407-434.

[20] S. Teleman, On the regular rings of John von Neumann, *Rev. roum. math. pures appl.*, XV, (1970), 735-742.

[21] S. Teleman, La représentation par faisceaux des modules sur les algèbres harmoniques, *Rev. roum. math. pures appl.*, XVI (to appear).

[22] A.B. Willcox, Some structure theorems for a class of Banach algebras, *Pacific J. of Math.*, 6, (1956), 177-192.

COLLOQUIA MATHEMATICA SOCIETATIS JÁNOS BOLYAI
5. HILBERT SPACE OPERATORS, TIHANY (HUNGARY), 1970

I. Cuculescu's proof for the commutant theorem

L. ZSIDÓ

Let M, N be von Neumann algebras in the Hilbert spaces H and K, respectively. $x \in B(H)$ and $y \in B(K)$ define in a usual way a bounded linear operator $x \bar{\otimes} y$ in the Hilbert tensor product $H \bar{\otimes} K$. The tensor product of M and N is defined as the weak closure of $\{x \bar{\otimes} y \mid x \in M, y \in N\}$ in $B(H \bar{\otimes} K)$. It is denoted by $M \bar{\otimes} N$.

The *commutant theorem* states that the commutant of a tensor product is the tensor product of the commutants, that is that

$$(M \bar{\otimes} N)' = M' \bar{\otimes} N'.$$

This theorem was proved by Tomita. Tomita's proof is presented in [5]. In [1], I. Cuculescu gives a simple proof of the theorem. Our purpose is to present a slightly modified version of Cuculescu's proof.

By lemma III. 1.7 and proposition 1.2.5 (ii) of [2], every von Neumann algebra has the form $\prod_{\iota \in I} (M_\iota \bar{\otimes} B(K_\iota))$, where M_ι are of countable type. Proposition I. 2.4

(iii) and theorem I.4.3 of [2] reduce our general problem to the case when M and N are of countable type. Considering faithful normal positive linear functionals on M and N and the representations defined by them, and using theorem I.4.3 of [2] we conclude that it is sufficient to consider the case when M and N have separating cyclic vector.

For the remaining part of the proof we use some elementary properties of densely defined closed operators in a Hilbert space (see for example [3], chapter I, §5).

Let $\xi \in H$ be a separating cyclic vector for a von Neumann algebra M and J the antilinear isometry of H onto H^* defined by $J\zeta = (\cdot\,|\,\zeta)$. We define the linear operator X_M from H into H^* by $X_M(x\xi) = Jx^*\xi$, $x \in M$.

Lemma 1. X_M *has closure and* $\bar{X}_M = JX^*_{M'}J$.

Proof. It is easy to verify that the graph Γ_{X_M} of X_M is contained in the graph $\Gamma_{JX^*_{M'}J}$ of $JX^*_{M'}J$. Hence X_M has closure.

Now let η be a vector in the domain of $X^*_{M'}J$. We associate to η the linear operator Y from H into H, defined by $Y(y\xi) = y\eta$, $y \in M'$. For every $y, z \in M'$ we have

$$(z\xi\,|\,Y(y\xi)) = (z\xi\,|\,y\eta) = (y^*z\xi\,|\,\eta) = (J\eta\,|\,Jy^*z\xi) =$$

$$= (J\eta\,|\,X_{M'}(z^*y\xi)) = (X^*_{M'}J\eta\,|\,z^*y\xi) = (zX^*_{M'}J\eta\,|\,y\xi).$$

Consequently Y^* is defined for every $z\xi$, $z \in M'$ by $Y^*(z\xi) = zX^*_{M'}J\eta$. This implies that Y^* is a densely defined closed operator which commutes with M'. Y has closure and $(1+Y^*\bar{Y})^{-1}$ is a bounded operator which commutes with M'. Hence $(1+Y^*\bar{Y})^{-1} \in M$. Let (e_n) be a sequence of spectral projections of $(1+Y^*\bar{Y})^{-1}$ such that $e_m \nearrow 1$ and such that there exists a sequence (x_n) in M with

$$(1+Y^*\bar{Y})^{-1}e_n x_n = e_n .$$

This equality implies that e_nH is contained in the domain of $\bar{Y}$, and since $\bar{Y}$ commutes with M', we have $\bar{Y}e_n \in M$.

Suppose that there exists in $\Gamma_{JX^*_{M'}J}$ an element $(\eta, JX^*_{M'}J\eta)$ orthogonal to Γ_{X_M}. This means that for every $x \in M$ we have

$$(\eta | x\xi) + (JX^*_{M'}J\eta | X_M x\xi) = 0 .$$

If Y is the operator associated to η, then $Y^*\xi = X^*_{M'}J\eta$, and so the above equality becomes the form

$$(\eta | x\xi) + (x^*\xi | Y^*\xi) = 0 .$$

In particular, for

$$(\eta | e_n\eta) + (e_n Y^*\xi | Y^*\xi) = 0 .$$

For $n \to \infty$ we get $\|\eta\|^2 + \|Y^*\xi\|^2 = 0$, so $\eta = 0$.

Hence $\bar{X}_M = JX^*_{M'}J$. q.e.d.

The following lemma is essentialy due to Sakai [4].

Lemma 2. *Let M, N be two von Neumann algebras in a Hilbert space H such that there exists a separating cyclic vector $\xi \in H$ for both algebras, furthermore let $M \subset N$, and $\bar{X}_M = \bar{X}_N$. Then $M = N$.*

Proof. Let $x_0 \in N$. Then $x_0\xi$ is in the domain of $\bar{X}_N = \bar{X}_M = JX^*_{M'}J$, and so we can associate to $x_0\xi$ an operator Y as in the proof of lemma 1. For every $y \in N'$ we have $\bar{Y}(y\xi) = yx_0\xi = x_0(y\xi)$. Hence $\bar{Y} = x_0$. But $\bar{Y}$ commutes with M', and so $x_0 = \bar{Y} \in M$. q.e.d.

Now by using our technic, we prove a known result.

Lemma 3. *Let X_1, X_2 be densely defined operators with closure, from H_1 into K_1 and from H_2 into K_2 respectively. Then $X_1 \otimes X_2$ defined by $(X_1 \otimes X_2)(\xi_1 \otimes \xi_2) = X_1\xi_1 \otimes X_2\xi_2$ for ξ_i in the domain of X_i has closure. If we denote this closure by $X_1 \bar{\otimes} X_2$, then*

$$(X_1 \bar{\otimes} X_2)^* = X_1^* \bar{\otimes} X_2^* .$$

Proof. If ξ_i is in the domain of X_i and η_i is in the domain of X_i^* we have

$$((X_1 \otimes X_2)(\xi_1 \otimes \xi_2) | \eta_1 \otimes \eta_2) = (\xi_1 \otimes \xi_2 | (X_1^* \otimes X_2^*)(\eta_1 \otimes \eta_2)).$$

Hence $X_1 \otimes X_2$ has closure and the graphe of $X_1^* \bar{\otimes} X_2^*$ is contained in the graph of $(X_1 \bar{\otimes} X_2)^*$. Let (η_0, ξ_0) be an element in the graph of $(X_1 \bar{\otimes} X_2)^*$, orthogonal to the graph of $X_1^* \bar{\otimes} X_2^*$. Hence, for every ξ_i in the domain of $\bar{X}_i$, we have

(i) $$(\bar{X}_1\xi_1 \otimes \bar{X}_2\xi_2 | \eta_0) = (\xi_1 \otimes \xi_2 | \xi_0),$$

and for every η_i in the domain of X_i^*,

(ii) $$(\eta_1 \otimes \eta_2 | \eta_0) + (X_1^*\eta_1 \otimes X_2^*\eta_2 | \xi_0) = 0.$$

For every ξ_i in the domain of $X_i^*\bar{X}_i$,

$$([1 + (X_1^*\bar{X}_1 \otimes X_2^*\bar{X}_2)](\xi_1 \otimes \xi_2) | \xi_0) =$$

$$= (\xi_1 \otimes \xi_2 | \xi_0) + (X_1^*\bar{X}_1\xi_1 \otimes X_2^*\bar{X}_2\xi_2 | \xi_0) = \text{(by (i))}$$

$$= (\bar{X}_1\xi_1 \otimes \bar{X}_2\xi_2 | \eta_0) + (X_1^*\bar{X}_1\xi_1 \otimes X_2^*\bar{X}_2\xi_2 | \xi_0) = \text{(by (ii))} = 0$$

holds.

Now $X_1^*\bar{X}_1 \bar{\otimes} X_2^*\bar{X}_2$ is densely defined closed and positive. Hence the range of $1 + (X_1^*\bar{X}_1 \bar{\otimes} X_2^*\bar{X}_2)$ is $H_1 \bar{\otimes} H_2$, and the above equality implies that $\xi_0 = 0$. From (ii) it results that $\eta_0 = 0$. q.e.d. Lemmas 1,2 and 3 imply immediately the commutant theorem.

REFERENCES

[1] I. Cuculescu, A proof of $(A \otimes B)' = A' \otimes B'$ for von Neumann algebras, *Revue Roum. de Math. Pures et Appl.* (to appear).

[2] J. Dixmier, *Les algèbres d'opérateurs dans l'espace hilbertien* (Paris, 1969).

[3] M.A. Naĭmark, *Normed rings* (in Russian) (Moscow, 1968).

[4] S. Sakai, On the tensor product of W^*-algebras, *Amer. J. Math.*, 90 (1968), 335-341.

[5] M. Takesaki, *Tomita's theory of modular Hilbert algebras and its applications*, Lecture Notes in Math., 128 (Springer Verlag, 1970).

CONTENTS

PREFACE 3
SCIENTIFIC PROGRAM 4
LIST OF PARTICIPANTS 8
Ju. M. Berezanskiĭ, Generalized power moment problems 11
F. A. Berezin – M. A. Šubin, Symbols of operators and quantization 21
J. Bognár, Involution as operator conjugation 53
F. F. Bonsall, Hermitian operators on Banach spaces 65
M. S. Brodskiĭ, On certain invariant subspaces of dissipative operators of exponential type 77
P. L. Butzer – U. Westphal, On the Cayley transform and semigroup operators . . 89
L. A. Coburn, C*-algebras generated by semi-groups of isometries 99
J. L. B. Cooper, Group representations and integral transforms 107
R. Denčev, On commutative self-adjoint extensions of differential operators 113
J. Dixmier – C. Foiaş, Sur le spectre ponctuel d'un opérateur 127
R. G. Douglas – J. L. Taylor, Wiener-Hopf operators with measure kernels 135
E. Durszt, A generalization of Schäffer's matrix 143
C. Foiaş, A classification of doubly cyclic operators 155
I. M. Gelfand – V. A. Ponomarev, Problems of linear algebra and classification of quadruples of subspaces in a finite-dimensional vector space 163
I. C. Gohberg – N. Ja. Krupnik, Banach algebras generated by singular integral operators 239
V. I. Gorbačuk, Self-adjoint extensions of some Hermitian operators in a space with indefinite metric 265
H. Helson, Invariant subspaces of the weighted shift 271
H. Helton, Operators with a representation as multiplication by x on a Sobolev space 279
I. S. Iohvidov, On the structure of the spectrum of G-selfadjoint and G-unitary operators in Hilbert space 289
H. Johnen, Best approximation on the unitary group 295
R. K. Juberg, On the boundedness of certain singular integral operators 305
M. A. Kaashoek, On the peripheral spectrum of an element in a strict closed semi-algebra 319
R. G. Kalisch, On the similarity of certain operators 333

G. E. Kisilevskiĭ, On the location of invariant subspaces of dissipative operators 347
I. Kovács, Power-bounded operators and finite type von Neumann algebras 351
M. G. Kreĭn – H. Langer, Über die verallgemeinerten Resolventen und die charakteristische Funktion eines isometrischen Operators im Raume $\Pi_\varkappa$ 353
G. W. Mackey, Products of subgroups and projective multipliers 401
P. Masani, Remarks on eigenpackets of self-adjoint operators 415
J. D. Pincus, The determining function method in the treatment of commutator systems 443
P. Rosenthal, Problems on invariant subspaces and operator algebras 479
Ju. A. Rozanov, On the structure of measurable linear operators 489
S. Sakai, On global type II_1 w^*-algebras 493
I. Suciu, Harnack inequalities for a functional calculus 499
B. Sz.-Nagy, Quasi-similarity of operators of class C_0 513
S. Teleman, Representations of von Neumann algebras by sheaves 519
L. Zsidó, I. Coculescu's proof for the commutant theorem 539
CONTENTS 543